The Practice of Research in Criminology and Criminal Justice

Third Edition

*In loving memory of two pioneer women
who instilled in me the importance
of education and lifelong learning:
To my great aunt, Martha Geiken Lund, 1912–2002
and
To my grandmother, Anna Geiken Bachman, 1907–1989*

R. B.

To Elizabeth and Julia

R. K. S.

The Practice of Research in Criminology and Criminal Justice

Third Edition

Ronet Bachman
University of Delaware

Russell K. Schutt
University of Massachusetts, Boston

SAGE Publications
Los Angeles • London • New Delhi • Singapore

For information:

SAGE Publications, Inc.
2455 Teller Road
Thousand Oaks, California 91320
E-mail: order@sagepub.com

SAGE Publications Ltd.
1 Oliver's Yard
55 City Road
London EC1Y 1SP
United Kingdom

SAGE Publications India Pvt. Ltd.
B 1/I 1 Mohan Cooperative Industrial Area
Mathura Road, New Delhi
India 110 044

SAGE Publications Asia-Pacific Pte. Ltd.
33 Pekin Street #02-01
Far East Square
Singapore 048763

Printed in the United States of America on acid-free paper.

Library of Congress Cataloging-in-Publication Data

Bachman, Ronet.
The practice of research in criminology and criminal justice / Ronet Bachman, Russell K. Schutt. — 3rd ed.
    p. cm.
Includes bibliographical references and index.
ISBN 978-1-4129-5032-9 (pbk.)
    SPSS Student Version 15.0: ISBN 978-1-4129-5749-2
    1. Criminology—Research. 2. Criminal justice, Administration of—Research.
I. Schutt, Russell K. II. Title.

HV6024.5.B33 2007
364.072—dc22

                                    2006034802

07   08   09   10   11   10   9   8   7   6   5   4   3   2   1

Acquisitions Editor:	Jerry Westby
Associate Editor:	Elise Smith
Editorial Assistant:	Kim Suarez
Production Editor:	Sanford Robinson
Copy Editor:	Ani L. Ayvazian
Proofreader:	Kevin Gleason
Typesetter:	C&M Digitals (P) Ltd.
Indexer:	Kathy Paparchontis
Cover Designer:	Michelle Kenny
Marketing Manager:	Jennifer Reed

Brief Contents

On the Student Study Site you will also find comprehensive study material that includes e-flashcards, Web exercises, practice self-tests, SAGE's online journal collection, and other interactive exercises.

Detailed Contents

2 The Process and Problems of Criminological Research   33

8 Qualitative Methods   257

11 Evaluation Research and Policy   361

ORGANIZATION OF THE BOOK

The way this book is organized reflects our beliefs in making research methods interesting, teaching students how to critique research, and viewing specific research techniques as parts of an integrated research strategy. Our concern with ethical issues in all types of research is underscored by the fact that we have a section on Ethics in every methodology chapter. The first two chapters introduce the why and how of research in general. Chapter 1 shows how research has helped us understand the magnitude of and the factors related to youth violence. Chapter 2 illustrates the basic stages of research with a series of experiments on the police response to intimate partner violence. Chapters 3 and 4 discuss how to evaluate the way researchers design their measures and draw their samples. Chapter 5 explores issues related to making causal connections.

Chapters 6 through 8 present the three most important methods of data collection: experiments, surveys, and qualitative methods (including participant observation, intensive interviews, and focus groups). Chapter 9 is a new chapter that reviews major analysis techniques that are used by researchers to identify and understand data collected in qualitative research investigations. Chapter 10 examines methodologies that rely on existing content and includes a discussion of secondary data analysis, historical and comparative research, content analysis, and crime mapping, along with a discussion of triangulating methods. Chapter 11 covers evaluation research and highlights the different alternatives to evaluation along with a discussion of the most appropriate methods to use for each evaluation question (e.g., process versus impact). In this chapter, you will see how various methods have been used to investigate the effects of several programs and policies, including problem-oriented policing, boot camps, rape-reform legislation, and mandatory sentencing laws.

Chapter 12 presents the basic statistical methods used to analyze the results of quantitative studies in a descriptive manner. We will work through an analysis of survey data on self-reported delinquency to see how these statistics are used to answer actual research questions. Chapter 13 provides an overview of the main types of written reports used in criminology and criminal justice. You will learn the required sections of a research proposal and a research article along with the peer-review process that accompanies such endeavors.

The substantive studies in each of these chapters show how each methodology has been used to improve our understanding of criminal justice–related issues, including the factors related to violence, how question wording affects estimates of victimization in surveys, how gang members perceive their world, how community police officers describe their role in comparison to regular patrol officers, the perceptions of jurors who have participated in a death penalty case, the effects of inmates' classification on institutional misconduct in prison, and the effects of war on violence in a cross-national comparison, to name just a few of the examples provided.

DISTINCTIVE FEATURES OF THE THIRD EDITION

The most distinctive feature of this text compared to others in the field continues to be the integration into each chapter of in-depth substantive research examples from the real world. Examples from the literature are not simply dropped here and there to keep students' attention. Rather, each chapter presents a particular research method in the context of a substantive research story. As such, this book's success is due in no small measure to the availability in the research literature of so many excellent examples. The following points are additional strengths of this text along with some new innovations in this edition:

A new chapter on causation and research design, which explores the difference between idiographic and nomothetic causal explanations as well as the criteria necessary to establish nomothetic causal relationships. Examples from literature in this chapter include an experimental design used to determine the effects of being exposed to violent media representations and actual violent behavior, a longitudinal analysis of delinquency offending to determine if the factors related to offending behavior are different for males and females, and an observational analysis of disorder and crime in urban neighborhoods.

A new chapter on qualitative data analysis, which focuses solely on the logic and design of qualitative data analysis. Examples from the literature in this chapter include a participant observational study of gang life in an urban residential housing facility, an observational study of community police officers, intensive interviews with violent female offenders, and a participant observational study of victimization and drug use in rave and hip hop clubs in Philadelphia.

A new appendix that provides step-by-step procedures for conducting literature reviews as well as finding information on the World Wide Web. A case study of a literature review on the deterrence value of arrest in cases of intimate partner assault is used to illustrate guidelines for effective literature searchers.

More integrated treatment of qualitative and quantitative methods. Qualitative and quantitative methods are contrasted in Chapter 1, and most other chapters include a special section on combining qualitative and quantitative approaches. Chapter 8 now focuses solely on the logic and design of qualitative research, whereas Chapter 9 provides an expanded treatment of qualitative data analysis. The value of multiple methods of investigation is now illustrated with examples and procedural discussions in many chapters.

Ethical concerns and ethical decision making. Every step in the research process raises ethical concerns, so ethics should be treated in tandem with the study of specific methods. While you will find ethics introduced in Chapter 2, every chapter reviews the relevant ethical issues in the context of each method of data collection, data analysis, and reporting. Discussion of Institutional Review Boards has been updated to include current policy and practice. In addition, there are new Ethics Questions at the end of each chapter.

Examples of criminological research as they occur in real-world settings. We include interesting studies taken from the literature on a variety of topics including the causes and correlates of violates, the efficacy of arrest for intimate partner assault, the perceptions of police officers regarding community policing, and an investigation into the lives of gang members, to name just a few. These real-world research examples illustrate the exigencies and complexities that shape the application of research methods.

End-of-chapter exercises. In addition to individual and group projects, each chapter includes exercises to give you experience in data analysis using SPSS, the statistical software for the social sciences. In addition, real data sets are provided to enhance your learning experience including subsets of the National Crime Victimization Survey, a state-level file containing crime rates and other variables that measure structural characteristics of each state, such as the poverty rate, the divorce rate, and so on. Each chapter also provides updated end-of-chapter Web exercises. In addition, a new section has been added to each chapter to guide you through the preparation of a research proposal and as we stated above, a new section posing various ethical dilemmas related to each chapter.

Aids to effective study. Lists of main points and key terms provide quick summaries at the end of each chapter. In addition, key terms are highlighted in boldface type when first introduced and defined in text. Definitions for these also can be found in the Glossary/Index at the end of the book. The instructor's manual includes more exercises that have been specially designed for collaborative group work inside and outside the classroom. Appendix A, "Conducting Literature Reviews and Finding Information," provides up-to-date information about using the Internet. The Student Study Site also provides invaluable tools for learning.

If this book communicates the excitement of research and the importance of evaluating carefully the methods we use in research, then we have succeeded in representing what social scientists interested in issues related to criminal justice and criminology do. We think it conveys the latest developments in research methodology and thereby demonstrates that researchers are committed to evaluating and improving their own methods of investigation.

We hope you enjoy learning how to investigate research questions related to criminal justice and criminology and perhaps do some research of your own along the way. We guarantee that the knowledge you develop about research methods will serve you well throughout your education, in your career, and in your community.

SUPPLEMENTS

Companion Student Study Site

www.sagepub.com/prccj3

This Web-based Student Study Site provides a variety of additional resources to enhance students' understanding of the book content and take their learning one step further. The site includes self-study quizzes, e-flashcards, a new "Learning from SAGE Research Articles" feature, Web exercises, GSS Datasets and Documentation, as well as additional appendices. It also contains interactive exercises with criminal justice and criminology tracks specifically designed to help students get into the latest research in the field.

Instructor's Resource CD-ROM

This CD offers the instructor a variety of resources to supplement the book material, including lecture outlines, PowerPoint® lecture slides, test questions with answers, and student project ideas. The CD also contains articles on teaching Criminal Justice Research Methods, film and software resources, as well as Web resources. An electronic test bank is also available so that instructors can create, deliver, and customize tests and study guides using Brownstone's Diploma test bank software.

A NOTE ABOUT USING SPSS AND HYPERRESEARCH

To carry out the SPSS exercises at the end of each chapter and in Appendix E, you must have SPSS installed on your computer. The Student Study Site includes several subsets of data, including data from the National Crime Victimization Survey and the Uniform Crime Reports. Appendix E will get you up and running with SPSS for Windows. You then may spend as much time as you like exploring the data sets provided, or may even use your own data. You also may carry out analyses of the General Social Survey at the University of California, Berkeley, Web site: http://sda.berkeley.edu:7502/archive.htm.

The study site to this text at http:www.sagepub.com/prccj3 contains a detailed discussion of what it is like to do qualitative research with the statistical software package HyperRESEARCH. Here, you will learn how to begin a simple project in HyperRESEARCH by creating and managing data and ideas, coding, linking, modeling, and asking questions about your narrative data.

Acknowledgments

We must first acknowledge our gratitude to Jerry Westby, whose hard work and guidance on this project are unrivaled. He has been more than an editor; he is an ideas man, a tenacious fact finder, a motivator, a therapist, and most important, a friend. We are also indebted to Elise Smith for her meticulous editing of the text and to Sanford Robinson, who made sure production was almost painless.

Gratitude also goes to the reviewers of this third edition, including Ira Sommers, California State University–Los Angeles; Kristy Holtfreter, Florida State University; Amy Craddock, Indiana State University; James R. Maupin, New Mexico State University; William Wells, Southern Illinois University–Carbondale; Gennifer Furst, The College of New Jersey; Lori Guevara, Fayetteville State University; Frank Cormier, University of Manitoba; Michael J. DeValve, Fayetteville State University; Brian Colwell, Stanford University; Susan B. Haire, University of Georgia; Lisa Anne Zilney, Montclair State University; and Stephen M. Haas, Marshall University. Reviewers of previous editions included Cathy Couglan, Texas Christian University; Lucy Hochstein, Radford University; Mark Winton, University of Central Florida; Stephen Haas, Marshall University; Hank J. Brightman, Saint Peter's College; Eric Metchick, Salem State College; Kristen Kuehnle, Salem State College; Wilson R. Palacios, University of South Florida; and Phyllis B. Gerstenfeld, California State University–Stanislaus. Andre Rosay also provided an extensive and invaluable review of the first edition of this text. We also thank Lindsay R. Reed and Hanna S. Scott for their diligence and hard work in writing the instructor's manual, Ann Dupuis and Sharlene Hesse-Biber for the appendix on HyperRESEARCH software, Lisa M. Gilman for the appendix on SPSS, and Kathryn Stoeckert and Heather Albertson for additional interactive exercises.

We continue to be indebted to the many students we have had an opportunity to teach and mentor, at both the undergraduate and graduate levels. In many respects, this book could not have been written without these ongoing reciprocal teaching and learning experiences. You inspire us to become better teachers!

Ronet is indebted to her terrific colleagues who are unwavering sources of support and inspire her. To a first-rate office staff who not only keep all the trains running on time in our department but also are wonderful friends: Linda Keen, Nancy Quillen, and Judy Watson. Ronet is also indebted to an amazing circle of friends who endured graduate school together and who get together for a retreat one weekend of the year (16 years and counting!) for guidance, support, therapy, and laughter: Dianne Carmody, Gerry King, Peggy Plass, and Barbara Wauchope. You are the most amazing women in the world, and I am so blessed to have you in my life. To Alex Alvarez and Michelle Meloy, my other kindred spirits for their support and guidance; to my mother, Jan, who remains my hero; and to my father, Ron, for his steadfast critical eye in all matters of life.

And most important, we both wish to thank our spouses, Raymond Paternoster and Elizabeth Schutt, for their love and support (and editorship!), and our children, John and Julia, for all the remarkable joy they have brought to our lives.

Science, Society, and Criminological Research

By now the scenario is common knowledge: On April 20, 1999, Eric Harris and Dylan Klebold turned Columbine High School, in suburban Colorado, into the scene of the deadliest school shooting in American history. After killing 12 students and a teacher, the youths apparently shot themselves in the head. In the end, 15 people were dead and 28 were injured. After the bloodbath, police found the school in Littleton littered with bombs and other booby traps.

Harris and Klebold were not typical terrorists. They were apparently bright young men who became social outcasts at their school, affiliated with a group known as the Trenchcoat Mafia, whose outward symbols included dark trenchcoats, dark sunglasses, and berets, part of a larger group described as "Goths." But signs of trouble were evident to many before the killings. Harris's Web pages were filled with images of fire, skulls, devils, and weapons and included recipes for and sketches of pipe bombs. The two young men also wrote poetry about death for their English class and made a video about guns for a video class.

The Littleton horror was simply the latest in a string of school shootings in 1998 and 1999. In the spring of 1998, in Jonesboro, Arkansas, two prepubescent boys decided that the best way to deal with girls who might have minds of their own was to shoot them. After pulling the school's fire alarm, the boys sat in the trees awaiting the targets. When the students filed out of the school, the boys opened fire, killing 4 students and a teacher. In the months that followed, 3 students were killed and 5 were injured in West Paducah, Kentucky, when another student shot them as they left a prayer meeting. In Edinboro, Pennsylvania, a student gunned down a teacher and wounded 2 classmates. And in Springfield, Oregon, a disgruntled student brought a rifle into the cafeteria at Thurston High School and fired indiscriminately into a crowd, killing 2 students and wounding 22 others. These were not inner-city schools riddled with economic deprivation and disorganization; they were all schools in quiet towns where violence, at least publicly displayed violence, was rare.

Each of these school shootings caused a media frenzy. Headlines such as "The School Violence Crisis" and "School Crime Epidemic" were plastered across national newspapers and weekly news journals. From a statistical standpoint, these headlines seemed a little late, particularly because the juvenile arrest rate for murder had been declining. In addition, the thousands of other shootings that occurred in less privileged communities during the same time rarely made the national news. For example, there were no such declarations when an 11-year-old boy was shot and killed in Chicago because he had allegedly shorted an 18-year-old on drug money. And where were such declarations the night 2 teenagers were shot outside a Baltimore nightclub when guns turned an argument into the final solution?

Many factors influence our beliefs about social phenomena, but the media play a large role in how we perceive both problems and solutions. What are your perceptions of violence committed by youth, and how did you acquire such perceptions? What do you believe are the causes of youth violence? Many factors have been blamed for youth violence in American society, including the easy availability of guns, the use of weapons in movies and television, the moral decay of our nation, poor parenting, unaware teachers, school and class size, racial prejudice, teenage alienation, the Internet and the World Wide Web, anti-Semitism, rap and rock music, and the Goth culture. When trying to make sense out of the Littleton incident, President Bill Clinton talked about hate, prejudice, community policing, conflict resolution, parental responsibility, and violence in the culture. Charlton Heston, spokesman for the National Rifle Association, blamed the absence of armed security guards in schools, even though one was present in Littleton. Heston also blamed the parents and the school for allowing kids to wear black.

Each of you probably has his or her own ideas about what factors may be related to violence in general and youth violence in particular. However, the factors you believe are important in explaining a phenomenon may not always be the ones supported by empirical research. In fact, the factors often touted by politicians and the media to be related to violence are not always supported by empirical evidence. In the rest of this chapter, you will learn how the methods of social science research go beyond stories in the popular media to help us answer questions like "What are the causes of youth violence?" By the chapter's end, you should know what is scientific in criminal justice and criminology and appreciate how the methods of social science can help us understand and answer research questions in this discipline.

REASONING ABOUT THE SOCIAL WORLD

The story of just one murderous youth raises many questions. Take a few minutes to read each of the following questions and jot down your answers. Don't ruminate about the questions or worry about your responses. This is not a test; there are no wrong answers.

- How would you describe Eric Harris?
- Why do you think Eric Harris wanted to kill other students?
- Was Eric Harris typical of other murderers under 18 years of age?
- In general, why do people become murderers?
- How have you learned about youth violence?

Now let us consider the possible answers to some of these questions. The information about Eric Harris is somewhat inconsistent (Duggan, Shear, & Fisher 1999). He was the 18-year-old son of middle-class professionals. He had an older brother who attended the University of Colorado. Harris apparently thought of himself as a white supremacist, but he also loved music by antiracist rock bands. On his Web page, he quoted from KMFDM, a German rock band whose song "Waste" includes these lyrics: "What I don't say I don't do. What I don't do I don't like. What I don't like I waste." Online, Harris referred to himself as "Darkness."

Do you have enough information now to understand why Eric went on a shooting rampage in his school?

A year before the shootings at Columbine High School, Harris was arrested on a felony count of breaking into a car. A juvenile court put him on probation, required him to perform community service and take criminal justice classes, and sent him to a school counseling program. He was described by one of his probation officers as a "very bright young man who is likely to succeed in life."

Now can you construct an adequate description of Eric Harris? Can you explain the reason for his murderous rampage? Or do you feel you need to know more about Eric Harris, about his friends and the family he grew up in? And how about his experiences in school and with the criminal justice system? We have attempted to investigate just one person's experiences, and already our investigation is spawning more and more questions.

Questions and Answers

When the questions concern not just one person but many people or general social processes, the possible questions and the alternative answers multiply. For example, consider the question of why incidents of school violence occur. Responses to a 1997 survey of mothers with children between the ages of 12 and 15 found that the majority of mothers considered the most important factors that contributed to school violence to be parents who did not teach children morals (84%), parents who supported aggressive behavior (78%), and gang or peer pressure (70%) (Kandakai et al. 1999). Compare these answers with the opinion you recorded earlier. Was your idea about the causes of youth violence one of the more popular ones held by this sample of mothers?

We cannot avoid asking questions about the actions and attitudes of others. We all try to make sense of the social world, which is a very complex place, and to make sense of our position in it, in which we have quite a personal stake. In fact, the more that you begin to "think like a social scientist," the more questions will come to mind.

But why does each question have so many possible answers? Surely our perspective plays a role. One person may see a homicide offender as a victim of circumstance, and another person may see the same individual as inherently evil. Answers to questions we ask in the criminological sciences also vary because individual life experiences and circumstances vary. The study of mothers' perceptions conducted by Kandakai et al. (1999), summarized in Exhibit 1.1, gives some idea of how opinions about the causes and cures for school violence vary. Even though previous research has found that the general public believes school violence is worse in urban schools than in suburban and rural schools, only 14% of the mothers who had children in urban schools believed their child's school was more violent than other schools. However, a higher percentage of urban mothers (37%) than suburban mothers (17%) reported that their child had been involved in one or more fights at school. Somewhat surprisingly, even though these mothers perceived parental support for violence as a major contributor to school violence, 4 in 10 of them believed it was acceptable for their child to fight in certain situations.

EXHIBIT 1.1 Mothers' Perceptions of the Factors Related to School Violence	
	Percentage Who Said *Factor Contributed a Great Deal*
Parents who do not teach their kids right from wrong	87
Parent support for aggressive behavior	78
Gang or peer pressure	70
Lack of family involvement	67
Sale of alcohol or cigarettes to children	61
Poor parent-teacher communication	60
Seeing violence in the community	53
Violent messages in rap music	47
Seeing violence in the media	44
Lack of a dress code	30

Everyday Errors in Reasoning

People give different answers to research questions for yet another reason: It is simply too easy to make errors in logic, particularly when we are analyzing the social world in which we ourselves are conscious participants. We can call some of these everyday errors, because they occur so frequently in the nonscientific, unreflective discourse about the social world that we hear on a daily basis.

For evidence of everyday errors, just listen to your conversations or the conversations of others for one day. At some point in the day, it is inevitable that you or someone you are talking with will say something like "Well, I knew a person who did X and Y happened." From this one piece of information, you therefore draw a conclusion about the likelihood of Y. Four general errors in everyday reasoning can be made: overgeneralization, selective or inaccurate observation, illogical reasoning, and resistance to change.

Overgeneralization

Overgeneralization, an error in reasoning, occurs when we conclude that what we have observed or what we know to be true for some cases is true for all cases. We are always drawing conclusions about people and social processes from our own interactions with them, but sometimes we forget that our experiences are limited. The social (and natural) world is, after all, a complex place. We have the ability (and inclination) to interact with just a small fraction of the individuals who inhabit the social world, especially in a limited span of time.

Selective or Inaccurate Observation

Selective observation is choosing to look only at things that are in accordance with our preferences or beliefs. When we are inclined to criticize individuals or institutions, it is all too easy to notice their every failing. For example, if we are convinced in advance that all kids who are violent are unlikely to be rehabilitated and will go on to commit violent offenses in adulthood, we will probably find many confirming instances. But what about other youths who have become productive and stable citizens after engaging in violence as adolescents? Or the child who was physically or sexually abused and joined a gang to satisfy the need for a family surrogate? If we acknowledge only the instances that confirm our predispositions, we are victims of our own selective observation. Exhibit 1.2 depicts the difference between selective observation and overgeneralization.

Recent research on cognitive functioning (how the brain works) helps to explain why our feelings so readily shape our perceptions (Seidman 1997). Emotional responses to external stimuli travel a shorter circuit in the brain than do reasoned responses (see Exhibit 1.3). The result, according to some cognitive scientists, is, "What something reminds us of can be far more important than what it is" (Goleman 1995:294–295). Our emotions can influence us even before we begin to reason about what we have observed.

Our observations also can simply be inaccurate. If a woman says she is *hungry* and we think she said she is hunted, we have made **an inaccurate observation**. If we think five people are standing on a street corner when seven actually are, we have made an

EXHIBIT 1.2   The Difference Between Overgeneralization and Selective Observation

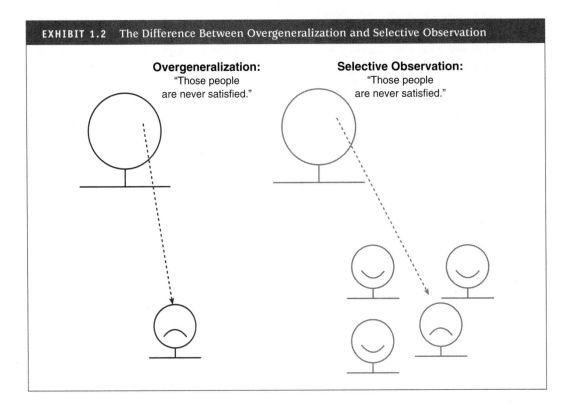

inaccurate observation. Such errors occur often in casual conservation and in everyday observation of the world around us. In fact, our perceptions do not provide a direct window onto the world around us, for what we think we have sensed is not necessarily what we have seen (or heard, smelled, felt, or tasted). Even when our senses are functioning fully, our minds have to interpret what we have sensed (Humphrey 1992). The optical illusion in Exhibit 1.4, which can be viewed as either two faces or a vase, should help you realize that perceptions involve interpretations. Different observers may perceive the same situation differently because they interpret it differently.

Illogical Reasoning

When we prematurely jump to conclusions or argue on the basis of invalid assumptions, we are using **illogical reasoning**. For example, it is not reasonable to propose that depictions of violence in media such as television and movies cause violence if evidence indicates that the majority of those who watch such programs do not become violent. However, it is also illogical to assume that media depictions of gratuitous violence have no effect on individuals. Of course, logic that seems impeccable to one person can seem twisted to another; the problem usually is reasoning from different assumptions rather than failing to "think straight."

EXHIBIT 1.3   Anatomy of an Emotional Hijacking

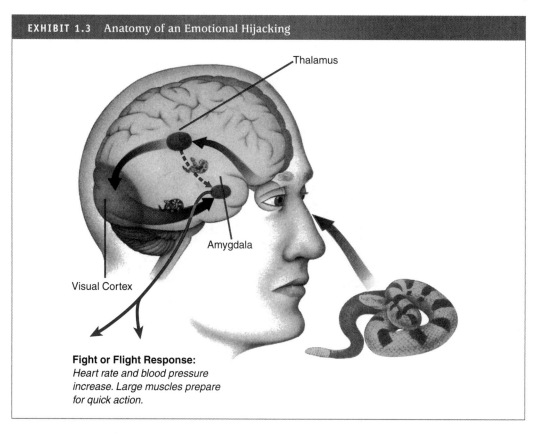

Thalamus

Amygdala

Visual Cortex

Fight or Flight Response:
*Heart rate and blood pressure
increase. Large muscles prepare
for quick action.*

Source: From Goleman, D., *Emotional Intelligence,* p. 19. Copyright © 1995. Used by permission of Bantam Books, a division of Bantam Doubleday Dell Publishing Group.

Resistance to Change

Resistance to change, the reluctance to change our ideas in light of new information, may occur for several reasons:

Ego-based commitments. We all learn to greet with some skepticism the claims by leaders of companies, schools, agencies, and so on that people in their organization are happy, that revenues are growing, that services are being delivered in the best possible way, and so forth. We know how tempting it is to make statements about the social world that conform to our own needs rather than to the observable facts. It also can be difficult to admit that we were wrong once we have staked out a position on an issue.

Excessive devotion to tradition. Some degree of devotion to tradition is necessary for the predictable functioning of society. Social life can be richer and more meaningful if it is allowed to flow along the paths charted by those who have preceded us. But too much

---

EXHIBIT 1.4   An Optical Illusion

---

devotion to tradition can stifle adaptation to changing circumstances. When we distort our observations or alter our reasoning so that we can maintain beliefs that "were good enough for my grandfather, so they're good enough for me," we hinder our ability to accept new findings and develop new knowledge. The consequences can be deadly, as residents of Hamburg, Germany, might have realized in 1892 (Freedman 1991). Until the last part of the 19th century, people believed that cholera, a potentially lethal disease, was due to minute, inanimate, airborne poison particles (miasmas). In 1850, English researcher John Snow demonstrated that cholera was, in fact, spread by contaminated water. When a cholera epidemic hit Hamburg in 1892, the authorities did what tradition deemed appropriate: digging up and carting away animal carcasses to prevent the generation of more miasmas. Despite their efforts, thousands died. New York City adopted a new approach based on Snow's discovery, which included boiling drinking water and disinfecting sewage. As a result, the death rate in New York City dropped to a tenth of what it had been in a previous epidemic.

Uncritical agreement with authority. If we do not have the courage to evaluate critically the ideas of those in positions of authority, we will have little basis for complaint if they exercise their authority over us in ways we do not like. And if we do not allow new discoveries to call our beliefs into question, our understanding of the social world will remain limited. An extreme example of this problem was the refusal of leaders in formerly Communist countries to acknowledge the decaying social and environmental fabric of their societies while they encouraged their followers to pay homage to the wisdom of Comrades Mao, Lenin, and Stalin. But we do not have to go so far afield to recognize that people often accept the beliefs of those in positions of authority without question.

Now take just a minute to reexamine the beliefs about youth violence that you recorded earlier. Did you grasp at a simple explanation even though reality was far more complex? Were your beliefs influenced by your own ego and feelings about your similarities to or differences from individuals prone to violence? Are your beliefs perhaps based on depictions of violence in the media or fiction? Did you weigh carefully the opinions of authority figures, including politicians, teachers, and even your parents, or just accept or reject those opinions out of hand? Could knowledge of research methods help to improve your own understanding of the factors related to violent behavior? By now, we hope that you will see some of the challenges faced by social scientists studying issues related to crime and the criminal justice system.

You do not have to be a scientist or use sophisticated research techniques to recognize and avoid these four errors in reasoning. If you recognize these errors for what they are and make a conscious effort to avoid them, you can improve your own reasoning. In the process, you will also be heeding the admonishments of your parents (or minister, teacher, or other adviser) to refrain from stereotyping people, to avoid jumping to conclusions, and to look at the big picture. These are the same errors that the methods of social science are designed to help criminologists avoid.

THE SOCIAL SCIENCE APPROACH    Page 9 notes

The social science approach to answering questions about the social world is designed to greatly reduce these potential sources of error in everyday reasoning. **Science** relies on logical and systematic methods to answer questions, and it does so in a way that allows others to inspect and evaluate its methods. In the realm of social research, these methods are not so unusual. After all, they involve asking questions, observing social groups, and counting people, which we often do in our everyday lives. However, social scientists develop, refine, apply, and report their understanding of the social world more systematically, or specifically, than Joanna Q. Public:

- **Social science** research methods can reduce the likelihood of overgeneralization by using systematic procedures for selecting individuals or groups to study that are representative of the individuals or groups that we wish to generalize.
- Social science methods can reduce the risk of selective or inaccurate observation by requiring that we measure and sample phenomena systematically.
- To avoid illogical reasoning, social researchers use explicit criteria for identifying causes and for determining if these criteria are met in a particular instance.
- Because they require that we base our beliefs on evidence that can be examined and critiqued by others, scientific methods lessen the tendency to develop answers about the social world from ego-based commitments, excessive devotion to tradition, and/or unquestioning respect for authority.

Science A set of logical, systematic, documented methods for investigating nature and natural processes; the knowledge produced by these investigations.

Social science The use of scientific methods to investigate individuals, societies, and social processes, including questions related to criminology and criminal justice; the knowledge produced by these investigations.

Motives for Criminological Research

Like you, social scientists read stories about incidents of violence committed by youth, observe this violence occasionally in their lives, and try to make sense of what they see. For most, that is the end of it. But for some social scientists, the problem of youth violence has become a major research focus. The motivations for selecting this particular research focus, as with any social science topic, can be anyone or some combination of the following:

Policy motivations. Many social service agencies and elected officials seek better assessments and descriptions of youth violence so they can identify needs and allocate responsibility among agencies that could meet these needs. For example, federal agencies such as the U.S. Department of Justice and the Centers for Disease Control and Prevention want to identify the magnitude of youth violence, and many state and local officials use social research to guide development of their social service budgets. Programs designed to rehabilitate young offenders often use research to learn more about the needs of their clientele. These policy guidance and program management needs have resulted in numerous research projects.

Academic motivations. Young offenders have been a logical focus for researchers interested in a number of questions ranging from how an individual's connection to parents and peers influences their behavior to how the social conditions under which they live, such as poverty, affect their behavior. For example, social scientists have long been concerned with the impact that social disorganization has on individual behavior. Early in this century, researchers at the University of Chicago were interested in the effects that residential mobility and immigration had on levels of crime and delinquency in urban neighborhoods. Today researchers are exploring similar questions concerning the impact of disintegrating economic bases in central cities and their relationship to crime and violence. Other researchers have focused on individual-level explanations such as neurological damage. Those who study social policy also have sought to determine whether correctional programs such as boot camps and other forms of shock incarceration serve to decrease the probability of juveniles reoffending in the future.

Personal motivations. Many who conduct research on youth violence feel that by doing so they can help to prevent it and/or ameliorate the consequences of this violence when it occurs. Some social scientists first volunteered with at-risk youth in such organizations as Big Sisters and Big Brothers, and only later began to develop a research agenda based on their experiences.

Social Criminological Research in Practice

Of course, youth violence did not first appear in the United States in the 1980s. In fact, youth violence always has been a popular topic of social science research. However, the sharp increase in this violence in the United States that began in the late 1980s was unprecedented. Predictably, whenever a phenomenon is perceived as an epidemic, numerous explanations emerge to explain it. Unfortunately, most of these explanations are based on the media and popular culture, not on empirical research. Despite the anecdotal information floating around in the mass media about the factors that may have contributed to increases in youth violence, social scientists interested in this phenomenon have amassed a substantial body of findings that have refined knowledge about the problem and shaped social policy (Tonry & Moore 1998). These studies fall into the four categories of purposes for social scientific research:

To test your understanding of the different purposes of social research, go to the Types of Research Interactive Exercises on the Student Study Site.

Descriptive research. Defining and describing social phenomena of interest is a part of almost any research investigation, but **descriptive research** is the primary focus of many studies of youth crime and violence. Some of the central questions used in these studies were: "How many people are victims of youth violence?" "How many youth are offenders?" "What are the most common crimes committed by youthful offenders?" and "How many youth are arrested and incarcerated each year for crime?" Measurement (see Chapter 3) and sampling (see Chapter 4) are central concerns in descriptive research.

Exploratory research. **Exploratory research** seeks to find out how people get along in the setting under question, what meanings they give to their actions, and what issues concern them. The goal is to answer the question "What is going on here?" and to investigate social phenomena without expectations. This purpose is associated with the use of methods that capture large amounts of relatively unstructured information. For example, researchers investigating the emergence of youth gangs in the 1980s were encountering a phenomenon with which they had no direct experience. Thus, an early goal was to find out what it was like to be a gang member and how gang members made sense of their situation. Exploratory research like this frequently involves qualitative methods (see Chapter 8).

Explanatory research. Many people consider explanation to be the premier goal of any science. **Explanatory research** seeks to identify causes and effects of social phenomena, to predict how one phenomenon will change or vary in response to variation in some other phenomenon. Researchers adopted explanation as a goal when they began to ask such questions as "Why do people become offenders?" and "Does the unemployment rate influence the frequency of youth crime?" Methods with which to identify causes and effects are the focus of Chapter 5.

Evaluation research. **Evaluation research** seeks to determine the effects of a social program or other types of intervention. It is a type of explanatory research, because it deals with cause and effect. However, evaluation research differs from other forms of explanatory research because evaluation research considers the implementation and effects of social policies and programs. These issues may not be relevant in other types of explanatory research. The increase of youth violence in the 1980s spawned many new government programs and, with them, evaluation research to assess the impact of these programs. Some of these studies are reviewed in Chapter 6, which covers experimental design, and in Chapter 11, which covers evaluation research.

We will now summarize one study in each of these four areas to give you a feel for the projects motivated by those different concerns.

Description: What Is the Magnitude of Youth Violence?

One large research project undertaken to examine the magnitude of youth violence (along with other risk-taking behavior such as taking drugs and smoking) is called the Youth Risk Behavior Survey, which has been conducted every 2 years in the United States since 1990. Respondents to this survey are a national sample of approximately 16,000 high school students in Grades 9 through 12. To determine trends and patterns in violence-related behavior, Brener et al. (1999) examined the data for the years 1991 and 1997 and published their findings in the prestigious *Journal of the American Medical Association*.

To measure the extent of youth violence, students were asked the following questions: "During the past 30 days, on how many days did you carry a weapon such as a gun, knife, or club?" "During the past 12 months, how many times were you in a physical fight?" "During the past 12 months, how many times were you in a physical fight in which you were injured and had to be seen by a doctor or nurse?" "During the past 30 days, how many times did you carry a weapon such as a gun, knife, or club on school property?" "During the past 12 months, how many times were you in a physical fight on school property?" and "During the past 12 months, how many times has someone threatened or injured you with a gun, knife, or club on school property?"

Before we tell you the results of this study, let us try a little experiment. You probably have your own ideas about the trends in youth violence and school-related violence, which may be based on your own experience or on media accounts. Do you believe that this violence increased or decreased from 1991 to 1997? Brener et al. (1999) found that the number of all students who carried a weapon within the 30 days preceding the survey actually decreased by 30% from 1991 to 1997. Similarly, the number of students who engaged in a physical fight one or more times during the 12 months preceding the survey decreased 14%, and the number injured in a physical fight decreased 20%. Trends in violence occurring on school property also decreased during this time period. The number of students who carried a gun on school property on one or more of the 30 days preceding the survey decreased 28%, and the number of students who engaged in a physical fight on school property one or more times during the 12 months preceding the survey decreased 9%.

This research is consistent with other studies that have found that official rates of youth violence, and youth homicide in particular, peaked in the early 1990s and declined thereafter (Cook & Laub 1998). What accounted for this unprecedented rise in youth violence during the late 1980s and early 1990s? To answer this question, explanatory research is necessary. After their in-depth descriptive analysis of youth violence during this time, Cook and Laub (1998) contend that the increasing homicide rates began with the introduction of crack cocaine and the conflict that surrounded its marketing. However, the primary factor postulated by Cook and Laub was the increased easy availability of firepower: "All of the increase in homicide rates was with guns, and it appears to be changing access [to] and use of guns, rather than a change in the character of the youths, that best accounts for their increased involvement in lethal violence" (p. 60).

Exploration: How Do Schools Respond to Gun Violence?

Research that is exploratory in nature is generally concerned with uncovering detailed information about a given phenomenon, learning as much as possible about particular people

Exploratory

and/or events. Asmussen and Creswell (1995) were interested in the responses to school shooting incidents. Because school shootings are relatively rare and there was virtually no empirical literature that addressed the topic, Asmussen and Creswell performed an in-depth qualitative case study of a shooting incident that occurred at a large public university. The incident occurred in a senior-level actuarial science class when a male graduate student arrived early for the class one Monday morning armed with a vintage Korean War semiautomatic military rifle loaded with a 30-round clip of .30-caliber ammunition. He carried another 30-round clip in his pocket. Twenty of the 34 students in the class were already there, and most of them were quietly reading the students' newspaper. The instructor was en route to class. The gunman pointed the rifle at the students, swept it across the room, and pulled the trigger. The gun jammed. Trying to unlock the rifle, he hit the butt of it on the desk and quickly tried firing it again. Again it did not fire. By this time, most students realized what was happening and dropped to the floor, overturned their desks, and tried to hide behind them. After about 20 seconds, one of the students shoved a desk into the gunman, students ran past him out into the hall and out of the building. The gunman hastily left the room and went out of the building to his parked car, which he had left running. He was captured by police within the hour.

Asmussen and Creswell (1995) were surprised to find that instead of seeking safety after leaving the classroom, all the students had stood together just outside the building. Although a few were openly emotional and crying, many were kidding about the incident, as if it had posed no real danger. This denial response is noted by the mental health literature, which finds that feelings of fear and anger usually follow an initial response of denial and disbelief. The researchers found that many people in addition to the students who witnessed the incident were traumatized by the incident. In fact, three distinct groups of people sought counseling services during the weeks following the event. The first group had some direct involvement with the assailant, either by seeing him the day of the gun incident or because they had known him personally. The second group were the "silent connection," comprising individuals who were indirectly involved and yet emotionally traumatized. A large number of this group were parents. The third group included people who had previously experienced a trauma and whose fears were retriggered by this incident.

Because the assailant's future was in the hands of the criminal justice system, the importance of information from this system became critical to feelings of safety. Two combined debriefing sessions by mental health counselors, the campus police chief, and two county attorneys were held for all those interested.

Another phenomenon observed on the campus was an increase in the number of professors and staff who were concerned with disruptive students or students who exhibited aberrant behavior in class. To deal with this, the Student Judiciary Office advised departments on various methods of dealing with students who exhibited abnormal behavior in class. In addition, plainclothes police officers were sent to sit outside classrooms and offices whenever faculty and staff indicated concerns.

To their surprise, Asmussen and Creswell (1995) found that the campus did not establish a special unit to manage future incidents. The only structural change made was the installation of emergency phones throughout the campus. In addition, no discussion was reported about formal linkages with community agencies that might assist in the event of a future tragedy. In the end, Asmussen and Creswell's exploratory research illustrates the complexity involved in responding to a campus shooting.

Explanation: What Factors Are Related to Youth Delinquency and Violence?

Felson et al. (1994) sought to determine whether individual and group differences in beliefs and attitudes about violence were related to committing acts of delinquency and interpersonal violence. The theoretical argument they were attempting to test is actually termed the "subculture of violence" thesis (Wolfgang & Ferracuti 1967). According to this thesis, some groups are more violent than others because they have a distinctive set of values that either support or tolerate violence.

Felson et al. (1994) used a data set called the Youth in Transition data, which was a panel study of high school boys. The first interviews were conducted with sophomore boys attending 87 randomly selected public high schools. The second interview was collected 18 months later at the end of the boys' junior year in high school. The researchers measured each boy's adherence to a subculture of violence by asking him about situations in which he would approve of aggression as a legitimate expression of grievances and as an appropriate response to personal attack. For example, each boy was presented a series of personal values and asked whether each was a "good thing for people to do": (1) turning the other cheek and forgiving others when they harm you, (2) replying to anger with gentleness, and (3) being kind to people even if they do things against one's own beliefs. These items each focus on the approval of nonaggressive responses to some type of provocation. Agreement with these statements therefore suggests that respondents disapprove of aggressive responses to personal attacks and wrongdoing. Disagreement with these statements would result in a high score on the subculture of violence. Each boy was asked the frequency with which he engaged in eight activities: got into a serious fight with a student in school, obtained something by telling a person something bad would happen to the person if he did not get what he wanted, hurt someone badly enough to need bandages or a doctor, hit a teacher, hit his father, hit his mother, took part in a group brawl in which his friends fought with another group, and used a knife or gun or some other thing to get something from a person. The respondents were also asked about other delinquent acts such as engaging in theft and vandalism.

The students' scores on the subculture of violence index were then used to predict scores on the interpersonal violence index, after controlling for the effects of other important variables such as socioeconomic status, race, attitudes toward academic achievement, and family stability. Felson et al. (1994) performed analyses using both individual and school-level data and found that in some schools, an aggressive response to a provocation was likely to meet with more peer approval than in other schools. They found that individuals who approved of aggressive responses to provocation were more likely to commit acts of interpersonal violence. In addition, individuals who went to schools where there was an atmosphere permissive of violence were more likely to act violently, a find that remained true even after controlling for individual attitudes. Felson and his colleagues attribute this to a social control process and explain, "If boys are expected to retaliate when provoked, it appears that they are more likely to engage in violence and other delinquent behavior, no matter what their personal values may be" (p. 164).

Evaluation: Do Violence Prevention Programs in Schools Work?

To reduce violence and create a safer atmosphere at schools across the country, literally thousands of schools have adopted some form of violence prevention training (Powell,

Muir-McClain, & Halasyamani 1995). These programs generally provide cognitive-behavioral and social skills training on various topics using a variety of methods. Such programs are commonly referred to as *conflict resolution* and *peer mediation training*. Many of these prevention programs are designed to improve interpersonal problem-solving skills among children and adolescents by training children in cognitive processing, such as identifying the interpersonal problem and generating nonaggressive solutions. There is limited evidence, however, that such programs are actually effective in reducing violence.

Grossman et al. (1997) assessed the efficacy of one such program for children in elementary school called "The Second Step: A Violence Prevention Curriculum." The program involved 30 lessons, each lasting about 35 minutes, taught once or twice a week. Each lesson consisted of a photograph accompanied by a social scenario that formed the basis for discussion, role playing, and conceptual activities. Lessons were arranged in three units: (1) empathy training, in which students identified their own feelings and those of others; (2) impulse control, in which students were presented a problem-solving strategy and behavioral skills for affecting solutions (e.g., apologizing or dealing with peer pressure); and (3) anger management, in which students were presented a coping strategy and behavioral skills for tense situations.

Twelve elementary schools in King County, Washington, were paired according to three criteria: the school district, the proportion of students receiving free or reduced-cost lunch, and the proportion of minority enrollment. After pairing, schools in each pair were randomly assigned either to receive the Second Step program (experimental groups) or to receive no violence prevention curriculum (control groups). Random assignment was necessary so the researchers could be more confident that any differences observed in aggression and violent behavior between the two groups after the program could be attributed to the program alone and not to some other factor. Violent and aggressive behavior was measured in three ways: teacher ratings of each child's behavior, parent ratings, and direct observation of students by trained observers in the classroom, playground, and cafeteria. Measures of aggression were taken at three time periods: before the start of the curriculum (baseline), 2 weeks after the conclusion of the curriculum, and 6 months after the curriculum.

To determine the effectiveness of the Second Step program, researchers examined the change in aggression between scores measured at baseline and those from the second and third periods of data collection. Grossman et al. (1997) found encouraging results: Observed physically aggressive behavior decreased significantly more among children who engaged in the curriculum than among children in the control group who were not exposed to the Second Step program. Moreover, prosocial behavior increased significantly among children in the Second Step program compared to the control group. Grossman et al. concluded, "This violence prevention curriculum appears to lead to modest reductions in levels of aggressive behavior and increases in neutral/prosocial behavior in school among second and third graders" (p. 1608).

STRENGTHS AND LIMITATIONS OF SOCIAL RESEARCH

These case studies are only four of the dozens of studies investigating youth violence, but they illustrate some of the questions criminological research can address, several different

SOCIAL RESEARCH PHILOSOPHIES

What influences the choice of a research strategy? The motive for conducting research is critical: An explanatory or evaluative motive generally leads a researcher to use quantitative methods, whereas an exploratory motive often results in the use of qualitative methods. Of course, a descriptive motive means choosing a descriptive research strategy.

Positivism and Postpositivism

A researcher's philosophical perspective on reality and on the appropriate role of the researcher also will shape her methodological preferences. Researchers with a **positivist** philosophy believe that there is an objective reality that exists apart from the perceptions of those who observe it; the goal of science is to better understand this reality.

> Whatever nature "really" is, we assume that it presents itself in precisely the same way to the same human observer standing at different points in time and space. . . . We assume that it also presents itself in precisely the same way across different human observers standing at the same point in time and space. (Wallace 1983:461)

This philosophy is traditionally associated with science (Weber 1949:72), with the expectation that there are universal laws of human behavior, and with the belief that scientists must be objective and unbiased to see reality clearly.

Postpositivism is a philosophy of reality that is closely related to positivism. Postpositivists believe that there is an external, objective reality, but are very sensitive to the complexity of this reality and the limitations of the scientists who study it—and, for social scientists, the biases they bring to the study of social beings like themselves (Guba & Lincoln 1994:109–111). As a result, they do not think scientists can ever be sure that their methods allow them to perceive objective reality; the goal of science can only be to achieve **intersubjective agreement** among scientists about the nature of reality (Wallace 1983:461). For example, postpositivists may worry that researchers' predispositions may bias them in favor of deterrence theory. Therefore, they will remain somewhat skeptical of results that support predictions based on deterrence until a number of researchers feel that they have found supportive evidence. The postpositivist retains much more confidence in the ability of the community of social researchers to develop an unbiased account of reality than in the ability of any individual social scientist to do so (Campbell & Russo 1999:144).

Positivist Research Guidelines

To achieve an accurate, or valid, understanding of the social world, a researcher operating within the positivist or postpositivist tradition must adhere to some basic guidelines about how to conduct research:

1. *Test ideas against empirical reality without becoming too personally invested in a particular outcome.* This guideline requires a commitment to "testing," as opposed to just reacting to events as they happen or looking for what we want to see (Kincaid 1996:51–54).

2. *Plan and carry out investigations systematically*. Social researchers have little hope of conducting a careful test of their ideas if they do not think through in advance how they should go about the test and then proceed accordingly.

3. *Document all procedures and disclose them publicly*. Social researchers should disclose the methods on which their conclusions are based so that others can evaluate for themselves the likely soundness of these conclusions. Such disclosure is a key feature of science. It is the community of researchers, reacting to each others' work, that provides the best guarantee against purely self-interested conclusions (Kincaid 1996).

4. *Clarify assumptions*. No investigation is complete unto itself; whatever the researcher's method, the research rests on some background assumptions. For example, research to determine whether arrest has a deterrent effect assumes that potential law violators think rationally, and that they calculate potential costs and benefits prior to committing crimes. By definition, research assumptions are not tested, so we do not know for sure whether they are correct. By taking the time to think about and disclose their assumptions, researchers provide important information for those who seek to evaluate the validity of research conclusions.

5. *Specify the meaning of all terms*. Words often have multiple or unclear meanings. "Recidivism," "self-control," "poverty," "overcrowded," and so on can mean different things to different people. In scientific research, all terms must be defined explicitly and used consistently.

6. *Maintain a skeptical stance toward current knowledge*. The results of any particular investigation must be examined critically, although confidence about interpretations of the social or natural world increases after repeated investigations yield similar results. A general skepticism about current knowledge stimulates researchers to improve the validity of current research results and expand the frontier of knowledge.

7. *Replicate research and build social theory*. No one study is definitive by itself. We cannot fully understand a single study's results apart from the larger body of knowledge to which it is related, and we cannot place much confidence in these results until the study has been replicated. Theories organize the knowledge accumulated by numerous investigations into a coherent whole and serve as a guide to future inquiries.

8. *Search for regularities or patterns*. Positivist and postpositivist scientists assume that the natural world has some underlying order of relationships, so that unique events and individuals can be understood at least in part in terms of general principles (Grinnell 1992:27–29).

Real investigations by social scientists do not always include much attention to theory, specific definitions of all terms, and so forth. But it behooves any social researcher to study these guidelines and to consider the consequences of not following any with which they do not agree.

A Positivist Research Goal: Advancing Knowledge

The goal of the traditional positivist scientific approach is to advance scientific knowledge. This goal is achieved when research results are published in academic journals or presented at academic conferences.

The positivist approach regards value considerations to be beyond the scope of science: "An empirical science cannot tell anyone what he should do—but rather what he can do—and under certain circumstances—what he wishes to do" (Weber 1949:54). The idea is that developing valid knowledge about how society *is* organized, or how we live our lives, does not tell us how society *should* be organized or how we *should* live our lives. The determination of empirical facts should be a separate process from the evaluation of these facts as satisfactory or unsatisfactory (Weber 1949:11).

The idea is not to ignore value considerations, because they are viewed as a legitimate basis for selecting a research problem to investigate. In addition, many scientists also consider it acceptable to encourage government officials or private organizations to act on the basis of a study's findings, after the research is over. During a research project, however, value considerations are to be held in abeyance.

Positivism The belief, shared by most scientists, that there is a reality that exists quite apart from our own perception of it, although our knowledge of this reality may never be complete.

Intersubjective agreement An agreement by different observers on what is happening in the natural or social world.

Postpositivism The belief that there is an empirical reality but that our understanding of it is limited by its complexity and by the biases and other limitations of researchers.

Interpretivism and Constructivism

Qualitative research is often guided by a different, **interpretivist** philosophy. Interpretive social scientists believe that social reality is socially constructed and that the goal of social scientists is to understand what meanings people give to reality, not to determine how reality works apart from these interpretations. This philosophy rejects the positivist belief that there is a concrete, objective reality that scientific methods help us to understand (Lynch & Bogen 1997); instead, interpretivists believe that scientists construct an image of reality based on their own preferences and prejudices and their interactions with others.

Here is the basic argument: All the empirical data we collect comes to us through our own senses and must be interpreted with our own minds. This suggests that we can never be sure that we have understood reality properly, or that we ever can, or that our own understandings can really be judged more valid than someone else's. Concerns like this have begun to appear in many areas of social science and have begun to shape some research methods. From this standpoint, the goal of validity becomes misleading: "Truth is a matter of the best-informed and most sophisticated construction on which there is consensus at a given time" (Schwandt 1994:128).

Searching for universally applicable social laws can distract from learning what people know and how they understand their lives. The interpretive social researcher examines meanings that have been socially constructed. . . . There is not one reality out there to be measured; objects and events are understood by different people differently, and those perceptions are the reality—or realities—that social science should focus on. (Rubin & Rubin 1995:35)

The **constructivist paradigm** extends interpretivist philosophy by emphasizing the importance of exploring how different stakeholders in a social setting construct their beliefs (Guba & Lincoln 1989:44–45). It gives particular attention to the different goals of researchers and other participants in a research setting and seeks to develop a consensus among participants about how to understand the focus of inquiry. The constructivist research report will highlight different views of the social program or other issue and explain how a consensus can be reached among participants.

Interpretivism The belief that reality is socially constructed and that the goal of social scientists is to understand what meanings people give to that reality. Max Weber termed the goal of interpretivist research *verstehen*, or "understanding."

Constructivist paradigm A perspective that emphasizes how different stakeholders in social settings construct their beliefs.

Constructivist inquiry uses an interactive research process, in which a researcher begins an evaluation in some social setting by identifying the different interest groups in that setting. The researcher goes on to learn what each group thinks, and then gradually tries to develop a shared perspective on the problem being evaluated (Guba & Lincoln 1989:42).

These steps are diagrammed as a circular process in Exhibit 1.5. In this process, called a **hermeneutic circle,**

the constructions of a variety of individuals—deliberately chosen so as to uncover widely variable viewpoints—are elicited, challenged, and exposed to new information and new, more sophisticated ways of interpretation, until some level of consensus is reached (although there may be more than one focus for consensus). (Guba & Lincoln 1989:180–181)

The researcher conducts an open-ended interview with the first respondent (R1) to learn about her thoughts and feelings on the subject of inquiry, her "construction" (C1). The researcher then asks this respondent to nominate a second respondent (R2), who feels very differently. The second respondent is then interviewed in the same way, but also is asked to comment on the themes raised by the previous respondent. The process continues until all major perspectives are represented, and then may be repeated again with the same set of respondents.

more precisely, it became clear that employee participation in production decisions had substantially increased overall productivity, whereas simple meeting attendance had not. This discovery would not have occurred without the active involvement of company employees in planning the research.

An Integrated Philosophy

It is tempting to think of positivism and postpositivism as representing an opposing research philosophy to interpretivism and constructivism. Then it seems that we should choose the one philosophy that seems closest to our own preferences and condemn the other as "unscientific," "uncaring," or perhaps just "unrealistic." But there are good reasons to prefer a research philosophy that integrates some of the differences between these philosophies (Smith 1991).

Society is a product of human action that in turn shapes how people act and think. The "sociology of knowledge" studies this process by which people make themselves as they construct society (Berger & Luckmann 1966). Individuals internalize the social order through the process of socialization, so that their own beliefs and actions are not entirely of their own making, but instead reflect the social order of which they are a part. This means that we should be very careful to consider how our research approaches and interpretations are shaped by our own social background, just as we are cautioned to do by interpretivist researchers.

When we peer below the surface of standardized research procedures, we also discover the importance of taking into account people's feelings and the meanings that they attach to these feelings. For example, Lavin and Maynard (2001) investigated how different survey research centers handle laughter by respondents during telephone surveys. The dilemma for the centers is this: When respondents laugh during an interview, it usually is an attempt to increase rapport with the interviewer, so turning down the "invitation" to laugh can make the interviewer seem unsympathetic. However, accepting this "invitation" to laugh injects an uncontrolled source of bias into what is supposed to be a standardized interview.

> As interviewers manage laughter in the interview, they artfully maneuver through the dilemma of adhering to standardization protocols while maintaining rapport with respondents that retains participation and continues to elicit answers. (P. 473)

As a result, what appear to be standardized interviews in fact vary in ways that are not apparent in the answers recorded by the interviewer. Recognition of this interpretive process can improve survey research conducted in the positivist tradition. Researchers cannot ignore the subjective aspects of human experience or expunge it entirely from the data collection process. This helps to explain why the debate continues between positivist and interpretivist philosophies and why research can often be improved by drawing on insights from both. In the words of Stephen P. Turner (1980), "The distinctive empirical concerns of 'interpretive' and 'statistical' sociologies, usually thought of as antithetical or mutually irrelevant, can be made to mesh" (p. 99).

And what about the important positivist distinction between facts and values in social research? Here, too, there is evidence that neither the "value-free" presumption of positivists nor the constructivist critique of this position is entirely correct. For example, Savelsberg, King, and Cleveland (2002) examined influences on the focus and findings of published criminal justice scholarship. They found that criminal justice research was more likely to be oriented to topics and theories suggested by the state when it was funded by government agencies. This reflects a political influence on scholarship. However, government funding did not have any bearing on the researchers' conclusions about the criminal justice processes they examined. This suggests that scientific procedures can insulate the research process itself from political pressure.

Which philosophy makes the most sense to you? Do you agree with positivists and post-positivists that scientific methods can help us understand the social world as it is, not just as we would like to think it is? Does the interpretivist focus on meanings sound like a good idea? Whatever your answers to these questions, you would probably agree that developing a valid understanding of the social world is not an easy task for social scientists and this is the subject we turn to next.

VALIDITY: THE GOAL OF SOCIAL RESEARCH

A scientist seeks to develop an accurate understanding of empirical reality, the reality we encounter firsthand, by conducting research that leads to valid knowledge about the world. But when is knowledge valid? In general, we have reached the goal of validity when our statements or conclusions about empirical reality are correct. If you look out your window and observe that it is raining, this is probably a valid observation, if your eyes and ears are to be trusted. However, if you pick up the newspaper and read that the majority of Americans favor the death penalty, this conclusion is of questionable validity because it is probably based on an interpretation of a social survey. As you will see in Chapter 7, attitudes toward the death penalty vary substantially depending on the wording of the questions asked.

To some of you, the goal of validity may sound a bit far-fetched. After all, how can we really be sure our understandings of empirical phenomena are correct when we can perceive the world only through the filter of our own senses? You need not worry. Such skepticism will help you remember the tenuousness of all knowledge and will keep you properly skeptical about new discoveries.

This book is about validity more than anything else, about how to conduct research that leads to valid interpretations of the social world. We will refer to validity repeatedly, and we ask you to register it in your brain now as the central goal of all the research conducted in our field. The goal of research conducted by social scientists investigating issues related to criminology and criminal justice is not to come up with conclusions that people will like or conclusions that suit their personal preferences. The goal is to determine the most valid answers through empirical research methods.

We must be concerned with three aspects of validity: **measurement validity**, **generalizability**, and **causal validity** (also known as internal validity). Each of these three aspects of validity is essential: Conclusions based on invalid measures, invalid generalizations, or invalid causal inferences will themselves be invalid.

Imagine that we survey a sample of 250 high school seniors and ask them two questions: "Do you have friends who have taken illegal drugs in the past 6 months?" (the measure of peer behavior) and "Have you taken illegal drugs in the past 6 months?" (individual use measure). We then compare the frequency of illegal drug use between students who have friends who have used illegal drugs and those whose friends have not used illegal drugs. We find that students who have friends who have used illegal drugs in the last 6 months are more likely to have used drugs themselves, and we conclude that drug use is, in part, due to the influence of peers.

But did our questions indeed tell us the frequency with which the students and their peers took illegal drugs? If they did, we achieved measurement validity. Do our results hold true of the larger adolescent population to which our conclusion referred? If so, our conclusion would satisfy the criterion for generalizability. Did the likelihood of students taking drugs actually increase if they had friends who also took drugs? If so, our conclusion is causally valid.

Measurement validity Exists when a measure measures what we think it measures.

Generalizability Exists when a conclusion holds true for the population, group, setting, or event that we say it does, given the conditions that we specify.

Causal validity (internal validity) Exists when a conclusion that A leads to or results in B is correct.

The goal in research is to achieve valid understandings of the social world by coming to conclusions that rest on valid measures and valid causal assertions and that are generalizable to the population of interest. Once we have learned how to develop studies that give us reasonably valid results and how to evaluate studies according to how well they meet this criterion, we will be well along on the road to becoming expert researchers. Chapters 3, 4, and 5 discuss these three aspects of validity and the specific techniques used to maximize the validity of our measures, our generalizations, and our causal assertions in greater detail.

CONCLUSION

We hope this first chapter has given you an idea of what to expect in the rest of this book. Our aim is to introduce you to social research methods by describing what social scientists have learned about issues in criminology and criminal justice as well as how they learned it. The substance of social science inevitably is more interesting than its methods, but the methods also become more interesting when they are not taught as isolated techniques. We have focused attention on research on youth violence and delinquency in this chapter; in subsequent chapters, we will introduce research examples from other areas.

Chapter 2 continues to build the foundation for our study of social research by reviewing the types of problems that criminologists study, the role of theory, the major steps in the research process, and other sources of information that may be used in social research.

We stress the importance of considering scientific standards in social research and review generally accepted ethical guidelines. Throughout the chapter, we use several studies of domestic violence to illustrate the research process.

KEY TERMS

Case report
Causal validity (internal validity)
Constructivist paradigm
Content analysis
Crime mapping
Descriptive research
Evaluation research
Experimental approach
Explanatory research
Exploratory research
External validity
Generalizability
Hermeneutic circle
Illogical reasoning
Inaccurate observation
Intensive interviewing
Interpretivism
Intersubjective agreement

Measurement validity
Overgeneralization
Participant observation
Participatory action research
Positivism
Postpositivism
Qualitative methods
Quantitative methods
Questionnaire
Resistance to change
Science
Secondary data analysis
Selective observation
Social science
Survey
Unobtrusive measures
Validity

HIGHLIGHTS

- Criminological research cannot resolve value questions or provide answers that will convince everyone and remain settled for all time.
- All empirically based methods of investigation are based on either direct experience or others' statements.
- Four common errors in reasoning are overgeneralization, selective or inaccurate observation, illogical reasoning, and resistance to change. Illogical reasoning is due to the complexity of the social world, self-interestedness, and human subjectivity. Resistance to change may be due to unquestioning acceptance of tradition or of those in positions of authority, or to self-interested resistance to admitting the need to change one's beliefs.
- Social science is the use of logical, systematic, documented methods to investigate individuals, societies, and social processes, as well as the knowledge produced by these investigations.
- Criminological research can be motivated by policy guidance and program management needs, academic concerns, and charitable impulses.
- Criminological research can be descriptive, exploratory, explanatory, or evaluative, or some combination of these.
- Quantitative methods record variation in social life in terms of categories that vary in amount. Qualitative methods are designed to capture social life as participants experience it, rather than in categories predetermined by the researcher.

- Positivism is the belief that there is a reality that exists quite apart from one's own perception of it that is amenable to observation.
- Intersubjective agreement is an agreement by different observers on what is happening in the natural or social world.
- Postpositivism is the belief that there is an empirical reality but that our understanding of it is limited by its complexity and by the biases and other limitations of researchers.
- Interpretivism is the belief that reality is socially constructed and the goal of social science should be to understand what meanings people give to that reality.
- The constructivist paradigm emphasizes the importance of exploring and representing the ways in which different stakeholders in a social setting construct their beliefs. Constructivists interact with research subjects to gradually develop a shared perspective on the issue being studied.
- Valid knowledge is the central concern of scientific research. The three components of validity are measurement validity, generalizability (both from the sample to the population from which it was selected and from the sample to other populations), and causal (internal) validity.

EXERCISES

1. What criminological topic or issue would you focus on if you could design a research project without any concern for costs? What are your motives for studying this topic? List at least four of your beliefs about this phenomenon. Try to identify the sources of each belief, for example, television, newspaper, parental influence.
2. Develop four research questions related to your chosen topic or issue, one for each of the four types of research (descriptive, exploratory, explanatory, and evaluative). Be specific.
3. Read the abstracts of each article in a recent issue of a major criminological journal. Identify the type of research conducted for each study.
4. Continue the debate between positivism and interpretivism with an in-class debate. Be sure to review the guidelines for these research philosophies and the associated goals. You might also consider whether an integrated philosophy is preferable.
5. Outline your own research philosophy. You can base your outline primarily on your reactions to the points you have read in this chapter, but try also to think seriously about which perspective seems more reasonable to you.

DEVELOPING A RESEARCH PROPOSAL

Will you develop a research proposal in this course? If so, you should begin to consider your alternatives.

1. What topic would you focus on if you could design a social research project without any concern for costs? What are your motives for studying this topic?
2. Develop four questions that you might investigate about the topic you just selected. Each question should reflect a different research motive: description, exploration, explanation, or evaluation. Be specific.
3. Which question most interests you? Would you prefer to attempt to answer that question with quantitative or qualitative methods? Why?

Student Study Site

The companion Web site for *The Practice of Research in Criminology and Criminal Justice*, Third Edition

http://www.sagepub.com/prccj3

Visit the Web-based Student Study Site to enhance your understanding of the chapter content and to discover additional resources that will take your learning one step further. You can enhance your understanding of the chapters by using the comprehensive study material, which includes e-flashcards, Web exercises, practice self-tests, and more. You will also find special features, such as Learning from Journal Articles, which incorporates SAGE's online journal collection.

WEB EXERCISES

1. You have been asked to prepare a brief presentation on a criminological topic or issue of interest to you. Go to the Bureau of Justice Statistics (BJS) Web site at www.ojp.usdoj.gov/bjs/. Browse the BJS publications for a topic that interests you. Write a short outline for a 5-to 10-minute presentation regarding your topic, including statistics and other relevant information.

2. Go to the Federal Bureau of Investigation (FBI) Web site at www.fbi.gov. Explore the types of programs and initiatives sponsored by the FBI. Discuss at least three of these programs or initiatives in terms of their purposes and goals. For each program or initiative examined, do you believe the program or initiative is effective? What are the major weaknesses? What changes would you propose the FBI make to more effectively meet the goals of the program or initiative?

3. There are many interesting Web sites that discuss philosophy of science issues. Read the summaries of positivism and interpretivism at http://www.rouncefield.homestead.com/theory.html (under the "Background" heading). What do these summaries add to your understanding of these philosophical alternatives?

ETHICS EXERCISES

Throughout the book, we will be discussing the ethical challenges that arise in research on crime and criminal justice. At the end of each chapter, we will ask you to consider some questions about ethical issues related to that chapter's focus. We introduce this critical topic formally in Chapter 2, but we will begin here with some questions for you to ponder.

1. You have now learned about Asmussen and Creswell's (1995) qualitative study of a school shooting incident. We think it provided important information for policymakers about the social dynamics in these tragedies. But what would *you* do if you were conducting a similar study in a high school and you learned that another student was planning to bring a gun to school to kill some other students? What if he was only thinking about it? Or just talking with his friends about how "neat" it would be? Can you suggest some guidelines for researchers?

2. Grossman et al. (1997) found that the "Second Step" program reduced aggressive behavior in schools and increased prosocial behavior. If you were David Grossman, would you announce your findings in a press conference and encourage schools to adopt this program? If you were a school principal who heard about this research, would you agree to let another researcher replicate (repeat) the Grossman et al. study in your school, with some classrooms assigned to receive the Second Step program randomly (on the basis of the toss of a coin) and others not allowed to receive the program for the duration of the study?

SPSS EXERCISES

The SPSS exercises at the end of each chapter use the data sets included on the companion Web site. In NCS.POR, a sample of 503 robbery incidents from the National Crime Victimization Survey, variable V2089 identifies the metropolitan statistical area (MSA) status (urban, suburban, or rural) of each robbery incident included in the data set.

1. Create a bar chart of V2089 using the graph procedure to show the percentage of robbery incidents occurring in urban, suburban, and rural areas. Be sure to select "options" and indicate that you want to leave out the missing values.

2. Write four research questions, one for each type of social research (descriptive, exploratory, explanatory, and evaluative), regarding the distribution of robbery incidents by MSA status.

3. Discuss the possible reasons (policy, academic, or personal) for conducting research on the location of robbery incidents.

The Process and Problems of Criminological Research

On July 4, 1997, the *Pathfinder* module landed on Mars after a 6-month trip from Earth. Within a day, a little vehicle was exploring Martian rocks and soil. Cameras were transmitting live pictures back to Earth, where millions of people were tuned in through their computers to special World Wide Web sites set up for the occasion. Once again, science was transforming our image of the universe and helping us transcend our natural physical and mental limits, just as computers, brain-imaging devices, and nuclear power had done before. It is no exaggeration to say that the physical and natural sciences have forever altered human life and continue to do so. Although social science has nothing like this impact, it

Social Science

does influence the design of social programs, the course of elections, the composition of juries, the strategies of business, and most important, our understanding of the social world.

Consider the impact of criminological research on the crime of domestic violence. Intimate partner violence (violence between spouses or intimates), sometimes referred to as domestic violence, is a major problem in our society, with police responding to between 2 million and 8 million complaints of assault by a spouse or lover yearly (Sherman 1992:6). Moreover, it is estimated from victimization surveys that many of these assaults are never reported to police (Bachman & Saltzman 1995; Tjaden & Thoennes 2000). Domestic violence is not just a frequent crime, it is also costly not only in terms of the injuries suffered by the parties involved but also in terms of shattered families. What to do about this major social problem, then, is an important policy question. One thing the police could do would be to respond to incidents of domestic violence in such a way that the offender would be less likely to be violent in the future. That is, proper police response could possibly prevent some acts of domestic violence. But what is the proper police response?

Policy Question

In 1981, the Police Foundation and the Minneapolis Police Department began an experiment to determine whether arresting accused spouse abusers on the spot would deter future offending incidents. In responding to police calls for service in misdemeanor domestic cases, the responding officer randomly assigned the case to be resolved by either arresting or not arresting the suspect on the scene. The experimental treatment, then, was whether the suspect was arrested, and the researchers wanted to know whether arrest was better than not arresting the suspect in reducing recidivism (subsequent assaults against the same victim). The study's results, which were widely publicized, indicated that arrest did have a deterrent effect. Partly as a result of the reported results of this experiment, the percentage of urban police departments that made arrest the preferred response to complaints of domestic violence rose from 10% in 1984 to 90% in 1988 (Sherman 1992:14). Six other cities then hosted studies like the Minneapolis experiment (collectively, this was called the Spouse Assault Replication Program [SARP]), but the results were not always so clear-cut as in the original study (Buzawa & Buzawa 1996; Hirschel, Hutchison, & Dean 1992; Pate & Hamilton 1992; Sherman 1992; Sherman & Berk 1984). In some cities (and for some people), arrest did seem to prevent future incidents of domestic assault; in other cities, it seemed only to make matters worse, contributing to additional assault; and in still other cities, arrest seemed to have no discernable effect. After these replications of the original Minneapolis experiment, people still wondered, "Just what is the effect of arrest in reducing domestic violence cases, and how should the police respond to such cases?" The answer was not clear. The Minneapolis Domestic Violence Experiment, the studies modeled after it, and the related controversies provide many examples for a systematic overview of the social research process.

sources in research is what to study

The first concern in criminological research (indeed in any research) is deciding what to study. That is, how does one go about selecting an issue, problem, or question to address with research? As you will learn in the next section, one source of the motivation to do research is criminological theory. In criminology, as in any other science, theory plays an important role as a basis for formulating research questions and later understanding the larger implications of one's research results. Another motivation for research is one's personal interests. There is nothing wrong with researching those issues that you find fascinating or interesting. There are other motivational sources for research that we will explore in this chapter, including helping to answer questions illuminated by earlier research. In this chapter we use the Minneapolis experiment and the SARP replication research to illustrate the three main research strategies: deductive, inductive, and descriptive research. In all three, theory and data are inextricably linked. The chapter ends with scientific and ethical guidelines that should be adhered to no matter what the research strategy, and shows how the Minneapolis experiment followed these guidelines. By the chapter's end, you should be ready to formulate a criminological research question, design a general strategy for answering this question, and critique previous studies that addressed this question. You can think of Chapter 1 as having introduced the "why" of criminological research; Chapter 2 introduces the "how."

CRIMINOLOGICAL RESEARCH QUESTIONS

How does a criminologist decide what to study and research? A **criminological research question** is a question about some aspect of crime or criminals that you seek to answer through the collection and analysis of firsthand, verifiable, empirical data. The types of questions that can be asked are virtually limitless. For example, "Are children who are violent more likely than nonviolent children to use violence as adults?" "Does the race of victim who is killed influence whether someone is sentenced to death rather than life imprisonment?" "Why do some kinds of neighborhoods have more crime than others? Is it due to the kinds of people who live there or characteristics of the neighborhood itself?" "Does community policing reduce the crime rate?" "Has the U.S. government's war on drugs done anything to reduce the use of illegal drugs?" So many research questions are possible in criminology that it is more of a challenge to specify what does not qualify as a social research question than to specify what does.

But that does not mean it is easy to specify a research question. In fact, formulating a good research question can be surprisingly difficult. We can break the process into three stages: identifying one or more questions for study, refining the questions, and then evaluating the questions.

Identifying Criminological Research Questions

Formulating a research question is often an intensely personal process in addition to being a scientific or professional one. Research questions may emerge from your "personal troubles," as Mills (1959) put it, or your personal experiences. These troubles or experiences could range from how you felt when you were picked up by the police and perhaps arrested when you were a teenager for something you did not do, to the awareness you may have that crime is not randomly distributed within a city but that there seem to be "good" or safe parts of town and "bad" or unsafe areas. You may find yourself asking questions such as "Do the police sometimes make things worse for people when they arrest them, worse in the sense that they are more likely to commit crime in the future?" or "Does victimization change a person's trust in others?" or "Does involvement in crime vary by one's age, social class, gender, or racial or ethnic group?" Can you think of other possible research questions that flow from your own experience in the world?

The experience of others is another fruitful source of research questions. Knowing a relative who was abused by a spouse, seeing a TV special about violence, or reading a gang member's autobiography can stimulate questions about general criminological processes. Can you draft a research question based on a relative's experiences, a TV show, or a book?

Sherman

Other researchers may also pose interesting questions for you to study. Most research articles end with some suggestions for additional research that highlight unresolved issues. For example, Sherman et al. (1992) concluded an article on some of the replications of the Minneapolis experiment on police responses to spouse abuse by suggesting that "deterrence may be effective for a substantial segment of the offender population. . . . However, the underlying mechanisms remain obscure" (p. 706). A new study could focus on the mechanisms: Why or under what conditions does the arrest of offenders who are employed deter them from future criminal acts? Exactly what occurs when someone is arrested for domestic violence that may lead him or her not to be violent against a spouse in the future? Is it the brute fear of being arrested and having to go to jail? Is it the fear that one's employer may find out and fire him or her? Is it the fear that members of the community may learn about the arrest and the offender does not want to lose his or her good standing in the neighborhood? Is it all these? Any issue of a journal in your field is likely to have comments that point toward unresolved issues.

Theory

The primary source of research questions for many criminologists is criminological theory. As you will soon learn, criminological theory provides an explanation as to why crime occurs, or occurs in some places and under some conditions but not others. Theory, then, is a very rich source of research ideas. Some researchers spend much of their careers conducting research intended to refine an answer to one central question. For example, you may find rational choice theory to be a useful approach to understanding diverse forms of social behavior, like crime, because you think people do seem to make decisions on the basis of personal cost-benefit calculations. So you may ask whether rational choice theory can explain why some people commit crimes and others do not, or why some people decide to quit committing crimes while others continue their criminal ways.

Rational Choice Theory

Finally, some research questions have very pragmatic sources. You may focus on a research question posed by someone else because doing so seems to be to your professional of financial advantage. Some criminologists conduct research on specific questions posed by a funding source in what is termed a request for proposals (RFP). (Sometimes the acronym RFA is used, meaning request for applications.) Or you may learn that the public defenders in your city are curious as to whether they are more successful in getting their clients acquitted of a criminal charge than private lawyers.

Refining Criminological Research Questions

As you have perhaps surmised by now, the problem is not so much coming up with interesting criminological questions for research as it is focusing on a problem of manageable size. We are often interested in much more than we can reasonably investigate with our limited time and resources (or the limited resources of a funding agency). Researchers may worry about staking a research project (and thereby a grant) on a particular problem and so address several research questions at once, often in a jumbled fashion. It may also seem risky to focus on a research question that may lead to results discrepant with our own cherished assumptions about the social world. In addition, the prospective commitment of time and effort for some research questions may seem overwhelming, resulting in a certain degree of paralysis (not that the authors have any experience with this!).

The best way to avoid these problems is to develop the research question one bit at a time with a step-by-step strategy. Do not keep hoping that the perfect research question will just spring forth from your pen. Instead, develop a list of possible research questions as you go along. At the appropriate time, you can look through this list for the research questions that appear more than once. Narrow your list to the most interesting, most workable candidates. Repeat this process as long as it helps to improve your research questions. Keep in mind that the research you are currently working on will likely generate additional research questions for you to answer.

Evaluating Criminological Research Questions   3rd Stage - Research question

In the third stage of selecting a criminological research question, you evaluate the best candidate against the criteria for good social research questions: feasibility given the time and resources available, social importance, and scientific relevance (King, Keohane, & Verba 1994).

The research question in the Minneapolis Domestic Violence Experiment, "Does the formal sanction of police arrest versus non-arrest inhibit domestic violence?" certainly meets the criteria of social importance and scientific relevance, but it would not be a feasible question for a student project because it would require you to try to get the cooperation of a police department. You might instead ask the question "Do people (students) think that arrest will inhibit domestic violence?" This is a question that you could study with an on-campus survey. Or perhaps you could work out an arrangement with a local battered women's shelter to study the question "What leads some women to call the police when they are the victims of domestic violence, and why do they sometimes not call?" A review of the literature, however, might convince you that this and other questions may not be scientifically relevant because they have been studied enough.

Feasibility

You must be able to conduct any study within the time frame and resources you have. If time is short, questions that involve long-term change may not be feasible, for example, "If a state has recently changed its law so that it now permits capital punishment for those convicted of murder, does it eventually see a reduction in the homicide rate over time?" This is an interesting and important question, but one that requires years of data collection and research. Another issue is what people or groups you can expect to gain access to. Although well-experienced researchers may be granted access to police or correctional department files to do their research, less seasoned and well-known researchers or students may not be granted such access. It is also often difficult for even the most experienced of researchers to be given full access to the deliberations of a criminal jury. For someone interested in white-collar crime, recording the interactions that take place in corporate boardrooms may also be taboo.

Then you must consider whether you will have any additional resources, such as other researchers to collaborate with or research funds. Remember that there are severe limits on what one person can accomplish. On the other hand, you might work in an organization that collects data on employees, customers, or clients that are relevant to your research interests. Or you may be able to piggyback your research onto a larger research project.

You also must be prepared to handle large amounts of quantitative data. A computer and the skills to use it will be essential. Also take into account the constraints you face due to your schedule and other professional or personal commitments and obligations.

The Minneapolis Domestic Violence Experiment shows how ambitious social research questions can be when a team of seasoned researchers secures the backing of influential groups. The project required hundreds of thousands of dollars, the collaboration of many social scientists and criminal justice personnel, and the volunteer efforts of 41 Minneapolis police officers. But don't worry, many worthwhile research questions can be investigated with much more limited resources. You will read in subsequent chapters about studies that addressed important research questions with much fewer resources than the Minneapolis social scientists commanded.

Social Importance

Criminological research is not a simple undertaking, so you must focus on a substantive area that you feel is important and that is either important to the discipline or important for public policy. You also need to feel personally motivated to carry out the study; there is little point in trying to answer a question that does not interest you.

In addition, you should consider whether the research question is important to other people. Will an answer to the research question make a difference for society? Again, the Minneapolis Domestic Violence Experiment is an exemplary case. If that study showed that a certain type of police response to domestic violence reduced the risk of subsequent victimization, a great deal of future violence could be prevented. But clearly, criminology and criminal justice are not wanting for important research questions. Social scientists deal with such important issues as the effect of get-tough "three strikes" laws, whether strict supervision of those on parole and probation is more successful than less strict (and less costly) supervision, whether the death penalty deters criminals, whether gangs or the drug trade contributed to the rise in youth violence, whether crime is more prevalent in the United States than in European countries, whether gun control legislation reduces violence; we could go on and on. Many criminological questions deal with very important and troubling issues for our society.

Scientific Relevance

Every research question in criminology should be grounded in the existing empirical literature. By *grounded* we mean the research we do must be informed by what others before us have done on the topic. Whether you formulate a research question because you have been stimulated by an academic article or because you want to investigate a current public policy problem, you must turn to the criminological literature to find out what has already been learned about this question. (Appendix A explains how to find information about previous research, using both printed and computer-based resources.) Even if your research topic has already been investigated by someone else, that does not necessarily mean it would be a bad idea for you to do research on the issue. It would be unreasonable to think of any criminological research question as being settled for all time. You can be sure that some prior study is relevant to almost any research question you can think of, and you can also think of better ways to do research than have been done in the past.

For example, the Minneapolis experiment was built on a substantial body of contradictory theorizing about the impact of punishment on criminality (Sherman & Berk 1984). Deterrence theory predicted that because it was a more severe penalty, arrest would better deter individuals from repeat offenses than not arresting them. Labeling theory, on the other hand, predicted that arrest would make repeat offenses more likely because it would stigmatize offenders. The researchers found one prior experimental study of this issue, but it was conducted with juveniles. Studies among adults and nonexperimental research had not yielded consistent findings about the effects of arrest on recidivism in domestic violence cases. Clearly, the Minneapolis researchers had good reason for another study. Prior research and theory also helped them develop the most effective research design.

THE ROLE OF CRIMINOLOGICAL THEORY   *Role of Theory*

We have already pointed out that criminological **theory** can be a rich source of research questions. What deserves more attention at this point is the larger role of theory in research. Criminological theories do many things:

- They help us explain or understand things like why some people commit crimes or more crimes than others; why some people quit and others continue; and what the expected effect of good families, harsh punishment, or other factors on crime might be.
- They help us make predictions about the criminological world: "What would be the expected effect on the homicide rate if we employed capital punishment rather than life imprisonment?" "What would be the effect on the rate of property crimes if unemployment were to substantially increase?"
- They help us organize and make sense of empirical findings in a discipline.
- They help guide research.
- They help guide public policy: "What should we do to reduce the level of domestic violence?"

Social scientists, such as criminologists, who connect their work to theories in their discipline can generate better ideas about what to look for in a study and develop conclusions with more implications for other research. Building and evaluating theory is therefore one of the most important objectives of a social science like criminology.

Theory A logically interrelated set of propositions about empirical reality. Examples of criminological theories are social learning, routine activities, labeling, general strain, and social disorganization theory.

Constructs

Differential Association

For centuries, scholars have been interested in developing theories about crime and criminals. Sometimes these theories involve very fanciful ideas that are not well developed or organized, whereas at other times they strike us as being very compelling and well organized. Theories usually contain what are called **theoretical constructs**. In criminology, these theoretical constructs describe what is important to look at to understand, explain, predict, and "do something about" crime. For example, an important theoretical construct in differential association theory is the notion of "definitions favorable and unfavorable to the violation of law." Theories usually link one or more theoretical constructs to others in what are called *relationship statements*. Differential association theory, for example, would link the theoretical construct of favorable or unfavorable definitions to the theoretical construct of involvement in crime to argue: "As one is exposed to more definitions favorable to the violation of law relative to definitions unfavorable to the violation of law, the more one is at risk for criminal behavior." This is a relationship statement that links two theoretical constructs; it states that as exposure to definitions favorable to the law increases, the risk of crime also increases. This is essentially a hypothesis that the theory of differential association entertains; if the theory is true, then the expected relationship should be true. The purpose of much criminological research is to examine the truth value, or empirical validity, of such theoretical relationship statements or hypotheses. Some criminological theories reflect a substantial body of research and the thinking of many social scientists; others are formulated in the course of one investigation. A few have been widely accepted, at least for a time; others are the subject of vigorous controversy, with frequent changes and refinements in response to criticism and new research.

Most criminological research is guided by some theory, although the theory may be only partially developed in a particular study or may even be unrecognized by the researcher. When researchers are involved in conducting a research project or engrossed in writing a research report, they may easily lose sight of the larger picture. It is easy to focus on accumulating or clarifying particular findings rather than considering how the study's findings fit into a more general understanding of the social world. Furthermore, as we shall soon see, just as theory guides research, research findings also influence the development of theory.

We can use the studies of the police response to domestic assault to illustrate the value of theory for social research. Even in this very concrete and practical matter, we must draw on social theories to understand how people act and what should be done about those actions. Consider the three action options that police officers have when they confront a domestic assault suspect (Sherman & Berk 1984:263). Fellow officers might urge forced separation to achieve short-term peace; police trainers might prefer mediation to resolve the underlying dispute; feminist groups might urge arrest to protect the victim. None of these recommendations is really a theory, but each suggests a different perspective on crime and legal sanctions. The traditional police perspective sees domestic violence as a family matter that should not be the object of formal legal action. The preference for medication reflects the view that domestic violence involves a family crisis that can be solved with special counseling. The pro-arrest position views domestic violence as a crime as serious as that between strangers and favors arrest for its presumed deterrent effect.

You will encounter these different perspectives if you read much of the literature on domestic violence, or even if you talk with your friends about it. As Exhibit 2.1 shows, each perspective reflects different assumptions about gender roles, about the sources of crime,

EXHIBIT 2.1	Bases for Three Perspectives on Intimate Partner Violence		
Police officers			"Temporary separation of ... couple will bring short-term peace."
Police trainers	Experiences with family conflict, police actions, the legal system, and so on	Assumptions about gender roles, source of crime, impact of punishment, and so on	"Mediation and counseling will resolve family problems." ·
Feminist and Victims' Rights activists			"Arresting spouse abusers will have a deterrent effect."

and about the impact of punishment. In turn, these assumptions reflect different experiences with family conflict, police actions, and the legal system. What we believe about one crime and the appropriate response to it relates to a great many other ideas we have about the social world. Recognizing these relationships is a first step toward becoming a theoretically guided social researcher and a theoretically informed consumer of social research.

Remember, however, that social theories do not provide the answers to the questions we confront as we formulate topics for research. Instead, social theories suggest the areas on which we should focus and the propositions that we should consider for a test. That is, theories suggest testable hypotheses about things and research verifies whether those hypotheses are true. In fact, one of the most important requirements of theory is that it be testable, or what philosophers of science call **falsifiable**; theoretical statements must be capable of being proven wrong. If a body of thought cannot be empirically tested, it is more likely philosophy than theory. For example, Sherman and Berk's (1984) domestic violence research was actually a test of predictions derived from two alternative theories of the impact of punishment on crime, deterrence theory and labeling theory:

Deterrence theory presumes that human beings are at least marginally rational beings who are responsive to the expected costs and benefits of their actions. Committing a crime nets certain benefits for offenders; therefore, if we want to inhibit crime, there must be a compensating cost; one cost is the criminal sanction (arrest, conviction, punishment). Deterrence theory expects punishment to inhibit crime in two ways. General deterrence occurs when people see that crime results in undesirable punishments for others, that "crime doesn't pay." The persons who are punished serve as examples for those who have not yet committed an offense but might be thinking of what awaits them should they engage in proscribed acts. Specific deterrence occurs when persons who are punished decide not to commit another offense so they can avoid further punishment (Lempert & Sanders 1986:86–87). Deterrence theory leads to the prediction that arresting spouse abusers will reduce the likelihood of their reoffending when compared with a less serious sanction (not being arrested but being warned or counseled).

Labeling theory distinguishes between primary deviance, the acts of individuals that lead to public sanctions, and secondary deviance, the deviance that occurs in response to public sanction (Hagan 1994:33). Arrest or some other public sanction for misdeeds labels the offender as deviant in the eyes of others. Once the offender is labeled, others will treat the offender as a deviant, and he or she is then more likely to act in a way that is consistent with the deviant label. Ironically, the act of punishment stimulates more of the very behavior that it was intended to eliminate (Tannenbaum 1938). This theory suggests that persons arrested for intimate partner violence are more likely to reoffend than those who are caught but not punished because the formal sanction of arrest is more stigmatizing than being warned or counseled. This prediction about the effect of formal legal sanctions is the reverse of the deterrence theory prediction.

Theorizing about the logic behind formal legal punishment also can help us draw connections to more general theories about social processes. Deterrence theory reflects the assumptions of rational choice theory, which assumes behavior is shaped by practical calculations: People break the law if the benefits of doing so exceed the costs. If crime is a rational choice for some people, then increasing the certainty or severity of punishment for crime should shift the cost-benefit balance away from criminal behavior. Labeling theory is rooted in symbolic interactionism, which focuses on the symbolic meanings that people give to behavior (Hagan 1994:40). Instead of assuming that some forms of behavior are deviant in and of themselves (Scull 1988:678), symbolic interactionists would view deviance as a consequence of the application of rules and sanctions to an offender (Becker 1963:9). Exhibit 2.2 summarizes how these general theories relate to the question of whether to arrest spouse abusers.

EXHIBIT 2.2 Two Social Theories and Their Predictions About the Effect of Arrest for Intimate Partner Assault

	Rational Choice Theory	*Symbolic Interactionism*
Theoretical assumption	People's behavior is shaped by calculations of the costs and benefits of their actions.	People give symbolic meanings to objects, behaviors, and other people
Criminological component	Deterrence theory: People break the law if the benefits of doing so outweigh the costs.	Labeling theory: People label offenders as deviant, promoting further deviance.
Prediction (effect of arrest for domestic assault)	Abusing spouse, having seen the costs of abuse (namely, arrest), decides not to abuse again.	Abusing spouse, having been labeled as "an abuser," abuses more often.

Does either deterrence theory or labeling theory make sense to you as an explanation for the impact of punishment? Do they seem consistent with your observations of social life? Over a decade after Sherman and Berk's (1984) study, Paternoster et al. (1997) decided to study punishment of domestic violence from a different perspective. They turned to a social psychological theory called procedural justice theory, which explains law-abidingness as resulting from a sense of duty or morality (Tyler 1990). People obey the law from a sense of obligation that flows from seeing legal authorities as moral and legitimate. From this perspective, individuals who are arrested seem less likely to reoffend if they are treated fairly, irrespective of the outcome of their case, because fair treatment will enhance their view of legal authorities as moral and legitimate. Procedural justice theory expands our view of the punishment process by focusing attention on how police act and how authorities treat subjects, rather than just on the legal decisions they make. Thus, it gives us a sense of the larger importance of the research question.

Are you now less certain about the likely effect of arrest for intimate partner violence? Will arrest decrease abuse because abusers do not wish to suffer from legal sanctions again? Will it increase abuse because abusers feel stigmatized by being arrested and thus are more likely to act like criminals? Or will arrest reduce abuse only if the abusers feel they have been treated fairly by the legal authorities? By suggesting such questions, social theory makes us much more sensitive to the possibilities and so helps us to design better research. Before, during, and after a research investigation, we need to keep thinking theoretically.

SOCIAL RESEARCH STRATEGIES

All social research, including criminological research, is the effort to connect theory and empirical data, the evidence we find in the real world. As Exhibit 2.3 shows, theory and data have a two-way, mutually reinforcing relationship. Research that begins with a theory implying that certain data should be found involves **deductive reasoning**, which moves from general ideas (theory) to specific reality (data). In contrast, **inductive reasoning** moves from the specific to the general.

Both deductive reasoning and inductive reasoning are essential to criminologists. We cannot test an idea fairly unless we use deductive reasoning, stating our expectations in advance and setting up a test in which our idea could be shown to be wrong (falsified). A theory that has not survived these kinds of tests can be regarded only as very tentative. Yet theories, no matter how cherished, cannot make useful predictions for every social situation or research problem that we seek to investigate. Moreover, we may find unexpected patterns in the data we collect, called **serendipitous findings** or **anomalous findings**. In either situation, we should reason inductively, making whatever theoretical sense we can of our unanticipated findings. Then, if the new findings seem sufficiently important, we can return to deductive reasoning and plan a new study to formally test our new ideas.

The Research Circle

This process of conducting research, moving from theory to data and back again, or from data to theory and back again, can be characterized as a **research circle**, Exhibit 2.4 depicts

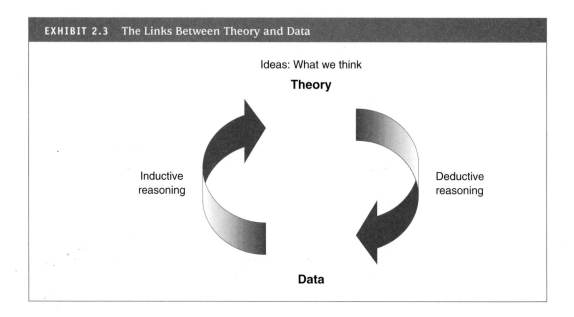

EXHIBIT 2.3   The Links Between Theory and Data

Ideas: What we think
Theory

Inductive
reasoning

Deductive
reasoning

Data

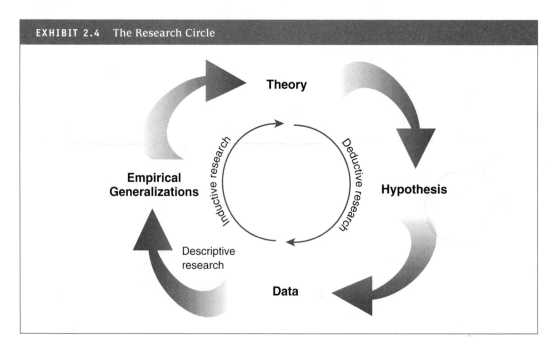

EXHIBIT 2.4   The Research Circle

Theory

Inductive research

Deductive research

**Empirical
Generalizations**

Hypothesis

Descriptive
research

Data

this circle. Note that it mirrors the relationship between theory and data shown in Exhibit 2.3 and that it comprises three main research strategies: deductive research, inductive research, and descriptive research.

Deductive Research

As Exhibit 2.4 shows, **deductive research** proceeds from theorizing to data collection and then back to theorizing. In essence, a specific expectation is deduced from a general premise and then tested.

Notice that a theory leads first to a **hypothesis**, which is a specific implication deduced from the more general theory. Researchers actually test a hypothesis, not the complete theory itself, because theories usually contain many hypotheses. As we stated earlier, a hypothesis proposes a relationship between two or more theoretical constructs or **variables**. A variable is a characteristic or property that can vary. A **constant** is a characteristic or a property that cannot vary. For example, if we were to conduct some research in a male adult penitentiary, the theoretical construct "type of crime committed" would be a variable because persons will have been incarcerated for different offenses (one person is in for armed robbery, another for rape, etc.). However, the theoretical construct "gender" would be a constant because every inmate in the penitentiary would be male; gender does not vary, it is constant. Would age be a variable or a constant in this group? Would "criminal status" (offender or nonoffender) be a variable or a constant?

Hypothesis A tentative statement about empirical reality, involving a relationship between two or more variables.

Example of a hypothesis: The higher the level of residential mobility in a community, the higher its rate of crime.

Variable A characteristic or property that can vary (take on different values or attributes).

Constant A characteristic or property that does not vary but takes on only one value.

Variables are of critical importance in research because in a hypothesis, variation in one variable is proposed to predict, influence, or cause variation in the other variable. The proposed influence is the **independent variable**; its effect or consequence is the **dependent variable**. After the researchers formulate one or more hypotheses and develop research procedures, they collect data with which to test the hypothesis.

To test your knowledge of hypotheses, go to the Variables and Hypotheses Interactive Exercises on the Student Study Site.

Independent variable A variable that is hypothesized to cause, or lead to, variation in another variable.

Example of an independent variable: Residential mobility (residents moving in and out of the community).

Dependent variable A variable that is hypothesized to vary depending on or under the influence of another variable.

Example of a dependent variable: The rate of crime in a community per 1,000 residents.

Hypotheses can be worded in several different ways, and identifying the independent and dependent variables is sometimes difficult. When in doubt, try to rephrase the hypothesis as an if-then statement: "If the independent variable increases (or decreases), then the dependent variable increases (or decreases)." Exhibit 2.5 presents several hypotheses with their independent and dependent variables and their if-then equivalents.

Exhibit 2.5 demonstrates another feature of hypotheses: **direction of association**. When researchers hypothesize that one variable increases as the other variable increases, the

---

EXHIBIT 2.5   Examples of Hypotheses

Original Hypothesis	Independent Variable	Dependent Variable	If-Then Hypothesis
1. The greater the social disorganization in a community, the higher the rate of crime.	Social disorganization	Crime rate	If social disorganization is higher, then the crime rate is higher.
2. As one's self-control gets stronger, the fewer delinquent acts one commits.	Self-control	Self-reported delinquency	If self-control is higher, then the number of delinquent acts is lower.
3. As the unemployment rate in a community decreases, the community rate of property crime decreases.	Unemployment rate	Rate of property crime	If the unemployment rate is lower, then the rate of property crime is lower.
4. As the discrepancy between one's aspirations and expectations increases, one's level of strain increases.	Discrepancy between one's aspirations and expectations	Strain	If the discrepancy between one's aspirations and expectations is high, then the level of strain is high.
5. Crime is lower in those communities where the police patrol on foot.	Presence of foot patrols	Crime	If a community has police foot patrols, then the level of crime is lower.

direction of the association is positive (Hypotheses 1 and 4 in the exhibit); when one variable decreases as the other variable decreases, the direction of association is also positive (Hypothesis 3). In positive relationships, then, the independent and dependent variables move in the same direction (as one increases, the other increases, or as one decreases, the other decreases). But when one variable increases as the other decreases, or vice versa, the direction of association is negative, or inverse (Hypothesis 2). In a negative relationship, then, the independent and dependent variables move in opposite directions (as one increases, the other decreases, or as one decreases, the other increases). Hypothesis 5 is a special case, in which the independent variable is categorical. The independent variable cannot be said to increase or decrease. In this case, the concept of direction of association does not apply, and the hypothesis simply states that one category of the independent variable is associated with higher values on the dependent variable.

You can get a better sense of what the direction of a relationship means by looking at Exhibits 2.6 through 2.10, which correspond to the five hypotheses shown in Exhibit 2.5. In these graphs, the independent variable is displayed as the x or horizontal axis and the dependent variable is displayed as the y or vertical axis. Exhibit 2.6 illustrates the hypothesis that social disorganization and a community's crime rate are positively related. The positive association implies that as social disorganization (the independent variable) increases (moves from being low to being high), the expected change is that the dependent variable will also increase (move from being low to high). This is the same thing as saying that as the level of social disorganization in a community moves from low to high, the level of crime also moves from low to high. In other words, the more social disorganization, the more crime. Notice that in a positive relationship, the independent and dependent variables are moving in the same direction; as social disorganization increases, so does crime. This positive relationship is shown in Exhibit 2.6 as an upward sloping line in the graph.

Exhibit 2.7 illustrates a graph for a negative association. This graph shows that as one's self-control becomes high (where more self-control means greater restraint over one's

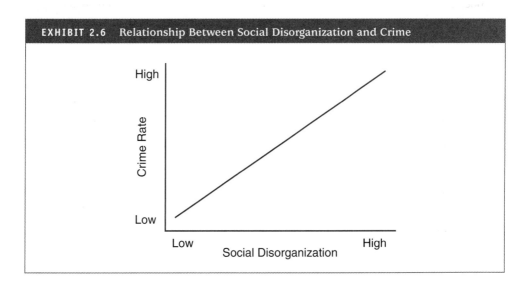

EXHIBIT 2.6   Relationship Between Social Disorganization and Crime

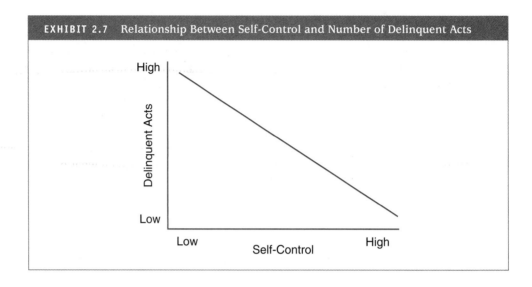

EXHIBIT 2.7   Relationship Between Self-Control and Number of Delinquent Acts

impulses), then the number of delinquent acts one commits declines or moves to low. This means that as self-control increases or strengthens, the frequency of delinquent behavior declines or decreases in number. Notice two things about this negative relationship. First, the independent and dependent variables move in opposite directions; as the independent variable (self-control) increases, the dependent variable (delinquency) decreases. Notice also that this negative relationship is shown as a downward sloping line in the graph.

Exhibit 2.8 illustrates a positive relationship but the hypothesis is stated differently. Hypothesis 3 in Exhibit 2.5 states that as the unemployment rate in a community decreases,

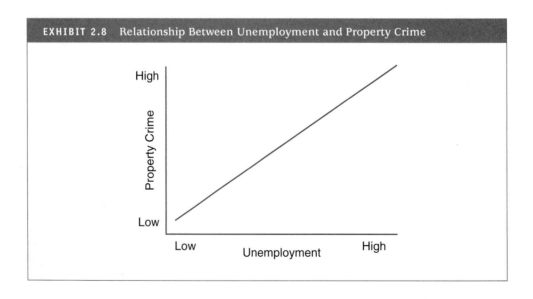

EXHIBIT 2.8   Relationship Between Unemployment and Property Crime

the rate of property crime in that community is also expected to decrease. Now, this is a positive relationship because the two variables move in the same direction; as the independent variable (unemployment) decreases, so does the dependent variable (the rate of property crime). As with Exhibit 2.6, this positive relationship is seen in an upward sloping line. Note that we could have expressed this hypothesis differently, and more similarly to Hypothesis 1, by stating, "As the level of unemployment in a community increases, the rate of property crime also increases." This, too, is a positive relationship, because the variables are again moving in the same direction.

Exhibit 2.9 corresponds to Hypothesis 4 in Exhibit 2.5, and illustrates another positive relationship. The hypothesis states that as the discrepancy between what one aspires to obtain and what one expects to obtain gets greater (moves from low to high), the level of strain the person feels also increases. Again, the independent and dependent variables move in the same direction, so the relationship or association between them is positive. Notice, we could have expressed this identical positive relationship slightly differently, by stating, "As the discrepancy between aspirations and expectations decreases, the level of strain decreases." Had we stated the hypothesis in this way, the graph would look exactly the same.

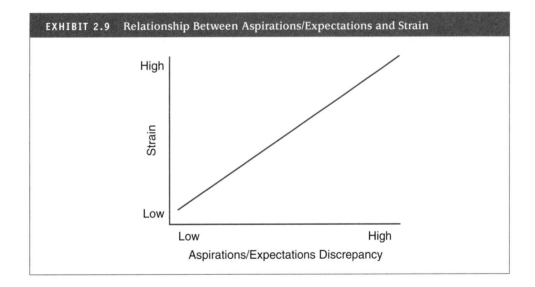

EXHIBIT 2.9   Relationship Between Aspirations/Expectations and Strain

Finally, Exhibit 2.10 illustrates a relationship with a graph that is slightly different from Exhibits 2.6 through 2.9. Here, the hypothesis (Hypothesis 5 in Exhibit 2.5) states that we expect to see lower crime in communities that have foot patrols compared with communities that do not have foot patrols. Unlike the other independent variables, the independent variable foot patrol does not vary from low to high: Either there are foot patrols in the community or there are not. If the hypothesis is true, what we should expect to see is that most of the communities that do not have foot patrols should have high rates of crime, whereas most of those communities that do have foot patrols should have lower rates of crime. This expectation is shown in Exhibit 2.10 with the clump of points (representing communities)

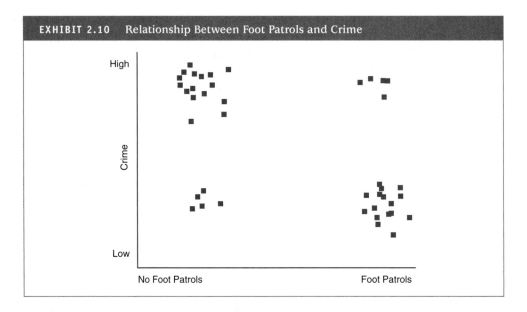

EXHIBIT 2.10   Relationship Between Foot Patrols and Crime

for the "no foot patrol" communities at the high range of crime and the clump of points for the "yes foot patrol" communities at the low range of crime.

The motives for deductive research include both explanation and evaluation (as described in Chapter 1). An example of explanatory deductive research is the Minneapolis Domestic Violence Experiment, in which Sherman and Berk (1984) sought to explain what sort of response by the authorities might keep a spouse abuser from repeating the offense. The researchers deduced from deterrence theory the expectation that arrest would deter domestic violence. They then collected data to test this expectation.

An example of evaluative deductive research is the study of the Second Step violence prevention curriculum, by Grossman et al. (1997), that was discussed in Chapter 1. The developers of the Second Step program had deduced from previous research in the violence prevention literature that violence in children could be inhibited if they learned and had practice with empathy, impulse control, and anger management. Based on this trinity of approaches, Second Step was devised to reduce the expressed violence among those children in the program. Grossman et al. collected data to test this expectation. They did find that physically aggressive behavior was lower among those children in the program than among a comparable group not exposed to the curriculum. Even though they began not with an explicit theory of violence prevention but with a mass of previous research findings, the developers of the Second Step program did begin with some ideas about what would work in inhibiting the violent behavior of young children. The evaluation by Grossman et al. confirmed those ideas.

In both explanatory and evaluative research, the statement of expectations for the findings and the design of the research to test these expectations strengthens the confidence we can place in the test. The deductive researcher shows her hand or states her expectations in advance and then designs a fair test of those expectations. Then "the chips fall where they may"; in other words, the researcher accepts the resulting data as a more or less objective picture of reality.

Inductive Research

In contrast to deductive research, **inductive research** begins at the bottom of the research circle and then works upward (see Exhibit 2.4). The inductive researcher begins with specific data, which are then used to develop (induce) a general explanation (a theory) to account for the data. The patterns in the data are then summarized in one or more **empirical generalizations** that can be compared to the hypothesis. If the empirical generalizations are those stated in the hypothesis, then the theory from which the hypothesis was deduced is supported. If the empirical generalizations are inconsistent with the hypothesis, then the theory is not supported (Wallace 1971:18).

The motive for inductive research is exploration. In Chapter 1, you read about an exploratory study of individuals' responses to an incident involving a school shooting (Asmussen & Creswell 1995). The incident took place at a public university where a gunman tried to shoot the students in his class. Fortunately, the gun jammed, and all the students escaped uninjured. Because there was very little previous work in this area, Asmussen and Creswell (1995) conducted in-depth interviews with the students and tried to classify typical responses to the situation. Although the researchers did not develop a theory from their work, they did develop a classification scheme or taxonomy of different responses to traumatic events.

In strictly inductive research, the researcher already knows what he has found when he starts theorizing, or attempting to explain what accounts for these findings. The result can be new insights and provocative questions. But the adequacy of an explanation formulated after the fact is necessarily less certain than an explanation presented prior to the collection of data. Every phenomenon can always be explained in some way. Inductive explanations are thus more trustworthy if they are tested subsequently with deductive research.

A Qualitative Exploration of the Response to Domestic Violence

Qualitative research is often inductive: The researchers begin by observing social interaction or interviewing social actors in depth and then developing an explanation for what has been found. The researchers often ask questions like "What is going on here?" "How do people interpret these experiences?" or "Why do people do what they do?" Rather than testing a hypothesis, the researchers are trying to make sense of some social phenomenon. Bennet, Goodman, and Dutton (1999) used this approach to investigate one of the problems that emerge when police arrest domestic batterers: The victims often decide not to press charges. Bennett et al. did not set out to test hypotheses with qualitative interviews (there was another, hypothesis-testing component in their research), but sought, inductively, to "add the voice of the victim to the discussion" and present "themes that emerged from [the] interviews" (p. 762).

Research assistants interviewed 49 victims of domestic violence in one court; Lauren Bennett also worked in the same court as a victim advocate. The researchers were able to cull from their qualitative data four reasons why victims became reluctant to press charges: Some were confused by the court procedures, others were frustrated by the delay, some were paralyzed by fear of retribution, and others did not want to send the batterer to jail.

Explanations developed inductively from qualitative research can feel authentic because we have heard what people have to say "in their own words" and we have tried to see the social world "as they see it." One victim interviewed by Bennett, Goodman, and Dutton

(1999) felt that she "was doing time instead of the defendant"; another expressed her fear by saying that she would like "to keep him out of jail if that's what it takes to keep my kids safe" (pp. 768–769). Explanations derived from qualitative research will be richer and more finely textured than those resulting from quantitative research, but they are likely to be based on fewer cases from a limited area. We cannot assume that the people studied in this setting are like others or that other researchers would develop explanations similar to ours to make sense of what was observed or heard. Because we do not initially set up a test of a hypothesis according to some specific rules, another researcher cannot come along and conduct just the same test.

Descriptive Research

You learned in Chapter 1 that description is one important motive for social research. Descriptive research can be considered a part of the research circle, even though such research does not involve connecting theory and data. As Exhibit 2.4 indicates, descriptive research starts with data and proceeds only to the stage of making empirical generalizations based on those data.

Valid description is important in its own right, but it is also critical in all research. Description of social phenomena can stimulate more ambitious deductive and inductive research. The Minneapolis Domestic Violence Experiment was motivated in part by a growing body of descriptive research indicating that spouse abuse is very common. You may recall from Chapter 1 that recent research on the magnitude of youth violence and other problem behaviors is heavily descriptive. The Youth Risk Behavior Survey is an ongoing study that attempts to measure the involvement of a national sample of youth in a wide sweep of antisocial behaviors. This descriptive work is valuable in documenting patterns and trends in violence by American youth throughout the 1990s. Theories will then be developed to account for these patterns and trends.

Domestic Violence and the Research Circle

The Sherman and Berk (1984) study of domestic violence that we have been discussing in this chapter is a good example of how the research circle works. In an attempt to determine ways to prevent the recurrence of spouse abuse, the researchers repeatedly linked theory and data, developing both hypotheses and empirical generalizations.

Phase 1: Deductive Research

The first phase of Sherman and Berk's (1984) study was designed to test a hypothesis. According to deterrence theory, punishment will reduce recidivism, or the propensity to commit further crimes. From this theory, Sherman and Berk deduced a specific hypothesis: "Arrest for spouse abuse reduces the risk of repeat offenses." In this hypothesis, arrest is the independent variable, and variation in the risk of repeat offenses is the dependent variable (it is hypothesized to depend on arrest).

Of course, in another study arrest might be the dependent variable in relation to some other independent variable. For example, in the hypothesis "The greater the rate of layoffs in a community, the higher the frequency of arrest," the dependent variable is frequency of arrest. Only within the context of a hypothesis, or a relationship between variables, does it make sense to refer to one variable as dependent and the other as independent.

Sherman and Berk (1984) tested their hypothesis by setting up an experiment in which the police responded to complaints of spouse abuse in one of three ways, one of which was to arrest the offender. When the researchers examined their data (police records for the persons in their experiment), they found that of those arrested for assaulting their spouse, only 13% repeated the offense, compared to a 26% recidivism rate for those who were separated from their spouse by the police without any arrest. This pattern in the data, or empirical generalization, was consistent with the hypothesis that the researchers deduced from deterrence theory. The theory thus received support from the experiment (see Exhibit 2.11).

EXHIBIT 2.11   The Research Circle: Minneapolis Domestic Violence Experiment

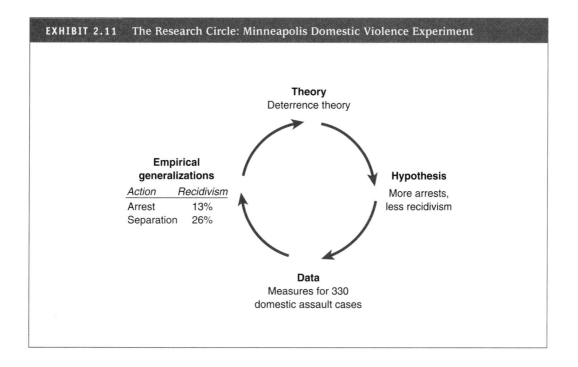

In designing their study, Sherman and Berk (1984) anticipated an important question: "How valid was the connection they [the researchers] were trying to make between theory and data?" The three dimensions of validity—measurement validity, generalizability, and causal validity—were at issue.

Determining whether spouses were assaulted after the initial police intervention was the key measurement concern. Official records of subsequent assaults by the suspect would provide one measure. But most spousal assaults are not reported to the police, and so research assistants also sought out the victims for interviews every 2 weeks during a 6-month follow-up period. Although fewer than half the victims completed all the follow-up interviews, the availability of the self-report measure allowed the researchers to shed some light on the validity of the official data. In general, the two measures yielded comparable results, although some discrepancies troubled critics.

The generalizability of the study's results was the researchers' greatest concern. Minneapolis is no more a typical U.S. city than any other, and we cannot assume that police policies that are effective in Minneapolis will be equally effective in cities with very different political histories, criminal justice agencies, and population characteristics. Sherman and Berk (1984) warned readers, "External validity will have to wait for replications" (p. 269); that is, for repetitions of the study using the same research methods to answer the same research question.

To test your knowledge of inductive and deductive research, go to the Research Circle Interactive Exercises on the Student Study Site.

Finally, Sherman and Berk's (1984) claims about the causal validity of their results rested primarily on the experimental design they used. The 330 domestic assault cases in the study were handled by the police in one of three ways: an arrest, an order that the offending spouse leave the house for 8 hours, or some type of verbal advice by the police officers. The officers were not allowed to choose which treatment to apply (except in extreme cases, such as when severe injury had occurred or when the spouse had demanded that an arrest be made). Instead, the treatments were carried out by police in random order, according to the color of the next report form on a pad that had been prepared by the researchers.

By insisting on the random assignment of cases to treatments, the researchers tried to ensure that police officers would not arrest just the toughest spouses or the spouses who seemed most obnoxious or the spouses they encountered late in the day. In other words, the random assignment procedure made it unlikely that arrested spouse abusers would differ, on average, from the other spouse abusers except for the fact that they were arrested (although, because of chance factors, the possibility of other differences cannot be completely ruled out). The researchers' conclusion that arrest caused a lower incidence of repeat offenses therefore seems valid.

Phase 2: Deductive Research

Because of their doubts about the generalizability of their results, Sherman, Berk, and new collaborators began to journey around the research circle again, with funding from the National Institute of Justice for **replications** (repetitions) of the experiment in six more cities. These replications used the same basic research approach but with some improvements. The random assignment process was tightened up in most of the cities so that police officers would be less likely to replace the assigned treatment with a treatment of their own choice. In addition, data were collected about repeat violence against other victims as well as against the original complainant. Some replications also examined different aspects of the arrest process, to see whether professional counseling helped and whether the length of time spent in jail after arrest mattered at all.

By the time results were reported from five of the cities in the new study, a problem was apparent. In three cities—Omaha, Nebraska; Charlotte, North Carolina; and Milwaukee, Wisconsin—researchers were finding long-term increases in domestic violence incidents among arrestees. But in Colorado Springs, Colorado, and Dade County, Florida, the predicted deterrent effects seemed to be occurring (Sherman et al. 1992).

Researchers had now traversed the research circle twice in an attempt to answer the original research question, first in Minneapolis and then in six other cities. But rather than leading to more confidence in deterrence theory, the research results were calling it into question. Deterrence theory now seemed inadequate to explain empirical reality, at least as the researchers had measured this reality. So the researchers began to reanalyze the follow-up data from several cities to try to explain the discrepant results, thereby starting around the research circle once again (Berk et al. 1992; Pate & Hamilton 1992; Sherman et al. 1992).

Phase 3: Inductive Research

At this point, the researchers' approach became more inductive, and they began trying to make sense of the differing patterns in the data collected in the different cities. Could systematic differences in the samples or in the implementation of arrest policies explain the differing outcomes? Or was the problem an inadequacy in the theoretical basis of their research? Was deterrence theory really the best way to explain the patterns in the data they were collecting?

Sherman et al. (1992) now turned to control theory (Toby 1957), yet another broad explanation for social behavior. It predicts that having a stake in conformity (resulting from inclusion in social networks at work or in the community) decreases a person's likelihood of committing crimes. The implication is that people who are employed and married are more likely to be deterred by the threat of arrest than those without such stakes in conformity. This is because an arrest for domestic violence could jeopardize one's job and one's marriage, thus making arrest more costly for the employed and married. This is indeed what a reexamination of the data revealed: Individuals who were married and employed were deterred from repeat offenses by arrest, but individuals who were unmarried and unemployed were actually more likely to commit repeat offenses if they were arrested. This was an important theoretical insight, for it suggested that one powerful way that formal sanctions work is that they can potentially trigger informal sanctions or costs (e.g., loss of respect from friends and family).

Now the researchers had traversed the research circle almost three times, a process perhaps better described as a spiral (see Exhibit 2.12). The first two times the researchers had traversed the research circle in a deductive, hypothesis-testing way: They started with theory and then deduced and tested hypotheses. The third time they traversed the research circle in a more inductive, exploratory way: They started with empirical generalizations from the data they had already obtained and then turned to a new theory to account for the unexpected patterns in the data. At this point they believed that deterrence theory makes correct predictions given certain conditions and that another theory, control theory, may specify what these conditions are.

After two and one-half cycles through the research circle, the picture became more complex but also conceptually richer. The researchers came closer to understanding how to inhibit domestic violence. But they cautioned us that their initial question, the research problem, was still not completely answered. Employment status and marital status alone do not measure the strength of social attachments; they also are related to how much people earn and the social standing of victims in court. So perhaps social ties are not really what makes arrest an effective deterrent to domestic violence. The real deterrent may be cost-benefit calculations ("If I have a higher income, jail is more costly to me") or perceptions about the actions of authorities ("If I am a married woman, judges will treat my complaint more seriously"). More research was still needed (Berk et al. 1992).

EXHIBIT 2.12   The Research Spiral: Minneapolis Domestic Violence Experiment

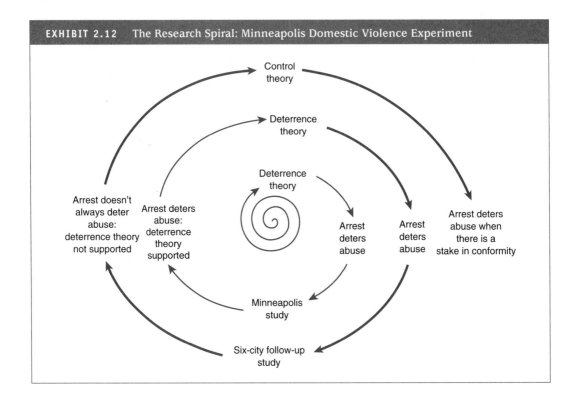

Phase 4: Deductive Research

In 1997, Paternoster et al. reexamined data from the Milwaukee Domestic Violence Experiment to test hypotheses derived from yet another theory, procedural justice theory. As explained earlier in this chapter, procedural justice theory predicts that people will comply with the law out of a sense of duty and obligation if they are treated fairly by legal authorities. In the Milwaukee sample, arrest had a criminogenic effect: Those who were arrested were subsequently more likely to abuse their spouses than those who were simply warned. Paternoster et al. (1997) thought that this effect might have been due to the way subjects were treated when they were arrested rather than simply to the fact that they were arrested. One of their hypotheses spells out the reasoning:

> Among those persons arrested for spouse assault, those who perceive themselves as being treated in a procedurally unfair manner will be more likely to commit acts of spouse assault in the future than those arrested persons who perceive themselves as being treated in a procedurally fair manner, net of other determinants of violence. (P. 173)

To carry out this study, Paternoster et al. (1997) reexamined data collected earlier in Milwaukee, where the findings had seemed anomalous. However, this reanalysis of the data qualifies as deductive research, because the hypotheses were derived from theory and then tested with the data, rather than being induced by the data.

The procedural justice hypotheses were supported: Persons who were arrested in the Milwaukee experiment became more likely to reoffend only if they perceived that they had been treated unfairly by the police. Otherwise, their rate of rearrest was similar to that for the persons who were not arrested. Thus, another element was added to our understanding of the effects of the police response to domestic violence.

Clearly our understanding of effective responses to domestic violence will never truly be complete, but research to date has greatly improved our understanding of this social problem. The future should yield an even better understanding, even though at times it may be hard to make sense out of conflicting findings from different studies. Science is an ongoing enterprise in which findings cumulate and eventually yield greater understanding or even radical revisions in our understanding. Needless to say, researchers do not need to worry about running out of work to do.

GUIDELINES FOR CRIMINOLOGISTS

Any effort to understand the social world is plagued by pitfalls, including (as you learned in Chapter 1) such everyday errors in reasoning as overgeneralization, selective or inaccurate observation, illogical reasoning, and resistance to change. Social scientists, including criminologists, cannot avoid these problems entirely, but they try to minimize their impact by adhering to certain guidelines.

The guidelines followed by social researchers fall into two categories: those that help keep research scientific and those that help keep research ethical. Both types of guidelines are essential for a field of inquiry that seeks empirical generalizations about human society. To point out their value, we use examples from the domestic violence research.

Scientific Guidelines

The following nine guidelines are applicable to any type of scientific research, but they are particularly useful to criminologists and to those who read about criminology and criminal justice. Adherence to these guidelines will reduce the temptation "to project on what is observed whatever [they] want the world to be for [their] own private purposes" (Hoover 1980:131).

1. *Test ideas against empirical reality without becoming too personally invested in a particular outcome.* This testing approach is reflected in the research process and is implicit in the goal of validity. It contrasts markedly with our everyday methods of figuring things out, in which we typically just react to events as they happen, without paying much attention to whether we really are putting our ideas to a test. Empirical testing requires a neutral and open-minded approach: The scientists are personally disinterested in the outcome and not swayed by the popularity or the social status of those who would prefer other outcomes. This does not mean that the researchers are not personally involved or interested in the research—they

must be—rather, the point is that they cannot have so much invested in a research project personally or professionally that they try in subtle or not-so-subtle ways to affect the outcome.

2. *Plan and carry out investigations systematically*. Social researchers have little hope of conducting a careful test of their ideas if they do not think through in advance how they should go about the test and then proceed accordingly. But a systematic approach is not always easy. For example, Sherman and Berk (1984) needed to ensure that spouse abusers were assigned to be either arrested or not on a random basis, rather than on the basis of the police officers' personal preferences. So the researchers devised an elaborate procedure using randomly sequenced report sheets in different colors. But the researchers found that police officers did not always follow this systematic procedure. Subsequently, in some replications of the study, the researchers ensured compliance with their research procedures by requiring police officers to call in to a central number to receive the experimentally determined treatment.

3. *Document all procedures, and disclose them publicly*. Social researchers who disclose the methods on which their conclusions rest allow others to evaluate for themselves the likely soundness of these conclusions. Such disclosure is a key feature of science. Again, Sherman and Berk (1984) provide a compelling example. In their research report, after describing the formal research plan, they described at length the apparent slippage from this plan, which occurred primarily because some police officers avoided implementing the random assignment procedure.

4. *Clarify assumptions*. No investigation is complete unto itself; whatever the researcher's method, the research rests on some background assumptions. Research to determine whether arrest has a deterrent effect assumes that potential law violators think rationally, that they calculate potential costs and benefits prior to committing crimes. When a researcher conducts an election poll, the assumption is that people actually vote for the candidate they say they will vote for. When government unemployment statistics are used to describe the state of the economy, the assumption is that those statistics reflect actual fluctuations in unemployment. By definition, research assumptions are not tested, so we do not know whether they are correct. In fact, researchers themselves do not always recognize the assumptions they are making. By taking the time to think about and to disclose their assumptions, researchers provide important information for those who seek to evaluate the validity of their conclusions.

5. *Specify the meaning of all terms*. Words often have multiple or unclear meanings. Strain, differential association, social disorganization, subculture of violence, problem-oriented policing, and so on can mean different things to different people. Thus, the terms used in scientific research must be defined explicitly and used consistently. For example, Sherman and Berk (1984) identified their focus as misdemeanor domestic assault, not just wife beating. They specified that their work concerned those cases of domestic assault in which severe injury was not involved and both partners were present when police arrived.

6. *Maintain a skeptical stance toward current knowledge.* Scientists may feel very confident about interpretations of the social or natural world that have been supported by repeated investigations, but the results of any particular investigation must be examined critically. A general skepticism about current knowledge stimulates researchers to improve the validity of current research results and expand the frontier of knowledge. For example, in response to questions raised about the Sherman and Berk (1984) study, Sherman and Cohn (1989) discussed 13 problems of the Minneapolis Domestic Violence Experiment in a published critique, weighing carefully the extent to which these problems might have affected its validity. This critique could then stimulate additional research designed to address the problematic aspects of the original research.

7. *Replicate research and accumulate knowledge.* No one study can be viewed as definitive in itself; usually at least some plausible threats to the validity of the conclusions exist. And no conclusion can be understood adequately apart from the larger body of knowledge to which the study is related. Scientific investigations may begin with a half-baked or off-the-wall idea, but a search of the literature for other relevant work must be conducted in short order. The other side of the coin is that the results of scientific research must be published, to serve as a foundation for others who seek to replicate or extend the research. Sherman (1992) reported that when he and his colleagues decided to attempt some replications of their own experiment, they found that another research team was already planning to do so. The process of extending knowledge gained in the Minneapolis experiment had already begun.

8. *Maintain an interest in theory.* Theories organize the knowledge accumulated by numerous investigations into a coherent whole and serve as a guide to future inquiries. Even though much research is purely descriptive, this research can still serve as a basis for others to evaluate different theories. The Minneapolis Domestic Violence Experiment was devised initially as a test of the competing predictions of deterrence and labeling theory, but the researchers extended their attention to control theory to help them explain unanticipated findings. These theoretical connections make the research much more relevant to other criminologists working to understand different types of crime and social control.

9. *Search for regularities or patterns.* Science is concerned with classes rather than with individuals (except inasmuch as individuals are representatives of a class). Scientists assume that the natural world has some underlying order of relationships, and that every event and individual is not so unique that general principles cannot be discerned (Grinnell 1992:27–29). Individuals are not unimportant to criminologists and other social scientists; Sherman (1992:162–164), for example, described the abuse histories of two men to provide greater insight into why arrest could have different effects for different people. But the goal of elaborating individual cases is to understand social patterns that characterize many individuals.

These general guidelines are only ideals for social research. No particular investigation will follow every guideline exactly. Real investigations by criminologists do not always include much attention to theory, specific definitions of all terms, and so forth. But any study that strays far from these guidelines cannot be considered scientific.

Ethical Guidelines

Every methodology (e.g., surveys, experiments, participant observation) has its own unique ethical considerations. As such, we will discuss ethics in all methodology chapters in this text. However, there are some general ethical guidelines that we would like to present here. Every scientific investigation, whether in the natural sciences or in the social sciences, such as criminology and criminal justice, has an ethical dimension. First and foremost, the scientific concern with validity requires that scientists be honest and reveal their methods. (How else could we determine whether the requirement of honesty has been met?) Scientists also have to consider the uses to which their findings will be put. In addition, because criminological research deals with people, such as criminals, criminal suspects, and incarcerated inmates, and controversial topics (involvement in crime), criminologists have some unique ethical concerns.

Honesty and Openness

Research distorted by political or personal pressures to find particular outcomes or to achieve the most marketable results is unlikely to be carried out in an honest and open fashion or to achieve valid results. For example, noted English biologist Sir Cyril Burt published fabricated evidence in 1961 that purported to show intelligence is determine primarily by heredity. In the 35 years before Burt's deliberate falsification was exposed, his study influenced much social science theory and research (and was in part responsible for Burt's knighthood) (Kamin 1974). Efforts to evaluate the influence of inherited characteristics on people were considerably set back by this fraud.

Being open about one's research is particularly important when the research is used or influential in public policy matters. For example, an economist, Professor Isaac Ehrlich (1975), who was interested in the study of crime conducted a study that examined whether capital punishment deters murder. Using national data for the time period 1933–1969, Ehrlich concluded that each execution prevented or deterred 7–8 homicides; in other words, Ehrlich claimed that executions saved 7–8 innocent crime victims. This was one of a very few studies at the time to find a deterrent effect for capital punishment. In fact, his results were cited in a U.S. Supreme Court case (*Gregg v. Georgia* 1976) as evidence in support of deterrence. His reported findings caused both a stir among proponents of capital punishment and skeptical hostility among its critics. Other researchers immediately wanted to reexamine Ehrlich's data to see how valid his results were. Unfortunately, Ehrlich was not immediately forthcoming with his data. The controversy surrounding his findings grew when other researchers tried to duplicate his findings with data they collected and found that when they analyzed slightly different years, with slightly different variables and sometimes slightly different statistical strategies, the effect of executions changed. In some analyses, executions seemed to deter murder; in some others, it seemed to increase murders (what was called a brutalization effect); and in others, it had no discernable effect at all. The controversy about Ehrlich's findings bubbled for years and might have been resolved much sooner had Ehrlich been more open and forthcoming about his data.

Openness about research procedures and results goes hand in hand with honesty in research design. Openness is also essential if researchers are to learn from the work of others. In spite of this need for openness, some researchers may hesitate to disclose their procedures or results to prevent others from building on their ideas and taking some of the credit. You may have heard of the long legal battle between a U.S. researcher and a French researcher about how credit should be allocated for discovering the AIDS virus. Although such public disputes are unusual, concerns with priority of discovery are common. Scientists are like other people in their desire to be first. Enforcing standards of honesty and encouraging openness about research is the best solution for these problems.

The Uses of Science

Scientists must also consider the extent to which they should publicize their research and the uses to which it is put. Although many scientists believe that personal values should be left outside the laboratory, some argue that it is proper, even desirable, for scientists in their role as citizens to attempt to influence public policy. In other words, if their research has something to say about how things *should be* done, some scientists feel that they should actively promote their work. Sometimes, however, there is controversy on this point.

Throughout this chapter, we have been talking about one criminological experiment, the Minneapolis Domestic Violence Experiment, which found that arresting domestic violence suspects was more effective in reducing subsequent violence than not arresting them. Although Sherman and Berk (1984) were generally cautious about the kinds of conclusions that should be drawn from their work, they were both active and passive in promoting their findings. Not only was their work published in scientific journals, but it also received widespread publicity in the national media: one of the authors and the Minneapolis chief of police wrote an editorial in the *Wall Street Journal* recommending that other jurisdictions pass laws similar to Minneapolis's, and one of the authors contracted with a Minneapolis television station to film a documentary on the research. In their published findings, they concluded,

> We favor a *presumption* of arrest: an arrest should be made unless there are good, clear reasons why an arrest would be counterproductive. We do not, however, favor *requiring* arrests in all misdemeanor domestic assault cases. (P. 270)

Sherman and Berk (1984) were criticized by other scholars on the basis that the findings of one study are not a sufficient empirical base on which to draw such an important public policy recommendation. In fact, Sherman and Berk did caution that their findings might not be replicated in other locations, and that their position is that police should be able to arrest domestic violence suspects, not that they should be required to. Nevertheless, because Sherman publicized the results of the research in the mass media, he was criticized by some social scientists for implicitly encouraging police departments to change their policies on the basis of preliminary evidence (the results of just one study in one city) (Binder & Meeker

1993; Lempert 1989). In part, the question was whether basing policy on partial information was preferable to waiting until the information was more complete (more studies conducted in different locations). Sherman (1992:150–153) later pointed out that in the Omaha follow-up study, arrest warrants were very effective in reducing repeat offenses among spouse abusers who had already left the scene when police arrived at the time of the initial complaint. Absent offenders had not been included in the initial study, and because the Omaha finding was not publicized, it did not become known to police chiefs or battered women's groups. As a consequence, Sherman suggested, some domestic violence that might have been prevented was not prevented. How much publicity is warranted, and at what point in the research is it warranted? He also argued that public policy is better guided by partial knowledge than no knowledge at all.

Criminologists who conduct research on behalf of organizations and agencies may face additional difficulties. When an organization contracts with a researcher to evaluate a program, identify community needs, or explore product potential, usually the organization, not the researcher, controls the final report and the publicity it receives. If organizational leaders decide that particular research results are inconsistent with their funding requests, community image, or employee relations, they may refuse to release the results or require changes that the researcher deems unacceptable. In a situation like this, a researcher's desire to have findings used appropriately and reported fully can conflict with contractual obligations.

Researchers often can anticipate such dilemmas in advance and resolve them when the contract for research is negotiated, or decline a particular research opportunity altogether if acceptable terms cannot be worked out. But often these problems come up after a report has been drafted, when the researcher finds out that the report is unacceptable to a top-level administrator or executive whom the researcher does not even know. In addition, a researcher's need to have a job or to maintain particular personal relationships may make it difficult to act in what the researcher thinks is the most ethical manner. A way to minimize these possibilities is to acknowledge the source of research funding in reports and to carefully scrutinize those research reports funded by organizations or agencies with a stake in the outcome.

Research on People

In physics or chemistry, research subjects (objects and substances) may be treated to extreme conditions and then discarded when they are no longer useful. However, social (and medical) scientists must concern themselves with the way their human subjects are treated in the course of research. This treatment may involve manipulations and deceptions in laboratory experiments, sensitive questions in survey research, observations in field studies, or analyses of personal data. Here we will briefly review current ethical standards for the treatment of human subjects and dilemmas in their application. In the chapters on data collection, Chapters 5 through 9, we will examine the specific ethical problems that may arise in the course of using particular research methods.

Contemporary standards for the treatment of human subjects are set by the federal government, by professional associations, by special university review boards, and in some cases by ethics committees in other organizations. Federal regulations require that the proposals of researchers seeking federal funds for research on human subjects be reviewed by an **institutional review board (IRB)** before they are submitted for fed universities and other agencies in turn apply ethics standards set by gov like the National Institutes of Health, and may develop more specific own. The American Society of Criminology (ASC), one of criminology's nizations, is in the process of redrafting its code of ethics, but advises mer ance from similar organizations including the American Sociological Asst⌐⌐⌐⌐⌐⌐ ⌐⌐⌐⌐⌐. ⌐⌐⌐ ASA's standards (ASA 1997) concerning the treatment of human subjects include federal regulations and ethics guidelines emphasized by most professional social science organizations:

- Research should cause no harm to subjects.
- Participation in research should be voluntary, and therefore subjects must give their informed consent to participate in the research.
- Researchers should fully disclose their identity.
- Anonymity or confidentiality must be maintained for individual research participants, unless it is voluntarily and explicitly waived.
- The benefits of a research project should outweigh any foreseeable risks.

As simple as these guidelines may seem, they are difficult to interpret in specific cases and harder yet to define in a way agreeable to all criminologists. For example, how should we interpret the admonition that no harm should be done to subjects? Does it mean that subjects should not be at all harmed psychologically as well as physically? That they should feel no anxiety or distress during the study or only after their involvement ends? Should the possibility of any harm, no matter how remote, deter research?

Consider the question of possible harm to the subjects of a well-known prison simulation study (Haney, Banks, & Zimbardo 1973). The study was designed to investigate the impact of social position on behavior, specifically, the impact of being either a guard or a prisoner in a prison, a total institution. The researchers selected 20 young men whom they judged to be the most stable and mature, and the least antisocial, of 75 applicants. The participants signed a contract agreeing to be either a guard or a prisoner in a simulated prison for 2 weeks, during which time they would be paid $15 daily and receive food, clothing, housing, and medical care. Some were randomly selected to be guards and were told to maintain order among the prisoners, who were then incarcerated in a makeshift basement prison. Within the first two days, marked differences in behavior emerged between the two groups. The prisoners acted passive and disorganized, and the guards became verbally and physically aggressive (although physical abuse was not allowed) and arbitrary. Five prisoners were soon released for depression, uncontrollable crying, fits of rage, and in one case a psychosomatic rash; on the sixth day the researchers terminated the experiment. Through discussions in special post-experiment encounter sessions, feelings of stress among the participants who played the role of prisoner seemed to the relieved; follow-up during the next year indicated no lasting negative effects on the participants and some benefits in the form of greater insight.

Would you ban such experiments because of the potential for harm to subjects? Does the fact that the experiment yielded significant insights into the effect of a situation on human behavior, insights that could be used to improve prisons, make any difference (Reynolds 1979:133–139)? Do you believe that this benefit outweighed the foreseeable risks?

The requirement of informed consent is also more difficult to define than it first appears. To be informed, consent must be given by persons who are competent to consent, have consented voluntarily, are fully informed about the research, and have comprehended what they have been told (Reynolds 1979). Can prisoners give informed consent? Can children or juveniles give their own consent? Can parents or guardians give consent on behalf of their children? Can students who are asked to participate in research by their professor give consent? Can participants in covert experiments do so?

Fully informed consent may alter participation in research and, because signing consent forms prior to participation may change participants' responses, produce biased results (Larson 1993:114). In addition, there is always the problem that those persons who give their support in a research project may be quite different from those who do not. This differential selection of participants has implications for both the validity of the reported results and the generalizability of the study. Experimental researchers whose research design requires some type of subject deception try to get around this problem by withholding some information before the experiment begins but then debriefing subjects at the end. In the **debriefing**, the researcher explains to the subject what happened in the experiment and why. However, even though debriefing can be viewed as a substitute in some cases for securing fully informed consent prior to the experiment, if the debriefed subjects disclose the nature of the experiment to other participants, subsequent results may still be contaminated (Adair, Dushenko, & Lindsay 1985).

Well-intentioned researchers also may fail to foresee all the potential problems. In the prison simulation, all the participants signed consent forms, but how could they have been fully informed in advance? The researchers themselves did not realize that the study participants would experience so much stress so quickly, that some prisoners would have to be released for severe negative reactions within the first few days, or that even those who were not severely stressed would soon be begging to be released from the mock prison. If this risk was not foreseeable, was it acceptable for the researchers to presume in advance that the benefits would outweigh the risks?

Maintaining confidentiality is another key ethical obligation. This standard, however, should be overridden if a health- or life-threatening situation arises and participants need to be alerted. Also, the standard of confidentiality does not apply to observation in public places and information available in public records.

The potential of withholding a beneficial treatment from some subjects is also cause for ethical concern. The Sherman and Berk (1984) experiment required the random assignment of subjects to treatment conditions and thus had the potential of causing harm to the victims of domestic violence whose batterers were not arrested. The justification for the study design, however, is quite persuasive: The researchers did not know prior to the experiment which response to a domestic violence complaint would be most likely to deter future incidents (Sherman 1992). The experiment provided clear evidence about the value of arrest, so it can be argued that the benefits outweighed the risks.

The evaluation of ethical issues in a research project should be based on a realistic assessment of the overall potential for harm to research subjects rather than an apparent inconsistency between any particular aspect of a research plan and a specific ethical guideline. For example, full disclosure of what is really going on in an experimental study is unnecessary if subjects are unlikely to be harmed. Nevertheless, researchers should make every effort to foresee all possible risks and to weigh the possible benefits of the research against these risks.

The extent to which ethical issues are a problem for researchers and their subjects varies dramatically with research design. Survey research, in particular, creates few ethical problems, unless the survey queries respondents about sensitive subject matter (e.g., victimization or offending experiences; see Chapter 7 for a more detailed discussion of this issue). In fact, researchers from Michigan's Institute for Survey Research interviewed a representative national sample of adults and found that 68% of those who participated in a survey were somewhat or very interested in participating in another; the more times respondents had been interviewed, the more willing they were to participate again. Presumably they would have felt differently if they had been treated unethically (Reynolds 1979:56–57). On the other hand, some experimental studies in the social sciences that have put people in uncomfortable or embarrassing situations have generated vociferous complaints and years of debate about ethics (Reynolds 1979; Sjoberg 1967).

CONCLUSION

Criminological researchers can find many questions to study, but not all questions are equally worthy. The ones that warrant the expense and effort of social research are feasible, socially important, and scientifically relevant.

The simplicity of the research circle presented in this chapter belies the complexity of the social research process. In the following chapters, we will focus on particular aspects of that process. Chapter 3 examines the interrelated processes of conceptualization and measurement, arguably the most important part of research. Measurement validity is the foundation for the other two aspects of validity. Chapter 4 reviews the meaning of generalizability and the sampling strategies that help us to achieve this goal. Chapter 5 introduces causal validity, the third aspect of validity, and illustrates different methods for achieving causal validity, with particular emphasis on experimental designs. The following four chapters then introduce different approaches to data collection, including surveys, qualitative research methods, secondary data analysis, and evaluation research, that help us, in different ways, to achieve validity.

As you encounter these specifics, do not lose sight of the basic guidelines that researchers need to follow to overcome the most common impediments to social research. Owning a large social science toolkit is no guarantee of making the right decisions about which tools to use and how to use them in the investigation of particular research problems. More important, our answers to research questions will never be complete or entirely certain. Thus, when we complete a research project, we should point out how the research could be extended and evaluate the confidence we have in our conclusions. Recall how the gradual elaboration of knowledge about the deterrence of domestic violence required sensitivity to research difficulties, careful weighing of the evidence, and identification of unanswered questions by several research teams.

Ethical issues also should be considered when evaluating research proposals and completed research studies. As the preceding examples show, ethical issues in social research are no less complex than the other issues that researchers confront. And it is inexcusable to jump into research on people without any attention to ethical considerations.

You are now forewarned about, and thus hopefully forearmed against, the difficulties that any scientists, but criminologists in particular, face in their work. We hope that you will return often to this chapter as you read the subsequent chapters, when you criticize the research literature, and when you design your own research projects. To be conscientious, thoughtful, and responsible is the mandate of every social scientist. If you formulate a feasible research problem, ask the right questions in advance, try to adhere to the research guidelines, and steer clear of the most common difficulties, you will be well along the road to fulfilling this mandate.

KEY TERMS

Anomalous finding	Independent variable
Constant	Inductive reasoning
Criminological research question	Inductive research
Debriefing	Institutional review board (IRB)
Deductive reasoning	Replication
Deductive research	Research circle
Dependent variable	Serendipitous finding
Direction of association	Theoretical construct
Empirical generalization	Theory
Falsifiable statement	Variable
Hypothesis	

HIGHLIGHTS

- Research questions should be feasible (within the time and resources available), socially important, and scientifically relevant.
- Building criminological theory is a major objective of criminological research. Investigate relevant theories before starting criminological projects, and draw out the theoretical implications of research findings.
- The type of reasoning in most criminological research can be described as primarily deductive or inductive. Research based on deductive reasoning proceeds from general ideas, deduces specific expectations from these ideas, and then tests the ideas with empirical data. Research based on inductive reasoning begins with specific data and then develops general ideas or theories to explain patterns in the data.
- It may be possible to explain unanticipated research findings after the fact, but such explanations have less credibility than those that have been tested with data collected for the purpose of the study.
- The scientific process can be represented as circular, with connections from theory to hypotheses to data to empirical generalizations. Research investigations may begin at different points along the research circle and traverse different portions of it. Deductive research begins at the point of theory; inductive research begins with data but ends

with theory. Descriptive research begins with data and ends with empirical generalizations.

- Replications of a study are essential to establish its generalizability in other situations. An ongoing line of research stemming from a particular question should include a series of studies that, collectively, traverse the research circle multiple times.

- Criminologists, like all social scientists, should structure their research so that their own ideas can be proved wrong, should disclose their methods for others to critique, and should recognize the possibility of error. Nine specific guidelines are recommended.

- Scientific research should be conducted and reported in an honest and open fashion. Contemporary ethical standards also require that social research not place subjects in any jeopardy, that research subjects be forewarned of any risk to them, that participation be voluntary as expressed in informed consent, that researchers fully disclose their identity, and that researchers fully and honestly report their research findings and sources of financial support.

EXERCISES

1. State a problem for research related to a criminological topic or issue of interest to you. Write down as many questions as you can about this topic.
   a. Considering your interest, opportunities, and the work for others, which of your research questions does not seem feasible or interesting?
   b. Pick out one question that seems feasible and that your other coursework suggests has been the focus of prior research or theorizing. Write this research question in one sentence. Elaborate on your question in a single paragraph. List at least three reasons for why it is a good research question to investigate.

2. Search the scholarly literature on your topic of interest. Refer to Appendix A for guidance on conducting the search if necessary.
   a. Copy at least 10 citations to recent articles reporting research relevant to your research question.
   b. Look up at least three of these articles. Write a brief description of each article, and evaluate its relevance to your research question. What additions or changes to your thoughts about the research question are suggested by these sources?
   c. Would you characterize the findings of these articles as largely consistent or inconsistent? How would you explain discrepant findings?
   d. How well did the authors summarize their work in their abstracts for the articles you consulted? What important points would you have missed if you had relied on only the abstracts?

3. Using one of the research articles you consulted for Exercise 2, identify and look up one of the cited articles, or Web sites. Compare the cited source to what was said about it in the original article or site. Was the discussion in the cited source accurate?

4. Using the same research article you focused on for Exercise 3, identify the stages of the research project corresponding to the points on the research circle. Did the research cover all four stages? Identify the theories and hypotheses underlying the study. What data were collected or utilized for the study? What were the findings (empirical generalizations)?

DEVELOPING A RESEARCH PROPOSAL

Now it is time to start writing the proposal. These next exercises are very critical first steps.

1. State a problem for research. If you have not already identified a problem for study, or if you need to evaluate whether your research problem is doable, a few suggestions should help to get the ball rolling and keep it on course:

   a. Jot down questions that have puzzled you in some area having to do with people and social relations, perhaps questions that have come to mind while reading textbooks or research articles or even while hearing news stories. Don't hesitate to jot down many questions, and don't bore yourself; try to identify questions that really interest you.

   b. Now take stock of your interests, your opportunities, and the work of others. Which of your research questions no longer seem feasible or interesting? What additional research questions come to mind? Pick out a question that is of interest and seems feasible and that your other coursework suggests has been the focus of some prior research or theorizing.

   c. Write out your research question in one sentence, and elaborate on it in one paragraph. List at least three reasons for why it is a good research question for you to investigate. Then present your proposal to your classmates and instructor for discussion and feedback.

2. Search the literature (and the Web) on the research question you identified. Refer to Appendix A for guidance on conducting the search. Copy down at least 10 citations to articles (with abstracts from *Sociology Abstracts* or *Psychology Abstracts*) and 5 Web sites reporting research that seems highly relevant to your research question; then look up at least 5 of these articles and 3 of the sites. Inspect the article bibliographies and the links in the Web site and identify at least one more relevant article and Web site from each source. Write a brief description of each article and Web site you consulted and evaluate its relevance to your research question. What additions or changes to your thoughts about the research question are suggested by the sources?

3. Propose at least two hypotheses that pertain to your research question. Justify these hypotheses in terms of the literature you have read.

4. Which standards for the protection of human subjects might pose the most difficulty for researchers on your proposed topic? Explain your answers and suggest appropriate protection procedures for human subjects.

Student Study Site

The companion Web site for *The Practice of Research in Criminology and Criminal Justice*, Third Edition

http://www.sagepub.com/prccj3

Visit the Web-based Student Study Site to enhance your understanding of the chapter content and to discover additional resources that will take your learning one step further. You can enhance your understanding of the chapters by using the comprehensive study material, which includes e-flashcards, Web exercises, practice self-tests, and more. You will also find special features, such as Learning from Journal Articles, which incorporates SAGE's online journal collection.

WEB EXERCISES

1. You have been assigned to write a paper on domestic violence and the law. To start, you would like to find out what the American Bar Association's stance is on the issue. Go to the American Bar Association Commission on Domestic Violence's Web site at www.abanet.org/domviol/mrdv/identify.html.

2. What is the American Bar Association's definition of domestic violence? How does it suggest one can identify a person as a victim of domestic violence? What does it identify as "basic warning signs"? Write your answers in a one- to two-page report.

3. Go to the Bureau of Justice Statistics (BJS) Web site at www.ojp.usdoj.gov/bjs/. Go to "Publications." Browse the list of publications for topics related to domestic violence. List the titles of all publications focusing on violence between intimate partners. Choose the most recent publication. How does the BJS define "intimate partners"? What are some of the characteristics of intimate partner violence? What trends are identified in the report?

ETHICS EXERCISES

1. Review the ethical guidelines adopted by the American Sociological Association (p. 63). Indicate whether you think each guideline was followed in the Sherman and Berk (1984) research on the policy response to domestic violence. If you find it hard to give a simple "yes" or "no" answer for each guideline, indicate the issues that make this evaluation difficult.

2. Concern with how research results are used is one of the hallmarks of ethical researchers, but deciding what form that concern should take is often difficult. You learned in this chapter about the controversy that occurred after Sherman and Berk (1984) encouraged police departments to adopt a pro-arrest policy in domestic abuse cases, based on findings from their Minneapolis study. Do you agree with the researchers' decision to suggest policy changes to police departments based on their study, in an effort to minimize domestic abuse? Several replication studies failed to confirm the Minneapolis findings. Does this influence your evaluation of what the researchers should have done after the Minneapolis study was completed? What about Larry Sherman's argument that failure to publicize the Omaha study finding of the effectiveness of arrest warrants resulted in some cases of abuse that could have been prevented? In one paragraph, propose a policy that researchers should follow about how much publicity is warranted and at what point in the research it should occur.

SPSS EXERCISES

Browse the variables in YOUTH.POR, a survey of high school youth regarding attitudes toward delinquency and delinquent behavior.

1. From these variables (excluding sex of respondent), write two hypotheses about levels of delinquency (DELINQ1) among high school youth.

2. Create a bar chart for at least one of the variables you hypothesize to be associated with levels of delinquency.

3. Compare the distribution of your chosen variable across gender groups. Select all males (SEX = 1) and request a bar chart of your chosen variable; then select all females (SEX = 2) and generate the bar chart again.

4. Compare the distributions between the two bar charts and formulate a hypothesis as to the relationship between the two variables. Is there a relationship between SEX and your chosen variable?

5. From these results, what do you hypothesize is the relationship between gender and level of delinquency?

CHAPTER 3

Conceptualization
and Measurement

Substance abuse is a social problem of remarkable proportions and is related to many forms of crime. For example, more than half of violent offenders incarcerated in state and federal prisons were under the influence of alcohol when they committed their crimes (Bureau of Justice Statistics 1998). Alcohol also is involved in about half of all fatal traffic crashes, and more than 1 million arrests are made annually for driving under the influence. As many as two-thirds of all persons arrested in urban areas test positive for drugs (Gruenewald et al.

Grunwald

1997). College presidents rate alcohol abuse as the number-one campus problem (Wechsler et al. 1994), and it is a factor in as many as two-thirds of on-campus sexual assaults (National Institute of Alcohol Abuse and Alcoholism [NIAAA], 1995). All told, the annual costs of prevention and treatment for alcohol and drug abuse exceed $4 billion (Gruenewald et al. 1997).

Whether your goal is to examine the factors related to criminal offending, to deliver useful services, or to design effective social policies, at some point you will probably need to read the research literature on substance abuse and perhaps even design your own study of it. Every time you begin to review or design relevant research, you will have to answer two questions: "What is meant by 'substance abuse' in this research?" (which concerns conceptualization) and "How was substance abuse measured?" (which concerns measurement). Both questions must be answered to evaluate the validity of substance abuse research. You cannot make sense of the results of a study until you know how the concepts were defined and measured. Nor are you ready to begin a research project until you have defined your concepts and constructed valid measures of them. Measurement validity is essential to successful research; in fact, without valid measures it is fruitless to attempt to achieve the other two aspects of validity: causal validity (see Chapter 5) and generalizability (see Chapter 4).

In this chapter, we first address the issue of conceptualization, using substance abuse and related concepts as examples. We also provide examples of the conceptualization process for other terms like "community policing" and "Street gangs." We then focus on measurement, reviewing first how measures of substance abuse have been constructed, using such operations as available data, questions, observations, and less direct and obtrusive measures. Then we explain how to assess the validity and reliability of these measures. The final topic is the level of measurement reflected in different measures. By chapter's end, you should have a good understanding of measurement, the first of the three legs on which a research project's validity rests.

CONCEPTS

A May 1997 *New York Times* article (Johnson 1997) reported that five U.S. colleges were participating in a pilot program to ban alcohol in their fraternities. Moreover, the article claimed that substance-free housing would soon become the norm on U.S. campuses. Do you know what the article means? Some of these **concepts**—*alcohol, colleges, campuses,* and *pilot program*—are widely understood and commonly used. However, do we all have the same thing in mind when we hear these terms? For example, are junior colleges subsumed within the term *college?* Does the concept of *on campus* extend to fraternity houses that are not physically on college property? Does *substance-free housing* mean banning tobacco products as well as alcohol?

Concept A mental image that summarizes a set of similar observations, feelings, or ideas.

Concepts such as *substance-free housing* require an explicit definition before they are used in research because we cannot be certain that all readers will share the same definition. It is even more important to define concepts that are somewhat abstract or

CONCEPTS: Poverty. Social Control, Strain (handwritten)

unfamiliar. When we refer to concepts such as *poverty* or *social control* or *strain*, we cannot be certain that others know exactly what we mean.

Clarifying the meaning of such concepts does not just benefit those unfamiliar with them; even experts often disagree about their meaning. But we need not avoid using these concepts; we just have to specify clearly what we mean when we use them, and we must expect others to do the same.

Specify concepts concepts clearly (handwritten)

(Conceptualization) in Practice

MUST HAVE A Definition (handwritten)

If we are to do an adequate job of **conceptualization**, we must do more than just think up some definition, any definition, for our concepts. We may need to distinguish subconcepts, or dimensions, of the concept. We also should ask how the concept's definition fits within the theoretical framework guiding the research and what assumptions underlie this framework.

Conceptualization The process of specifying what we mean by a term. In deductive research, conceptualization helps to translate portions of an abstract theory into testable hypotheses involving specific variables. In inductive research, conceptualization is an important part of the process used to make sense of related observations.

Defining Youth Gangs

Do you have a clear image in mind when you hear the term *youth gangs?* Although this is a very ordinary term, social scientists' attempts to define precisely the concept, youth gang, have not yet succeeded: "Neither gang researchers nor law enforcement agencies can agree on a common definition . . . and a concerted national effort . . . failed to reach a consensus" (Howell 2003:75). Exhibit 3.1 lists a few of the many alternative definitions of youth gangs.

What is the basis of this conceptual difficulty? Howell (2003:27–28) suggests that defining the term *youth gangs* has been difficult for four reasons:

- Youth gangs are not particularly cohesive.
- Individual gangs change their focus over time.
- Many have a "hodgepodge of features," with diverse members and unclear rules.
- There are many incorrect but popular "myths" about youth gangs.

Concepts (handwritten)

In addition, youth gangs are only one type of social group, and it is important to define youth gangs in a way that distinguishes them from these other types of groups, for example, childhood play groups, youth subculture groups, delinquent groups, and adult criminal organizations. You can think of "social group" as a broader concept that has multiple dimensions, one of which is youth gangs. In the same way, you can think of substance abuse as a concept with three dimensions: alcohol abuse, drug abuse, and polysubstance abuse. Whenever you define a concept, you need to consider whether the concept is unidimensional or multidimensional. If it is multidimensional, your job of conceptualization is not complete until you have specified the related subconcepts that belong under the umbrella of the larger concept.

→ Unidimensional / Multidimensional (handwritten)

EXHIBIT 3.1   Alternative Definitions of Youth Gangs

The term gang tends to designate collectivities that are marginal members of mainstream society, loosely organized, and without a clear, social purpose. (Ball & Curry 1995:227)

The gang is an interstitial group (between childhood and maturity) originally formed spontaneously, and then integrated through conflict. (Thrasher 1927:18)

[A gang is] any denotable adolescent group of youngsters who a) are generally perceived as a distinct aggregation by others in the neighborhood, b) recognize themselves as a denotable group (almost invariably with a group name), and c) have been involved in a sufficient number of delinquent incidents to call forth a consistently negative response from neighborhood residents and/or law enforcement agencies. (Klein 1971:13)

A youth gang is a self-formed association of peers united by mutual interests with identifiable leadership and internal organization who act collectively or as individuals to achieve specific purposes, including the conduct of illegal activity and control of a particular territory, facility, or enterprise. (Miller 1992:21)

[A gang is] an age-graded peer group that exhibits some permanence, engages in criminal activity, and has some symbolic representation of membership. (Decker & Van Winkle 1996:31)

[A gang is] a self-identified group of kids who act corporately, at least sometimes, and violently, at least sometimes. (Kennedy, Piehl, & Braga 1996:158)

A Criminal Street Gang is any ongoing organization, association, or group of three or more persons, whether formal or informal, having as one of its primary activities the commission of criminal acts. (Street Terrorism Enforcement and Prevention Act, 1988, California Penal Code sec. 186.22[f])

Source: Based on Howell 2003:76.

Defining Substance Abuse

What observations or images should we associate with the concept *substance abuse?* Someone leaning against a building with a liquor bottle, barely able to speak coherently? College students drinking heavily at a party? Someone in an Alcoholics Anonymous group drinking one beer? A 10-year-old boy drinking a small glass of wine in an alley? A 10-year-old boy drinking a small glass of wine at the dinner table in France? Do all these images share something in common that we should define as substance abuse for the purposes of a particular research study? Do some of them? Should we take into account cultural differences? Social situations? Physical tolerance for alcohol? Individual standards?

Many researchers now use the definition of substance abuse contained in the *Diagnostic and Statistical Manual of Mental Disorders* (1994) of the American Psychiatric Association (Mueser et al. 1990): "repeated use of a substance to the extent that it interferes with adequate social, vocational, or self-care functioning" (p. 33). In contrast, substance dependence is defined as "development of tolerance to a substance such that the person requires larger dosages to achieve the same psychoactive effect, and the experience of withdrawal symptoms and craving after a period of abstinence from the substance." Note that these definitions rely on behavioral and biological criteria rather than social expectations or cultural norms.

Not correct of journal

Clear and Precise language that should minimize differences in

We cannot judge the *DSM-IV* definition of substance abuse as correct or incorrect. Each researcher has the right to conceptualize as he or she sees fit. However, we can say that the *DSM-IV* definition of substance abuse is useful, even good, in part because it has been very widely adopted. If we conceptualize substance abuse in the same way that the *DSM-IV* does, many others will share our definition and understand what we are talking about. This definition of substance abuse has two other attractive features: It is stated in clear and precise language that should minimize differences in interpretation and maximize understanding; and it can clearly be distinguished from the more specific concept of substance dependence.

One caution is in order. The definition of any one concept rests on a shared understanding of the terms used in the definition. So if our audience does not already have a shared understanding of terms such as *adequate social functioning*, *self-care functioning*, and *repeated use*, we must also define these terms before we are finished with the process of defining substance abuse.

Interpretation and Maximize Understd

Defining Community Policing

Most of us believe that the police are supposed to prevent crime, not just respond to it after it occurs. However, early research demonstrated that there was little difference in rates of crime between jurisdictions where there was a strong police presence in the form of visible patrols and those jurisdictions where there were virtually no patrols (Bayley 1994). In the 1980s, police began to acknowledge the limitations of basic patrols and seek more effective prevention strategies. David Bayley (1994) describes three insights that helped shape the new crime-prevention strategies that emerged: (1) that police could not prevent crime without community help, (2) that police must do more than react to crime, and (3) that simply patrolling an area is too passive. Most people label the crime-prevention efforts that emerged from these insights as "community policing." Unfortunately, there is still a large amount of variation and disagreement among practitioners and researchers *alike* on what exactly community policing really is. As Bayley describes,

> I have heard police describe community policing as being foot patrol, aggressive enforcement of minor ordinances, electronic surveillance of shopping malls, enhanced traffic enforcement, and any police action that instills public confidence. . . . One police chief thought that any contact between the police and the public was community policing. (P. 104)

After examining the evolution of police practices from the common practice of basic patrolling to the new community approaches, Bayley (1994) observed that four elements were common: consultation, adaptation, mobilization, and problem solving. He conceptualized each of these elements as follows:

> *Consultation:* Establishing new mechanisms for discussing police priorities and strategies with their communities.

> *Adaptation:* Because crime and order needs vary from place to place, adaptation involves reshaping command structures so that local police commanders can use resources more flexibly.

Mobilization: Rather than relying on just their own efforts to prevent crime, mobilization involves police developing programs that enlist the active assistance of the public (e.g., Neighborhood Watches).

Problem Solving: Instead of simply responding to crimes after they have occurred, problem solving involves police studying conditions that lead to calls for their services, drawing up plans to correct these conditions, and taking the lead in evaluating and implementing remedial actions.

As you can see, community policing involves several different dimensions. There is wide variability in the extent to which police departments have adopted these different dimensions, and the future of community policing, at least in the United States, is unclear. However, in other industrialized countries, including Japan and Singapore, community policing has become entrenched (Bayley 1994).

Defining Poverty

Decisions about how to define a concept reflect the theoretical framework that guides the researchers. For example, the concept *poverty* has always been somewhat controversial, because different notions of what poverty is shape estimates of how prevalent it is and what can be done about it.

Most of the statistics that you see in the newspaper about the poverty rate reflect a conception of poverty that was formalized by Mollie Orshansky of the Social Security Administration in 1965 and subsequently adopted by the federal government and many researchers (Putnam 1977). She defined poverty as an *absolute* standard, based on the amount of money required to purchase an emergency diet that is estimated to be nutritionally adequate for about 2 months. The idea is that people are truly poor if they can barely purchase the food they need and other essential goods. This poverty standard is adjusted for household size and composition (number of children and adults), and the minimal amount needed for food is multiplied by 3 because a 1955 survey indicated that poor families spend about one-third of their incomes on food (Orshansky 1977).

Some social scientists disagree with the absolute standard and have instead urged adoption of a *relative* poverty standard or income inequality (Blau & Blau 1982). The idea behind this relative conception is that poverty should be defined in terms of what is normal in a given society at a particular time.

Some social scientists prefer yet another conception of poverty. With the *subjective* approach, poverty is defined as what people think would be the minimal income they need to make ends meet. Of course, many have argued that this approach is influenced too much by the different standards that people use to estimate what they need (Ruggles 1990:20–23).

Which do you think is a more reasonable approach to defining poverty: an absolute standard, a relative standard, or a subjective standard? Our understanding of the concept of poverty is sharpened when we consider the theoretical ramifications of these alternative definitions.

Defining Strain

Some concepts have multiple dimensions, bringing together several related concepts under a larger conceptual umbrella. One such concept is *strain*, first defined by Merton (1938) and later by Cloward and Ohlin (1960). In its simplest form, strain theory contends that delinquency and other forms of criminality and/or deviance are the result of frustrated needs or wants; this frustration is generally believed to result from a breakdown in the relationship between socially induced aspirations or goals and socially approved ways of achieving these goals. According to strain theory, delinquency is a response to actual or anticipated failure to achieve socially induced needs or goals (e.g., status, wealth, power, social acceptance). To determine the relationship between perceived strain and delinquency in a sample of 13- through 19-year-olds from the National Youth Survey (NYS), Elliott, Huizinga, and Ageton (1985) conceptualized strain as goal expectation discrepancies in both the home and school contexts. For example, strain would be increased at the individual level if being a good student was important to someone, but he or she did not perceive that teachers thought of him or her as a good student.

Relying on Merton's (1938) earlier conceptualization, Messner and Rosenfeld (1994) utilized the macro-level analogue of strain called *anomie*. Anomie, in its most general formulation, is defined as a weakening in the normative regulation of behavior. Anomie theory posits that diminished normative regulation results from an overemphasis on cultural goals (e.g., wealth, prestige) relative to the legitimate means to achieve them (e.g., availability of a college education, high status jobs), or from the differential distribution of opportunities to achieve highly valued goals (some groups in society may not have the same opportunities to achieve these goals as others). According to Messner and Rosenfeld, there are four cultural values within American society that contribute to anomie and therefore to crime in general: achievement, individualism, universalism, and materialism. To fully define anomie, Messner and Rosenfeld thus had to define these concepts. For example, to describe what they meant by the value of materialism, they state,

> In American culture, success is signified in a profoundly significant way: by the accumulation of monetary rewards. Money is awarded special priority in American culture. The point to emphasize here is not that Americans are uniquely materialistic, for a strong interest in material well-being can be found in most societies. Rather, the distinctive feature of American culture is the preeminent role of money as the "metric" of success. Monetary success is inherently open-ended. It is always possible in principle to have more money. Hence, the American Dream offers no final stopping point. It requires never-ending achievement. The pressure to accumulate money is therefore relentless, which entices people to pursue their monetary goals by any means necessary. (P. 71)

Concepts and Variables

After we define the concepts in a theory, we can identify variables corresponding to the concepts and develop procedures to measure them. This is an important step. Consider the concept of social control, which Donald Black (1984) defines as "all of the processes by which people define and respond to deviant behavior." What variables can represent this conceptualization of social control? Proportion of persons arrested in a community? Average length of sentences for crimes? Types of bystander reactions to public intoxication? Some combination of these?

Although we must proceed carefully to specify what we mean by a concept like social control, some concepts are represented well by the specific variables in the study and need not be defined so carefully. We may define binge drinking as heavy episodic drinking and measure it, as a variable, by asking people how many drinks they consumed in succession during some period (see Wechsler et al. 1994). That is pretty straightforward.

Be aware that not every concept in a study is represented by a variable. For example, if the term *tolerance of drinking* is defined as the absence of rules against drinking in a fraternity, it brings to mind a phenomenon that varies across different fraternities at different colleges. But if we study social life at only those fraternities that prohibit drinking, tolerance of drinking would not be a variable: All the fraternities studied have the same level of tolerance, and thus tolerance of drinking is a constant and not a variable. Of course, the concept of tolerance of drinking would still be important for understanding social life in the "dry" fraternities.

MEASUREMENT OPERATIONS

After we have defined our concepts in the abstract—that is, after conceptualizing—and after we have specified the specific variables we want to measure, we must develop our measurement procedures. The goal is to devise **operations** that actually measure the concepts we intend to measure, in other words, to achieve measurement validity.

Operation The procedure for actually measuring the concepts we intend to measure, identifying the value of a variable for each case.

Operationalization The process of specifying the operations that will indicate the value of a variable for each case.

Exhibit 3.2 represents the **operationalization** process in three studies. The first researcher defines her concept (binge drinking) and chooses one variable (frequency of heavy episodic drinking) to represent it. This variable is then measured with responses to a single question, or indicator: "How often within the past two weeks did you consume five or more drinks containing alcohol in a row?" The second researcher defines her concept, poverty, as having two aspects or dimensions, subjective poverty and absolute poverty. Subjective poverty is measured with responses to a survey question: "Do you consider yourself poor?" Absolute poverty is measured by comparing family income to the poverty threshold. The third researcher decides that her concept, social class, can be indicated with three measured variables: income, education, and occupational prestige. The values of these three variables for each case studied are then combined into a single **indicator**.

Good conceptualization and operationalization can prevent confusion later in the research process. For example, a researcher may find that substance abusers who join a self-help group are less likely to drink again than those who receive hospital-based substance abuse treatment. But what is it about these treatment alternatives that is associated with successful abstinence? Level of peer support? Beliefs about the causes of alcoholism? Financial

EXHIBIT 3.2   Concepts, Variables, and Indicators

Concept	Variable	Indicator
Binge drinking	Frequency of heavy episodic drinking	"How often within the past two weeks did you consume five or more drinks containing alcohol in a row?"
Poverty	Subjective poverty	"Would you say you are poor?"
	Absolute poverty	Family income ÷ poverty
Social class	Income	
	Education	Income + education + prestige
	Occupational prestige	

What measures are based on.

investment in the treatment? If the researcher had considered such aspects of the concept of substance abuse treatment before collecting her data, she might have been able to measure different aspects of treatment and so figure out which, if any, were associated with differences in abstinence rates. Because she did not measure these variables, she will not contribute as much as she might have to our understanding of substance abuse treatment.

Social researchers have many options for operationalizing their concepts. Measures can be based on activities as diverse as asking people questions, reading judicial opinions, observing social interactions, coding words in books, checking census data, enumerating the contents of trash receptacles, or drawing urine and blood samples. We focus here on the operations of using published data, asking questions, observing behavior, and using unobtrusive means of measuring people's behavior and attitudes.

Using Available Data

Government reports are rich and readily accessible sources of criminal justice data, as are datasets available from nonprofit advocacy groups, university researchers, and some private businesses. For example, law-enforcement and health statistics provide several community-level indicators of substance abuse (Gruenewald et al. 1997). Statistics on arrests for the sale and possession of drugs, drunk driving arrests, and liquor law violations (such as sales to minors) can usually be obtained on an annual basis, and often quarterly, from local police departments or state crime information centers.

Indicators like these cannot be compared across communities or over time without reviewing carefully how they were constructed. The level of alcohol in the blood that is legally required to establish intoxication can vary among communities, creating the appearance of different rates of substance abuse even though drinking and driving practices may be identical. Enforcement practices can vary among police jurisdictions and over time (Gruenewald et al. 1997:14). We also cannot assume that available data are accurate, even when they appear to measure the concept in which we are interested in a way that is consistent across communities. "Official" counts of homeless persons have been notoriously unreliable because of the difficulty of locating homeless persons on the streets, and government agencies have, at times, resorted to "guesstimates" by service providers (Rossi 1989). Even available data for such seemingly straightforward measures as cause of death can contain a surprising amount of error. For example, between 30% and 40% of death certificates incorrectly identify the cause of death (Altman 1998).

Government statistics that are generated through a central agency like the U.S. Bureau of the Census are often of high quality, but caution is warranted when using official data collected by local levels of government. For example, the Uniform Crime Reports (UCR) program administered by the Federal Bureau of Investigation imposes standard classification criteria, with explicit guidelines and regular training at the local level, but data are still inconsistent for many crimes. Consider only a few of the many sources of inconsistency between jurisdictions: variation in the classification of forcible rape cases due to differences in what is considered to be "carnal knowledge of a female"; different decisions about what is considered "more than necessary force" in the definition of "strong-arm" robberies; whether offenses in which threats were made but no physical injury occurred are classified as aggravated or simple assaults (Mosher, Miethe, & Phillips 2002:66). A new National Incident-Based Reporting System (NIBRS) corrects some of the problems with the UCR, but it requires much more training and documentation and has not yet been widely used (Mosher, Miethe, & Phillips 2002:70).

In some cases, problems with an available indicator can be lessened by selecting a more precise indicator. For example, the number of single-vehicle nighttime crashes, whether fatal or not, is a more specific indicator of the frequency of drinking and driving than just the number of single-vehicle fatal accidents (Gruenewald et al. 1997:40–41). Focusing on a different level of aggregation may also improve data quality, because procedures for data collection may differ between cities, counties, states, and so on (Gruenewald et al. 1997:40–41). It is only after such factors as legal standards, enforcement practices, and measurement procedures have been taken into account that comparisons among communities become credible.

Constructing Questions

Asking people questions is the most common and probably the most versatile operation for measuring social variables. Most concepts about individuals can be defined in such a way that measurement with one or more questions becomes an option. In this section, we introduce some options for writing single questions; in Chapter 7, we explain why single questions can be inadequate measures of some concepts, and then examine approaches that rely on multiple questions to measure a concept.

Measuring variables with single questions is very popular. Public opinion polls based on answers to single questions are reported frequently in newspaper articles and TV newscasts: "Do you favor or oppose U.S. policy in . . . ?" "If you had to vote today, for which candidate would you vote?" Criminal justice surveys also rely on single questions to measure many variables: "Overall, how satisfied are you with the police in your community?" "How would you rate your current level of safety?"

Single questions can be designed with or without explicit response choices. The question that follows is a **closed-ended (fixed-choice) question** because respondents are offered explicit responses to choose from. It has been selected from the Core Alcohol and Drug Survey distributed by the Core Institute (1994), Southern Illinois University, for the Fund for the Improvement of Postsecondary Education (**FIPSE**) Core Analysis Grantee Group (Presley, Meilman, & Lyerla 1994).

> Compared to other campuses with which you are familiar, this campus's use of alcohol is. . . . (*Mark one*)
>
> _____ Greater than other campuses
>
> _____ Less than other campuses
>
> _____ About the same as other campuses

Response choices should be mutually exclusive and exhaustive, so that every respondent can find one and only one choice that applies to him or her (unless the question is of the "Check all that apply" format). To make response choices exhaustive, researchers may need to offer at least one option with room for ambiguity. For example, a questionnaire asking college students to indicate their school status should not use freshman, sophomore, junior, senior, and graduate student as the only response choices. Most campuses also have students in a "special" category, so you might add "Other (please specify)" to the five fixed responses to this question. If respondents do not find a response option that corresponds to their answer to the question, they may skip the question entirely or choose a response option that does not indicate what they are really thinking.

Most surveys of a large number of people contain primarily fixed-choice questions, which are easy to process with computers and analyze with statistics. With fixed-choice questions, respondents are also more likely to answer the question that the researcher really wants them to answer. Including response choices reduces ambiguity and makes it easier for respondents to answer. However, fixed-response choices can obscure what people really think if the choices do not match the range of possible responses to the question; many studies show that some respondents will choose response choices that do not apply to them simply to give some sort of answer (Peterson 2000:39). We will discuss question wording and response options in greater detail in Chapter 7.

Open-ended questions—questions without explicit response choices, to which respondents write in their answers—are preferable when the range of responses cannot adequately be anticipated, namely, questions that have not previously been used in surveys and questions that are asked of new groups. Open-ended questions can also lessen confusion about the meaning of responses involving complex concepts. The next question is an open-ended version of the earlier fixed-choice question:

> How would you say alcohol use on this campus compares to that on other campuses?

Making Observations

Observations can be used to measure characteristics of individuals, events, and places. The observations may be the primary form of measurement in a study, or they may supplement measures obtained through questioning. Reiss (1971a) developed a careful method of observing phenomena that he termed **systematic social observation** (SSO). In his study of police interaction with the public, Reiss's SSO method involved riding in police squad cars, observing police-citizen interactions and recording features of these interactions characteristics on a form.

Sampson and Raudenbush (1999) refined the SSO technique in their study of neighborhood disorder and crime. Teams drove in "a sport utility vehicle at a rate of five miles per hour down every street" in a sample of Chicago neighborhoods (Raudenbush and Sampson 1999). Two video cameras recorded people and activities on both sides of the vehicle, while a trained observer completed a log for each block. The resulting 23,816 observer logs contained information about building conditions and land use, while the videotapes were coded to measure features of streets, buildings, businesses, and social interaction on 15,141 blocks. Direct observation is often the method of choice for measuring behavior in natural settings, as long as it is possible to make the requisite observations.

Collecting Unobtrusive Measures

Unobtrusive measures allow us to collect data about individuals or groups without their direct knowledge or participation. In their classic book (now revised), Webb et al. ([1966] 2000) identified four types of unobtrusive measures: physical trace evidence, archives (available data), simple observation, and contrived observation (using hidden recording hardware or manipulation to elicit a response). Let us consider the first two types in more detail: physical trace evidence and archives.

The physical traces of past behavior are one type of unobtrusive measure that is most useful when the behavior of interest cannot be directly observed (perhaps because it is hidden or occurred in the past) and has not been recorded in a source of available data. To measure the prevalence of drinking in college dorms or fraternity houses, we might count the number of empty bottles of alcoholic beverages in the surrounding dumpsters. However, you can probably see that care must be taken to develop trace measures that are useful for comparative purposes. For instance, comparison of the number of empty bottles in dumpsters outside different dorms can be misleading; at the very least, you would need to take into account the number of residents in the dorms, the time since the last trash collection, and the accessibility of each dumpster to passersby.

Unobtrusive measures can also be created from such diverse forms of media as newspaper archives or magazine articles, TV or radio talk shows, legal opinions, historical documents, personal letters, or e-mail messages. An investigation of the drinking climate on campuses might include a count of the amount of space devoted to ads for alcoholic beverages in a sample of issues of the student newspaper. Campus publications also might be coded to indicate the number of times that statements discouraging substance abuse appear. With this tool, you could measure the frequency of articles reporting substance abuse–related crimes, the degree of approval of drinking expressed in TV shows or songs, or the relationship between region of the country and amount of space devoted in the print media to drug usage.

Combining Measurement Operations

Using available data, asking questions, making observations, and using unobtrusive indicators are interrelated measurement tools, each of which may include or be supplemented by the others. From people's answers to survey questions, the U.S. Bureau of the Census develops widely consulted census reports containing available data on people, firms, and geographic units in the United States. Data from employee surveys may be supplemented by information available in company records. Interviewers may record observations about those whom they question. Researchers may use insights gleaned from questioning participants to make sense of the social interaction they have observed. Unobtrusive indicators can be used to evaluate the honesty of survey responses.

Questioning can be a particularly poor approach for measuring behaviors that are very socially desirable, such as voting or attending church, or that are socially stigmatized or illegal, such as abusing alcohol or drugs. **Triangulation**, the use of two or more different measures of the same variable, can strengthen measurement considerably (Brewer & Hunter 1989:17). When we achieve similar results with different measures of the same variable, particularly when they are based on such different methods as survey questions and field-based observations, we can be more confident in the validity of each measure. If results diverge with different measures, it may indicate that one or more of these measures are influenced by more measurement error than we can tolerate. Divergence between measures could also indicate that they actually operationalize different concepts. An interesting example of this interpretation of divergent results comes from research on crime. Official crime statistics only indicate those crimes that are reported to and recorded by the police; when surveys are used to measure crimes with self-reports of victims, many "personal annoyances" are included as if they were crimes (Levine 1976). We will talk more about triangulation in Chapter 10.

EVALUATION OF MEASURES

The issue of measurement error is very important. Do the operations developed to measure our concepts actually do so; are they valid? If we have weighed our measurement options, carefully constructed our questions and observational procedures, and carefully selected indicators from the available data, we should be on the right track. But we cannot have much confidence in a measure until we have empirically evaluated its validity.

Measurement Validity

We can consider **measurement validity** the first concern in establishing the validity of research results, because without having measured what we think we measured, we really do not know what we are talking about.

Measurement validity The extent to which measures indicate what they are intended to measure.

As an example of measurement validity, consider the following question: "How prevalent is youth violence and delinquency in the United States; how many juveniles are involved in delinquency?" Data on the extent of juvenile delinquency come from two primary sources: official statistics and unofficial statistics. Official statistics are based on the aggregate records of juvenile offenders and offenses processed by agencies of the criminal justice system: police, courts, and corrections. One primary source of official statistics on juvenile delinquency is the *Uniform Crime Reports (UCR)*, produced by the Federal Bureau of Investigation (FBI). Unofficial statistics are data produced by people or agencies outside the criminal justice system such as victimization surveys and self-report studies. The validity of official statistics for measuring the extent of juvenile delinquency is a heated debate among criminologists. Although some researchers believe official reports are a valid measure of serious delinquency, others contend that UCR data say more about the behavior of the police than about delinquency. These criminologists think the police are predisposed against certain groups of people or certain types of crimes.

Unquestionably, official reports underestimate the actual amount of delinquency because a great deal of delinquent behavior never comes to the attention of police (Mosher, Miethe, & Phillips 2002). Sometimes delinquent acts are committed and not observed; or they are observed and not reported. There is also evidence that the UCRs often reflect the political climate and police policies as much as they do criminal activity. Take the United States' "War on Drugs," which heated up in the 1980s. During this time, arrest rates for drug offenses soared, giving the illusion that drug use was increasing at an epidemic pace. However, self-report surveys that asked citizens directly about their drug use behavior during this time period found that use of most illicit drugs was actually declining (Regoli & Hewitt 1994). In your opinion, then, which measure of drug use, the UCR or self-report surveys, was more valid? Before you answer this question, let us continue with the delinquency example.

Despite the limitations of official statistics for measuring delinquency, these data were relied on by criminologists and used as a valid measure of the prevalence of delinquency for many decades. As such, delinquency and other violent offenses were thought to primarily involve minority populations and/or disadvantaged youth. In 1947, however, Wallerstein and Wyle (1947) surveyed a sample of 700 juveniles and found that 91% admitted to having committed at least one offense that was punishable by one or more years in prison, and 99% admitted to at least one offense for which they could have been arrested had they been caught. In 1958, Short and Nye (1957–1958) reported the results from the first large-scale self-report study involving juveniles from a variety of locations. In their research, Short and Nye concluded that delinquency was widespread throughout the adolescent population and that youth from high-income families were just as likely to engage in delinquency as youth from low-income families. Contemporary studies using self-report data from the National Youth Survey (NYS) indicate that the actual amount of delinquency is much greater than that reported by the UCR and that unlike these official data, where nonwhites are overrepresented, self-report data indicate that white juveniles report nearly the identical number of delinquencies as nonwhites, but fewer of them are arrested (Elliott & Ageton 1980). This is just one example of measurement validity and should convince you that we must be very careful in designing our measures and in subsequently evaluating how well they have performed.

The extent to which measures indicate what they are intended to measure can be assessed with one or more of four basic approaches: face validation, content validation, criterion validation, and construct validation. Whatever the approach to validation, no one measure will be valid for all times and places. For example, the validity of self-report measures of substance abuse varies with such factors as whether the respondents are sober or intoxicated at the time of the interview, whether the measure refers to recent or lifetime abuse, and whether the respondents see their responses as affecting their chances at receiving housing, treatment, or some other desired outcome (Babor, Stephens, & Marlatt 1987). In addition, persons with severe mental illness are, in general, less likely to respond accurately (Corse, Hirschinger, & Zanis 1995). These types of possibilities should always be considered when evaluating measurement validity.

Face Validity

To test your understanding of evaluating measures, go to the Valid and Reliable Measures Interactive Exercises on the Student Study Site.

Researchers apply the term **face validity** to the confidence gained from careful inspection of a concept to see if it is appropriate "on its face," simply whether it appears to measure what it intends. For example, measuring people's favorite color seems unlikely on its face to tell us much about their alcohol consumption patterns. A measure with greater face validity would be a count of how many drinks they had consumed in the past week.

Although every measure should be inspected in this way, face validation in itself does not provide very convincing evidence of measurement validity. The question "How much beer or wine did you have to drink last week?" may look valid on its face as a measure of frequency of drinking, but people who drink heavily tend to underreport the amount they drink. So the question would be an invalid measure in a study that includes heavy drinkers.

Content Validity

Content validity establishes that the measure covers the full range of the concept's meaning. To determine that range of meaning, the researcher may solicit the opinions of experts and review literature that identifies the different aspects of the concept.

An example of a measure that covers a wide range of meaning is the Michigan Alcoholism Screening Test (MAST). The MAST includes 24 questions representing the following subscales: recognition of alcohol problems by self and others; legal, social, and work problems; help seeking; marital and family difficulties; and liver pathology (Skinner & Sheu 1982). Many experts familiar with the direct consequences of substance abuse agree that these dimensions capture the full range of possibilities. Thus, the MAST is believed to be valid from the standpoint of content validity.

Criterion Validity

When people drink an alcoholic beverage, the alcohol is absorbed into their blood and then gradually metabolized (broken down into other chemicals) in their liver (NIAAA 1997). The alcohol that remains in their blood at any point, unmetabolized, impairs both thinking and behavior (NIAAA 1994). As more alcohol is ingested, cognitive and behavioral consequences multiply. These biological processes can be identified with direct measures of alcohol concentration in the blood, urine, or breath. Questions about drinking behavior, on the other hand, can be viewed as attempts to measure indirectly what biochemical tests measure directly.

Criterion validity is established when the scores obtained on one measure can be accurately compared to those obtained with a more direct or already validated measure of the same phenomenon (the criterion). A measure of blood-alcohol concentration or a urine test could serve as the criterion for validating a self-report measure of drinking, as long as the questions we ask about drinking refer to the same period. Observations of substance use by friends or relatives could also, in some circumstances, serve as a criterion for validating self-report substance use measures.

Criterion validation studies of substance abuse measures have yielded inconsistent results. Self-reports of drug use agreed with urinalysis results for about 85% of the drug users who volunteered for a health study in several cities (Weatherby et al. 1994). On the other hand, the post-treatment drinking behavior self-reported by 100 male alcoholics was substantially less than the drinking behavior observed by the alcoholics' friends or relatives (Watson et al. 1984). Such inconsistent findings can occur because of differences in the adequacy of a measures across settings and populations. This underscores our point that you cannot assume that a measure that was validated in one study is also valid in another setting or with a different population.

An attempt at criterion validation is well worth the effort because it greatly increases confidence that the measure is measuring what was intended. However, often no other variable might reasonably be considered a criterion for feelings or beliefs or other subjective states. Even with variables for which a reasonable criterion exists, the researcher may not be able to gain access to the criterion, as would be the case with a tax return or employer document as criterion for self-reported income.

Construct Validity

Measurement validity also can be established by showing that a measure is related to a variety of other measures as specified in a theory. This validation approach, known as **construct validity**, is commonly used in social research when no clear criterion exists for validation purposes. For example, in one study of the validity of the Addiction Severity Index (ASI), A. Thomas McLellan et al. (1985) compared subject scores on the ASI to a number of indicators that they felt from prior research should be related to substance abuse: medical problems, employment problems, legal problems, family problems, and psychiatric problems. They could not use a criterion validation approach because they did not have a more direct measure of abuse, such as laboratory test scores or observer reports. However, their extensive research on the subject had given them confidence that these sorts of problems were all related to substance abuse, and thus their measures seemed to be valid from the standpoint of construct validity. Indeed, the researchers found that individuals with higher ASI ratings tended to have more problems in each of these areas, giving us more confidence in the ASI's validity as a measure.

A somewhat different approach to construct validation is termed **discriminant validity**. In this approach, scores on the measure to be validated are compared to scores on another measure of the same variable and to scores on variables that measure different but related concepts. Discriminant validity is achieved if the measure to be validated is related most strongly to its comparison measure and less to the measures of other concepts. McLellan et al. (1985) found that the ASI passed this test, too: The ASI's measures of alcohol and drug problems were related more strongly to other measures of alcohol and drug problems than they were to measures of legal problems, family problems, medical problems, and the like.

The distinction between criterion and construct validation is not always clear. Opinions can differ about whether a particular indicator is indeed a criterion for the concept that is to be measured. For example, if you need to validate a question-based measure of sales ability for applicants to a sales position, few would object to using actual sales performance as a criterion. But what if you want to validate a question-based measure of the amount of social support that people receive from their friends? Should you just ask people about the social support they have received? Could friends' reports of the amount of support they provided serve as a criterion? Are verbal accounts of the amount of support provided adequate? What about observations of social support that people receive? Even if you could observe people in the act of counseling or otherwise supporting their friends, can an observer be sure that the interaction is indeed supportive? There isn't really a criterion here, just related concepts that could be used in a construct validation strategy. Even biochemical measures of substance abuse are questionable as criteria for validating self-reported substance use. Urine test results can be altered by ingesting certain substances, and blood tests vary in their sensitivity to the presence of drugs over a particular period of time.

What construct and criterion validation have in common is the comparison of scores on one measure to scores on other measures that are predicted to be related. It is not so important that researchers agree that a particular comparison measure is a criterion rather than a related construct. But it is very important to think critically about the quality of the comparison measure and whether it actually represents a different measure of the same phenomenon. For example, it is only a weak indication of measurement validity to find that scores on a new self-report measure of alcohol use are associated with scores on a previously used self-report measure of alcohol use.

Reliability

Reliability means that a measurement procedure yields consistent scores when the phenomenon being measured is not changing (or that the measured scores change in direct correspondence to actual changes in the phenomenon). If a measure is reliable, it is affected less by random error, or chance variation, than if it is unreliable. Reliability is a prerequisite for measurement validity; we cannot really measure a phenomenon if the measure we are using gives inconsistent results.

Reliability A measure is reliable when it yields consistent scores or observations of a given phenomenon on different occasions. Reliability is a prerequisite for measurement validity.

There are four possible methods for measuring the reliability of a measure: test-retest reliability, interitem reliability, alternate-forms reliability, and interobserver reliability.

Test-Retest Reliability

When researchers measure a phenomenon that does not change between two points separated by an interval of time, the degree to which the two measurements yield comparable, if not identical, values is the **test-retest reliability** of the measure. If you take a test of your math ability and then retake the test 2 months later, the test is performing reliably if you receive a similar score both times, presuming that nothing happened during the 2 months to change your math ability. Of course, if events between the test and the retest have changed the variable being measured, then the difference between the test and retest scores should reflect that change.

When ratings by an observer, rather than ratings by the subjects themselves, are being assessed at two or more points in time, test-retest reliability is termed **intraobserver reliability** or **intrarater reliability**.

To test your understanding of evaluating measures, go to the Valid and Reliable Measures Interactive Exercises on the Student Study Site.

One example of how test-retest reliability may be assessed is a study by Sobell et al. (1988) of alcohol abusers' past drinking behavior (using the Lifetime Drinking History questionnaire) and life changes (using the Recent Life Changes questionnaire). All 69 subjects in the study were patients in an addiction treatment program. They had not been drinking prior to the interview (determined by a breath test). The two questionnaires were administered by different interviewers about 2 or 3 weeks apart, both times asking the subjects to recall events 8 years prior to the interviews. Reliability was high: 92% of the subjects reported the same life events both times, and at least 81% of the subjects were classified consistently at both interviews as having had an alcohol problem or not. When asked about their inconsistent answers, subjects reported that in the earlier interview they had simply dated an event incorrectly, misunderstood the question, evaluated the importance of an event differently, or forgotten an event. Answers to past drinking questions were less reliable when they were very specific, apparently because the questions exceeded subjects' capacities to remember accurately.

Interitem Reliability (Internal Consistency)

When researchers use multiple items to measure a single concept, they are concerned with **interim reliability** (or internal consistency). For example, if we are to have confidence that a set of questions (such as those in Exhibit 3.3) reliably measures attitudes toward violence, the answers to the questions should be highly associated with one another. The stronger the association among the individual items, and the more items that are included, the higher the reliability of the index.

Alternate-Forms Reliability

Researchers are testing **alternate-forms reliability** when they compare subjects' answers to slightly different versions of survey questions (Litwin 1995:13–21). A researcher may reverse the order of the response choices in an index or modify the question wording in minor ways and then readminister that index to subjects. If the two sets of responses are not too different, alternate-forms reliability is established.

A related test of reliability is the **split-halves reliability** approach. A survey sample is divided in two by flipping a coin or using some other random assignment method. These two halves of the sample are then administered the two forms of the questions. If the responses of the two halves of the sample are about the same, the measurer's reliability is established.

Interobserver Reliability

When researchers use more than one observer to rate the same persons, events, or places, **interobserver reliability** is their goal. If observers are using the same instrument to rate the same thing, their ratings should be very similar. If they are similar, we can have much more confidence that the ratings reflect the phenomenon being assessed rather than the orientations of the observers.

EXHIBIT 3.3   Questions Used in the Violent Defensive Values Index

Would you approve of a man punching a stranger who had hit the man's child after the child accidentally damaged the stranger's car?

_____ No (0)

_____ I don't know or not sure (1)

_____ Yes (2)

Would you approve of a man punching a stranger who was beating up a woman and the man saw it?

_____ No (0)

_____ I don't know or not sure (1)

_____ Yes (2)

Would you approve of a man punching a stranger who had broken into the man's house?

_____ No (0)

_____ I don't know or not sure (1)

_____ Yes (2)

Source: Cao, Adams & Jensen 1997.

Ways to Improve Reliability and Validity

We must always assess the reliability of a measure if we hope to able to establish its validity. Remember that a reliable measure is not necessarily a valid measure, as Exhibit 3.4 illustrates. This discrepancy is a common flaw of self-report measures of substance abuse. The multiple questions in self-report indexes of substance abuse are answered by most respondents in a consistent way, so the indexes are reliable. However, a number of respondents will not admit to drinking, even though they drink a lot. Their answers to the questions are consistent, but they are consistently misleading. So the indexes based on self-report are reliable but invalid. Such indexes are not useful and should be improved or discarded. Unfortunately, many measures are judged to be worthwhile on the basis only of a reliability test.

The reliability and validity of measures in any study must be tested after the fact to assess the quality of the information obtained. But then if it turns out that a measure cannot be considered reliable and valid, little can be done to save the study. Hence it is supremely important to select in the first place measures that are likely to be reliable and valid. In studies that use interviewers or observers, careful training is often essential to achieving a consistent approach. In most cases, however, the best strategy is to use measures that have been used before and whose reliability and validity have been established in other contexts. But the

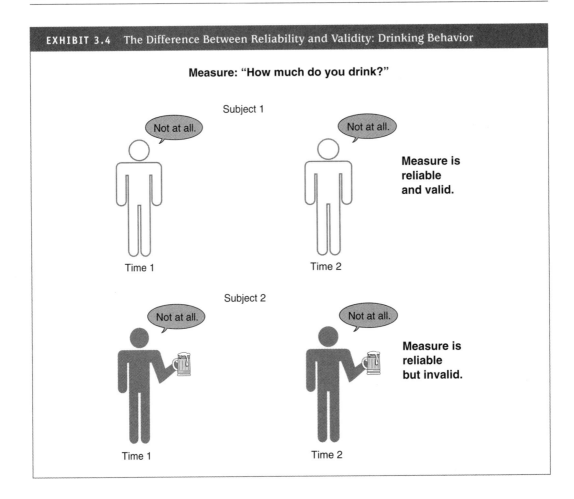

EXHIBIT 3.4 The Difference Between Reliability and Validity: Drinking Behavior

selection of "tried and true" measures still does not absolve researchers from the responsibility of testing the reliability and validity of the measure in their own studies.

The process of evaluating the reliability and validity of measures about individuals is termed **psychometrics**. Measures of individuals that range from tests you take in school to personality assessments you complete on the job are advertised as "psychometrically valid" after multiple studies have demonstrated their reliability and validity. The process of evaluating the reliability and validity of measures about organizations, neighborhoods, or other collective units is termed **ecometrics**, a term coined by Raudenbush and Sampson (1999). For example, Raudenbush and Sampson's ecometric evaluation of their observational measures of Chicago neighborhoods included a test of the consistency of ratings by multiple observers of the same neighborhoods (p. 7).

It may be possible to improve the reliability and validity of measures in a study that already has been conducted if multiple measures were used. For example, in our study of housing for homeless mentally ill persons, funded by the National Institute of Mental Health, we assessed substance abuse with several different sets of direct questions as well as with reports from subjects' case managers and others (Goldfinger et al. 1996). We found that the observational reports were often inconsistent with self-reports and that different self-report measures were not always in agreement and were thus unreliable. A more reliable measure of substance abuse was initial reports of lifetime substance abuse problems. This measure was extremely accurate in identifying all those who subsequently abused substances during the project. We concluded that the lifetime measure was a valid way to identify persons at risk for substance abuse problems.

LEVELS OF MEASUREMENT

To test your knowledge, go to the Levels of Measurement interactive exercises on the Student Study Site.

When we know a variable's **level of measurement**, we can better understand how cases vary on that variable and so understand more fully what we have measured. Level of measurement also has important implications for the types of statistics that can be used with the variable, as you will learn in Chapter 12. There are four levels of measurement: nominal, ordinal, interval, and ratio. Exhibit 3.5 depicts the differences among these four levels.

Level of measurement The complexity of the mathematical means that can be used to express the relationship between a variable's values. The nominal level of measurement, which is qualitative, has no mathematical interpretation; the quantitative levels of measurement (ordinal, interval, and ratio) are progressively more complex mathematically.

Nominal Level of Measurement

The **nominal level of measurement** (also called the categorical or qualitative level) identifies variables whose values have no mathematical interpretation; they vary in kind or quality but not in amount. In fact, it is conventional to refer to the values of nominal variables as attributes instead of values. Gender is one example. The variable "gender" has two attributes (or categories or qualities): male and female. We might indicate male with the value 1 and female with the value 2, but these numbers do not tell us anything about the difference between male and female except that they are different. Female is not one unit more of "gender" than male, nor is it twice as much "gender." Ethnicity, occupation, religious affiliation, and region of the country are also measured at the nominal level. A person may be Spanish or Portuguese, but one ethnic group does not represent more ethnicity than another, just a different ethnicity. A person may be a doctor or a truck driver, but one does not represent three units more occupation than the other.

Although the attributes of categorical variables do not have a mathematical meaning, they must be assigned to cases with great care. The attributes we use to measure, or categorize, cases must be mutually exclusive and exhaustive:

- A variable's attributes or values are **mutually exclusive attributes** if every case can have only one attribute.
- A variable's attributes or values are **exhaustive attributes** when every case can be classified into one of the categories.

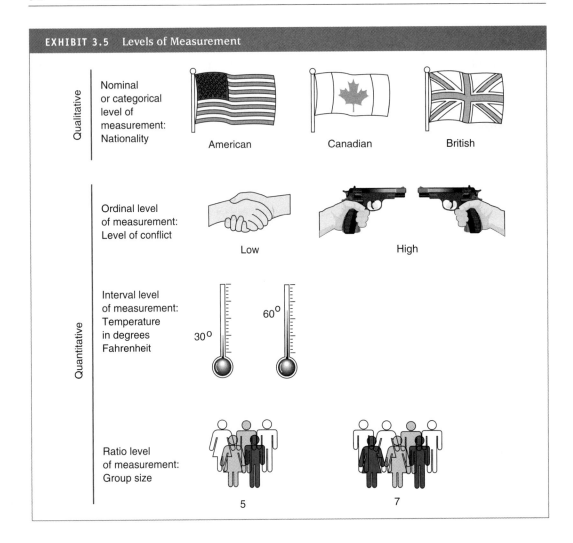

EXHIBIT 3.5   Levels of Measurement

When a variable's attributes are mutually exclusive and exhaustive, every case corresponds to one, and only one, attribute. Imagine the challenge of coming up with an exhaustive set of attributes when a variable with a large number of attributes is being studied.

Ordinal Level of Measurement

The first of the three quantitative levels is the **ordinal level of measurement**. At this level, the numbers assigned to cases specify only the order of the cases, permitting greater-than and less-than distinctions; absolute mathematical distinctions cannot be made between categories. The Core Alcohol and Drug Survey (Core Institute 1994) measures substance abuse with a series of questions that permit ordinal distinctions (see Exhibit 3.6). Although these

EXHIBIT 3.6 Example of Ordinal Measures: Core Alcohol and Drug Survey

Within the last year, about how often have you used . . . (mark one for each line)	Did Not Use	Once a Year	6 Times a Year	Once a Month	Twice a Month	Once a Week	3 Times a Week	5 Times a Week	Every Day
a. Tobacco (smoke, chew, snuff)									
b. Alcohol (beer, wine, liquor)									
c. Marijuana (pot, hash, hash oil)									
d. Cocaine (crack, rock, freebase)									
e. Amphetamines (diet pills, speed)									
f. Sedatives (downers, ludes)									
g. Hallucinogens (LSD, PCP)									
h. Opiates (heroin, smack, horse)									
i. Inhalants (glue, solvents, gas)									
j. Designer drugs (ecstasy, MDMA)									
k. Steroids									
l. Other illegal drugs									

Source: Core Institute 1994.

categories may seem at first to be mathematically distinct, you can really only determine the extent to which a respondent who checks one category consumes more or less than another respondent.

The properties of variables measured at the ordinal level are illustrated in Exhibit 3.5 by the contrast between the level of conflict in two groups. The first group, symbolized by two people shaking hands, has a low level of conflict. The second group, symbolized by two persons using guns against each other, has a higher level of conflict. The third group, symbolized by two people pointing guns at each other, has an even higher level of conflict. To measure conflict, we would put the groups "in order" by assigning the number 1 to the low-conflict group, the number 2 to the group using fists, and the number 3 to the high-conflict group using guns. The numbers thus indicate only the relative position or order of the cases. Although low level of conflict is represented by the number 1, it is not mathematically two less units of conflict than the high level of conflict, which is represented by the number 3. These numbers really have no mathematical qualities; they are just used to represent relative rank in the measurement of conflict.

As with nominal variables, the different values of a variable measured at the ordinal level must be mutually exclusive and exhaustive. They must cover the range of observed values and allow each case to be assigned no more than one value.

Interval Level of Measurement

The numbers indicating the values of a variable at the **interval level of measurement** represent fixed measurement units but have no absolute, or fixed, zero point. This level of measurement is represented in Exhibit 3.5 by the difference between two Fahrenheit temperatures. Although 60 degrees is 30 degrees hotter than 30 degrees, 60 in this case is not twice as hot as 30. Why not? Because heat does not begin at 0 degrees on the Fahrenheit scale.

An interval-level measure is created by a scale that has fixed measurement units but no absolute, or fixed, zero point. The numbers can therefore be added and subtracted, but ratios are not meaningful. Again, the values must be mutually exclusive and exhaustive.

Social scientists often treat indexes that were created by combining responses to a series of variables measured at the ordinal level as interval-level measures. An index of this sort could be created with responses to the Core Institute's (1994) questions about friends' disapproval of substance use (see Exhibit 3.7). The survey has 13 questions on the topic, each of which has the same three response choices. If Don't disapprove is valued at 1, Disapprove is valued at 2, and Strongly disapprove is valued at 3, the summed index of disapproval would range from 12 to 36. The average could then be treated as a fixed unit of measurement. So a score of 20 could be treated as if it were 4 more units than a score of 16.

Ratio Level of Measurement

The numbers indicating the values of a variable at the **ratio level of measurement** represent fixed measuring units and an absolute zero point (zero means absolutely no amount of whatever the variable indicates). On a ratio scale, 10 is two points higher than 8 and is also two times greater than 5. Ratio numbers can be added and subtracted, and because the

EXHIBIT 3.7 Ordinal Level Variables Can Be Added to Create an Index With Interval-Level Properties: Core Alcohol and Drug Survey

How do you think your close friends feel (or would feel) about you . . . (mark one for each line)	Do Not Disapprove	Disapprove	Strongly Disapprove
a. Trying marijuana once or twice			
b. Smoking marijuana occasionally			
c. Smoking marijuana regularly			
d. Trying cocaine once or twice			
e. Taking cocaine regularly			
f. Trying LSD once or twice			
g. Taking LSD regularly			
h. Trying amphetamines once or twice			
i. Taking amphetamines regularly			
j. Taking one or two drinks of an alcoholic beverage (beer, wine, liquor) nearly every day			
k. Taking four or five drinks nearly every day			
1. Having five or more drinks in one sitting			
m. Taking steroids for bodybuilding or improved athletic performance			

Source: Core Institute 1994.

numbers begin at an absolute zero point, they can be multiplied and divided (so ratios can be formed between the numbers). For example, people's ages can be represented by values ranging from 0 years (or some fraction of a year) to 120 or more. A person who is 30 years old is 15 years older than someone who is 15 years old (30 – 15 = 15) and is twice as old as that person (30/15 = 2). Of course, the numbers also are mutually exclusive and exhaustive, so that every case can be assigned one and only one value.

Exhibit 3.5 displays an example of a variable measured at the ratio level. The number of people in the first group is 5, and the number in the second group is 7. The ratio of the two groups' sizes is then 1.4, a number that mirrors the relationship between the sizes of the groups. Note that there does not actually have to be any group with a size of 0; what is important is that the numbering scheme begins at an absolute zero, in this case, the absence of any people. The number of days a convicted felon was sentenced to prison would represent a ratio level of measurement because sentence length begins with an absolute 0 point. The number of days an addict stays clear after treatment, too, has a ratio level of measurement.

The Case of Dichotomies

Dichotomies, variables having only two values, are a special case from the standpoint of levels of a measurement. Although variables with only two categories are generally thought of as nominally measured, we can also think of a dichotomy as indicating the presence or absence of an attribute. Suppose, for example, we were interested in differences between individuals who had never used illegal drugs in the last year compared to those who had used at least one illegal drug in the last year. We could create a variable that indicated this dichotomous distinction by coding those individuals who said they did not use any of the substances listed as 0, and all others as 1. Viewed in this way, there is an inherent order to the two values: In one group the attribute of consuming illegal substances is absent (those coded 0), and in another it is present (those coded 1).

Comparison of Levels of Measurement

Exhibit 3.8 summarizes the types of comparisons that can be made with different levels of measurement, as well as the mathematical operations that are legitimate. All four levels of measurement allow researchers to assign different values to different cases. All three quantitative measures allow researchers to rank cases in order.

An important thing to remember is that researchers *choose* levels of measurement in the process of operationalizing the variables; the level of measurement is not inherent in the variable itself. Many variables can be measured at different levels, with different procedures.

EXHIBIT 3.8   Properties of Measurement Levels

		Relevant Level of Measurement			
Examples of Comparison Statements	*Appropriate Math Operations*	*Nominal*	*Ordinal*	*Interval*	*Ratio*
A is equal to (not equal to) *B*	= (≠)	✓	✓	✓	✓
A is greater than (less than) *B*	> (<)		✓	✓	✓
A is three more than (less than) *B*	+ (−)			✓	✓
A is twice (half) as large as *B*					✓

For example, The Core Alcohol and Drug Survey (Core Institute 1994) identifies binge drinking by asking students, "Think back over the last two weeks. How many times have you had five or more drinks at a sitting?" You might be ready to classify this as a ratio-level measure, but you must first examine the fixed response options given to respondents. This is a **closed-ended question**, and students are asked to indicate their answer by checking None, Once, Twice, 3 to 5 times, 6 to 9 times, or 10 or more times. Use of these categories makes the level of measurement ordinal. The distance between any two cases cannot be clearly determined. A student with a response in the "6 to 9 times" category could have binged just one more time than a student who responded 3 to 5 times, or they could have binged 4 more times. With these response categories, you cannot mathematically distinguish the number of times a student binged, only the relative amount of binging behavior.

The more information available, the more ways you have to compare cases. You also have more possibilities for statistical analysis with quantitative than with qualitative variables. Thus, it often is a good idea to try to measure variables at the highest level of measurement possible, if doing so does not distort the meaning of the concept that is to be measured. For example, measure age in years, if possible, rather than in categories. You can always recategorize the responses into categories that compare teenagers to young adults, but it is impossible to obtain the actual age in years when the question is asked using an ordinal response format.

Be aware, however, that other considerations may preclude measurement at a high level. For example, many people are very reluctant to report their exact incomes, even in anonymous questionnaires. So asking respondents to report their income in categories (such as under $10,000, $10,000–19,999, $20,000–29,999, etc.) will result in more responses, and thus more valid data, than asking respondents for their income in dollars.

CONCLUSION

Remember always that measurement validity is a necessity for social research. Gathering data without careful conceptualization or conscientious efforts to operationalize key concepts often is a wasted effort.

The difficulties of achieving valid measurement vary with the concept being operationalized and the circumstances of the particular study. The examples in this chapter of difficulties in achieving valid measures of substance abuse should sensitize you to the need for caution, particularly when the concepts you wish to measure are socially stigmatized and/or illegal.

Planning ahead is the key to achieving valid measurement in your own research; careful evaluation is the key to sound decisions about the validity of measures in others' research. Statistical tests can help to determine whether a given measure is valid after data have been collected, but if it appears after the fact that a measure is invalid, little can be done to correct the situation. If you cannot tell how key concepts were operationalized when you read a research report, do not trust the findings. And if a researcher does not indicate the results of tests used to establish the reliability and validity of key measures, remain skeptical.

KEY TERMS

Alternate-forms reliability
Closed-ended question
Concept
Conceptualization
Construct validity
Content validity
Criterion validity
Dichotomy
Discriminant validity
Ecometrics
Exhaustive attributes
Face validity
Indicator
Interitem reliability
Interobserver reliability
Interval level of measurement
Intraobserver reliability (intrarater reliability)

Level of measurement
Measurement validity
Mutually exclusive attributes
Nominal level of measurement
Open-ended questions
Operation
Operationalization
Ordinal level of measurement
Psychometrics
Ratio level of measurement
Reliability
Split-halves reliability
Systematic social observation
Test-retest reliability
Triangulation
Unobtrusive measures

HIGHLIGHTS

- Conceptualization plays a critical role in research. In deductive research, conceptualization guides the operationalization of specific variables; in inductive research, it guides efforts to make sense of related observations.

- Concepts are operationalized in research by one or more indicators, or measures, which may derive from observation, self-report, available records or statistics, books and other written documents, clinical indicators, discarded materials, or some combination.

- The validity of measures should always be tested. There are four basic approaches: face validation, content validation, criterion validation, and construct validation. Criterion validation provides the strongest evidence of measurement validity, but there often is no criterion to use in validating social science measures.

- Measurement reliability is a prerequisite for measurement validity, although reliable measures are not necessarily valid. Reliability can be assessed through test-retest procedures, in terms of interitem consistency, through a comparison of responses to alternate forms of the test, or in terms of consistency among observers.

- Level of measurement indicates the type of information obtained about a variable and the type of statistics that can be used to describe its variation. The four levels of measurement can be ordered by complexity of the mathematical operations they permit: nominal (least complex), ordinal, interval, ratio (most complex). The measurement level of a variable is determined by how the variable is operationalized. Dichotomies are a special case of measurement.

EXERCISES

1. Are important concepts in criminological research always defined clearly? Are they defined consistently? Search the literature for six articles that focus on "violent crime," "domestic violence," or some other concept suggested by your instructor. Is the concept

defined clearly in each article? How similar are the definitions? Write what yo found in a short report.

2. What are some of the research questions you could attempt to answer with av statistical data? Visit your library and ask for an introduction to the governmen documents collection. Inspect the volumes from the Federal Bureau of Investigation (FBI) Uniform Crime Report (UCR) or the *Sourcebook for Criminal Justice Statistics*, both of which report statistics on crimes committed by offender characteristics. List 10 questions you could explore with such data.

3. Develop a plan for evaluating the validity of a measure. Your instructor will give you a copy of a questionnaire actually used in a study. Choose one question and define the concept that you believe it is intended to measure. Then develop a construct validation strategy involving other measures in the questionnaire that you think should be related to the question of interest; that is, if it measures what you think it measures.

DEVELOPING A RESEARCH PROPOSAL

At this point you can begin the processes of conceptualization and operationalization. You will need to assume that your primary research method will be conducting a survey.

1. List at least 10 variables that will be measured in your research. No more than two of these should be sociodemographic indicators like race or age. The inclusion of each variable should be justified in terms of theory or prior research that suggests it would be an appropriate independent or dependent variable, or will have some relation to either of these.

2. Write a conceptual definition for each variable. Whenever possible, this definition should come from the existing literature, either a book you have read for a course or the research literature that you have been searching. Ask two class members for feedback on your definitions.

3. Develop measurement operations for each variable. Several measures should be single questions and indexes that were used in prior research (search the Web and the journal literature in *Soc Abstracts* or *Pscyh Abstracts*). Make up a few questions and one index yourself. Ask classmates to answer these questions and give you feedback on their clarity.

4. Propose tests of reliability and validity for four of the measures.

Student Study Site

The companion Web site for *The Practice of Research in Criminology and Criminal Justice*, Third Edition

http://www.sagepub.com/prccj3

Visit the Web-based Student Study Site to enhance your understanding of the chapter content and to discover additional resources that will take your learning one step further. You can enhance your understanding of the chapters by using the comprehensive study material, which includes e-flashcards, Web exercises, practice self-tests, and more. You will also find special features, such as Learning from Journal Articles, which incorporates SAGE's online journal collection.

WEB EXERCISES

1. Use the Web to find information regarding alcohol consumption and crime. Write a short report on your findings. How is "alcohol consumption" conceptualized and measured in the various sources you find?

2. How would you define "rape"? Write a brief definition. Based on this conceptualization, what circumstances constitute rape? Describe a method of measurement that would be valid for a study of rape (as you define it). Now go to the Rape Victim Advocates' Web site at www.RapeVictimAdvocates.org. Go to "Myths & Facts." Discuss some facts about rape that you were previously unaware of or some myths you believed. Rewrite your definition of rape based on your new knowledge. What additional circumstances constitute rape based on your new conceptualization?

ETHICS EXERCISES

1. In order to measure disorder in Chicago neighborhoods, Sampson and Raudenbush (1999) recorded the street scene with video cameras in a van with darkened windows. Do you judge this measurement procedure as ethical? Refer to each of the guidelines in chapter 2. How could the guideline about "informed consent" be interpreted to permit this type of observational procedure?

2. Both some Homeland Security practices and inadvertent releases of Web searching records have raised new concerns about the use of unobtrusive measures of behavior and attitudes. If all identifying information is removed, do you think criminologists should be able to study who is stopped by police for traffic violations? What types of books are checked out in libraries in different communities? The extent of use of pornography in different cities by analyzing store purchases? How much alcohol different types of people use by linking credit card records to store purchases?

SPSS EXERCISES

HOMICIDE.POR contains a sample of homicide defendants from a sample of 33 U.S. counties for the year 1988.

1. Obtain a frequency distribution for the variables INTIMATE, NUMVICT, and PRIMTIME. At what levels (nominal or categorical, ordinal, interval, ratio) are each of these variables measured?

2. What conclusions do you make about the victim-offender relationship, number of victims, and length of sentence received based on these frequency distributions?

CHAPTER 4

Sampling

When a jury in Albany, New York, acquitted the four Caucasian police officers who had
fired 41 bullets and killed a young African immigrant named Amadou Diallo, reporters went
to the streets to uncover the public sentiment about the verdict. Reporters from the
Associated Press went to the people gathered outside Diallo's Bronx home; reporters were
also gathered to hear reaction from the rally at Reverend Al Sharpton's National Action
Network headquarters in Harlem. Speakers at both rallies were not just average citizens but
also politicians running for office or seeking reelection. For example, the Associated Press
paraphrased U.S. Representative Charles Rangel's contention that officers never would have
shot Diallo if he had been Caucasian. Other citizens' anger and frustration were revealed
by these statements: "Murderers!" they shouted, "Racist cops!" "We want justice!" Providing

quotations from individuals within news stories is a common technique in journalism. These quotes help to create a human face to go with the story.

The problem with these journalistic methods, however, is that we do not know whether the opinions selected to be printed by journalists represent the opinions of the larger population. For example, is the sentiment of Representative Rangel similar to that of all U.S. representatives from New York? All Democratic U.S. representatives? Do the quotes from citizens represent the opinion of everyone living in Harlem? How about the Bronx, or New York City as a whole? In other words, we do not know how generalizable these comments are, and if we do not have confidence in their generalizability, their validity is suspect. In reality, of course, we have no idea whether these opinions are widely shared or unique.

In this chapter, we first review the rationale for using sampling in social research and consider two alternatives to sampling. We then turn to the topic of specific sampling methods and when they are most appropriate, using a variety of examples from the criminological research literature. We also briefly introduce the concept of a sampling distribution and explain how it helps in estimating our degree of confidence in statistical generalizations. By the chapter's end, you should know which questions you need to ask to evaluate the generalizability of a study as well as what choices you need to make to design a sampling strategy.

SAMPLE PLANNING

You have encountered the problem of generalizability in each of the studies you read about in this book. For example, Felson et al. (1994) generalized their sample-based explanation of adolescent aggression and violence to the population of high school students in the United States; Powell, Muir-McClain, and Halasyamani (1995) generalized their evaluation findings of violence prevention programs from an elementary school in King County, Washington, to all public elementary schools; and Lawrence Sherman and Richard Berk (1984) and others (Berk et al. 1992; Dunford, Huizinga, & Elliott 1990; Garner, Fagan, & Maxwell 1995; Hirschel, Hutchison, & Dean 1992; Pate & Hamilton 1992; Sherman et al. 1992) tried to determine the generalizability of findings from the original study of domestic violence in Minneapolis. Whether we, like these other researchers, are designing a sampling strategy or evaluating the generalizability of someone else's findings, we have to understand how and why researchers decide to sample. Sampling is very common in social research, but sometimes it is not necessary.

Define Sample Components and the Population

Let us say that we are designing a study of a topic that involves a lot of people (or other entities such as cities, or countries), which are the elements in our study. We do not have the time or resources to study the entire **population**—all the elements in which we are interested—and so we resolve to study a **sample**, a subset of this population.

> ✱ **Population** The entire set of elements (e.g., individuals, cities, states, countries, prisons, schools) in which we are interested.

> ✱ **Sample** A subset of elements from the larger population.

Sampling Units Primary - Secondary

We may collect our data directly from the elements in our sample. Some studies are not so simple, however. The entities we can easily reach to gather information are not the same as the entities about whom we really want information. So we may collect information about the elements from another set of entities called the **sampling units**. For example, if we interview mothers to learn about their families, the families are the elements and the mothers are the sampling units. If we survey prison wardens to learn about prisons, the prisons are the elements and the wardens are the sampling units.

- Single Stage Sample Study -

A single-stage sample is a study in which individual people are sampled and are the focus of the study; the sampling units are the same as the elements. However, if a sample is selected in two or more stages, the units selected—let's say groups and individuals—at each stage within the groups are sampling units, but it may be that only one sampling unit is the study's element (see Exhibit 4.1). For example, a researcher might sample families for a survey about household theft victimizations and then interview only one adult representative of the family, such as a parent (a sample), to obtain information about any victimizations. The families are the primary sampling units (and are the elements in the study), and the parents are secondary sampling units (but they are not elements, because they provide information about the family).

One key issue with selecting or evaluating sample components is understanding exactly what population they represent. In a survey of adult Americans, the general population may be reasonably construed as all residents of the United States who are at least 21 years old. But always be alert to ways the population may have been narrowed by the sample selection procedures. Perhaps only English-speaking adult residents of the continental United States were actually sampled. The population for a study is the aggregation of elements that we actually focus on and sample from, not a larger aggregation that we really wish we could have studied.

Some populations cannot be easily identified by a simple criterion such as a geographic boundary or an organizational membership. Let us say you were interested in victimizations experienced by the homeless population. In this case, a clear definition of the homeless population is difficult, but quite necessary. In research, anyone should be able to determine what population was actually studied. However, studies of homeless persons in the early 1980s "did not propose definitions, did not use screening questions to be sure that the people they interviewed were indeed homeless, and did not make major efforts to cover the universe of homeless people" (Burt 1996:15). For example, some studies relied on homeless persons in only one shelter. The result was "a collection of studies that could not be compared" (Burt 1996:15). Several studies of homeless persons in urban areas addressed the problem by employing a more explicit definition of the population: People are homeless if they have no home or permanent place to stay of their own (renting or owning) and no regular arrangement to stay at someone else's place (Burt 1996).

EXHIBIT 4.1   Sample Components in a Two-Stage Study

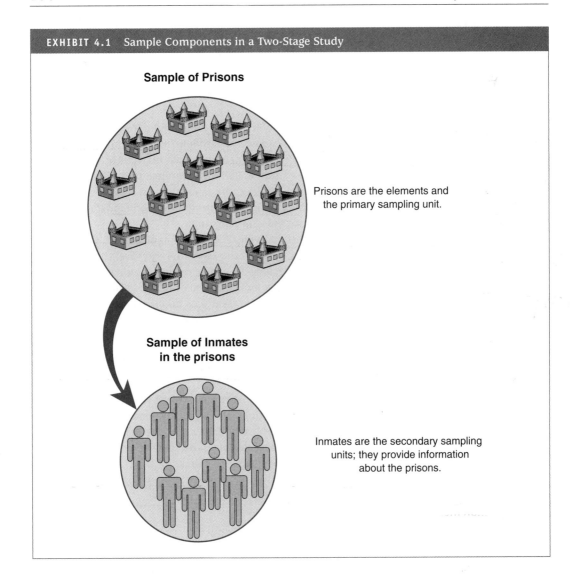

Sample of Prisons

Prisons are the elements and the primary sampling unit.

Sample of Inmates in the prisons

Inmates are the secondary sampling units; they provide information about the prisons.

Even this more explicit definition still leaves some questions unanswered: What is a regular arrangement? How permanent does a permanent place have to be? In a study of homeless persons in Chicago, Sosin, Colson, and Grossman (1988) answered these questions in their definition of the population of interest:

> We define the homeless as those currently residing for at least one day but for less than fourteen with a friend or relative, not paying rent, and not sure that the length of stay will surpass fourteen days; those currently residing in a shelter, whether overnight or transitional; those currently without normal, acceptable

shelter arrangements and thus sleeping on the street, in doorways, in abandoned buildings, in cars, in subway or bus stations, in alleys, and so forth; those residing in a treatment center for the indigent who have lived at the facility for less than 90 days and who claim that they have no place to go, when released. (P. 22)

This definition accurately reflects these researchers' concept of homelessness and allows researchers in other locations or at other times to develop procedures for studying a comparable population. The more complete and explicit the definition of the population from which a sample is selected, the more precise our generalizations from a sample to that population can be.

Evaluate Generalizability

After we clearly define the population we will sample, we need to determine the scope of the generalizations we will seek to make from our sample. Let us say we were interested in the extent to which high school youth are fearful of being attacked or harmed at school or going to and from their schools. It would be easy to go down to the local high school and hand out a survey asking students to report their levels of fear in these situations. But what if my local high school were located in a remote and rural area of Alaska? Would this sample reflect levels of fear perceived by suburban youth in California or urban youth in New York City? Obviously not. Oftentimes, regardless of the sample utilized, researchers will go on to talk about "this percentage of high school students are fearful" or "freshman students are more fearful than seniors," as if their study results represented all high school students. Many social researchers and criminologists (and most everyone else, for that matter) are eager to draw conclusions about all individuals they are interested in, not just their samples. Generalizations make their work (and opinions) sound more important. If every high school student were like every other one, generalizations based on observations of one high school student would be valid. But of course, that is not the case.

Generalizability has two aspects. **Sample generalizability** refers to the ability to generalize from a sample, or subset, of a larger population to that population itself. This is the most common meaning of generalizability. **Cross-population generalizability** refers to the ability to generalize from findings about one group, population, or setting to other groups, populations, or settings (see Exhibit 4.2). In this book, we use the term **external validity** to refer only to cross-population generalizability, not to sample generalizability.

> **Sample generalizability** Exists when a conclusion based on a sample, or subset, of a larger population holds true for that population.
>
> **Cross-population generalizability** Exists when findings about one group, population, or setting hold true for other groups, populations, or settings. Also called *external validity*.

Sample generalizability is a key concern in survey research, whether you are interested in measuring the extent of violent victimization, perceived levels of fear, or any other issue. A good example of the importance of sample generalizability comes from legitimate

EXHIBIT 4.2   Sample and Cross-Population Generalizability

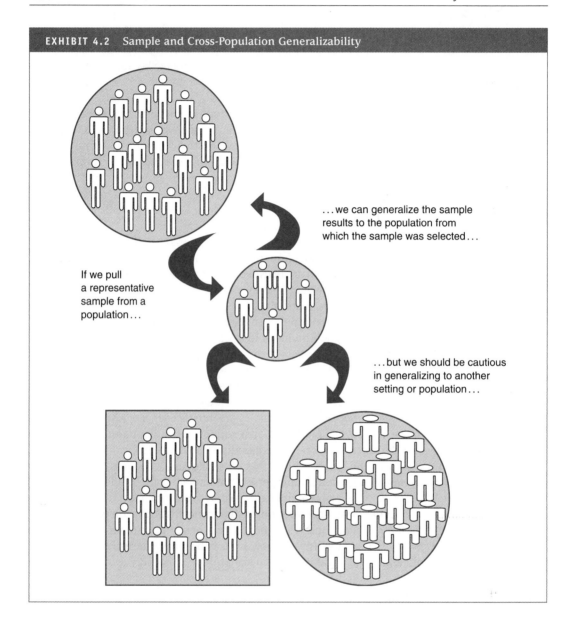

...we can generalize the sample results to the population from which the sample was selected...

If we pull a representative sample from a population...

...but we should be cautious in generalizing to another setting or population...

pollsters such as Gallup and Roper. Say we were interested in the likely winner of the next presidential election. To determine this, these pollsters might study a sample of likely voters, for example, and then generalize their findings to the entire population of likely voters. No one would be interested in the results of political polls if they represented only the tiny sample that actually was surveyed rather than the entire population.

Cross-population generalizability occurs to the extent that the results of a study hold true for multiple populations; these populations may not all have been sampled, or they may be represented as subgroups within the sample studies. For example, our cross-population generalizability would be called into question if we said all high school students perceived low levels of fear while going to and from school and we only had our rural Alaskan sample.

Generalizability is a key concern in research design. We rarely have the resources to study the entire population that is of interest to us, so we have to select cases to study that will allow our findings to be generalized to the population of interest. We can never be sure that our propositions will hold under all conditions, so we should be cautious in generalizing to populations that we did not actually sample.

This chapter primarily focuses on the problem of sample generalizability: Can findings from a sample be generalized to the population from which the sample was drawn? This is the most basic question to ask about a sample, and social research methods provide many tools to address it.

Sample generalizability depends on sample quality, which is determined by the amount of **sampling error**. Sampling error can generally be defined as the difference between the characteristics of a sample and the characteristics of the population from which it was selected. The larger the sampling error, the less representative the sample, and thus the less generalizable the findings. To assess sample quality when you are planning or evaluating a study, ask yourself these questions:

- From what population were the cases selected?
- What method was used to select cases from this population?
- Do the cases that were studied represent, in the aggregate, the population from which they were selected?

Sampling error Any difference between the characteristics of a sample and the characteristics of the population from which it was drawn. The larger the sampling error, the less representative the sample is of the population.

In reality, researchers often project their theories onto groups or populations much larger than, or simply different from, those they have actually studied. The **target population** is a set of elements larger than or different from the population that was sampled, and to which the researcher would like to generalize any study findings. When we generalize findings to target populations, we must be somewhat speculative. We must carefully consider the claim that the findings can be applied to other groups, geographic areas, cultures, or times.

Because the validity of cross-population generalizations cannot be tested empirically, except by conducting more research in other settings, we do not focus much attention on this problem here. We will return to the problem of cross-population generalizability in Chapter 5, which addresses experimental research.

Assess Population Diversity

Sampling is unnecessary if all the units in the population are identical. Physicists do not need to select a representative sample of atomic particles to learn about basic physical processes. They can study a single atomic particle, because it is identical to every other particle of its type. Similarly, biologists do not need to sample a particular type of plant to determine whether a given chemical has toxic effects on it. The idea is, "If you've seen one, you've seen 'em all."

What about people? Certainly all people are not identical; nor are animals in many respects. Nonetheless, if we are studying physical or psychological processes that are the same among all people, sampling is not needed to achieve generalizable findings. Psychologists and social psychologists often conduct experiments on college students to learn about processes that they think are identical for all individuals. They believe that most people will have the same reactions as the college students if they experience the same experimental conditions. Field researchers who observe group processes in a small community sometimes make the same assumption.

Milgram

There is a potential problem with this assumption, however. There is no way to know if the processes being studied are identical for all people. In fact, experiments can give different results depending on the type of people studied or the conditions for the experiment. Milgram's (1965) classic experiments on obedience to authority, among the most replicated experiments in the history of social psychological research, illustrate this point very well. The Milgram experiments tested the willingness of male volunteers in New Haven, Connecticut, to comply with instructions from an authority figure to give electric shocks to someone else, even when these shocks seemed to harm the person receiving them. In most cases, the volunteers complied. Milgram concluded that people are very obedient to authority.

Were these results generalizable to all men, to men in the United States, or to men in New Haven? Similar results were obtained in many replications of the Milgram experiments when the experimental conditions and subjects were similar to those studied by Milgram. Other studies, however, showed that some groups were less likely to react so obediently. Given certain conditions, such as another subject in the room who refused to administer the shocks, subjects were likely to resist authority.

So what do the experimental results tell us about how people will react to an authoritarian movement in the real world, when conditions are not so carefully controlled? In the real social world, people may be less likely to react obediently. Other individuals may argue against obedience to a particular leader's commands or people may see, on TV, the consequences of their actions. But alternatively, people may be even more obedient to authority than the experimental subjects, as they get swept up in mobs or are captivated by ideological fervor. Milgram's research gives us insight into human behavior, but there is no guarantee that what he found with particular groups in particular conditions can be generalized to the larger population (or to any particular population) in different settings.

Generalizing the results of experiments and of participant observation is risky, because such research often studies a small number of people who do not represent a particular population. Researchers may put aside concerns about generalizability when they observe the social dynamics of specific clubs, or college dorms, or a controlled experiment that tests the effect of, say, a violent movie on feelings for others. But we should be cautious about generalizing the results of such studies.

The important point is that social scientists rarely can skirt the problem of demonstrating the generalizability of their findings. If a small sample has been studied in an experiment or field research project, the study should be replicated in different settings or, preferably, with a **representative sample** of the population for which the generalizations are sought (see Exhibit 4.3).

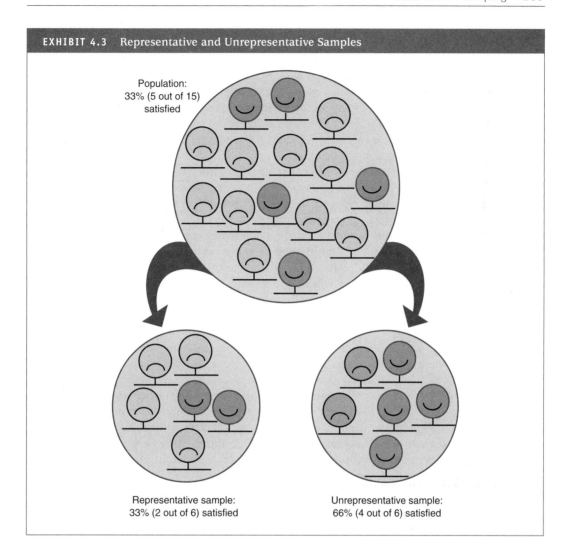

EXHIBIT 4.3   Representative and Unrepresentative Samples

Population:
33% (5 out of 15)
satisfied

Representative sample:
33% (2 out of 6) satisfied

Unrepresentative sample:
66% (4 out of 6) satisfied

The social world and the people in it are just too diverse to be considered identical units. Social psychological experiments and small field studies have produced good social science, but they need to be replicated in other settings, with other subjects, to claim any generalizability. Even when we believe that we have uncovered basic social processes in a laboratory experiment or field observation, we should be very concerned with seeking confirmation in other samples and other research.

Representative sample A sample that looks like the population from which it was selected in all respects that are potentially relevant to the study. The distribution of characteristics among the elements of a representative sample is the same as the distribution of those characteristics among the total population. In an unrepresentative sample, some characteristics are overrepresented or underrepresented.

Consider a Census

In some circumstances, it may be feasible to skirt the issue of generalizability by conducting a **census** studying the entire population of interest rather than drawing a sample. The federal government tries to do this every 10 years with the U.S. Census. A census can also include studies of all the employees (or students) in small organizations (or universities), studies comparing all 50 states, and studies of the entire population of a particular type of organization in a particular area. However, in all these instances, except for the U.S. Census, the population studied is relatively small.

Social scientists do not often attempt to collect data from all the members of some large population simply because doing so would be too expensive and time-consuming. Some social scientists do conduct research with data from the U.S. Census, but it is the government that collects the data and your tax dollars that pay for the effort. For the 2000 census, the Bureau of the Census used half a million temporary workers just to follow up on the 34% of households that did not return their census form in the mail.

Even if the population of interest for a survey is a small town of 20,000 or students in a university of 10,000, researchers will have to sample. The costs of surveying just thousands of individuals far exceed the budgets for most research projects. In fact, even the U.S. Bureau of the Census cannot afford to have everyone answer all the questions that should be covered in the census. So it draws a sample. Every household must complete a short version of the census (it had 7 questions in 2000), and a sample consisting of one in six households must complete a long form (with 53 additional questions) (Rosenbaum 2000).

Another costly fact is that it is hard to get people to complete a survey. Even the U.S. Bureau of the Census (1999) must make multiple efforts to increase the rate of response in spite of the federal law requiring all citizens to complete their census questionnaire. After spending $167 million on publicity (Forero 2000), the Bureau still planned up to six attempts to contact each household that did not respond by mail (U.S. Bureau of the Census 2000a). And then it planned an even more intensive sample survey to learn about the characteristics of those who still had not been contacted (U.S. Bureau of the Census, 2000b). Even relatively small rates of nonresponse in a survey can skew the results, so efforts to decrease nonresponse are essential. As the U.S. 2000 census progressed, concerns arose about under-representation of minority groups (Kershaw 2000), impoverished cities (Zielbauer 2000), well-to-do individuals in gated communities and luxury buildings (Langford 2000), and even college students (Abel 2000). The average survey has far less legal and financial backing, and so it is essential to limit the total number in the sample so there are more resources for follow-up procedures.

SAMPLING METHODS

As you can probably guess, the most important feature to know about a sample is whether it is truly representative of the population from which it was selected. The most important distinction made about samples is whether they are based on a probability or a nonprobability sampling method. Sampling methods that allow us to know in advance how likely it is that any element of a population will be selected for the sample are **probability sampling methods**. Sampling methods that do not let us know the likelihood in advance are **nonprobability sampling methods**.

Probability sampling methods rely on a random selection procedure. In princip
is the same as flipping a coin to decide which person wins and which one loses. Hea
tails are equally likely to turn up in a coin toss, so both persons have an equal chance to win.
That chance, their **probability of selection**, is 1 out of 2, or .5.

✸ *Probability of selection* The likelihood that an element will be selected from the
population for inclusion in the sample. In a census of all the elements of a population,
the probability that any particular element will be selected is 1.0 because everyone will
be selected. If half the elements in the population are sampled on the basis of chance
(say, by tossing a coin), the probability of selection for each element is one-half, or
0.5. When the size of the sample as a proportion of the population decreases, so does the
probability of selection.

Flipping a coin is a fair way to select one of two people, because the selection process
harbors no systematic bias. You might win or lose the coin toss, but you know that the outcome was due simply to chance, not to bias (unless your opponent tossed a two-headed
coin!). For the same reason, rolling a six-sided die is a fair way to choose one of six possible outcomes (the odds of selection are 1 out of 6, or .17). Dealing out a hand after shuf-
fling a deck of cards is a fair way to allocate sets of cards in a poker game (the odds of each
person getting a particular outcome, such as a full house or a flush, are the same). Similarly,
state lotteries use a random process to select winning numbers. Thus, the odds of winning
a lottery, the probability of selection, are known even though they are very small (perhaps
1 out of 1 million) compared to the odds of winning a coin toss. As you can see, the fundamental strategy in probability sampling is the **random selection** of elements into the
sample. When a sample is randomly selected from the population, every element has a
known and independent chance of being selected into the sample.

✗ *Random selection* The fundamental element of probability samples. The essential characteristic of random selection is that every element of the population has a known and independent chance of being selected into the sample.

There is a natural tendency to confuse the concept of probability, in which cases are
selected only on the basis of chance, with a haphazard method of sampling. On first impression, leaving things up to chance seems to imply not exerting control over the sampling
method. But to ensure that nothing but chance influences the selection of cases, the researcher
must proceed very methodically, leaving nothing to chance except the selection of the cases
themselves. The researcher must carefully follow controlled procedures if a purely random
process is to occur. In fact, when reading about sampling methods, do not assume that a random sample was obtained just because the researcher used a random selection method at
some point in the sampling process. Look for these two particular problems: selecting elements from an incomplete list of the total population and failing to obtain an adequate
response rate (say, only 45% of the people who were asked to participate actually agreed).
The list from which the elements of the population are selected is the **sampling frame.**
If this list is incomplete, a sample selected randomly from the list will obviously not be a

random sample of the population because not everyone in the population was represented on the list. Thus, you should always consider the adequacy of the sampling frame. Even for a simple population like a university's student body, the registrar's list is likely to be at least a bit out of date at any given time. For example, some students will have dropped out, but their status will not yet be officially recorded. Although you may judge the amount of error introduced in this particular situation to be negligible, the problems are greatly compounded for a larger population. The sampling frame for a city, state, or nation is always likely to be incomplete because of constant migration into and out of the area. Even unavoidable omissions from the sampling frame can bias a sample against particular groups within the population.

A very inclusive sampling frame may still yield systematic bias if many sample members cannot be contacted or refuse to participate. **Nonresponse** is a major hazard in survey research because individuals who do not respond to a survey are likely to differ systematically from those who take the time to participate. You should not assume that findings from a randomly selected sample will be generalizable to the population from which the sample was selected if the rate of nonresponse is considerable (certainly not if it is much above 30%).

Probability Sampling Methods

Probability sampling methods are those in which the probability of selection is known and is not zero (so there is some chance of selecting each element). These methods randomly select elements and therefore have no systematic bias; nothing but chance determines which elements are included in the sample. This feature of probability samples, sometimes referred to as random samples, makes them much more desirable than non-probability samples when the goal is to generalize your findings to a larger population.

Even though a random sample has no systematic bias, it certainly will have some sampling error due to chance. The probability of selecting a head is .5 in a single toss of a coin, and in 20, 30, or however many tosses of a coin you like. But it is perfectly possible to toss a coin twice and get a head both times. The random sample of the two sides of the coin is selected in an unbiased fashion, but it still is unrepresentative. Imagine randomly selecting a sample of 10 people from a population comprising 50 men and 50 women. Just by chance it is possible that your sample of 10 people will include 7 women and only 3 men. Fortunately, we can determine mathematically the likely degree of sampling error in an estimate based on a random sample (as you will see later in this chapter), assuming that the sample's randomness has not been destroyed by a high rate of nonresponse or by poor control over the selection process.

In general, both the size of the sample and the homogeneity (sameness) of the population affect the degree of error due to chance; the proportion of the population that the sample represents does not. To elaborate:

The larger the sample, the more confidence we can have in the sample's representativeness of the population from which it was drawn. If we randomly pick five people to represent the entire population of our city, our sample is unlikely to be very representative of the entire population in terms of age, gender, race, attitudes, and so on. But if we randomly pick 100 people, the odds of having a representative sample are much better; with a random sample of 1,000, the odds become very good indeed.

The more homogeneous the population, the more confidence we can have in the representativeness of a sample of any particular size. Let us say we plan to draw samples of 50 from each of two communities to estimate mean family income. One community is very diverse, with family incomes ranging from $12,000 to $85,000. In the more homogeneous community, family incomes are concentrated in a narrower range, from $41,000 to $64,000. The estimated average family income based on the sample from the homogeneous community is more likely to be representative than is the estimate based on the sample from the more heterogeneous community. With less variation, fewer cases are needed to represent the homogeneous community.

The fraction of the total population that a sample contains does not affect the sample's representativeness, unless that fraction is large. We can regard any sampling fraction under 2% with about the same degree of confidence (Sudman 1976:184). In fact, sample representativeness is not likely to increase much until the sampling fraction is quite a bit higher. Other things being equal, a sample of 1,000 from a population of 1 million (with a sampling fraction of 0.001, or 0.1%) is much better than a sample of 100 from a population of 10,000 (although the sampling fraction is 0.01, or 1%, which is 10 times higher). The size of a sample is what makes representativeness more likely, not the proportion of the whole that the sample represents.

Polls that predict presidential election outcomes illustrate both the value of random sampling and the problems that it cannot overcome. In most presidential elections, pollsters have accurately predicted the outcomes of the actual vote by using random sampling and, these days, phone interviewing to learn which candidate voters intend to choose. Exhibit 4.4 shows how close these sample-based predictions have been in the last 12 contests. The big exception was the 1980 election, when a third-party candidate had an unpredicted effect. Otherwise, the small discrepancies between the votes predicted through random sampling and the actual votes can be attributed to random error.

The Gallup poll did quite well in predicting the result of the remarkable 2000 presidential election. The final Gallup prediction was that George W. Bush would win with 48% (Al Gore was predicted to receive only 46%, and Green Party nominee Ralph Nader was predicted to secure 4%). Although the race turned out much closer, with Gore actually winning the popular vote (before losing in the electoral college), Gallup accurately noted that there appeared to have been a late-breaking trend in favor of Gore (Newport 2000). In 2004, the final Gallup prediction of 49% for Bush was within 2 percentage points of his winning total of 51%.

Despite accurate sampling, election polls have produced some major errors in prediction. The reasons for these errors illustrate how unintentional systematic bias can influence sample results. In 1936, a *Literary Digest* poll predicted that Alfred M. Landon would defeat President Franklin Delano Roosevelt in a landslide, but instead Roosevelt took 63% of the popular vote. The problem? The *Digest* mailed out 10 million mock ballots to people listed in telephone directories, automobile registration records, voter lists, and so on. But in 1936, the middle of the Great Depression, only relatively wealthy people had phones and cars, and they were likely to vote Republican. Furthermore, only 2,376,523 completed ballots were returned, creating a response rate of only 24%, which left much room for error. Of course, this poll was not designed as a random sample, so the appearance of systematic bias is not surprising. Gallup was able to predict the 1936 election results accurately with a randomly selected sample of just 3,000 (Bainbridge 1989:43–44).

EXHIBIT 4.4   Election Outcomes: Predicted[a] and Actual

Winner (Year)	Polls (in %)	Result (in %)
Kennedy (1960)	51	50
Johnson (1964)	64	61
Nixon (1968)[b]	43	43
Nixon (1972)	62	62
Carter (1976)	48	50
Reagan (1980)[b]	47	51
Reagan (1984)	59	59
Bush (1988)	56	54
Clinton (1992)[b]	49	43
Clinton (1996)[b]	52	50
Bush, G. W. (2000)[b]	48	50
Bush, G. W. (2004)[b]	49	51

Source: Gallup Poll Accuracy Record, 12-13-00, www.gallup.com/poll/trends/ptaccuracy.asp.

a. Final Gallup Poll prior to the election.
b. There was also a third-party candidate.

The year 1980 was the only year since 1948 that pollsters had the wrong prediction for the winner of a presidential election in the week prior to the election. With Jimmy Carter ahead of Ronald Reagan in the polls by 45% to 42%, Gallup predicted a race too close to call. The outcome: Reagan 51%, Carter 42%. The problem? A large bloc of undecided voters, an unusually late debate with a strong performance by Reagan, and the failure of many pollsters to call back voters whom interviewers had failed to reach on the first try (these harder-to-reach voters were more likely to be Republican-leaning) (Dolnick 1984; Loth 1992). In this case, the sample was systematically biased against voters who were harder to reach and those who were influenced by the final presidential debate. The presence of many undecided voters in the sample was apparently an accurate representation of sentiment in the general population, so the problem would not be considered sample bias; but it did make measuring voting preferences all the more difficult.

Because they do not disproportionately exclude or include particular groups within the population, random samples that are successfully implemented avoid systematic bias. Random sampling error can still be considerable, however, and different types of random samples vary in their ability to minimize it. The four most common methods for drawing random samples are simple random sampling, systematic random sampling, stratified random sampling, and cluster sampling.

Simple Random Sampling

Simple random sampling requires a procedure that generates numbers or identifies cases strictly on the basis of chance. As you know, flipping a coin and rolling a die can be used to identify cases strictly on the basis of chance, but these procedures are not very efficient tools for drawing a sample. A **random number table**, such as the one in Appendix D, simplifies the process considerably. The researcher numbers all the elements in the sampling frame and then uses a systematic procedure for picking corresponding numbers from the random number table. (Exercise 2 at the end of the chapter explains the process step by step.) Alternatively, a researcher may use a lottery procedure. Each case number is written on a small card, and then the cards are mixed up and the sample selected from the cards.

When a large sample must be generated, these procedures are very cumbersome. Fortunately, a computer program can easily generate a random sample of any size. The researcher must first number all the elements to be sampled (the sampling frame) and then run the computer program to generate a random selection of the numbers within the desired range. The elements represented by these numbers are the sample.

Organizations that conduct phone surveys often draw random samples with another automated procedure called **random digit dialing**. A machine dials random numbers within the phone prefixes corresponding to the area in which the survey is to be conducted. Random digit dialing is particularly useful when a sampling frame is not available. The researcher simply replaces any inappropriate numbers (e.g. those no longer in service or for businesses) with the next randomly generated phone number.

The probability of selection in a true simple random sample is equal for each element. If a sample of 500 is selected from a population of 17,000 (i.e., a sampling frame of 17,000), then the probability of selection for each element is 500/17,000, or .03. Every element has an equal and independent chance of being selected, just like the odds in a toss of a coin (1/2) or a roll of a die (1/6). Thus, sample random sampling is an equal probability of selection method (EPSEM).

Simple random sampling can be done either with or without replacement sampling. In **replacement sampling**, each element is returned to the sampling frame from which it is selected so that it may be sampled again. In sampling without replacement, each element selected for the sample is then excluded from the sampling frame. In practice it makes no difference whether sampled elements are replaced after selection, as long as the population is large and the sample is to contain only a small fraction of the population.

Tjaden and Thoennes (2000) utilized simple random sampling when they sought to determine the magnitude of violent victimization in the United States using a telephone survey. They used random digit dialing to contact adult (18 years and older) residential household members in the continental United States and District of Columbia; 72% of the females and 69% of the males selected for the sample participated in the survey. The final sample contained 8,000 adult men and 8,000 adult women. In their final report to the National Institute of Justice, Tjaden and Thoennes provide a comparison of demographic characteristics of men

EXHIBIT 4.5   Stratified Random Sampling

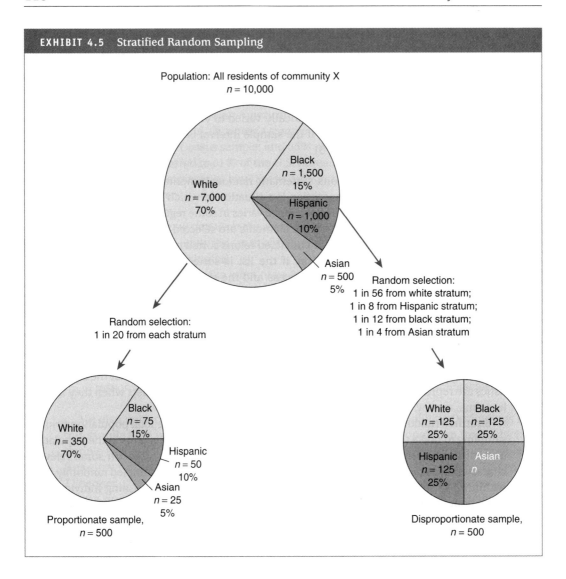

and 125 Caucasians (25%). In this type of sample, the probability of selection of every case is known but unequal between strata. You know what the proportions are in the population, and so you can easily adjust your combined sample statistics to reflect these true proportions. For instance, if you want to combine the ethnic groups and estimate the average income of the total population, you would have to weight each case in the sample. The weight is a number you multiply by the value of each case based on the stratum it is in. For example, you would multiply the incomes of all African Americans in the sample by 0.6 (75/125), the incomes of all Hispanics by 0.4 (50/125), and so on. Weighting in this way reduces the influence of the oversampled strata and increases the influence of the undersampled strata to just what they would have been if pure probability sampling had been used.

Why would anyone select a sample that is so unrepresentative in the first place? The most common reason is to ensure that cases from smaller strata are included in the sample in sufficient numbers. Only then can separate statistical estimates and comparisons be made between strata (e.g., between African Americans and Caucasians). Remember that one determinant of sample quality is sample size. The same is true for subgroups within samples. If a key concern in a research project is to describe and compare the incomes of people from different racial and ethnic groups, then it is important that the researchers base the mean income of each group on enough cases to be a valid representation. If few members of a particular minority group are in the population, they need to be oversampled. Such disproportionate sampling may also result in a more efficient sampling design if the costs of data collection differ markedly between strata or if the variability (heterogeneity) of the strata differs.

Cluster Sampling

Although stratified sampling requires more information than usual prior to sampling (about the size of strata in the population), **cluster sampling** requires less prior information. Specifically, cluster sampling can be useful when a sampling frame is not available, as often is the case for large populations spread across a wide geographic area or among many different organizations. In fact, if we wanted to obtain a sample from the entire U.S. population, there would be no list available. Yes, there are lists in telephone books of residents in various places who have telephones, lists of those who have registered to vote, lists of those who hold driver's licenses, and so on. However, all these lists are incomplete; some people do not list their phone number or do not have a telephone, some people are not registered to vote, and so on. Using incomplete lists such as these introduces bias into a sample.

In such cases, the sampling procedures become a little more complex. We usually end up working toward the sample we want through a series of steps, first by extracting a random sample of groups or clusters of elements that are available and then by randomly sampling the individual elements of interest from within these selected clusters. So what is a cluster? A **cluster** is a naturally occurring, mixed aggregate of elements of the population, with each element appearing in one and only one cluster. Schools could serve as clusters for sampling students, blocks could serve as clusters for sampling city residents, counties could serve as clusters for sampling the general population, and businesses could serve as clusters for sampling employees.

Drawing a cluster sample is at least a two-stage procedure. First, the researcher draws a random sample of clusters. A list of clusters should be much easier to obtain than a list of all the individuals in each cluster in the population. Next, the researcher draws a random sample of elements within each selected cluster. Because only a fraction of the total clusters are involved, obtaining the sampling frame at this stage should be much easier.

In a cluster sample of city residents, for example, blocks could be the first-stage clusters. A research assistant could walk around each selected block and record the addresses of all occupied dwelling units. Or in a cluster sample of students, a researcher could contact the schools selected in the first stage and make arrangements with the registrars to obtain lists of students at each school. Cluster samples often involve multiple stages (see Exhibit 4.6).

How many clusters and how many individuals within clusters should be selected? As a general rule, cases in the sample will be closer to the true population value if the researcher maximizes the number of clusters selected and minimizes the number of individuals

EXHIBIT 4.6   Cluster Sampling

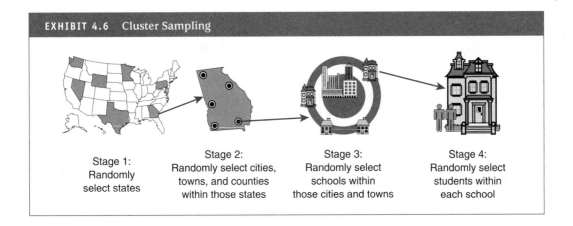

| Stage 1: Randomly select states | Stage 2: Randomly select cities, towns, and counties within those states | Stage 3: Randomly select schools within those cities and towns | Stage 4: Randomly select students within each school |

within each cluster. Unfortunately, this strategy also maximizes the cost of the sample. The more clusters selected, the higher the travel costs. It also is important to take into account the homogeneity of the individuals within clusters; the more homogeneous the clusters, the fewer cases needed per cluster. This should make intuitive sense, as it is more likely that any selected element will represent other elements if they are alike. Although cluster sampling is a very popular method among survey researchers, it is has one drawback: Sampling error is greater in a cluster sample than in a simple random sample. And as a general rule, this error increases as the number of clusters decreases.

Many professionally designed surveys use multistage cluster samples or even combinations of cluster and stratified probability sampling methods. For example, the U.S. Justice Department's National Crime Victimization Survey (NCVS) is an excellent example of a cluster sample. In the NCVS, the first stage of clusters selected are referred to as primary sampling units (PSUs), and represent a sample of rural counties and large metropolitan areas. From these PSUs, another stage of sampling involves the selection of geographic districts within each of the PSUs that have been listed by the U.S. Bureau of the Census population census. Finally, a probability sample of residential dwelling units is selected from these geographic districts. These dwelling units, or addresses, represent the final stage of the multistage sampling. Anyone who resides at a selected address who is 12 years of age or older and is a U.S. citizen is eligible for the NCVS sample. Approximately 50,500 housing units or other living quarters are designated for the NCVS each year and are selected in this manner.

How would we evaluate the NCVS sample, using the sample evaluation questions?

- *From what population were the cases selected?* The population was clearly defined for each cluster.
- *What method was used to select cases from this population?* The random selection method was carefully described.
- *Do the cases that were studied represent, in the aggregate, the population from which they were selected?* The unbiased selection procedures make us reasonably confident in the representativeness of the sample.

Nonprobability Sampling Methods

Unlike probability samples, when collecting a sample using nonprobability sampling techniques, elements within the population do not have a known probability of being selected into the sample. Thus, because the chance of any element being selected is unknown, we cannot be certain the selected sample actually represents our population. Why, you may be asking yourself right now, would we want to use such a sample if we cannot generalize our results to a larger population? Well, because these methods are useful for several purposes, including those situations in which we do not have a population list, when we are exploring a research question that does not concern a large population, or when we are doing a preliminary or exploratory study. Suppose, for example, that we were interested in the crime of shoplifting and wanted to investigate the rationalization used by shoplifters. It would be hard to define a population in this case because we do not have a list of shoplifters from which to randomly select. There may be lists of convicted shoplifters, but of course, they only represent those shoplifters who were actually caught.

There are four nonprobability sampling methods frequently used in criminological research: availability sampling, quota sampling, purposive sampling, and snowball sampling.

Availability Sampling

The definition of **availability sampling** is straightforward: elements are selected because they are available or easy to find. Consequently, this sampling method is also known as haphazard, accidental, or convenience sampling. News reporters often use person-on-the-street interviews-availability samples, to inject a personal perspective into a news story and show what ordinary people may think of a given topic. Availability samples are also used by university professors and researchers all the time. Have you ever been asked to complete a questionnaire in one of your classes? If so, you have been selected for inclusion in an availability sample.

Even though they are not generalizable, availability samples are often appropriate in research; for example, when a field researcher is exploring a new setting and trying to get some sense of prevailing attitudes or when a survey researcher conducts a preliminary test of a questionnaire. There are a variety of ways to select elements for an availability sample: standing on street corners and talking to anyone walking by, asking questions of employees who come to pick up their paychecks at a personnel office, or distributing questionnaires to an available and captive audience such as a class or a group meeting. Availability samples are also frequently used in field work studies interested in obtaining detailed information about a particular group. For example, when Bourgois, Lettiere, and Quesada (1997) wanted to understand homeless heroin addicts, they immersed themselves in a community of addicts living in a public park in San Francisco. These addicts thus became the availability sample.

Another classic example of an availability sample used in research is found in John Irwin's *The Felon* (1970), a study of the post-prison careers of a group of California criminal felons. To conduct his study, Irwin did not take a simple or systematic random sample of all felons released from the California Department of Corrections. Instead, because he resided and worked in the San Francisco area, Irwin studied offenders released on parole from penal institutions in San Francisco and Oakland parole districts who volunteered for his study.

When such samples are used, it is necessary to explicitly describe the sampling procedures used in the methodology section of research reports to acknowledge the nonrepresentativeness of the sample. For example, in a study conducted by Bachman, Paternoster, and Ward (1992) to investigate the factors related to sexual assault offending, a sample of male college students was utilized:

> The respondents in this study consisted of 94 male undergraduate students enrolled in introductory social science courses at a state university in New England. The students were given extra credit for their participation in the study, and were administered the data collection instrument during nonclass hours. (P. 412)

To test your understanding of the different sampling methods, go to the Identifying Sampling Techniques Interactive Exercises on the Student Study Site.

This availability sample was partially justified for two reasons: The study was exploratory in nature, and college women have been shown to be at extremely high risk of becoming the victims of sexual assault perpetrated by their male peers. However, it was still a large leap of faith to assume that the college males from one university used in this sample were representative of males in general, or even of male college students in general.

How do the answers to the sample evaluation questions differ for this type of sample from those obtained for probability samples? Well, there is no clearly definable population from which the respondents were drawn, and no systematic technique was used to select the respondents. Consequently, there is not much likelihood that the sample is representative of any target population; the problem is that we can never be sure.

Availability sampling often masquerades as a more rigorous form a research. Popular magazines and Internet sites periodically survey their readers by asking them to fill out questionnaires. Follow-up articles then appear in the magazine or on the site, displaying the results under such titles as "What You Think About the Death Penalty for Teenagers." If the magazine's circulation is large, a large sample can be achieved in this way. The problem is that usually only a tiny fraction of readers fill out the questionnaire, and these respondents are probably unlike other readers who did not have the interest or time to participate. So the survey is based on an availability sample. Even though the follow-up article may be interesting, we have no basis for thinking that the results describe the readership as a whole, much less the population at large. In fact, Richard Morin (1999), a columnist for the *Washington Post*, implored his readers, "Promise me this: While I'm gone, please, oh please, don't take Internet polls seriously. For several years now, your Wiz has condemned those ghastly web pseudo-surveys. But not everybody's paying attention" (p. B5). Morin went on to chastise ABC's *Good Morning America* for grabbing a story based on a survey that Internet users filled out on ABC.com. Based on this survey, *Good Morning America*, and other journalists including those at the *Boston Herald*, ran stories stating that nearly 6% of Internet users were addicted to the Internet. Not only was the sample not selected randomly, but electronic surveys also remain very susceptible to electronic ballot stuffing; few sites restrict access, so you can answer the posted questions as many times as you want. The 6% estimate is obviously not generalizable to the population of all Internet users. Because of their nonscientific basis, many Internet sites that conduct such polls, including ABC.com, now add this disclaimer to the online poll's question of the day: "Not a scientific poll; for entertainment only."

Quota Sampling

Quota sampling is intended to overcome the most obvious flaw of availability sampling, that the sample will just consist of who or what is available, without any concern for its

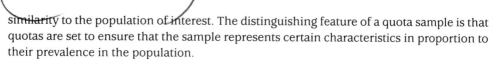

similarity to the population of interest. The distinguishing feature of a quota sample is that quotas are set to ensure that the sample represents certain characteristics in proportion to their prevalence in the population.

Quota samples are similar to stratified probability samples, but they are generally less rigorous and precise in their selection procedures. Quota sampling simply involves designating the population into proportions of some group that you want to be represented in your sample. Similar to stratified samples, in some cases, these proportions may actually represent the true proportions observed in the population. At other times, these quotas may represent predetermined proportions of subsets of people you deliberately want to oversample.

Suppose we were interested in investigating the crime of personal larceny with contact, which generally involves pocket picking and purse snatching. Because elderly citizens are disproportionately victimized by this crime, we would want to make sure we had enough elderly victims in our sample to make comparisons to their younger cohorts. For a study such as this, we may decide that we want 50% of the victims in our sample to consist of individuals 65 years of age or older and 50% to consist of those younger than 65 years of age. When these proportions are decided, they represent our *quotas*. Elements of the population are then collected until each quota is filled. In our example, we would select the sample until we have exactly 50% of the elements younger than 65 and 50% 65 years of age or older. Quota sampling may be more complex, in that we can add quotas to fill. For example, in addition to age, we may want to make sure that within each age group we have a certain proportion of African American men and women, a certain proportion of Hispanic men and women, and a certain proportion of Caucasian men and women.

In other cases, you may want to represent the distribution of a characteristic (e.g., gender) within your population. For example, suppose you know that the distribution of a city's residents regarding unemployment is as follows: 60% employed, 35% out of the labor force, 5% unemployed. If you wanted your sample to be representative of this distribution, and you were going to sample 500 people, 300 of them would need to be employed (60% of 500), 175 of them would need to out of the workforce (e.g., retired), and 25 would need to be unemployed (not employed but looking for a job). With this quota list in hand, you and your research team would then go out into the community looking for the right number of people in each quota category.

The problem is that even when we know that a quota sample is representative of the particular characteristics for which quotas have been set, we have no way of knowing if the sample is representative in terms of any other characteristics. In Exhibit 4.7, for example, quotas have been set for gender only. Under the circumstances, it's no surprise that the sample is representative of the population only in terms of gender, not in terms of race. Interviewers are only human; they may avoid potential respondents with menacing dogs in the front yard, or they could seek out respondents who are physically attractive or who look like they would be easy to interview. Realistically, researchers can set quotas for only a small fraction of the characteristics relevant to a study, so a quota sample is really not much better than an availability sample (although following careful, consistent procedures for selecting cases within the quota limits always helps).

This last point leads to another limitation of quota sampling: You must know the characteristics of the entire population to set the right quotas. In most cases, researchers know what the population looks like in terms of no more than a few of the characteristics relevant to their

Snowball Sampling

For **snowball sampling**, you identify one member of the population and speak to him or her, then ask that person to identify others in the population and speak to them, then ask them to identify others, and so on. The sample thus snowballs in size. This technique is useful for hard-to-reach or hard-to-identify, interconnected populations where at least some members of the population know each other, such as drug dealers, prostitutes, practicing criminals, gang leaders, and informal organizational leaders.

In their study of juvenile gangs in St. Louis, Decker and Van Winkle (1996) utilized the technique of snowball sampling. Specifically, the snowball began with an earlier fieldwork project involving active residential burglars (Wright & Decker 1994). The young members from this sample, along with contacts a field ethnographer had with several active street criminals, started the referral process. The initial interviewees then nominated other gang members as potential interview subjects.

One problem with this technique is that the initial contacts may shape the entire sample and foreclose access to some members of the population of interest. Because Decker and Van Winkle (1996) wanted to interview members from several gangs, they had to restart the snowball sampling procedure many times to gain access to a large number of gangs. One problem, of course, was validating whether individuals claiming to be gang members, so-called wannabes, actually were legitimate members. Over 500 contacts were made before the final sample of 99 was complete.

More systematic versions of snowball sampling can also reduce the potential for bias. The most sophisticated version, *respondent-driven sampling*, gives financial incentives, also called gratuities, to respondents to recruit peers (Heckathorn 1997). Limitations on the number of incentives that any one respondent can receive increase the sample's diversity. Targeted incentives can steer the sample to include specific subgroups. When the sampling is repeated through several waves, with new respondents bringing in more peers, the composition of the sample converges on a more representative mix of characteristics. Exhibit 4.9 shows how the sample spreads out through successive recruitment waves to an increasingly diverse pool (Heckathorn 1997:178). As with all nonprobability sampling techniques, however, researchers using even the most systematic versions of snowball sampling cannot be confident that their sample is representative of the population of interest.

Lessons About Sample Quality

Some lessons are implicit in our evaluations of the samples in this chapter:

- We cannot evaluate the quality of a sample if we do not know what population it is supposed to represent. If the population is unspecified because the researchers were never clear about just what population they were trying to sample, then we can safely conclude that the sample itself is no good.
- We cannot evaluate the quality of a sample if we do not know exactly how cases in the sample were selected from the population. If the method was specified, we then need to know whether cases were selected in a systematic fashion or on the

EXHIBIT 4.9   Respondent-Driven Sampling—A Version of Snowball Sampling

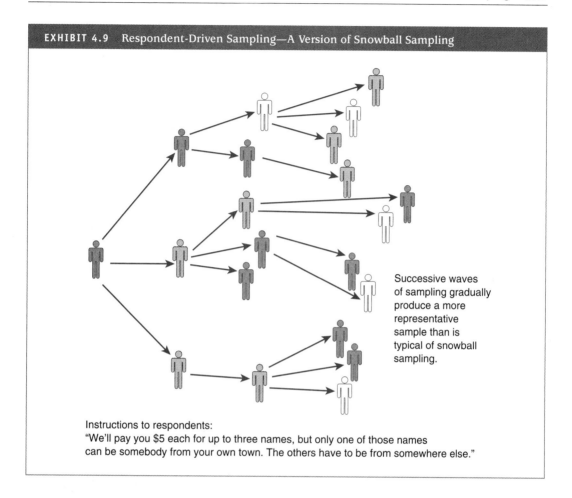

Successive waves of sampling gradually produce a more representative sample than is typical of snowball sampling.

Instructions to respondents:
"We'll pay you $5 each for up to three names, but only one of those names can be somebody from your own town. The others have to be from somewhere else."

basis of chance. In any case, we know that a haphazard method of sampling (as in person-on-the-street interviews) undermines generalizability.

- Sample quality is determined by the sample actually obtained, not just by the sampling method itself. If many of the people (or other elements) selected for our sample do not respond or participate in the study, even though they have been selected for the sample, the quality of our sample is undermined even if we chose the sample in the best possible way.

- We need to be aware that even researchers who obtain very good samples may talk about the implications of their findings for some group that is larger than or just different from the population they actually sampled. For example, findings from a representative sample of students in one university often are discussed as if they tell us about university students in general. And maybe they do; the problem is, we just don't know.

SAMPLING DISTRIBUTIONS

A well-designed probability sample is one that is likely to be representative of the population from which it was selected. But as you have seen, random samples still are subject to sampling error due only to chance. To deal with that problem, social researchers take into account the properties of a **sampling distribution**, a hypothetical distribution of a statistic across an infinite number of random samples that could be drawn from a population. Any single random sample can be thought of as just one of an infinite number of random samples that, in theory, could have been selected from the population. If we had the finances of Bill Gates and the patience of Job, and were able to draw an infinite number of samples, and we calculated the same type of statistic (e.g., percentage of those in favor of the three strikes laws) for each of these samples, we would then have a sampling distribution. Understanding sampling distributions is the foundation for understanding how researchers can estimate sampling error.

Sampling distribution A hypothetical distribution of a statistic (e.g., proportion, mean) across an infinite number of random samples that could be drawn from a population.

What does a sampling distribution look like? Because a sampling distribution is based on some statistic calculated for different samples, we need to choose a statistic. Let us focus on the arithmetic average, or mean. We will explain the calculation of the mean in Chapter 12, but you may already be familiar with it: It is simply the arithmetic average of a variable distribution and is obtained by adding up the values of all the cases and then dividing this sum by the total number of cases. Let us say you draw a random sample of 500 families and find that their average (mean) family income is $36,239. This $36,239 would be the sample statistic for this one sample. But imagine that you then drew another random sample, and another, and another, and so on, and calculated the mean from each of these samples. Then imagine marking these means on graph paper. The resulting graph would be a sampling distribution of the mean.

Exhibit 4.10 demonstrates what happens when we do something very similar to what we just described. We did not take an infinite number of samples from a large population but used the 1996 General Social Survey (GSS) sample as if it were a population. We first drew 50 different random samples, each consisting of 30 cases, from the 1996 GSS. (The standard notation for the number of cases in each sample is $n = 30$.) Then we calculated for each random sample the approximate mean family income (approximate because the GSS does not record actual income in dollars). We then graphed the means of the 50 samples. Each column in Exhibit 4.10 shows how many samples had a particular family income. The mean for the population (the total sample) is $38,249, and you can see that the sampling distribution clusters around this value. Also notice, however, that although many of our selected sample means are close to the population mean, some are quite far from it. If you had calculated the mean from only one sample, it could have fallen anywhere in this distribution, from the low of $25,000 to the high of $47,000. From probability theory, however, we know that the one mean we actually obtain from our sample is unlikely to be far from the true population mean if, in fact, we have drawn a true random sample.

EXHIBIT 4.10   Partial Sampling Distribution: Mean Family Income

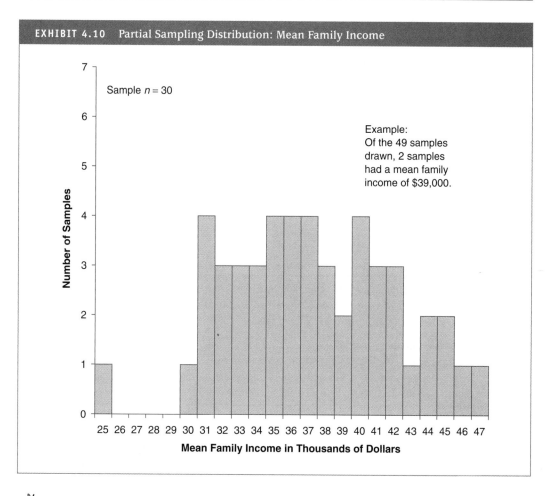

Sample $n = 30$

Example:
Of the 49 samples
drawn, 2 samples
had a mean family
income of $39,000.

Y-axis: Number of Samples
X-axis: Mean Family Income in Thousands of Dollars

Estimating Sampling Error

In reality, we do not actually observe sampling distributions in real research; it would take too much time and money. Instead, researchers just draw the best sample they can. (Now you understand why it is important to have a sample that is representative of the population.) A sampling distribution usually remains hypothetical or a theoretical distribution. We can use the properties of sampling distributions, however, to calculate the amount of sampling error that was likely with the random sample used in a study. The tools for calculating sampling error are called **inferential statistics**.

Inferential statistics Mathematical tools for estimating how likely it is that a statistical result based on data from a random sample is representative of the population from which the sample was selected.

Sampling distributions for many statistics, including the mean, have a normal shape. A graph of a **normal distribution** looks like a bell, with one hump in the middle, centered around the population mean, and the number of cases tapering off on both sides of the mean. Note that a normal distribution is symmetrical: If you fold it in half at its center (at the population mean), the two halves will match perfectly. This shape is produced by **random sampling error**. The value of the statistic varies from sample to sample because of chance, so higher and lower values are equally likely. A good example of a normally shaped distribution is the grades that will be received in the class for which you are reading this text. The majority of students in the class will receive C's, fewer will receive B's and D's, and fewer still will receive A's and F's.

The partial sampling distribution in Exhibit 4.10 does not have a completely normal shape because it involves only a small number of samples (50), each of which has only 30 cases. Exhibit 4.11 shows what the sampling distribution of family incomes would look like if it formed a perfectly normal distribution if, rather than 50 random samples, we had selected thousands of random samples.

Systematic sampling error Overrepresentation or underrepresentation of some population characteristics in a sample due to the method used to select the sample. A sample shaped by systematic sampling error is a biased sample.

Random sampling error Differences between the population and the sample that are due only to chance factors (random error), not to systematic sampling error. Random sampling error may or may not result in an unrepresentative sample. The magnitude of sampling error due to chance factors can be estimated statistically.

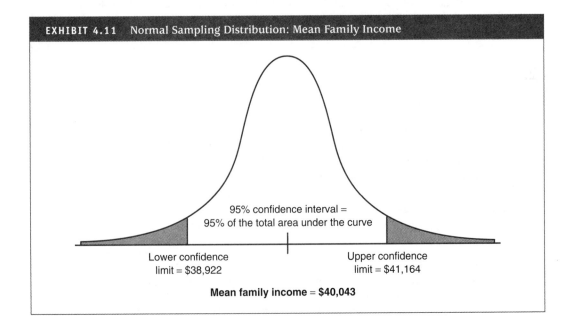

EXHIBIT 4.11 Normal Sampling Distribution: Mean Family Income

95% confidence interval = 95% of the total area under the curve

Lower confidence limit = $38,922

Upper confidence limit = $41,164

Mean family income = $40,043

As we have stated earlier, in research our goal is to generalize what we observe in our sample to the population of interest. In statistical terminology, a **sample statistic** is an estimate of the population parameter we want to estimate. The properties of a sampling distribution facilitate this process of statistical inference. In the sampling distribution, the most frequent value of the sample statistic (such as the mean) computed from sample data is identical to the **population parameter**, the statistic computed for the entire population. In other words, we can have a lot of confidence that the value at the peak of the bell curve represents the norm for the entire population.

Sample statistic An estimated statistic (e.g., proportion, mean) from the one sample we actually selected from a population.

In a normal distribution, a predictable proportion of cases also falls within certain ranges under the curve. Inferential statistics take advantage of this feature and allow us to estimate how likely it is that, given a particular sample, the true population value will be within some range around the statistic. For example, a statistician might conclude from a sample of 30 families that we can be 95% confident that the true mean family income in the total population is between $23,012 and $38,120. The interval from $23,012 to $38,120 would then be called the confidence interval for the mean. The upper ($38,120) and lower ($23,012) bounds of this interval are the confidence limits. Exhibit 4.11 marks such confidence limits, indicating the range that encompasses 95% of the area under the normal curve; in theory, then, 95% of all sample statistics would fall within this range.

Although all normal distributions have these same basic features, they differ in the extent to which they cluster around the mean. A sampling distribution is more compact when it is based on larger samples. Stated another way, we can be more confident in estimates based on large random samples because we know that a larger sample creates a more compact sampling distribution. This should make intuitive sense. If you want to estimate the average family income of residents in New York City and you only select a sample of 30 families, the mean you estimate from this sample of 30 is less likely to reflect New York City residents' annual family income than a sample of, say, 3,000.

Let us illustrate the importance of sample size for you. Compare the two sampling distributions of mean family income in Exhibit 4.12. Both depict the results for 50 samples. However, in one study each sample comprised 100 families, and in the other study each sample comprised only 5 families. Clearly, the larger samples ($n = 100$) result in a sampling distribution that is much more tightly clustered around the true mean of the population compared to the smaller samples ($n = 5$). The 95% confidence interval for mean family income for the entire GSS sample of 1,368 cases was $29,421 to $31,711, an interval only about $2,300 wide. But the 95% confidence interval for the mean family income in one GSS subsample of 100 cases was much wider, with limits of $25,733 and $35,399. And for a subsample of only 5 cases, the 95% confidence interval was very broad: $14,104 to $47,028. This wide interval reflects the fact that we are less confident in this sample estimate. Because the confidence we have in statistics obtained from such small samples is so low, they give us very little useful information about the population.

In most social science disciplines, including ours, researchers typically rely on 95% or 99% confidence intervals. In fact, every time you read the results of an opinion poll in the

involved only aggregate data about cities, and he explained his research approach as, in part, a response to the failure of other researchers to examine this problem at the structural, aggregate level. Moreover, Sampson argued that the rates of joblessness and family disruption in communities influence community social processes, not just the behavior of the specific individuals who are unemployed or who grew up without two parents. Yet Sampson suggests that the experience of joblessness and poverty is what tends to reduce the propensity of individual men to marry and that the experience of growing up in a home without two parents in turn increases the propensity of individual juveniles to commit crimes. These conclusions about the behavior of individuals seem consistent with the patterns Sampson found in his aggregate, city-level data; so it seems unlikely that he is committing an ecological fallacy.

The solution is to know what the units of analysis and units of observation were in a study, and to take these into account when weighing the credibility of the researcher's conclusions. The goal is not to reject conclusions that refer to a level of analysis different from what was actually studied. Instead, the goal is to consider the likelihood that an ecological fallacy or a reductionist fallacy has been made when estimating the causal validity of the conclusions.

CONCLUSION

Sampling is the fundamental starting point in criminological research. Probability sampling methods allow a researcher to use the laws of chance, or probability, to draw samples from population parameters that can be estimated with a high degree of confidence. A sample of just 1,000 or 1,500 individuals can be used to reliably estimate the characteristics of the population of a nation comprising millions of individuals.

But researchers do not come by representative samples easily. Well-designed samples require careful planning, some advance knowledge about the population to be sampled, and adherence to systematic selection procedures so that the selection procedures are not biased. And even after the sample data are collected, the researcher's ability to generalize from the sample findings to the population is not completely certain. The best the researcher can do is to perform additional calculations that state the degree of confidence that can be placed in the sample statistic.

The alternatives to random, or probability-based, sampling methods are almost always much less palatable, even though they typically are much cheaper. Without a method of selecting cases likely to represent the population in which the researcher is interested, research findings will have to be carefully qualified. Unrepresentative samples may help researchers understand which aspects of a social phenomenon are important, but questions about the generalizability of this understanding are left unanswered.

Social scientists often seek to generalize their conclusions from the population that they studied to some larger target population. The validity of generalizations of this type is necessarily uncertain, for having a representative sample of a particular population does not at all ensure that what we find will hold true in other populations. Nonetheless, the accumulation of findings from studies based on local or otherwise unrepresentative populations can provide important information about broader populations.

KEY TERMS

Availability sampling
Census
Cluster
Cluster sampling
Confidence interval
Confidence limits
Cross-population generalizability
Disproportionate stratified sampling
Ecological fallacy
External validity
Inferential statistics
Nonprobability sampling method
Nonresponse
Normal distribution
Periodicity
Population
Population parameter
Probability of selection
Probability sampling method
Proportionate stratified sampling
Purposive sampling
Quota sampling
Random digit dialing

Random number table
Random sampling error
Random selection
Reductionist fallacy (reductionism)
Replacement sampling
Representative sample
Sample
Sample generalizability
Sample statistic
Sampling distribution
Sampling error
Sampling frame
Sampling interval
Sampling unit
Simple random sampling
Snowball sampling
Stratified random sampling
Systematic random sampling
Systematic sampling error
Target population
Units of analysis
Units of observation

HIGHLIGHTS

- Sampling theory focuses on the generalizability of descriptive findings to the population from which the sample was drawn. It also considers whether statements can be generalized from one population to another.

- Sampling is usually necessary except in two conditions: (1) when the elements that would be sampled are identical, which is almost never the case, and (2) when you have the option of conducting a complete census of a population.

- Nonresponse undermines sample quality: It is the obtained sample, not the desired sample, that determines sample quality.

- Probability sampling methods rely on a random selection procedure to ensure there is no systematic bias in the selection of elements. In a probability sample, the odds of selecting elements are independent, equal, and known, and the method of selection is carefully controlled.

- A sampling frame (a list of elements in the population) is required in most probability sampling methods. The adequacy of the sampling frame is an important determinant of sample quality.

- Simple random sampling and systematic random sampling are equivalent probability sampling methods in most situations. However, systematic random sampling is inappropriate for sampling from lists of elements that have a regular, periodic structure.

- Stratified random sampling uses prior information about a population to make sampling more efficient. Stratified sampling may be either proportionate or disproportionate. Disproportionate stratified sampling is useful when a research question focuses on a stratum or on strata that make up a small proportion of the population.

- Cluster sampling is less efficient than simple random sampling but is useful when a sampling frame is unavailable. It is also useful for large populations spread out across a wide area or among many organizations.

- Nonprobability sampling methods can be useful when random sampling is not possible, when a research question does not concern a larger population, and when a preliminary exploratory study is appropriate. However, the representativeness of nonprobability samples cannot be determined.

- The likely degree of error in an estimate of a population characteristic based on a probability sample decreases as the size of the sample increases. Sampling error also decreases if the population from which the sample was selected is homogeneous. Sampling error is not affected by the proportion of the population that is sampled, except when that proportion is large. The degree of sampling error affecting a sample statistic can be estimated from the characteristics of the sample and knowledge of the properties of sampling distributions. Through these inferential statistics, we can estimate the degree of confidence we have in our findings.

- Invalid generalizations about causal relationships may occur when relationships between variables measured at the group level are assumed to apply at the individual level (the ecological fallacy) and when relationships between variables measured at the level of individuals are assumed to apply at the group level (the reductionist fallacy). Nonetheless, many research questions point to relationships at multiple levels and may profitably be answered by studying different units of analysis.

EXERCISES

1. Locate one or more newspaper articles reporting the prevalence of crime or drugs or alcohol. What information does the article provide on the sample that was selected? What additional information do you need to determine whether the sample was a representative one?

2. Select a random sample using the table of random numbers in Appendix D. Compute a statistic based on your sample, and compare it to the corresponding figure for the entire population:

   a. First, select the crime statistics of your choice at the city or state level from the most recent FBI Uniform Crime Report (UCR).
   b. Next, create your sampling frame, a numbered list of all elements in the population. If you are using a complete listing of all elements, such as state-level crime statistics, the sampling frame is the same as the list. Just number the elements (states). If your population is composed of cities, your sampling frame will be those cities that meet certain criteria (e.g., all cities with a population of 250,000 or greater); you may need to refer to another source, such as the U.S. Bureau of the Census, to determine which cities meet your criteria. Identify these cities, and then number them sequentially, starting with 1.
   c. Decide on a method of picking numbers out of the random number table in Appendix D, such as taking every number in each row, row by row, or by moving down or diagonally

across the columns. Use only the first (or last) digit in each number if yo[...]
1 to 9 cases; select the first (or last) two digits if you want fewer than 10C[...]

   d. Pick a starting location in the random number table. It is important to [...]
point in an unbiased way, perhaps by closing your eyes and then point[...]
of the page.

   e. Record the numbers you encounter as you move from the starting loca[...] [...] [...]
direction you decided in advance, until you have recorded as many random numbers
as the number of cases you need in the sample. If you are selecting states, 10 might be
a good number. Ignore numbers that are too large (or small) for the range of numbers
used to identify the elements in the population. Discard duplicate numbers.

   f. Calculate the mean value (mathematical average) in your sample for the crime statistic
you selected. Calculate the mean by adding up the values of all the elements in the
sample and dividing by the number of elements in the sample.

   g. Go back to the sampling frame and calculate this same average for all the elements in
the list. How close is the sample average to the population average?

3. Select five scholarly journal articles that describe criminological research using a
sample drawn from some population. Identify the type of sample used in each
study, and note any strong and weak points in how the sample was actually drawn.
Did the researchers have a problem due to non-response? Considering the sample, how
confident are you in the validity of generalizations about the population based on the
sample? Do you need any additional information to evaluate the sample? Do you think
a different sampling strategy would have been preferable? What larger population
were the findings generalized to? Do you think these generalizations were warranted?
Why or why not?

DEVELOPING A RESEARCH PROPOSAL

Consider the possibilities for sampling.

1. Propose a sampling design that would be appropriate if you were to survey students on
your campus only. Define the population, identify the sampling frame(s), and specify the
elements and any other units at different stages. Indicate the exact procedure for selecting
people to be included in the sample.

2. Propose a different sampling design for conducting your survey in a larger population,
such as your city, state, or the entire nation.

Student Study Site

The companion Web site for *The Practice of Research in Criminology and Criminal Justice*, Third
Edition

   http://www.sagepub.com/prccj3

   Visit the Web-based Student Study Site to enhance your understanding of the chapter content and to discover additional resources that will take your learning one step further. You can
enhance your understanding of the chapters by using the comprehensive study material,
which includes e-flashcards, Web exercises, practice self-tests, and more. You will also find special features, such as Learning from Journal Articles, which incorporates SAGE's online journal
collection.

---

because of Compstat, the city's computer program that identifies to police where crimes are clustering (Dewan 2004b:A1; Kaplan 2002:A3)? Or should credit be given to New York's "Safe Streets, Safe Cities" program, which increased the ranks of police officers (Rashbaum 2002)? What about better emergency room care causing a decline in homicides (Harris et al. 2002)? And what about the decline in usage of crack cocaine on the streets of New York City (Dewan 2004b:C16)? To determine which of these possibilities could contribute to the increase or decline of serious crime, we must design our research strategies carefully.

In this chapter, we first discuss the meaning of causation from two different perspectives—nomothetic and idiographic—and then review the criteria for achieving causally valid explanations. During this review, we give special attention to several key distinctions in research design that are related to our ability to come to causal conclusions: the use of an experimental or nonexperimental design, and reliance on a cross-sectional or longitudinal design. By the end of the chapter, you should have a good grasp of the different meanings of causation and be able to ask the right questions to determine whether causal inferences are likely to be valid. You also may have a better answer about the causes of crime and violence.

CAUSAL EXPLANATION

A cause is an explanation for some characteristic, attitude, or behavior of groups, individuals, or other entities (such as families, organizations, or cities) or for events. Most social scientists seek causal explanations that reflect tests of the types of hypotheses with which you are familiar (see Chapter 3): The independent variable is the presumed cause, and the dependent variable is the potential effect. For example, the study by Sampson and Raudenbush (2001) tested whether disorder in urban neighborhoods (the independent variable) leads to crime (the dependent variable). (As you know, they concluded that it did not, at least not directly.) This type of causal explanation is termed *nomothetic*.

A different type of cause is the focus of some qualitative research (see Chapter 8) and our everyday conversations about causes. In this type of causal explanation, termed *idiographic*, individual events or the behaviors of individuals are explained with a series of related, prior events. For example, you might explain a particular crime as resulting from several incidents in the life of the perpetrator that resulted in a tendency toward violence, coupled with stress resulting from a failed marriage, and a chance meeting.

Nomothetic Causal Explanation

A **nomothetic causal explanation** is one involving the belief that variation in an independent variable will be followed by variation in the dependent variable, when all other things are equal (*ceteris paribus*). In this perspective, researchers who claim a causal effect have concluded that the value of cases on the dependent variable differs from what their value would have been in the absence of variation in the independent variable. For instance, researchers might claim that the likelihood of committing violent crimes is higher for individuals who were abused as children than it would be if these same individuals had not been abused as children. Or, researchers might claim that the likelihood of committing violent crimes is higher for individuals exposed to media violence than it would be if these same individuals

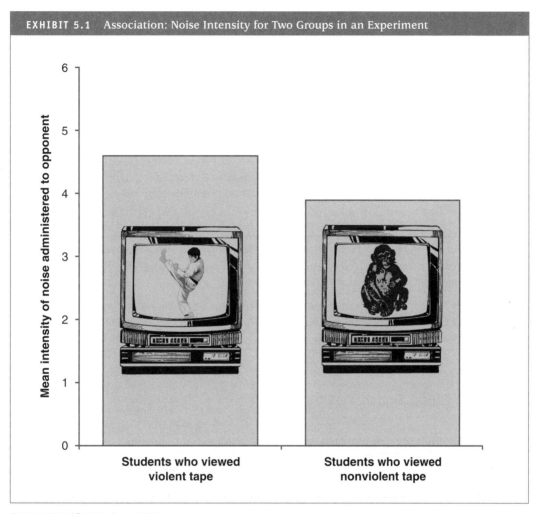

EXHIBIT 5.1 Association: Noise Intensity for Two Groups in an Experiment

Source: Adapted from Bushman 1995.

had not been exposed to media violence. The situation as it would have been in the absence of variation in the independent variable is termed the **counterfactual** (see Exhibit 5.1).

Of course, the fundamental difficulty with this perspective is that we never really know what would have happened at the same time to the same people (or groups, cities, and so on) if the independent variable had not varied, because it did. We cannot rerun real-life scenarios (King, Keohane, & Verba 1994). We could observe the aggressiveness of people's behavior before and after they were exposed to media violence. But this comparison involves an earlier time period, when, by definition, the people and their circumstances were not exactly the same.

But we do not need to give up hope! Far from it. We can design research to create conditions that are comparable indeed, so that we can confidently assert our conclusions *ceteris paribus*, other things being equal. We can examine the impact on the dependent variable of variation in the independent variable alone, even though we will not be able to compare the same people at the same time in exactly the same circumstances except for the variation in the independent variable. And by knowing the ideal standard of comparability, we can improve our research designs and strengthen our causal conclusions even when we cannot come so close to living up to the meaning of *ceteris paribus*.

Quantitative researchers seek to test nomothetic causal explanations with either experimental or nonexperimental research designs. However, the way in which experimental and nonexperimental designs attempt to identify causes differs quite a bit. It is very hard to meet some of the criteria for achieving valid nomothetic causal explanations using a nonexperimental design. Most of the rest of this chapter is devoted to a review of these causal criteria and a discussion of how experimental and nonexperimental designs can help to establish them.

Causal effect (nomothetic perspective) When variation in one phenomenon, an independent variable, leads to or results, on average, in variation in another phenomenon, the dependent variable.

Example of a nomothetic causal effect: Individuals arrested for domestic assault tend to commit fewer subsequent assaults than do similar individuals who are accused in the same circumstances but not arrested.

Idiographic Causal Explanation

The other meaning of the term *cause* is one that we have in mind very often in everyday speech. This is **idiographic causal explanation**: the concrete, individual sequence of events, thoughts, or actions that resulted in a particular outcome for a particular individual or that led to a particular event (Hage & Meeker 1988). An idiographic explanation also may be termed an *individualist* or a *historicist explanation*.

Causal effect (idiographic perspective) When a series of concrete events, thoughts, or actions result in a particular event or individual outcome.

Example of an idiographic causal effect: An individual is neglected by his parents. He comes to distrust others, has trouble maintaining friendships, has trouble in school, and eventually gets addicted to heroin. To support his habit, he starts selling drugs and is ultimately arrested and convicted for drug trafficking.

A causal explanation that is idiographic includes statements of initial conditions and then relates a series of events at different times that led to the outcome, or causal effect. This narrative, or story, is the critical element in an idiographic explanation, which may therefore be classified as narrative reasoning (Richardson 1995:200–201). Idiographic explanations focus on particular social actors, in particular social places, at particular social times

(Abbott 1992). Idiographic explanations are also typically very concerned with context, with understanding the particular outcome as part of a larger set of interrelated circumstances. Idiographic explanations thus can be termed holistic.

Idiographic explanation is deterministic, focusing on what caused a particular event to occur or what caused a particular case to change. As in nomothetic explanations, idiographic causal explanations can involve counterfactuals, by trying to identify what would have happened if a different circumstance had occurred. But unlike nomothetic explanations, in idiographic explanations the notion of a probabilistic relationship, an average effect, does not really apply. A deterministic cause has an effect in every case under consideration.

Anderson's (1990) field research in a poor urban community produced a narrative account of how drug addiction can result in a downward slide into residential instability and crime:

> When addicts deplete their resources, they may go to those closest to them, drawing them into their schemes. . . . The family may put up with the person for a while. They provide money if they can. . . . They come to realize that the person is on drugs. . . . Slowly the reality sets in more and more completely, and the family becomes drained of both financial and emotional resources. . . . Close relatives lose faith and begin to see the person as untrustworthy and weak. Eventually the addict begins to "mess up" in a variety of ways, taking furniture from the house [and] anything of value. . . . Relatives and friends begin to see the person . . . as "out there" in the streets. . . . One deviant act leads to another. (Pp. 86–87)

An idiographic explanation like Anderson's (1990) pays close attention to time order and causal mechanisms. Nonetheless, it is difficult to make a convincing case that one particular causal narrative should be chosen over an alternative narrative (Abbott 1992). Does low self-esteem result in vulnerability to the appeals of drug dealers, or does a chance drug encounter precipitate a slide in self-esteem? The prudent causal analyst remains open to alternative explanations.

RESEARCH DESIGNS AND CRITERIA FOR CAUSAL EXPLANATIONS

In the movie *Money Train*, two men spray the inside of a subway token booth with a flammable liquid, blowing up the toll booth and killing the collector. In 1995, while the movie was still showing in theaters, a similar incident actually occurred in a New York City subway. The toll collector was hospitalized with widespread third-degree burns. The media violence, it was soon alleged, had caused the crime. How would you evaluate this claim? What evidence do we need to develop a valid conclusion about a hypothesized causal effect? Imagine a friend saying, after reading about the *Money Train* incident, "See, media violence causes people to commit crimes." Of course, after reading Chapter 1 you would not be so quick to jump to such a conclusion. "Don't overgeneralize," you would remind yourself. When your friend insists, "But I recall that type of thing happening before," you might even suspect selective observation. As a blossoming criminological researcher, you now know that if we want to have confidence in the validity of our causal statements, we must meet a higher standard.

To test your understanding of criteria of causal explanations, go to the Concerns with Causal Reasoning Interactive Exercises on the Student Study Site.

How research is designed influences our ability to draw causal conclusions. In this section, we will introduce the features that need to be considered in a research design in order to evaluate how well it can support nomothetic causal conclusions.

Five criteria must be considered when deciding whether a causal connection exists. When a research design leaves one or more of the criteria unmet, we may have some important doubts about causal assertions the researcher may have made. The first three of the criteria are generally considered the necessary and most important basis for identifying a nomothetic causal effect: empirical association, appropriate time order, and nonspuriousness. The other two criteria, identifying a causal mechanism and specifying the context in which the effect occurs, can also considerably strengthen causal explanations although many do not consider them as requirements for establishing a causal relationship.

Conditions necessary for determining causality:

1. empirical association

2. appropriate time order

3. nonspuriousness

Conditions important in specifying causal relationships:

1. mechanism

2. context

We will use Brad Bushman's (1995) experiment on media violence and aggression to illustrate the five criteria for establishing causal relationships. Bushman's study focused in part on this specific research question: Do individuals who view a violent videotape act more aggressively than individuals who view a nonviolent videotape?

Undergraduate psychology students were recruited to watch a 15-minute videotape in a screening room, one student at a time. Half of the students watched a movie excerpt that was violent (from *Karate Kid III*), and half watched a nonviolent movie excerpt (from *Gorillas in the Mist*). After viewing the videotape, the students were told that they were to compete with another student, in a different room, on a reaction-time task. When the students saw a light cue, they were to react by trying to click a computer mouse faster than their opponent. On a computer screen, the students set a level of radio static that their opponents would hear when the opponents reacted more slowly. The students themselves heard this same type of noise when they reacted more slowly than their opponents, at the intensity level supposedly set by their opponents.

Each student in the study participated in 25 trials, or competitions, with the unseen opponent. Their aggressiveness was operationalized as the intensity of noise that they set for their opponents over the course of the 25 trials. The louder the noise level they set, the more aggressively they were considered to be behaving toward their opponents. The question that we will focus on first is whether students who watched the violent video behaved more aggressively than those who watched the nonviolent video.

Association

The results of Bushman's (1995) experiment are represented in Exhibit 5.1. The average intensity of noise administered to the opponent was indeed higher for students who watched the violent videotape than for those who watched the nonviolent videotape. But is Bushman justified in concluding from these results that viewing a violent videotape increased aggressive behavior in his subjects? Would this conclusion have any greater claim to causal validity than the statement that your friend made in response to the *Money Train* incident? Perhaps it would.

If for no other reason, we can have greater confidence in Bushman's (1995) conclusion because he did not observe just one student who watched a violent video and then acted aggressively, as was true in the *Money Train* incident. Instead, Bushman observed a number of students, some of whom watched a violent video and some of whom did not. So his conclusion is based on finding an **association** between the independent variable (viewing of a violent videotape) and the dependent variable (likelihood of aggressive behavior).

Time Order

Association is a necessary criterion for establishing a causal effect, but it is not sufficient. Suppose you find in a survey that most people who have committed violent crimes have also watched the movie *Money Train*, and that most people who have not committed violent crimes have not watched the movie. You believe you have found an association between watching the movie and committing violent crimes. But imagine you learn that the movie was released after the crimes were committed. Thus, those people in your survey who said they had seen the movie had actually committed their crimes before the movie characters committed their crimes. Watching the movie, then, could not possibly have led to the crimes. Perhaps the criminals watched the movie because committing violent crimes made them interested in violent movies.

This discussion points to the importance of the criterion of **time order**. To conclude that causation was involved, we must see that cases were exposed to variation in the independent variable *before* variation in the dependent variable. Bushman's (1995) experiment satisfied this criterion because he controlled the variation in the independent variable: All the students saw the videotape excerpts (which varied in violent content) before their level of aggressiveness was measured.

Nonspuriousness

Even when research establishes that two variables are associated and that variation in the independent variable precedes variation in the dependent variable, we cannot be sure we identified a causal relationship between the two variables. Have you heard the old adage "Correlation does not prove causation"? It is meant to remind us that an association between two variables might be caused by something else. If we measure children's shoe sizes and their academic knowledge, for example, we will find a positive association. However, the association results from the fact that older children have larger feet as well as more academic knowledge. Shoe size does not cause knowledge or vice versa.

Before we conclude that variation in an independent variable causes variation in a dependent variable, we must have reason to believe that the relationship is nonspurious. **Nonspuriousness** is a relationship between two variables that is not due to variation in a third variable. When this third variable, an **extraneous variable**, causes the variation, it is said to have created a **spurious relationship** between the independent and dependent variables. We must design our research so that we can see what happens to the dependent variable when only the independent variable varies. If we cannot do this, there are other statistical methods we must use to control the effects of other variables we also believe are related to our dependent variable. (You will be relieved to know that a discussion of these statistical techniques is way beyond the scope of this text!)

In reality, then, the fact that someone blew up a toll booth after seeing the movie *Money Train* might be related to the fact that he was already feeling enraged against society. This led him to seek out a violent movie for entertainment purposes (see Exhibit 5.2). Thus, seeing the violent movie itself in no way led him to commit the crime. We must be sure that all three conditions of association, time order, and nonspuriousness are met before we make such claims.

Does Bushman's (1995) claim of a causal effect rest on any stronger ground? To evaluate nonspuriousness, you need to know about one more feature of his experiment. He assigned students to watch either the violent video or the nonviolent video randomly, that is, by the toss of a coin. Because he used **random assignment**, the characteristics and attitudes that students already possessed when they were recruited for the experiment could

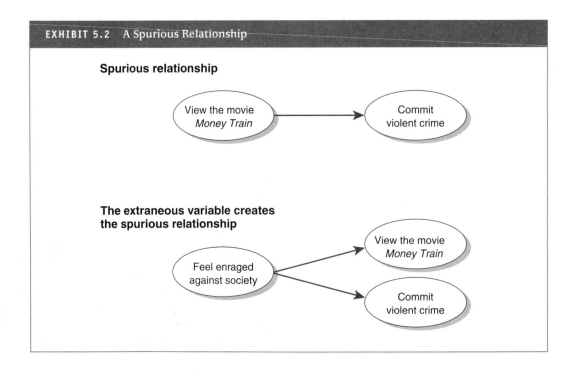

EXHIBIT 5.2   A Spurious Relationship

not influence either of the two videos they watched. As a result, the students' characteristics and attitudes could not explain why one group reacted differently from the other after watching the videos. In fact, because Bushman used 296 students in his experiment, it is highly unlikely that the violent video group and the nonviolent video group differed in any relevant way at the outset, even on the basis of chance. This experimental research design meets the criterion of nonspuriousness. Bushman's conclusion that viewing video violence causes aggressive behavior thus rests on firm ground indeed.

Causal (internal) validity is achieved by meeting the criteria of association, time order, and nonspuriousness. Others, however, believe that two additional criteria should also be considered: mechanism and context.

Mechanism

Confidence in a conclusion that two variables have a causal connection will be strengthened if a **mechanism**, some discernable means of creating a connection, can be identified (Cook & Campbell 1979:35; Marini & Singer 1988). Many social scientists (and scientists in other fields) argue that a causal explanation is not adequate until a causal mechanism is identified. What process or mechanism actually is responsible for the relationship between the independent and dependent variables?

Bushman (1995) did not empirically identify a causal mechanism in his experiment, but he did suggest a possible causal mechanism for the effect of watching violent videos. Before we can explain this causal mechanism, we have to tell you about one more aspect of his research. He was not interested simply in whether viewing violent films resulted in aggressive behavior. Actually, his primary hypothesis was that individuals who are predisposed to aggression before the study began would be more influenced by a violent film than individuals who were not aggressive at the outset. And that is what happened: Individuals who were predisposed to aggression became more aggressive after watching Bushman's violent video, but individuals who were not predisposed to aggression did not become more aggressive.

After the experiment, Bushman (1995) proposed a causal mechanism to explain why aggressive individuals became even more aggressive after watching the film:

> High trait aggressive individuals [people predisposed to aggression] are more susceptible to the effects of violent media than are low trait aggressive individuals because they possess a relatively large network of aggressive associations that can be activated by violent cues. Habitual exposure to television violence might be partially responsible. (P. 959)

Note that this explanation relies more on speculation than on the actual empirical evidence from this particular experiment. Nonetheless, by proposing a reasonable causal mechanism that connects the variation in the independent and dependent variables, Bushman (1995) strengthens the argument for the causal validity of his conclusions.

It is often possible to go beyond speculation by designing research to test one or more possible causal mechanisms. Perhaps other researchers will design a new study to measure directly the size of individuals' networks of aggressive associations that Bushman (1995) contends are part of the mechanism by which video violence influences aggressive behavior.

Context

In the social world, it is virtually impossible to claim that one and only one independent variable is responsible for causing or affecting a dependent variable. Stated another way, no cause can be separated from the larger **context** in which it occurs. A cause is really only one of a set of interrelated factors required for the effect (Hage & Meeker 1988; Papineau 1978). When relationships among variables differ across geographic units like counties or across other social settings, researchers say there is a **contextual effect**. Identification of the context in which a causal relationship occurs can help us to understand that relationship.

Some researchers argue that we do not fully understand the causal effect of media violence on behavioral aggression unless we have identified these other related factors. As we have just seen, Bushman (1995) proposed at the outset of his research at least one other condition: Media violence would increase aggression only among individuals who were already predisposed to aggression.

Identification of the context in which a causal effect occurs is not a criterion for a valid causal conclusion. Some contextual factors may not turn out to be causes of the effect being investigated. The question for researchers is, "How many contexts should we investigate?" In a classic study of children's aggressive behavior in response to media violence, Bandura, Ross, and Ross (1963) examined several contextual factors. They found that effects varied with the children's gender and with the gender of the opponent toward whom they acted aggressively, but not with whether they saw a real (acted) or filmed violent incident. For example, children reacted more aggressively after observing men committing violent acts than after observing women committing these same acts. But Bandura et al. did not address the role of violence within the children's families or the role of participation in sports or many other factors that could be involved in children's responses to media violence. Bandura et al. strengthened their conclusions by focusing on a few likely contextual factors.

Specifying the context for a causal effect helps us to understand that effect, but it is a process that can never really be complete. We can always ask what else might be important: In which country was the study conducted? What are the ages of the study participants? We need to carefully review the results of prior research and the implications of relevant theory to determine what contextual factors are likely to be important in a causal relationship. Our confidence in causal conclusions will be stronger when we know these factors are taken into account.

In summary, before researchers can infer a causal relationship between two variables, three criteria are essential: empirical association, appropriate time order, and nonspuriousness. After these three conditions have been met, two other criteria are also important: causal mechanism and context.

RESEARCH DESIGNS AND CAUSALITY

How research is designed influences our ability to draw causal conclusions. Obviously, if you conclude that playing violent video games causes violent behavior after watching your

8-year-old nephew playing a violent video game and then hitting his 4-year-old brother, you would be on shaky empirical ground. In this section, we will introduce features that need to be considered in a research design in order to evaluate how well it can support nomothetic causal conclusions.

True Experiments

In a true experiment, the time order is determined by the researcher. The experimental design provides the most powerful design for testing causal hypotheses about the effect of a treatment or some other variable whose values can be manipulated by the researchers. It is so powerful for testing causal hypotheses because it allows us to establish the three criteria for causality with a great deal of confidence. The Bushman (1995) study we examined in the last section was a true experiment.

True experiments must have at least three things:

1. Two comparison groups, one receiving the experimental condition (e.g., treatment or intervention) termed the experimental group and the other receiving no treatment or intervention or another form thereof, termed the control group.
2. Random assignment to the two (or more) comparison groups.
3. Assessment of change in the dependent variable for both groups after the experimental condition has been received.

The combination of these features permits us to have much greater confidence in the validity of causal conclusions than is possible in other research designs. Confidence in the validity of an experiment's findings is further enhanced by identification of the causal mechanism and control over the context of an experiment. We will discuss experimental designs in more detail in the next chapter (see Chapter 6). For now, we want to highlight how true experimental designs lend themselves to meeting the criteria necessary for causality.

Causality and True Experimental Designs

A prerequisite for meeting each of the three criteria to identify causal relations is maintaining control over the conditions subjects are exposed to after they are assigned to the experimental and comparison groups. If these conditions begin to differ, the variation between the experimental and comparison groups will not be what was intended. Even a subsequent difference in the distribution of cases on the dependent variable will not provide clear evidence of the effect of the independent variable. Such unintended variation is often not much of a problem in laboratory experiments where the researcher has almost complete control over the conditions and can ensure that these conditions are nearly identical for both groups. But control over conditions can become a very big concern for experiments that are conducted in the field in real-world settings, such as Sherman and Berk's (1984) study of the deterrent effects of arrest on intimate partner assaults.

Let us examine how well true experiments meet the criteria necessary for establishing causality in greater detail:

Association between the hypothesized independent and dependent variables. As you have seen, experiments can provide unambiguous evidence of association by randomly assigning subjects to experimental and comparison groups.

Time order of effects of one variable on the others. Unquestionably, the independent variable (treatment of condition) preceded the posttest measures in the experiments described so far. For example, arrest for partner abuse preceded recidivism in the Sherman and Berk (1984) study, and the exposure to media violence preceded the aggression in the Bushman (1995) experiment. In experiments with a pretest, time order can be established by comparing posttest to pretest scores. In experiments with random assignment of subjects to the experimental and comparison groups, time order can be established by comparison of posttest scores only.

Nonspurious relationships between variables. Nonspuriousness is difficult to establish; some would say it is impossible to establish in nonexperimental designs. The random assignment of subjects to experimental and comparison groups makes true experiments powerful designs for testing causal hypotheses. Random assignment controls a host of possible extraneous influences that can create misleading, spurious relationships in both experimental and nonexperimental data. If we determine that a design has used randomization successfully, we can be much more confident in the causal conclusions.

Mechanism that creates the causal effect. The features of true experiment do not, in themselves, allow identification of causal mechanisms; as a result there can be some ambiguity about how the independent variable influenced the dependent variable and the causal conclusions.

Context in which change occurs. Control over conditions is more feasible in many experimental designs than it is in nonexperimental designs, but it is often difficult to control conditions in field experiments. In the next chapter, you will learn how the lack of control over experimental conditions can threaten internal validity.

Nonexperimental Designs

Nonexperimental research designs can be either cross-sectional or longitudinal. In a **cross-sectional research design**, all data are collected at one point in time. Identifying the **time order** of effects—what happened first, and so on—is critical for developing a causal analysis, but can be an insurmountable problem with a cross-sectional design. In **longitudinal research designs**, data are collected at two or more points in time, and so identification of the time order of effects can be quite straightforward. An experiment, of course, is a type of longitudinal design because subjects are observed at 2 or more points in time.

Cross-Sectional Designs

Much of the research you have encountered so far in this text has been cross-sectional. Although each survey and interview takes some time to carry out, if they measure the actions, attitudes, and characteristics of respondents at only one time, they are considered cross-sectional. The name comes from the idea that a snapshot from a cross-section of the population is obtained at one point in time.

As you learned in chapter 3, Sampson and Raudenbush (1999) used a very ambitious cross-sectional design to study the effect of visible public social and physical disorder on the crime rate in Chicago neighborhoods. Their theoretical framework focused on the concept of informal social control: the ability of residents to regulate social activity in their neighborhoods through their collective efforts according to desired principles. They believed that informal social control would vary between neighborhoods, and they hypothesized that it was the strength of informal social control that would explain variation in crime rates rather than just the visible sign of disorder. They contrasted this prediction to the "broken windows" theory: the belief that signs of disorder themselves cause crime. Their findings supported their hypothesis: both visible disorder and crime were consequences of low levels of informal social control (measured with an index of "collective efficacy"). One did not cause the other (see Exhibit 5.3).

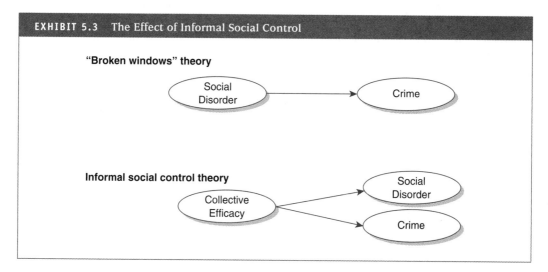

EXHIBIT 5.3   The Effect of Informal Social Control

"Broken windows" theory

Social Disorder → Crime

Informal social control theory

Collective Efficacy → Social Disorder

Collective Efficacy → Crime

Source: Based on Sampson & Raudenbush 1999:635.

In spite of these compelling findings (see Exhibit 5.4), Sampson and Raudenbush's (1999) cross-sectional design could not establish directly that the variation in the crime rate occurred after variation in informal social control. Maybe it was a high crime rate that led residents to stop trying to exert much control over deviant activities in the neighborhood, perhaps because of fear of crime. It is difficult to discount such a possibility when only cross-sectional data are available.

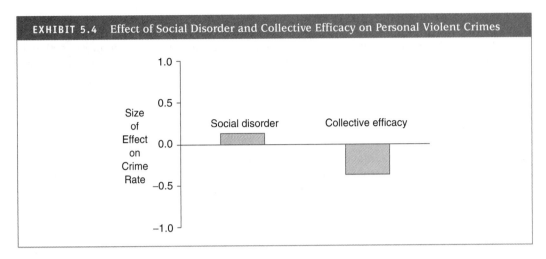

EXHIBIT 5.4 Effect of Social Disorder and Collective Efficacy on Personal Violent Crimes

Source: Adapted from Sampson & Raudenbush 1999.

There are four special circumstances in which we can be more confident in drawing conclusions about time order on the basis of cross-sectional data. Because in these special circumstances the data can be ordered in time, they might even be thought of as longitudinal designs (Campbell 1992).

The independent variable is fixed at some point prior the variation in the dependent variable. So-called demographic variables that are determined at birth—such as sex, race, and age—are fixed in this way. So are variables like education and marital status, if we know when the value of cases on these variables was established and if we know that the value of cases on the dependent variable was set some time afterward. For example, say we hypothesize that educational opportunities in prison affect recidivism rates. Let us say we believe those inmates who are provided with greater educational and vocational opportunities in prison will be less likely to reoffend after release from prison. If we know that respondents completed their vocational or other educational training before leaving prison, we would satisfy the time order requirement even if we were to measure education at the same time we measure recidivism after release. However, if some respondents possibly went back to school after prison release, the time order requirement would not be satisfied.

We believe that respondents can give us reliable reports of what happened to them or what they thought at some earlier point in time. Horney, Osgood, and Marshall (1995) provide an interesting example of the use of such retrospective data. The researchers wanted to identify how criminal activity varies in response to changes in life circumstances. They interviewed 658 newly convicted male offenders sentenced to a Nebraska state prison. In a 45- to 90-minute interview, they recorded each inmate's report of his life circumstances and of his criminal activities for the preceding 2 to 3 years. They then found that criminal involvement was related strongly to adverse changes in life circumstances, such as marital separation or drug use. Retrospective data are often inadequate for measuring variation in past psychological states or behaviors, however, because what we recall about our feeling or actions in the past is likely to be influenced by what we feel in the present. For example, retrospective reports by both adult alcoholics and their parents appear to greatly overestimate the frequency of childhood problems (Vaillant 1995). People cannot report reliably the frequency and timing of many past events, from hospitalization to hours worked. However, retrospective data tend to be reliable when it concerns major, persistent experiences in the past, such as what type of school someone went to or how a person's family was structured (Campbell 1992).

Our measures are based on records that contain information on cases in earlier periods. Government, agency, and organizational records are an excellent source of time-ordered data after the fact. However, sloppy record keeping and changes in data-collection policies can lead to inconsistencies, which must be taken into account. Another weakness of such archival data is that they usually contain measures of only a fraction of the variables that we think are important.

We know that the value of the dependent variable was similar for all cases prior to the treatment. For example, we may hypothesize that an anger management program (independent variable) improves the conflict resolution abilities (dependent variable) of individuals arrested for intimate partner assault. If we know that none of the arrested individuals could employ verbal techniques for resolving conflict prior to the training program, we can be confident that any subsequent variation in their ability to do so did not precede exposure to the training program. This is one way that traditional experiments establish time order: Two or more equivalent groups are formed prior to exposing one of them to some treatment.

Longitudinal Designs

In longitudinal research, data are collected at 2 or more points in time and, as such, data can be ordered in time. By measuring the value of cases on an independent variable and a dependent variable at different times, the researcher can determine whether variation in the independent variable precedes variation in the dependent variable.

In some longitudinal designs, the same sample (or panel) is followed over time; in other designs, sample members are rotated or completely replaced. The population from which the sample is selected may be defined broadly, as when a longitudinal survey of the general population is conducted. Or the population may be defined narrowly, as when members of a specific age group are sampled at multiple points in time. The frequency of follow-up measurement can vary, ranging from a before-and-after design with just one follow-up to studies in which various indicators are measured every month for many years.

Certainly it is more difficult to collect data at two or more points in time than at one time. Quite frequently researchers simply cannot, or are unwilling to, delay completion of a study for even 1 year in order to collect follow-up data. But think of the many research questions that really should involve a much longer follow-up period: Does community-oriented policing decrease rates of violent crime? What is the impact of job training in prison on recidivism rates? How effective are batterer-treatment programs for individuals convicted of intimate partner assault? Do parenting programs for young mothers and fathers reduce the likelihood of their children becoming delinquent? It is safe to say that we will never have enough longitudinal data to answer many important research questions. Nonetheless, the value of longitudinal data is so great that every effort should be made to develop longitudinal research designs when they are appropriate for the research question asked. The

EXHIBIT 5.5   Three Types of Longitudinal Design

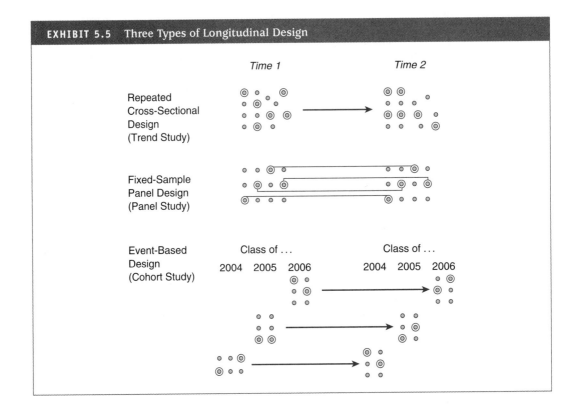

following discussion of the three major types of longitudinal designs will give you a sense of the possibilities (see Exhibit 5.5).

Repeated Cross-Sectional Designs

Studies that use a **repeated cross-sectional design**, also known as **trend studies**, have become fixtures of the political arena around election time. Particularly in presidential election years, we have all become accustomed to reading weekly, even daily, reports on the percentage of the population that supports each candidate. Similar polls are conducted to track sentiment on many other social issues. For example, a 1993 poll reported that 52% of adult Americans supported a ban on the possession of handguns, compared to 41% in a similar poll conducted in 1991. According to pollster Louis Harris, this increase indicated a "sea change" in public attitudes (cited in Barringer 1993). Another researcher said, "It shows that people are responding to their experience [of an increase in handgun-related killings]" (cited in Barringer 1993:A14).

Repeated cross-sectional design (trend study) A type of longitudinal study in which data are collected at two or more points in time from different samples of the same population.

Repeated cross-sectional surveys are conducted as follows:

1. A sample is drawn from a population at Time 1, and data are collected from the sample.
2. As time passes, some people leave the population and others enter it.
3. At Time 2, a different sample is drawn from this population.

These features make the repeated cross-sectional design appropriate when the goal is to determine whether a population has changed over time. Has racial tolerance increased among Americans in the past 20 years? Are prisons more likely to have drug-treatment programs available today than they were in the 1950s? These questions concern changes in the population as a whole, not changes in individuals within the population. We want to know whether racial tolerance increased in society, not whether this change was due to migration that brought more racially tolerant people into the country or to individual U.S. citizens becoming more tolerant. We are asking whether state prisons overall are more likely to have drug-treatment programs available today than they were a decade or two decades ago, not whether any such increase was due to an increase in prisoner needs or to individual prisons changing their program availability. When we do need to know whether individuals in the population changed, we must turn to a panel design.

Fixed-Sample Panel Designs

Panel designs allow us to identify changes in individuals, groups, or whatever we are studying. This is the process for conducting **fixed-sample panel designs**:

1. A sample (called a panel) is drawn from a population at Time 1, and data are collected from the sample.
2. As time passes, some panel members become unavailable for follow-up, and the population changes.
3. At time 2, data are collected from the same people as at Time 1 (the panel), except for those people who cannot be located.

Fixed-sample panel design (panel study) A type of longitudinal study in which data are collected from the same individuals—the panel—at two or more points in time. In another type of panel design, panel members who leave are replaced with new members.

Because a panel design follows the same individuals, it is better than a repeated cross-sectional design for testing causal hypotheses. For example, Sampson and Laub (1990) used a fixed-sample panel design to investigate the effect of childhood deviance on adult crime. They studied a sample of white males in Boston when the subjects were between 10 and 17 years old and then followed up when the subjects were in their adult years. Data were collected from multiple sources, including the subjects themselves and criminal justice records. Sampson and Laub (p. 614) found that children who had been committed to a correctional school for persistent delinquency were much more likely than other children in the study to commit crimes as adults: 61% were arrested between the ages of 25 and 32, compared to 14% of those who had not been in correctional schools as juveniles. In this study, juvenile delinquency unquestionably occurred before adult criminality. If the researchers had used a cross-sectional design to study the past of adults, the juvenile delinquency measure might have been biased by memory lapses, by self-serving recollections about behavior as juveniles, or by loss of agency records.

If you now wonder why not every longitudinal study is designed as a panel study, you have understood the advantages of panel designs. However, remember that this design does not in itself establish causality. Variation in both the independent variable and the dependent variables may be due to some other variable, even to earlier variation in what is considered the dependent variable. In the example in Exhibit 5.6, there is a hypothesized association between delinquency in the 11th grade and grades obtained in the 12th grade (the dependent variable). The time order is clear. However, both variables are consequences of grades obtained in the 7th grade. The apparent effect of 11th-grade delinquency on 12th-grade grades is spurious because of variation in the dependent variable (grades) at an earlier time.

Panel designs are also a challenge to implement successfully, and often are not even attempted, because of two major difficulties:

Expense and attrition. It can be difficult, and very expensive, to keep track of individuals over a long period, and inevitably the proportion of panel members who can be located for follow-up will decline over time. Panel studies often lose more than one quarter of their members through attrition (Miller 1991:170), and those who are lost are often not

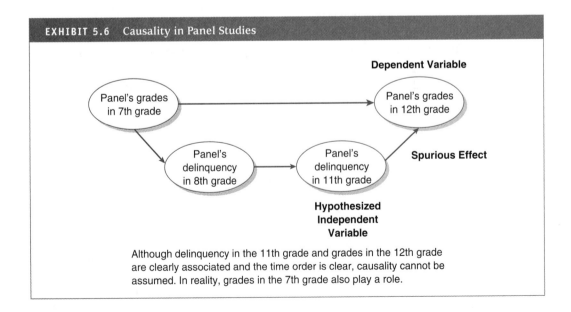

EXHIBIT 5.6   *Causality in Panel Studies*

Dependent Variable

Panel's grades in 7th grade

Panel's grades in 12th grade

Panel's delinquency in 8th grade

Panel's delinquency in 11th grade

Spurious Effect

Hypothesized Independent Variable

Although delinquency in the 11th grade and grades in the 12th grade are clearly associated and the time order is clear, causality cannot be assumed. In reality, grades in the 7th grade also play a role.

necessarily like those who remain in the panel. As a result, a high rate of subject attrition may mean that the follow-up sample will no longer be representative of the population from which it was drawn and may no longer provide a sound basis for estimating change. Subjects who were lost to follow-up may have been those who changed the most, or the least, over time. For example, between 5% and 66% of subjects are lost in substance abuse prevention studies, and the dropouts typically had begun the study with higher rates of tobacco and marijuana use (Snow, Tebes, & Arthur 1992:804).

It does help to compare the baseline characteristics of those who are interviewed at follow-up with characteristics of those lost to follow-up. If these two groups of panel members were not very different at baseline, it is less likely that changes had anything to do with characteristics of the missing panel members. Even better, subject attrition can be reduced substantially if sufficient staff can be used to keep track of panel members. In their panel study, Sampson and Laub (1990) lost only 12% of the juveniles in the original sample (8% if you do not count those who had died).

Subject fatigue. Panel members may grow weary of repeated interviews and drop out of the study, or they may become so used to answering the standard questions in the survey that they start giving stock answers rather than actually thinking about their current feelings or actions (Campbell 1992). This is called the problem of **subject fatigue**. Fortunately, subjects do not often seem to become fatigued in this way, particularly if the research staff have maintained positive relations with the subjects. For example, at the end of an 18-month-long experimental study of housing alternatives for persons with mental illness who had been homeless, only 3 or 4 individuals (out of 93 who could still be located) refused to participate in the fourth and final round of interviews. The interviews took a total of about 5 hours to complete, and participants received about $50 for their time (Schutt, Goldfinger, & Penk 1997).

Because panel studies are so useful, social researchers have developed increasingly effective techniques for keeping track of individuals and overcoming subject fatigue. But when resources do not permit use of these techniques to maintain an adequate panel, repeated cross-sectional designs usually can be employed at a cost that is not a great deal higher than that of a one-time-only cross-sectional study. The payoff in explanatory power should be well worth the cost.

Event-Based Designs

In an **event-based design**, often called a *cohort study*, the follow-up samples (at one or more times) are selected from the same **cohort**: people who all have experienced a similar event or a common starting point. Examples include the following:

- *Birth cohorts*: those who share a common period of birth (those born in the 1940s, 1950s, 1960s, etc.)
- *Seniority cohorts*: those who have worked at the same place for about 5 years, about 10 years, and so on
- *School cohorts*: freshmen, sophomores, juniors, and seniors

EXHIBIT 5.7 The Use of Statistical Control to Reduce Spuriousness

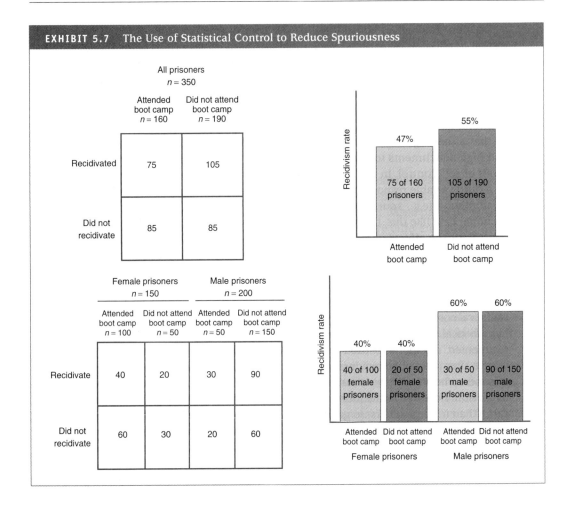

Statistical control A technique used in nonexperimental research to reduce the risk of spuriousness. One variable is held constant so the relationship between two or more other variables can be assessed without the influence of variation in the control variable.

Example of statistical control: Sampson (1987) found that the relationship between rates of family disruption and violent crimes held true for cities with similar levels of joblessness (the control variable). So the rate of joblessness could not have caused the association between family disruption and violent crime.

Our confidence in causal conclusions based on nonexperimental research also increases with identification of a causal mechanism. These mechanisms are called **intervening variables** in nonexperimental research, and help us to understand how variation in the independent variable results in variation in the dependent variable. For example, in a study that reanalyzed data from Glueck and Glueck's (1950) pathbreaking study of juvenile delinquency, Sampson and

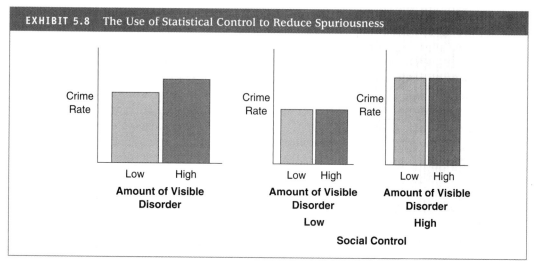

EXHIBIT 5.8   The Use of Statistical Control to Reduce Spuriousness

Source: Based on Sampson & Raudenbush 1999.

Laub (1994) found that children who grew up with such structural disadvantages as family poverty and geographic mobility were more likely to become juvenile delinquents. Why did this occur? Their analysis indicated that these structural disadvantages led to lower levels of informal social control in the family (less parent-child attachment, less maternal supervision, and more erratic or harsh discipline). Lower levels of informal social control resulted in a higher probability of delinquency (see Exhibit 5.9). Informal social control intervened in the relationship between structural context and delinquency.

Of course, identification of one (or two or three) intervening variables does not end the possibilities for clarifying the causal mechanisms. You might ask why structural

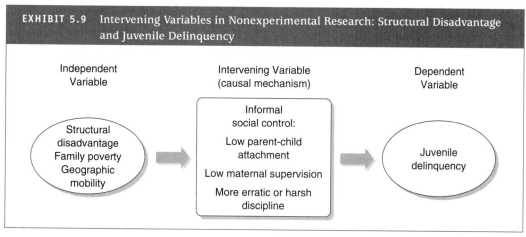

EXHIBIT 5.9   Intervening Variables in Nonexperimental Research: Structural Disadvantage and Juvenile Delinquency

Source: Based on Sampson & Laub 1994.

disadvantage tends to result in lower levels of family social control or how family social control influences delinquency. You could then conduct research to identify the mechanisms that link, for example, family social control and juvenile delinquency. (Perhaps the children feel they are not cared for, so they become less concerned with conforming to social expectations.) This process could go on and on. The point is that identification of a mechanism through which the independent variable influences the dependent variable increases our confidence in the conclusion that a causal connection does indeed exist.

When you think about the role of variables in causal relationships, do not confuse variables that cause spurious relationships with variables that intervene in causal relationships, even though both are third variables that do not appear in the initial hypothesis. Intervening variables help explain the relationship between the independent variable (juvenile delinquency) and the dependent variable (adult criminality).

Nonexperimental research can be a very effective tool for exploring the context in which causal effects occur. Administering surveys in many different settings and to different types of individuals is usually much easier than administering various experiments. The difficulty of establishing nonspuriousness does not rule out using nonexperimental data to evaluate causal hypotheses. In fact, when enough nonexperimental data are collected to allow tests of multiple implications of the same causal hypothesis, the results can be very convincing (Freedman 1991).

In any case, nonexperimental tests of causal hypotheses will continue to be popular because the practical and ethical problems in randomly assigning people to different conditions preclude the test of many important hypotheses with an experimental design. Just remember to carefully consider possible sources of spuriousness and other problems when evaluating causal claims based on individual nonexperimental studies.

CONCLUSION

In this chapter, you have learned about two alternative meanings of causation (nomothetic and idiographic). You have studied the five criteria used to evaluate the extent to which particular research designs may achieve causally valid findings. You have learned how our ability to meet these criteria is shaped by research design features including the use of true experimental designs, the use of cross-sectional or longitudinal designs, and the use of statistical control to deal with the problem of spuriousness. You have also seen why the distinction between experimental and nonexperimental designs has so many consequences for how, and how well, we are able to meet nomothetic criteria for causation.

It is important to remember that the results of any particular study are part of an always-changing body of empirical knowledge about social reality. Thus, our understandings of causal relationships are always partial. Researchers always wonder whether they have omitted some relevant variables from their controls, whether their experimental results would differ if the experiment were conducted in another setting, or whether they have overlooked a critical historical event. But by using consistent definitions of terms and maintaining clear standards for establishing the validity of research results, and by expecting the same of others who do research, social researchers can contribute to a growing body of knowledge that can reliably guide social policy and social understanding.

When you read the results of a social scientific study, you should now be able to evaluate critically the validity of the study's findings. If you plan to engage in social research, you should now be able to plan an approach that will lead to valid findings. And with a good understanding of three dimensions of validity (measurement validity, generalizability, and causal (internal validity) under your belt, you are ready to focus on the major methods of data collection used by social scientists.

KEY TERMS

Association
Ceteris paribus
Cohort
Cohort study
Context
Contextual effect
Counterfactual
Cross-sectional research design
Event-based design (cohort study)
Extraneous variable
Fixed-sample panel design (panel study)
Idiographic causal explanation

Intervening variable
Longitudinal research design
Mechanism
Nomothetic causal explanation
Nonspuriousness
Random assignment
Repeated cross-sectional design (trend study)
Spurious relationship
Statistical control
Subject fatigue
Time order
Trend study

HIGHLIGHTS

- Causation can be defined in either nomothetic or idiographic terms. Nomothetic causal explanations deal with effects on average. Idiographic causal explanations deal with the sequence of events that led to a particular outcome.

- The concept of nomothetic causal explanation relies on a comparison. The value of cases on the dependent variable is measured after they have been exposed to variation in an independent variable. This measurement is compared to what the value of cases on the dependent variable would have been if they had not been exposed to the variation in the independent variable (the counterfactual). The validity of nomothetic causal conclusions rests on how closely the comparison group comes to the ideal counterfactual.

- From a nomothetic perspective, three criteria are generally viewed as necessary for identifying a causal relationship: association between the variables, proper time order, and nonspuriousness of the association. In addition, the basis for concluding that a causal relationship exists is strengthened by identification of a causal mechanism and the context for the relationship.

- Association between two variables is in itself insufficient evidence of a causal relationship. This point is commonly made with the expression "Correlation does not prove causation."

- Experiments use random assignment to make comparison groups as similar as possible at the outset of an experiment in order to reduce the risk of spurious effects due to extraneous variables.

- Nonexperimental designs use statistical controls to reduce the risk of spuriousness. A variable is controlled when it is held constant so that the association between the independent and dependent variables can be assessed without being influenced by the control variable.

- Ethical and practical constraints often preclude the use of experimental designs.

- Idiographic causal explanations can be difficult to identify, because the starting and ending points of particular events and the determination of which events act as causes in particular sequences may be ambiguous.

- Longitudinal designs are usually preferable to cross-sectional designs for establishing the time order of effects. Longitudinal designs vary in terms of whether the same people are measured at different times, how the population of interests is defined, and how frequently follow-up measurements are taken. Fixed-sample panel designs provide the strongest test for the time order of effects, but they can be difficult to carry out successfully because of their expense as well as subject attrition and fatigue.

EXERCISES

1. Review articles in several newspapers, copying down all causal assertions. These might range from assertions that community policing was related to decreasing rates of violence, that the stock market declined because of uncertainty in the Middle East, or to explanations about why a murder was committed. Inspect the articles carefully, noting all evidence used to support the causal assertions. Are the explanations nomothetic, idiographic, or a combination of both? Which criteria for establishing causality in a nomothetic framework are met? How satisfactory are the idiographic explanations? What other potentially important influences on the reported outcome have been overlooked?

2. Select several research articles in professional journals that assert, or imply, that they have identified a causal relationship between two or more variables. Are all the criteria for establishing the existence of a causal relationship met? Find a study in which subjects were assigned randomly to experimental and comparison groups to reduce the risk of spurious influences on the supposedly causal relationship. How convinced are you by the study? Find a survey study that makes causal assertions based on the relationships, or correlations, among variables. What variables have been statistically controlled? List other variables that might be influencing the relationship but that have not been controlled. How convinced are you by the study?

3. Search *Sociological Abstracts* or another index to the social literature for several articles on studies using any type of longitudinal design. You will be searching for article titles that use words like "longitudinal," "panel," "trend," or "over time." How successful were the researchers in carrying out the design? What steps did the researchers who used a panel design take to minimize panel attrition? How convinced are you by those using repeated cross-sectional designs that they have identified a process of change in individuals? Did any researchers use retrospective questions? How did they defend the validity of these measures?

5. Go to the book's Study Site, http://www.sagepub.com/prccj3, and choose two research articles that include some attention to causality (as indicated by a check in that column

of the article matrix). Describe the approach taken in each article to establishing causality. How do the approaches differ from each other? Which approach seems stronger to you?

DEVELOPING A RESEARCH PROPOSAL

How will you try to establish the causal effects you hypothesize?

1. Identify at least one hypothesis involving what you expect is a causal relationship.

2. Identify key variables that should be controlled in your survey design in order to increase your ability to avoid arriving at a spurious conclusion about the hypothesized causal effect. Draw on relevant research literature and social theory to identify these variables.

3. Add a longitudinal component to your research design. Explain why you decided to use this particular longitudinal design.

4. Review the criteria for establishing a nomothetic causal effect and discuss your ability to satisfy each one.

Student Study Site

The companion Web site for *The Practice of Research in Criminology and Criminal Justice*, Third Edition
   http://www.sagepub.com/prccj3
   Visit the Web-based Student Study Site to enhance your understanding of the chapter content and to discover additional resources that will take your learning one step further. You can enhance your understanding of the chapters by using the comprehensive study material, which includes e-flashcards, Web exercises, practice self-tests, and more. You will also find special features, such as Learning from Journal Articles, which incorporates SAGE's online journal collection.

WEB EXERCISES

1. Go to the Disaster Center Web site, http://www.disastercenter.com/crime/. Review the crime rate nationally, and, by picking out links to state reports, compare the recent crime rates in two states. Report on the prevalence of the crimes you have examined. Propose a causal explanation for variation in crime between states, over time, or both. What research design would you propose to test this explanation? Explain.

2. Go to Crime Stoppers USA's Web site at http:/www.crimestopusa.com/. Check out "About Us" and "Crime Stoppers." How is CSUSA "fighting crime"? What does CSUSA's approach assume about the cause of crime? Do you think CSUSA's approach to fighting crime is based on valid conclusions about causality? Explain.

3. What are the latest trends in crime? Write a short statement after inspecting the FBI's Uniform Crime Reports at www.fbi.gov (go to the "statistics" section under "reports and publications"). You will need to use Adobe Acrobat Reader to access some of these reports (those in PDF format). Follow the instructions on the site if you are not familiar with this program.

Experimental Designs

$\mathbf{A}$s of June 30, 2005, there were 2,186,230 prisoners held in federal or state prisons or in local jails in the United States. This represents a 2.6% increase from the same time in 2004. The rate of incarceration is also increasing. In 2005, there were an estimated 488 prison inmates per 100,000 U.S. residents compared to 411 in 1995. The number of women under jurisdiction of state and federal prison authorities is also increasing, and increasing at a faster rate than men (Bureau of Justice Statistics 2006). The rising prison population coupled with lawsuits has prompted most correctional institutions to begin to classify inmates into differing levels of security (e.g., minimum, maximum) based on objective criteria like severity of their offenses, previous records and so forth instead of more subjective criteria. Obviously, the security level of an institution in which an inmate is classified will affect their incarceration experience. Someone who is assigned to a maximum security prison instead of a lower level of security will also have differential access to things like mental health services, drug treatment programs, and vocational training (Brennan & Austin 1997). But is the classification of inmates also related to their behavior while incarcerated? Do those assigned to maximum security prisons engage in more misconduct compared to inmates assigned to less secure facilities? How could you answer this question? If you compared rates of misconduct across prison settings, you would not have the answer because the inmates may have been very different at their time of incarceration. As such, any differences you observe could be attributable to these "before incarceration" differences, not to the type of facility in which they were housed.

In the last chapter, you learned that experimental research provides the most powerful design for testing nomothetic causal hypotheses. This chapter examines experimental methodology in more detail: You will learn to distinguish different types of experimental designs (which include true experiments and quasi-experiments), to evaluate the utility of particular designs for reaching causally valid conclusions, to identify problems of generalizability with experiments, and to consider ethical problems in experimentation. We will also return to the relationship between prison security level and inmate misconduct later in the chapter.

TRUE EXPERIMENTS

Recall from the last chapter that true experiments must have at least three things:

1. Two comparison groups, one receiving the experimental condition (e.g., treatment or intervention) and the other receiving no treatment or intervention or another form thereof.
2. Random assignment to the two (or more) comparison groups.
3. Assessment of change in the dependent variable for both groups after the experimental condition has been received.

Experimental and Comparison Groups

True experiments must have at least one **experimental group** (subjects who receive some treatment) and at least one **control or comparison group** (subjects to whom the experimental group can be compared). The control group differs from the experimental group by one or more independent variables, whose effects are being tested. In other words, the difference between the two groups is that the experimental group receives the treatment or condition we are interested in, whereas the comparison group does not.

Experimental group In an experiment, the group of subjects that receives the treatment or experimental manipulation.

Control or comparison group The group of subjects that is either exposed to a different treatment from the experimental group or that receives no treatment at all.

An experiment can have more than one experimental group if the goal is to test several versions of the **treatment** (the independent variable) or several combinations of different treatments. An experiment also may have more than one comparison group, as when outcome scores for the treatment group need to be compared to more than one comparison group.

Pretest and Posttest Measures

All true experiments have a **posttest**, that is, measurement of the outcome in both groups after the experimental group has received the treatment. In fact, we might say that any hypothesis-testing research involves a posttest. The dependent variable is measured after the independent variable has had its effect, if any. True experiments also may have **pretests** that measure the dependent variable prior to the experimental intervention. A pretest is exactly the same as the posttest, just administered before the treatment. Strictly speaking, however, an experiment does not require a pretest. When researchers use random assignment, the groups' initial scores on the dependent variable and on all other variables will most likely be similar; we can be confident in this because of the random assignment. Any difference in outcome between the experimental group and the comparison group is likely caused by only the intervention or experimental condition.

Having pretest scores, however, has many advantages. They provide a direct measure of how much the experimental and comparison groups changed over time. They allow the researcher to verify that randomization was successful (that chance factors did not lead to an initial difference between the groups). In addition, by identifying subjects' initial scores on the dependent variable, a pretest provides a more complete picture of the conditions in which the intervention had (or did not have) an effect (Mohr 1992:46–48).

An experiment may have multiple posttests and perhaps even multiple pretests. Multiple posttests can identify just when the treatment has its effect and for how long. They are particularly important for treatments delivered over time (Rossi & Freeman 1989:289–290).

Random Assignment

Random assignment, sometimes referred to as randomization, is what makes the comparison group in a true experiment such a powerful tool for identifying the effects of the treatment. If subjects are randomly assigned to either the experimental group or the comparison group, then a researcher can assume that the only difference between the two groups is that the experimental group received the treatment or intervention and the comparison group did not. Thus, a randomly assigned comparison group can provide a good estimate of the **counterfactual**, the outcome that would have occurred if the subjects had not been exposed to the treatment but had otherwise equal experiences (Mohr 1992:3; Rossi & Freeman 1989:229). If the comparison group differed from the experimental group in any way besides not receiving the treatment (or receiving a different treatment), a researcher would not be able to determine the unique effects of the treatment.

Assigning subjects randomly to the experimental and comparison groups ensures that systematic bias does not affect the assignment of subjects to groups. But of course random assignment cannot guarantee that the groups are perfectly identical at the start of the experiment. Similar to random selection in sampling, random assignment removes bias from the assignment process but only by relying on chance, which itself can result in some intergroup differences. Fortunately, researchers can use statistical methods to determine the odds of ending up with groups that differ on the basis of chance alone; these odds are low even for groups of moderate size.

Although similar in procedure, it is important to note here that random assignment of subjects to experimental and comparison groups is not the same as random sampling of individuals from a larger population (see Exhibit 6.1). In fact, random assignment (randomization) does not help to ensure that the research subjects are representative of some larger population; that is the goal of random sampling. What random assignment does

EXHIBIT 6.1 Random Sampling Versus Random Assignment

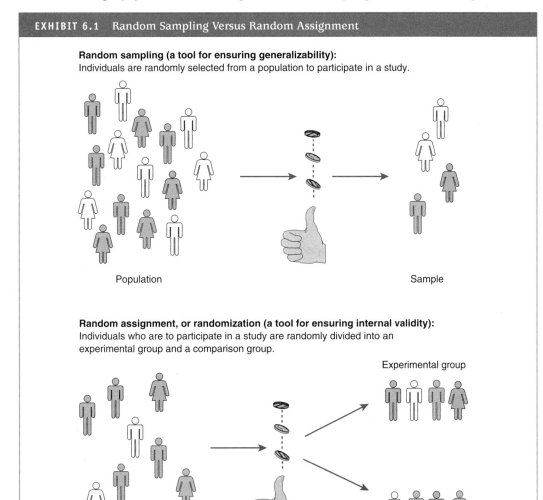

Random sampling (a tool for ensuring generalizability):
Individuals are randomly selected from a population to participate in a study.

Population

Sample

Random assignment, or randomization (a tool for ensuring internal validity):
Individuals who are to participate in a study are randomly divided into an experimental group and a comparison group.

Experimental group

Study participants

Comparison group

by creating two or more equivalent groups is useful for ensuring internal validity, not generalizability.

Random assignment does share with random sampling the use of a chance selection method. In random assignment, a random procedure is used to determine the group into which each subject is placed. In random sampling, a random procedure is used to determine which cases are selected for the sample. The random procedure of tossing a coin, using a random number table, or generating random numbers with a computer can be similar in both random assignment and random sampling.

Matching is another procedure sometimes used to equate experimental and comparison groups, but by itself it is a poor substitute for randomization. Matching of individuals in a treatment group with those in a comparison group might involve pairing persons on the basis of similarity of gender, age, year in school, or some other characteristic. The basic problem is that, as a practical matter, individuals can be matched on only a few characteristics; unmatched differences between the experimental and comparison groups may still influence outcomes. However, matching combined with randomization can reduce the possibility of differences due to chance. For example, if individuals are matched by gender and age, then the members of each matched pair are randomly assigned to the experimental and comparison groups and the possibility of differences due to chance in the gender and age composition of the groups is eliminated (see Exhibit 6.2). Matching is also used in some quasi-experimental designs when randomization is not possible, as you will see later in this chapter.

We have already discussed two true experimental designs in this text: the Bushman (1995) experiment on media violence and the Sherman and Berk (1984) experiment on the

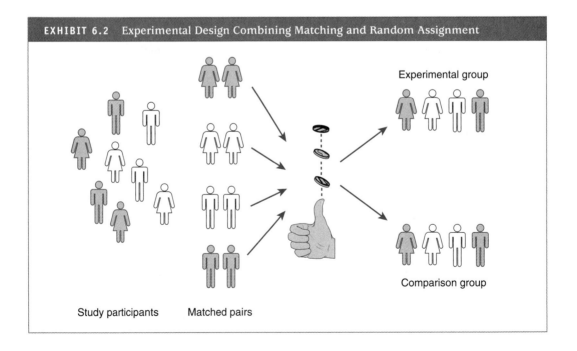

EXHIBIT 6.2 Experimental Design Combining Matching and Random Assignment

Study participants   Matched pairs

Experimental group

Comparison group

effects of arrest on intimate partner violence. Both are graphically displayed for you now in Exhibit 6.3. As you can see, in the Bushman experiment of media violence and aggression, the experimental design included a pretest. In this experiment, undergraduate college students were randomly assigned to experimental and control groups.

Recall from the last chapter that subjects in both groups were required to perform a learning task with a fake partner. In the task, subjects were to administer noises as a form of punishment to their partners when their partners gave an incorrect answer. The noise level was the operationalization of aggression for the experiment. The louder the noises administered, the higher the aggression score. Both experimental and control group subjects thus received a pretest score on their aggressiveness.

Subjects in both groups then watched a videotape; experimental group subjects watched a violent video, whereas control group subjects watched a nonviolent video. After viewing of the tapes, aggression levels were then similarly measured in the posttests. Using this design, Bushman (1995) was not only able to measure differences in aggression after the treatment (viewing the videotapes), but he was also able to determine whether aggression levels for the experimental and control subjects differentially changed from pretest to posttest. This illustrates the advantage of utilizing pretest measures in an experimental design.

The bottom panel of Exhibit 6.3 depicts the Sherman and Berk (1984) experiment. The subjects for the Sherman and Berk experiment were all reported offenders of misdemeanor intimate-partner assault in Minneapolis during the time of the experiment. The experimental treatment was arrest and the control treatment was separation and no arrest. The key to this experimental design was that incidents were randomly assigned to the arrest and nonarrest conditions. That is, when an officer arrived at the scene, he or she randomly assigned offenders to the treatment based on a randomly color-coded police record sheet. For example, if they arrived at the scene and the next sheet in their book was pink, they would make an arrest; if the sheet was white, they would not make an arrest. The posttest measurement was recidivism. If those offenders who were arrested were less likely than nonarrested offenders to have assaulted their partners 6 months after the incident, this decreased rate of assault could be attributed to the treatment of arrest. This, of course, is what happened.

Throughout the remainder of this chapter, you will learn more about each of these key features of experimental design. To help you understand the variability across different experimental designs, we will incorporate the use of simple diagrams. These diagrams also show at a glance just how well suited any experiment is to identifying causal relations, by indicating whether it has a comparison group, a pretest and a posttest, and random assignment. We will now turn to the question that we asked at the beginning of the chapter, "Does the type of security classification inmates are given affect their behavior?"

Prison Classification and Inmate Behavior

There is wide variability in the criteria used to classify prisoners across States. Regardless of how these classifications are made, once these labels are assigned, they have the effect that all labels have: they attach various expectations to prisoners. Bench and Allen (2003) state,

that people residing in the southern region of the United States are more violent than those residing in the north. Nisbett and Cohen (1996) examined one hypothesized cause of male violence related to regional differences, namely a male's reputation for strength and toughness. What does this have to do with geographic regions? Well, along with other researchers, Nisbett and Cohen contend that the South has developed what has been termed a culture of honor. One property of this culture is the necessity for men to appear strong and unwilling to tolerate an insult. "[An individual] must constantly be on guard against affronts that could be construed by others as disrespect. When someone allows himself to be insulted, he risks giving the impression that he lacks the strength to protect what is his" (p. xv).

Nisbett and Cohen (1996) developed an experiment that would determine whether southerners became more upset by insults and affronts and therefore would be more likely to take aggressive action than would northerners. In this experiment, one group of southern students and one group of northern students were brought to a laboratory and told they would be participating in a study concerning the effect of response-time constraints on judgment. After an initial introduction to the study, the subjects were asked to fill out a brief demographic questionnaire and take it to a table at the end of a long, narrow hallway. Subjects from both southern and northern groups were randomly assigned to a control and an experimental condition. In the control condition, nothing happened as the subject returned the questionnaire. However, in the experimental condition, as the subject walked down the hall, a confederate of the experimenter walked out of a door marked "Photo Lab" and began working at a file cabinet in the hall. The confederate had to push the file drawer in to allow the subject to pass by him and drop his paper off at the table. As the subject returned seconds later and walked down the hall toward the experiment room, the confederate, who had just reopened the file drawer, slammed it shut on seeing the subject approach, bumped into him with his shoulder, and called the subject an asshole. The confederate then walked back into the photo lab. Exhibit 6.5 illustrates this experiment.

Two observers, who appeared to be working on homework, were stationed in the hall to monitor the subjects' verbal, nonverbal, and/or physical response to the insult. When the subjects returned to the laboratory they were asked to conduct several tasks, including a word completion task in which they were given a string of letters (e.g., __ight or gu__) that they could complete in either a hostile way (fight or gun) or a nonhostile way (light or gum). The subjects were also asked to do a scenario completion task in which the subjects were required to fill in the beginning or ending of a story. One story was neutral, in which a man was rescued by an ambulance; the other story was presented as clearly confrontational, in which a woman at a party is telling her boyfriend that another man has made several passes at her and even attempted to kiss her.

Ratings by the staged observers revealed that for northerners, the overwhelmingly dominant emotional reaction to the insult was to show more amusement than anger. Conversely, the overwhelmingly dominant reaction for southerners was to show as much or more anger than amusement. For the word-completion task and story completion of the neutral scenario (involving the man and the ambulance), Nisbett and Cohen (1996) found no differences between northerners and southerners in their hostile responses. However, for the scenario describing the attempted pass at the girlfriend, the southerners who were insulted in the hall were much more likely to end this scenario with violence than were northerners who were also insulted.

EXHIBIT 6.5   Experimental and Control Groups in One of Nisbett and Cohen's (1996)
Experiments on Hostile Responses to an Insult

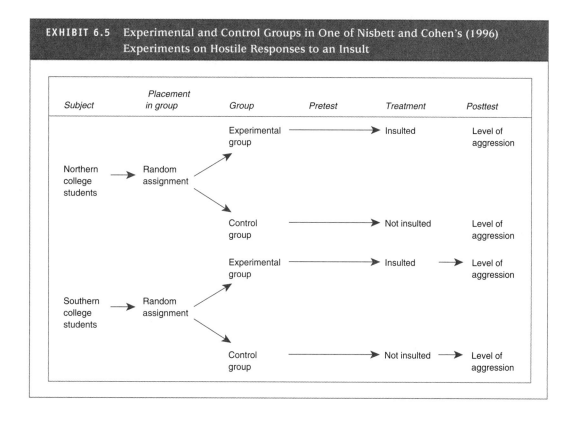

The four experiments we have discussed so far in this text, the Sherman and Berk (1984) experiment on arrest and domestic assault, the Bushman (1995) experiment on media violence, the Bench and Allen (2003) experiment on inmate classification, and the Nisbett and Cohen (1996) experiment on reactions to insults are all true experiments. Take a minute now and think about how these studies meet the criteria necessary for determining causality that were discussed in the last chapter (see Chapter 5). In the next section, we discuss quasi-experimental designs, which are experiments in which at least one of the conditions of a true experiment are not met.

QUASI-EXPERIMENTS

Testing a hypothesis with a true experimental design often is not feasible with the desired subjects and in the desired setting. Such a test may be too costly, take too long to carry out, be inappropriate for the particular research problem, or presume ability to manipulate an intervention that already has occurred. To overcome these problems, yet still benefit from the logic of the experimental method, researchers may use designs that retain several components of experimental design but differ in other important details.

Usually the best alternative to an experimental design is a **quasi-experimental design**, maximizing internal validity. Although not defined consistently by all experts, in a quasi-experimental design the comparison group is predetermined to be comparable to the treatment group in critical ways, such as eligibility for the same services or attendance at the same school cohort (Rossi & Freeman 1989:313). However, because subjects are not randomly assigned to the comparison and experimental groups, these research designs are "quasi"-experimental. In other quasi-experimental designs, there may be just one group that is monitored before and after a treatment is delivered and no comparison group. In both cases we cannot be as confident in their causal conclusions as we can with true experimental designs.

The following are the three major types of quasi-experimental designs (others can be found in Cook & Campbell 1979; Mohr 1992):

- **Nonequivalent control group designs** have experimental and comparison groups designated before the treatment occurs and are not created by random assignment.
- **Before-and-after designs** have a pretest and posttest but no comparison group. In other words, the subjects exposed to the treatment served, at an earlier time, as their own controls.
- **Ex post facto control group designs**, like nonequivalent control group designs, have experimental and comparison groups that are not created by random assignment. But unlike the groups in nonequivalent control group designs, the groups in ex post facto designs are designated after the treatment or intervention has occurred. For this reason, some researchers consider this design to be nonexperimental, not even quasi-experimental.

Nonequivalent Control Group Designs

In this type of quasi-experimental design, a comparison group is selected to be as comparable as possible to the treatment group. Two selection methods can be used:

Individual matching. Individual cases in the treatment group are matched with similar individuals in the comparison group. A Mexican American male about 20 years old who is assigned to the treatment group may be matched with another Mexican American male about 20 years old who is assigned to the comparison group. The problem with this method is determining in advance which variables should be used for matching. It is also unlikely that a match can be found for all cases. However, in some situations matching can create a comparison group that is very similar to the experimental group. For example, some studies of the effect of Head Start, the government program that prepares disadvantaged toddlers for school, used participants' siblings as the comparison group. Therefore, they were similar to the experimental group.

Aggregate matching. A comparison group is identified that matches the treatment group in the aggregate rather than trying to match individual cases. In most situations when random assignment is not possible, this method of matching makes sense. Matching in the aggregate means finding a group with similar distributions on key variables such as the same average age, the same percentage of females, and so on. For this design to be considered even quasi-experimental, individuals cannot choose which group to join or where to seek services. In other words, the subjects cannot opt for or against the experimental treatment.

Roy Watson's (1986) study of the deterrent effect of police action on violations of a seat-belt law illustrates a quasi-experimental nonequivalent control group design. Watson's quasi-experimental design is diagrammed in Exhibit 6.6. This study used aggregate matching and a pretest. It was conducted in British Columbia, Canada, where a mandatory seat-belt law had been enacted but had not elicited high rates of compliance. Watson selected two communities of comparable size where police enforcement of the seat-belt law was low. The units of analysis in the study were the drivers, and seat-belt usage was the dependent variable.

In a pretest, Watson (1986) measured seat-belt usage in both communities. Then, in the experimental community, he instituted a media campaign to increase seat-belt usage, followed by increased police enforcement of the seat-belt law. A posttest followed both the media campaign and the increased police enforcement. In the comparison community, one posttest was conducted at about the same time as a final posttest was conducted in the experimental community. Because the two communities were geographically distant and in different media markets, it seemed unlikely that the experiment in one community would have affected the other community while the study was in progress. As Exhibit 6.7 shows, the experimental program had a marked effect on seat-belt use.

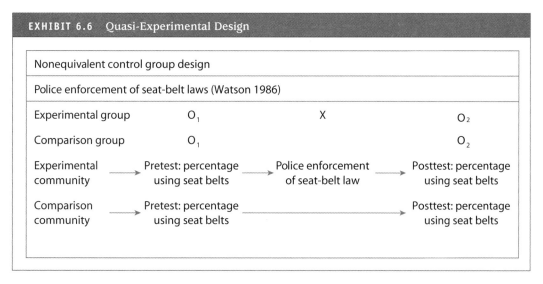

EXHIBIT 6.6 Quasi-Experimental Design

Nonequivalent control group design

Police enforcement of seat-belt laws (Watson 1986)

Experimental group	O_1	X	O_2
Comparison group	O_1		O_2

Experimental community	→ Pretest: percentage using seat belts	→ Police enforcement of seat-belt law	→ Posttest: percentage using seat belts
Comparison community	→ Pretest: percentage using seat belts		Posttest: percentage using seat belts

Note: O = Observation (pretest or posttest); X = Experimental treatment.

EXHIBIT 6.7   Driver Use of Seat Belts in Two Communities

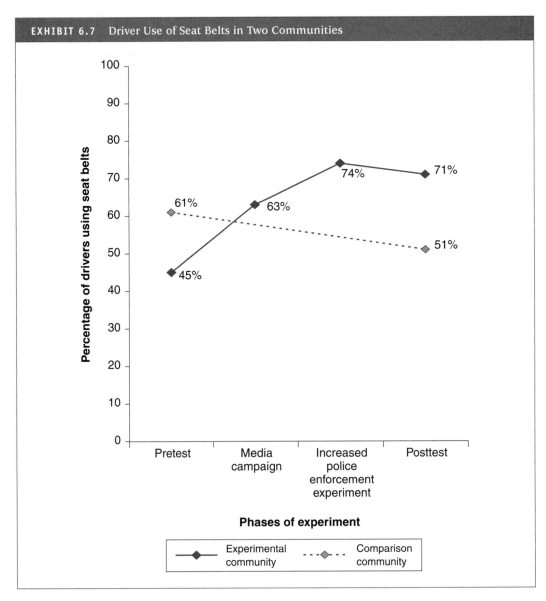

Source: Watson 1986.

Before-and-After Designs

The common feature of before-and-after designs is the absence of a comparison group. All cases are exposed to only the experiment treatment. The basis for comparison is provided by the pretreatment measures in the experimental group. These designs are useful for studies of interventions that are experienced by virtually every element in some population.

The simplest type of before-and-after design is the fixed-sample panel design. A panel design involves only one pretest and one posttest. It does not qualify as a quasi-experimental design because comparing subjects to themselves at only one earlier point in time does not provide an adequate comparison group. Many influences other than the experimental treatment may affect a subject following the pretest. Consequently, if there is a change in the dependent variable from pre- to post-measures, we cannot be certain that the treatment itself is the sole cause of this change.

Horney and Spohn's (1991) study of the effect of rape law reforms on the number of actual rapes reported to police and processed in six urban jurisdictions illustrates a more powerful before-and-after design sometimes called a **time series design** or a **repeated measures panel design**. This design typically includes many pretest and posttest observations that allow the researcher to study the process by which an intervention or treatment has an impact over time. Horney and Spohn identified six urban jurisdictions that implemented some type of rape reform legislation: Detroit, Chicago, Philadelphia, Atlanta, Houston, and Washington, DC. The overall purpose of the reforms was to treat rape like other crimes, focusing not on the behavior and reputation of the victim but on the unlawful acts of the offender. Advocates of the reformed laws anticipated that by improving the treatment of rape victims, the reforms would lead to an increase in the number of reports of rape to police. Advocates also expected the reforms to remove legal barriers to effective prosecution and would make arrest, prosecution, and conviction for rape more likely. Horney and Spohn tracked several dependent variables thought to measure the success of these reforms, including reports of rape to police. These data were monitored for several years before and after reforms were implemented in each jurisdiction. An example of this design for one jurisdiction is presented in Exhibit 6.8.

Horney and Spohn (1991) performed experiments for three different variables for each of the six jurisdictions. As you can see, this design allows researchers to determine the magnitude of change in a given variable before and after an intervention or event (in this case rape law reform) has taken place. It appears from Exhibit 6.8 that reports of rape to police did increase in the post-reform time period as compared to the time period before rape reform legislation was implemented. Of course, special statistics are required to analyze time series data. The intent is to identify a trend in the dependent variable up to the date of the intervention or event whose effect is being studied and then to project the trend into the postintervention period. This projected trend is then compared to the actual trend of the dependent variable after the intervention. A substantial disparity between the actual trend and the projected trend is evidence that the intervention or event had an impact (Rossi & Freeman 1989:260–261, 358–363). This time series design is particularly useful for studying the impact of new laws or social programs that affect everyone and can be readily assessed by ongoing measurement.

Ex Post Facto Control Group Designs

A design in which the treatment and comparison groups are designated after the treatment is administered is an ex post facto (after the fact) control group design. This design perhaps should be considered nonexperimental rather than quasi-experimental, because the comparison group may not be comparable to the treatment group. Carefully designed ex post facto studies, however, can result in comparisons between the treatment and control

EXHIBIT 6.8   A Repeated Measures Panel Design Used to Examine the Efficacy
of Rape Law Reforms

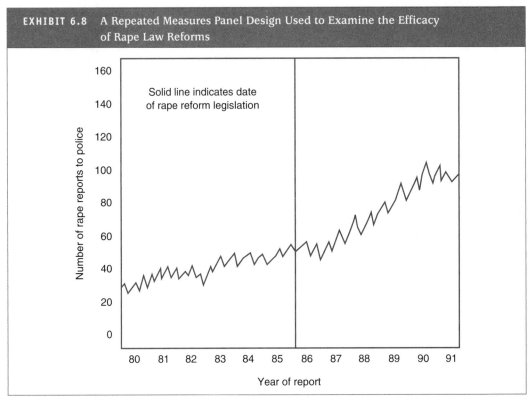

Source: Horney & Spohn 1991.

groups that give us almost as much confidence in the validity of their causal conclusions as we can have in the causal conclusions from a quasi-experimental design (Rossi & Freeman 1989:343–344).

Farrington's (1977) classic study of the deviance amplification process is an excellent example of an ex post facto control group design. Farrington was interested in testing the main hypothesis about deviance amplification from labeling theory. Becker (1963:31) describes deviance amplification thus: "One of the most crucial steps in the process of building a stable pattern of deviant behavior is likely to be the experience of being caught and publicly labeled as deviant." Accordingly, Farrington tested the hypothesis that individuals who were publicly labeled as deviant would increase their deviant behavior compared to those who were not so labeled. To measure the process of being publicly labeled, he used the criteria of whether individuals had been found guilty of a delinquent act in court.

To use a true experimental design to test this hypothesis, individuals who engaged in criminal behavior would have to be randomly assigned to either a group found guilty in court or a group not taken to court. It would then be possible to conclude that any subsequent differences between the groups resulted from the presence or absence of public labeling (being convicted), and did not reflect pre-existing differences or the influence of any

EXHIBIT 6.9   Ex Post Facto Control Group Design: Farrington's (1977) Test of Deviance Amplification

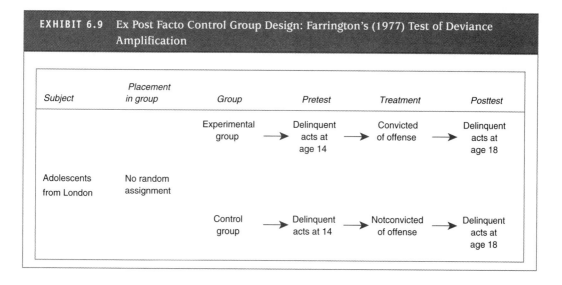

Subject	Placement in group	Group	Pretest	Treatment	Posttest
		Experimental group →	Delinquent acts at age 14 →	Convicted of offense →	Delinquent acts at age 18
Adolescents from London	No random assignment				
		Control group →	Delinquent acts at 14 →	Notconvicted of offense →	Delinquent acts at age 18

other factor. Because of ethical considerations, of course, such an experiment is difficult to carry out.

Instead, Farrington (1977) used the Cambridge Study of Delinquent Development that interviewed adolescents from London at three different ages: 14, 16, and 18. (The Cambridge Study in Delinquent Development actually maintained intermittent contact with 400 London working-class youths from ages 8 to 18. In this study, only the interviews at ages 14, 16, and 18 were used). At all three times, the youth were asked about their participation in delinquency. For example, the questionnaire given at age 18 included seven delinquent acts: breaking and entering, stealing from vehicles, stealing from slot machines, shoplifting, joy-riding, damaging property, and buying stolen goods. The dependent variable, then, was the youth's self-reported delinquency score. The treatment or intervention was whether the youth was convicted of an offense by the Criminal Record Office. Pretest measures of self-reported delinquency were taken at age 14, and posttest scores were taken at age 18. Individuals convicted of an offense between the times of these interviews were considered the experimental publicly labeled group. Individuals who were not convicted were considered the control group (see Exhibit 6.9).

Farrington (1977) actually conducted many hypothesis tests with these data, trying to rule out the effects of other variables and to investigate the intervening processes or causal mechanisms linking public labeling and deviant acts. In conclusion, Farrington appropriately included a caveat about his methodology along with directions for future research:

> The results obtained with the quasi-experimental matching design, although less conclusive than would have been the case in a true experiment, were in agreement with the hypothesis that public labeling (measured by criminal convictions) increases deviant behavior (measured by self-reported delinquency scores). As a next step, it would be useful to establish the generality of this finding by investigating whether deviance amplification still occurred with different operational definitions of public labeling and deviant behavior, and with different samples of people in different places and at different times. (P. 123).

VALIDITY IN EXPERIMENTS

Like any research design, experimental designs must be evaluated for their ability to yield valid conclusions. True experiments are particularly well suited to producing valid conclusions about causality (internal validity), but they are less likely to fare well in achieving generalizability. Quasi-experiments may provide more generalizable results than true experiments, but they are more prone to problems of internal invalidity (although some allow the researcher to rule out almost as many potential sources of internal invalidity as does a true experiment). Measurement validity is also a central concern, but an experimental design does not in itself offer any special tools or particular advantages or disadvantages in measurement. In this section you will learn more about the ways in which experiments help (or do not help) to resolve potential problems of internal validity and generalizability.

Causal (Internal) Validity

An experiment's ability to yield valid conclusions about causal effects is determined by the comparability of its experimental and comparison groups. First, of course, a comparison group must be created. Second, this comparison group must be so similar to the experimental group that it will show what the experimental group would be like if it did not receive the experimental treatment, if the independent variable was not varied. For example, the only difference between the two groups in Bushman's (1995) study was that one group watched a violent movie and the other group did not.

There are five basic sources of internal invalidity.

1. *Selection bias*. When characteristics of the experimental and comparison group subjects differ.
2. *Endogenous change*. When the subjects develop or change during the experiment as part of an ongoing process independent of the experimental treatment.
3. *External events or history effects*. When something occurs during the experiment, other than the treatment, that influences outcome scores.
4. *Contamination*. When either the experimental group of the comparison group is aware of the other group and is influenced in the posttest as a result (Mohr 1992).
5. *Treatment misidentification*. When variation in the independent variable (the treatment) is associated with variation in the observed outcome, but the change occurs through a process that the researcher has not identified.

Selection Bias

You may already realize that the composition of the experimental and comparison groups in a true experiment is unlikely to be affected by their difference. If it were affected, it would cause **selection bias**. Random assignment equates the groups' characteristics, though with some possibility for error due to chance. The likelihood of difference due to chance can be identified with appropriate statistics.

But in field experiments, what is planned as a random assignment process may deteriorate when it is delegated to front-line program staff. The staff may not always follow the process guidelines. This problem occurred in the Sherman and Berk (1984) domestic violence experiment in Minneapolis. Police officers sometimes violated the random assignment plan because they thought the circumstances warranted arresting a suspect who had been randomly assigned to receive only a warning. In several of the follow-up studies, the researchers maintained closer supervision over the assignment process so the randomization could be maintained.

Even when the random assignment plan works, the groups can differ over time because of **differential attrition**, or what can be thought of as deselection. That is, the groups become different because for various reasons subjects are more likely to drop out of groups. This is not a likely problem for a laboratory experiment that occurs in one session, but for laboratory experiments that occur over time, differential attrition may become a problem.

To test your understanding of internal validity, go to the Sources of Internal Invalidity Interactive Exercises on the Student Study Site.

Differential attrition is quite common in field experiments that evaluate the impact of social programs (Cook & Campbell 1979:359–366). Subjects who receive some advantageous program benefit are more likely to stay in the experiment (making themselves available for measurement in the posttest); subjects who do no receive program benefits are more likely to drop out. Another possibility is that individuals who may be more in need of a program, such as severe alcoholics in a detoxification study, are more likely to drop out (Rossi & Freeman 1989:236). The problem this creates is that the two groups are no longer equivalent, despite the fact that subjects were originally randomly assigned.

In evaluation research that monitors program impact over an extended period, the possibility of differential attrition can be high. When the treatment group is receiving some service or benefit that the control group is not, the individuals in the control group will be more likely to drop out of the study.

When subjects are not randomly assigned to treatment and comparison groups, as in nonequivalent control group designs, there is a serious threat of selection bias. Even if the researcher selects a comparison group that matches the treatment group on important variables, there is no guarantee that the groups were similar initially in terms of the dependent variable or another characteristic that ultimately influences posttest scores. However, a pretest helps the researchers to determine whether selection bias exists and to control for it. Most variables that might influence outcome scores will also influence scores on the pretest. Statistically controlling for the pretest scores also serves to control for unmeasured variables that influence the pretest scores.

Endogenous Change

The type of problem considered an **endogenous change** occurs when natural developments in the subjects, independent of the experimental treatment, account for some or all the observed change between pretest and posttest. Endogenous change includes three specific threats to internal validity:

Testing. Taking the pretest can in itself influence posttest scores. Subjects may learn something or be sensitized to an issue by the pretest and as a result respond differently when they are asked the same questions in the posttest.

Maturation. Changes in outcomes scores during experiments that involve a lengthy treatment period may be due to maturation. Subjects may age, or gain experience in school, or grow in knowledge, all as part of a natural maturational experience, and thus respond differently on the posttest from the way they responded on the pretest.

Regression. People experience cyclical or episodic changes that result in different posttest scores, a phenomenon known as a **regression effect**. Subjects who are chosen for a study because they received very low scores on a test may show improvement in the posttest, on average, simply because some of the low scorers were having a bad day. On the other hand, individuals selected for an experiment because they are suffering from tooth decay will not show improvement in the posttest because a decaying tooth is not likely to improve in the natural course of things. It is hard, in many cases, to know whether a phenomenon is subject to naturally occurring fluctuations, so the possibility of regression effects should be considered whenever subjects are selected because of their extremely high or low values on the outcome variable (Mohr 1992:56, 71–79).

Testing, maturation, and regression effects are generally not a problem in true experiments. Both the experimental group and the comparison group take the pretest, so even if this experience itself leads to a change in posttest scores, the comparison between the experimental and comparison groups will not be affected. Of course, in experiments with no pretest, testing effects are not a problem. Similarly, both the experimental and comparison groups are equally subject to maturation and regression, since they are the same at the outset (unless chance factors led to a difference). These endogenous changes may affect outcome scores for both groups, but both groups should be equally affected. As a result, the estimates of treatment effects should not be biased.

Endogenous change is, however, a major problem with quasi-experimental designs, particularly before-and-after designs. In panel designs with a single pretest and a single posttest, any change may be due to testing, maturation, or regression. But a study that is completed within a short time would have little concern with maturation effects, and use of a diverse subject group will reassure those concerned about regression effects. Unfortunately, the possibility of endogenous change accounting for pretest-posttest change cannot be eliminated with before-and-after designs. Repeated-measures panel studies and time series designs are more effective because they allow the researcher to trace the pattern of change or stability in the dependent variable up to and after the treatment. Ongoing effects of maturation and regression can thus be identified and taken into account.

External Events

External events, sometimes referred to as the **history effect** during the experiment—things that happen outside the experiment—can also change the subjects' outcome scores. An example of this is a newsworthy event that has to do with the focus of an experiment and the major disasters to which subjects are exposed. What if researchers were evaluating the effectiveness of a mandatory arrest policy in decreasing incidents of intimate partner assault, and an event such as the murder trial of O. J. Simpson occurred during the experiment? This would clearly be a historical event that might compromise the results. The media were obsessed with the trial of O. J. Simpson, and as a result, intimate partner assault

To test your understanding of internal validity, go to the Sources of Internal Invalidity Interactive Exercises on the Student Study Site.

and homicide were given a tremendous amount of media attention. Because of this increased awareness, many victims of intimate partner violence reported their victimizations during this time; police agencies and women's shelters were flooded with calls. If a researcher had been using calls to police in a particular jurisdiction as an indicator of the incidence of intimate partner assault, this historical event would have seriously jeopardized the internal validity of his or her results. Why? Because the increase in police calls would have had more to do with the trial than with any recent change in arrest policies.

Causal conclusions can be invalid in true experiments because of the influence of external events, but not every experiment is affected by them. The specific features of an experiment must be considered carefully to evaluate the possibility of problems due to external events. For example, in an experiment in which subjects go to a particular location for the treatment and the control group subjects do not, something at the location unrelated to the treatment could influence the experimental subjects.

The more carefully controlled the conditions for experimental and comparison groups, the less likely external events will invalidate the causal conclusions of an experiment. External events seldom affect laboratory experiments. In addition, when using nonequivalent control group designs, a comparison group can minimize the effect of external events if the two groups are exposed to the same environment during the experiment. For example, the relatively constant level of seat-belt use in the comparison community in the Watson (1986) study suggests that a national series of TV announcements to encourage compliance is not what resulted in the changes in seat-belt use observed in the experimental community (p. 294).

Before-and-after designs are much less prone to problems of external events, and become quasi-experimental when the researchers make multiple before-and-after comparisons or when they apply multiple pretest and posttest measures to the same group in successive time intervals. If **multiple-group before-and-after designs** yield comparable results, the observed pretest-posttest change is less likely to be due to unique external events that would have affected only some of the comparisons.

Contamination

Contamination occurs in an experiment when the comparison group is in some way affected by, or affects, the treatment group. This problem basically arises from failure to adequately control the conditions of the experiment. When comparison group members become aware that they are being denied some advantage, they may increase their efforts to compensate, creating a problem called **compensatory rivalry**, or the **John Henry effect** (Cook & Campbell 1979:55). (Saretsky, 1972, coined the term *John Henry effect* in honor of the steel driver who, when he learned that his output was to be compared to that of a steam drill, worked so hard that he outperformed the drill and died of overexertion.)

On the other hand, comparison group members may become demoralized if they feel that they have been left out of some valuable treatment and may perform worse than they would have outside the experiment. The treatment may seem, in comparison, to have a more beneficial effect than it actually did. Both compensatory rivalry and demoralization thus distort the impact of the experimental treatment. Similar problems could occur if members of the experimental group become aware of the control group and feel differently as a result.

Although the components of a true experiment guard against internal invalidity due to history and selection, the threat of contamination is always present. Careful inspection of the research design can determine whether contamination is likely to be a problem in a particular experiment. If the experiment is conducted in a laboratory, if members of the experimental group and the comparison group have no contact while the study is in progress, and if the treatment is relatively brief, contamination is not likely to be a problem. To the degree that these conditions are not met, the likelihood of contamination will increase.

The threat of contamination is not necessarily any different in a quasi-experimental design than it is in a true experiment. Some nonequivalent control group designs minimize the threat of contamination by using a comparison group whose members will have no contact with the treatment group or awareness of the treatment. For example, Watson (1986) chose two communities for his seat-belt compliance study that were far enough apart to be in different media markets.

Treatment Misidentification

Treatment misidentification occurs when the treatment itself does not cause the outcome but does cause some intervening process the researcher is not aware of and has not identified. This term can also refer to unknown concomitants or consequences for subjects in the control group that have to do with the experimental manipulation. In either case, the subjects experienced something other than, or in addition to, what the researchers believed they would experience. Treatment misidentification has at least three sources:

1. *Expectancies of experimental staff.* Change among experimental subjects may be due to the positive **expectancies of the experimental staff** who are delivering the treatment rather than due to the treatment itself. This type of treatment misidentification can occur even in randomized experiments when well-trained staff convey their enthusiasm for an experimental program to the subjects in subtle ways. Such positive staff expectations create a self-fulfilling prophecy. Because social programs are delivered by human beings, such expectancy effects can be very difficult to control in field experiments. However, in some experiments on the effects of treatments such as medical drugs, **double-blind procedures** can be used. Staff will deliver the treatments without knowing which subjects are getting the treatment and which are receiving a placebo, something that looks like the treatment but has no effect. These procedures can be used in other experiments as well. In fact, the prison experiment discussed at the beginning of this chapter used a double-blind procedure to randomly assign inmates. In the experiment, only the executive director of the Corrections Department and the director of classification were aware of the research. Correctional staff, other individuals who worked with the inmates, and the inmates themselves were unaware of the study. In this way, any expectancies that the staff may have had were unlikely to affect inmate behavior.

2. *Placebo effect.* Treatment misidentification may occur when subjects receive a treatment that they consider likely to be beneficial, then improve because of that expectation rather than the treatment itself. In medical research the placebo is

often a chemically inert substance that looks like the experimental drug but actually has no medical effect. Research indicates that the **placebo effect** produces positive health effects in two-thirds of patients suffering from relatively mild medical problems (Goleman 1993a:C3). Placebo effects can also occur in social science research. The only way to reduce this threat to internal validity is to treat the comparison group with something similar.

3. *Hawthorne effect*. Members of the treatment group may change in terms of the dependent variable because their participation in the study makes them feel special. This problem can occur when treatment group members compare their situation to that of the control group members, who are not receiving the treatment. In this case, this is a type of contamination effect. But experimental group members could feel special simply because they are in the experiment. The **Hawthorne effect** is named after a famous productivity experiment at the Hawthorne electric plant outside Chicago. Workers were moved to a special room for a study of the effects of lighting intensity and other work conditions on their productivity. After this move, the workers began to increase their output no matter what change was made in their working conditions, even when the conditions became worse. The researchers concluded that the workers felt they should work harder because they were part of a special experiment. By the time the study was over, the researchers were convinced that it was social interaction among workers, not physical arrangements, that largely determined their output. Most management historians believe that the human relations school of management began with this insight.

In a **process analysis**, periodic measures are taken throughout the experiment to assess whether the treatment is being delivered as planned. To avoid treatment misidentification, field researchers are now using process analysis more often than in the past. Process analysis is also a necessary component of evaluation studies, when researchers need to be very confident that the effects they observe are caused by the policy or program change they are studying (Hunt 1985:272–274).

Generalizability

The need for generalizable findings can be thought of as the Achilles heel of the true experimental design. The design components that are essential for a true experiment and minimize the threats to causal (internal) validity make it more difficult to achieve sample generalizability, the ability to apply the findings to a clearly defined, larger population. In contrast, cross-population generalizability is no more or less achievable with experiments than with other research designs. As you learned in Chapter 1, the extent to which treatment effects can be generalized across subgroups and to other populations and settings is external validity.

Sample Generalizability

Subjects who can be recruited for a laboratory experiment, randomly assigned to a group, and kept under carefully controlled conditions for the study's duration are unlikely to be a representative sample of any large population of interest to social scientists. In fact, most are recruited from college populations. Can they be expected to react to the experimental treatment in the same way as members of the larger population? The more artificial the experimental arrangements, the greater the problem (Campbell & Stanley 1966:20–21). As you learned in Chapter 4, randomly selecting a sample from a population is the best way to ensure sample generalizability (this has nothing to do with randomly assigning the sample to the control or experimental groups). A large percentage of laboratory experiments, including those performed by Nisbett and Cohen (1996), utilize availability samples such as college students. As Nisbett and Cohen clearly state,

> When we refer to northerners and southerners who participated in the experiments, we use those words as a shorthand way of referring to male students from the North and the South attending the University of Michigan who are Caucasian, non-Hispanic, and non-Jewish. Our sample is certainly not a representative one—even of Caucasian, non-Hispanic, non-Jewish males. On average, the students came from families that were financially well-off, with the median income for northerners about $85,000 and for southerners, $95,000. The southerners might also be unusual in that they chose to leave the South at least temporarily and come to school in the North. (Pp. 41–42)

It is not only the characteristics of the subjects themselves that determine the generalizability of the experimental results. The generalizability of the treatment and of the setting for the experiment also must be considered (Cook & Campbell 1979:73–74). However, researchers can take steps both before and after an experiment to increase a study's generalizability. Field experiments are likely to yield findings that are more generalizable to broader populations than are laboratory experiments using subjects who must volunteer. Sherman and Berk's (1984) experimental study of arrest and domestic violence and Watson's (1986) quasi-experimental study of seat-belt use were both conducted with people and police engaged in their normal activities in real communities, greatly increasing our confidence in the studies' generalizability.

In a few field experiments, participants can be randomly selected from the population of interest, so the researchers can achieve results generalizable to that population. The Sherman and Berk (1984) arrest study did not use random selection. Because it was to include all actual domestic assault cases processed by police in two Minneapolis precincts during a certain period, the resulting sample of cases should represent at least the two precincts.

But in most experiments, neither random selection from the population nor selection of the entire population is possible. Potential subjects must make a conscious decision to participate. This undoubtedly increases the possibility that the resulting sample is an unrepresentative pool of volunteers. Even in the Sherman and Berk (1984) study, many police officers did not participate, so the attempt to include all domestic assault cases that met study criteria did not succeed. The results of the study may not have been generalizable even to the two precincts.

When random selection is not feasible, the researchers may be able to increase generalizability by selecting several sites for conducting the experiments that offer obvious contrasts in the key variables of the population. The follow-up studies to Sherman and Berk's

(1984) work, for example, were conducted in cities that differed from Minneapolis, the original site, in social class and ethnic composition. As a result, although the findings are not statistically generalizable to a larger population, they do give some indication of the study's general applicability (Cook & Campbell 1979:76–77).

Factorial Surveys

Factorial surveys embed the features of true experiments into a survey design in order to maximize generalizability. In the most common type of factorial survey, respondents are asked for their likely responses to one or more vignettes about hypothetical situations. The content of these vignettes is varied randomly among survey respondents so as to create "treatment groups" that differ in terms of particular variables reflected in the vignettes.

Bachman, Paternoster, and Ward (1992) conducted a factorial survey designed to investigate the factors that constrained male college students from committing a sexual assault. Each respondent read 5 different vignettes in which a female was forced to have sex against her will by a male. The experimental component of the survey was created by randomly varying the vignettes along a number of important dimensions including the victim's situation prior to the assault, the victim/offender relationship, the victim's initial response, the offender's response, and harm to the victim (see Exhibit 6.10). After reading each vignette, respondents were asked to estimate the likelihood that they would act as the vignette male acted under the same set of circumstances. In addition, respondents were asked to estimate the likelihood that the victim would report the incident to police, the likelihood that the police would make an arrest if it was reported, how morally wrong respondents thought the act was, the probability of friends and family losing respect for the vignette male if they found out about the incident, and other questions intended to measure factors thought to deter would-be offenders.

Among other things, Bachman, Paternoster, and Ward (1992) found support for rational choice theory; males who thought that the victim would likely report the incident and that police would likely arrest the offender were less likely to say they would do the same thing compared to males who thought the likelihood of formal sanctions was low. However, there is still an important limitation to the generalizability of factorial surveys like this: A factorial survey research design only indicates what respondents say they would do in situations that have been described to them. It is impossible to determine what the males in this sample would actually do in these situations. So factorial surveys do not completely resolve the problems caused by the difficulty of conducting true experiments with representative samples. Nonetheless, by combining some of the advantages of experimental and survey designs, factorial surveys can provide stronger tests of causal hypotheses than surveys and more generalizable findings than experiments conducted under some laboratory conditions.

External Validity

Researchers are often interested in determining whether the treatment effects identified in an experiment hold true for subgroups of subjects or across different populations. Of course, determining that a relationship between the treatment and the outcome variable holds true for certain subgroups does not establish that the same relationship also holds true for these subgroups in the larger population, but it does suggest that the relationship might be **externally valid**.

EXHIBIT 6.10   Conditions Manipulated in a Factorial Survey Design to Examine Sexual Assault

CONSTANT: Lori B. is a 20-year-old female.

DIMENSION I: Victim's Situation Prior to Assault

1. When returning to her apartment from shopping
2. When returning to her apartment from a party
3. When returning to her apartment from a party where she had too much to drink

CONSTANT: She was approached by Tom, a 22-year-old male.

DIMENSION II: Victim/Offender Relationship

1. Who Lori had been dating
2. Who Lori had met through a mutual friend. He accompanied her home.
3. Who Lori did not know. He asked her if he could use her phone, and she let him in.

CONSTANT: After Tom was inside the apartment, he told her that he wanted to have sex with her.

DIMENSION III: Victim's Initial Response

1. She said no and told him to leave, but he didn't.
2. She told him to leave, but he didn't.
3. She said no and tried to push him out of the apartment, but she couldn't.
4. She tried to push him out of the apartment, but she couldn't.

DIMENSION IV: Offender's Response

1. Tom ignored her.
2. Tom argued with her and tried to persuade her.
3. Tom threatened to beat her with his fists if she didn't.
4. Tom threatened to cut her with a knife if she didn't.
5. Tom beat her with his fists.
6. Tom cut her with a knife.

DIMENSION V: Victim's Second Response

1. Lori allowed Tom to start kissing her. She later said no again, but Tom continued and had sexual intercourse with her anyway.
2. Lori allowed Tom to start kissing and fondling her. She later said no again, but Tom continued and had sexual intercourse with her anyway.
3. Lori was too intimidated and frightened to protest, and Tom had sexual intercourse with her.
4. Lori started screaming and crying, but Tom had sexual intercourse with her anyway.
5. Lori started fighting back by hitting and kicking, but Tom had sexual intercourse with her anyway.

DIMENSION VI: Harm to Victim

1. Tom then left.
2. Tom then left. Lori was bruised.
3. Tom then left. Lori had cuts and bruises.
4. Tom then left. Lori required psychological counseling.

Source: Bachman, Paternoster & Ward 1992.

External validity The applicability of a treatment effect (or noneffect) across subgroups within an experiment and/or across different populations, times, or settings.

Example: Sherman et al. (1992) found that arrest reduced repeat offenses for employed subjects but not for unemployed subjects. The effect of arrest thus varied with employment status, so a conclusion that arrest deters recidivism would not be externally valid.

Imagine a field experiment in which subjects were randomly sampled from the population of interest but in which the treatment effect occurred only for some subgroups. Suppose, for example, that you conduct an experiment on the effects of alcohol on violent behavior and find that alcohol increases the probability for males to act violently, but not females. The finding about the overall effect of alcohol, then, would not apply to all groups within the population. In this instance, the external validity of the conclusions would be equivalent to their cross-population generalizability.

Ultimately, the external validity of experimental results will increase with the success of replications taking place at different times and places, using different forms of the treatment. As indicated by the replications of the Sherman and Berk (1984) study of arrest for domestic violence, the result may be a more detailed, nuanced understanding of the hypothesized effect.

Interaction of Testing and Treatment

A variation of the problem of external validity occurs when the experimental treatment is effective only when particular conditions created by the experiment occur. For example, if subjects have had a pretest it may sensitize them to a particular issue, so when they are exposed to the treatment, their reaction is different from what it would have been if they had not taken the pretest. In other words, testing and treatment interact to produce the outcome.

Suppose you were interested in the effects of a diversity training film on prejudicial attitudes. After answering questions in a pretest about their attitudes on various topics related to diversity (e.g., racial or sexual prejudice), the subjects generally became more sensitive to the issue of prejudice without seeing the training film. On the posttest, then, their attitudes may be different from pretest attitudes simply because they have become sensitized to the issue of diversity through pretesting. In this situation, the treatment may actually have an effect, but it would be difficult to determine how much of the effect was attributable to the sensitizing pretest and how much was due to seeing the film.

Researchers sometimes deliberately take advantage of the interaction of testing and treatment to increase external validity. One design for doing so, the **Solomon four-group design**, randomly assigns subjects to at least two experimental groups and at least two comparison groups. One experimental group and one comparison group will have a pretest, and the other two groups will not have a pretest (see Exhibit 6.11). If testing and treatment do interact, the difference in outcome scores between the experimental and comparison groups will differ between subjects who took the pretest and those who did not.

that their behavior was entirely normal and that their feelings of conflict or tension were shared by other participants. To determine long-term trauma, participants were also interviewed after one year. No evidence of lingering trauma was uncovered, and almost all participants said they strongly supported the research.

Reynolds (1979:130–133) makes several additional points about Milgram's classic experiment:

- The knowledge resulting from these experiments could be of substantial benefit to the general public if it were used to reduce conflict in organizations and to encourage individuals to question authority. On the other hand, there was some risk that organizational leaders might use the knowledge to manipulate members.
- The research had no direct major benefits to participants while clearly producing stress for individuals during the experiment and possibly some self-doubt afterward. However, the follow-up interviews indicated that research subjects interpreted the experiment in a way that avoided a harmful level of self-blame.
- The research subjects were not a disadvantaged group, and their rights and welfare were taken adequately into account through careful monitoring of the experiment (it was stopped for subjects in extreme stress), full debriefing, and follow-up interviews.

Overall, Reynolds (1979) suggests, the costs of the research were minor relative to its benefits:

The effects for the individuals cannot be considered benign: they suffered extreme stress. . . . But participants appeared to have received no lasting negative effects. . . . The value of the project can be judged only on the basis of its contribution to knowledge and possible contributions to the improvement of society. (Pp. 132–133)

Today researchers must usually follow more stringent standards set by current federal ethical guidelines and administered by institutional review boards (IRBs). It is unlikely an experiment like Milgram's would pass the requirements of an IRB today. At minimum, the current guidelines would require informed consent to be obtained from the subjects. Informed consent, in turn, would require subjects to be told that another human being would not be harmed in the experiment. And since the experimental design required deceiving subjects about this precise point, informed consent could not have been carried out. Clearly, the need to adhere to ethical standards can conflict with the investigation of important social issues.

Selective Distribution of Benefits

Although all experiments, whether conducted in the laboratory or in the field, can involve issues of informed consent (Hunt 1985:275–276), one ethical issue somewhat unique to field experiments is the **distribution of benefits**. How much are subjects harmed by the way benefits are distributed as part of an experiment? This issue is particularly salient for treatments that may hold some positive benefit for the experimental subjects and deny the same benefits to subjects in the control group. Although this issue poses more ethical dilemmas for medical research investigating the effectiveness of new drugs, it is also important for many social and criminological experiments. For example, Sherman and Berk's (1984) experiment, and its successors, required police to make arrests in domestic violence cases primarily on the basis of a random process. When arrests were not made, did the subjects' abused spouses suffer?

Is it ethical to give some potentially advantageous or disadvantageous treatment to people on a random basis? As a general rule, random distribution of benefits is justified when the researchers do not know whether some treatment actually is beneficial or not and, of course, it is the goal of the experiment to find out. In this case, random assignment is the only way to assess whether one treatment is more beneficial than another or than no treatment at all. Also, if insufficient resources are available to fully fund a benefit for every eligible person, distribution of the benefit on the basis of chance to equally needy persons is ethically defensible.

Let us examine the experiment of Sherman and Berk (1984). Victims who called police to report an incident of misdemeanor assault perpetrated by an intimate during the course of the experiment (and who did not insist on arrest) did not know they were in a research study and had no choice about whether their assailants were arrested or received a warning. Perhaps it seems unreasonable to let a random procedure determine how police resolve cases of domestic violence. And indeed it would be unreasonable if this procedure were regular police practice. The Sherman and Berk experiment and its successors do pass ethical muster, however, when seen for what they were: a way of learning how to increase the effectiveness of police response to this all-too-common crime. The initial Sherman and Berk findings encouraged police departments to make many more arrests for these crimes, and the follow-up studies resulted in a better understanding of when arrests are not likely to be effective. The implications of this research may be complex and difficult to implement, but the research provides a much stronger factual basis for policy development.

CONCLUSION

True experiments play two critical roles in criminological research. First, they are the best research design for testing causal hypotheses. Even when conditions preclude the use of a true experimental design, many research designs can be improved by adding experimental components. Second, true experiments provide a comparison point for evaluating the ability of the other research designs to achieve causally valid results.

In spite of their obvious strengths, true experiments are used infrequently to study many research problems related to criminology and criminal justice. There are three basic reasons for this:

1. The experiments required to test many important hypotheses require more resources than most social scientists can access.

2. Most research problems of interest to social scientists simply are not amenable to experimental designs, for reasons ranging from ethical considerations to the

limited possibilities for randomly assigning people to different conditions in the real world.

3. The requirements of experimental design usually preclude large-scale studies and so limit generalizability to a degree that is unacceptable to many social scientists.

And just because it is possible to test a hypothesis with an experiment, it may not always be desirable to do so. When a program is first developed and its elements are in flux, it is not a good idea to begin a large evaluation study that cannot possibly succeed unless the program design remains constant. Researchers should wait until the program design stabilizes somewhat. It also does not make sense for evaluation researchers to test the impact of programs that cannot actually be implemented or that are unlikely to be implemented in the real world due to financial or political problems (Rossi & Freeman 1989:304–307).

Even laboratory experiments are inadvisable when they do not test the real hypothesis of interest but instead test a limited version amenable to laboratory manipulation. The intersecting complexity of societies, social relationships, and social beings of people and the groups to which they belong is so great that it often defies reduction to the simplicity of a laboratory or restriction to the requirements of experimental design. Yet the virtues of experimental designs indicate they should always be considered when explanatory research is planned.

When a true experimental design is not feasible, researchers may instead use a quasi-experimental design, including nonequivalent control group designs, before-and-after designs, and ex post facto control group designs. These research designs are termed "quasi"-experimental because the units under study are not randomly assigned to the comparison and experimental groups. As a result, we cannot be as confident in the comparability of the groups in these designs. However, there are several steps that can be taken, such as matching certain individual or group characteristics across the two groups that allow us to have more confidence in the results of quasi-experimental research.

KEY TERMS

Association	Event- or cohort-based design
Before-and-after design	Ex post facto control group design
Causal effect	Expectancies of the experimental staff
Ceteris paribus	Experimental group
Cohort	External event
Compensatory rivalry (John Henry effect)	External validity
Contamination	Extraneous variable
Context	Factorial survey design
Control or comparison group	Fixed-sample panel design
Cross-sectional research design	Hawthorne effect
Debriefing	History effect
Differential attrition	Intervening variables
Distribution of benefits	Longitudinal research design
Double-blind procedure	Matching
Endogenous change	Mechanism
Evaluation research	Multiple-group before-and-after design

Nonequivalent control group design
Nonexperimental approach to
  establishing causality
Nonspuriousness
Placebo effect
Posttest
Pretest
Process analysis
Quasi-experimental design
Random assignment
Regression effect
Repeated cross-sectional design
Repeated measures panel design

Selection bias
Self-fulfilling prophecy
Solomon four-group design
Spurious relationship
Statistical control
Subject fatigue
Time order
Time series design
  (repeated measures panel design)
Treatment
Treatment misidentification
True experiment

HIGHLIGHTS

- The independent variable in an experiment is represented by a treatment or other intervention. Experiments have a comparison group that represents what subjects are like on the dependent variable without the treatment. In true experiments, subjects are randomly assigned to comparison and experimental groups.

- Experimental research designs have three essential components: use of at least two groups of subjects for comparison, measurement of the change that occurs as a result of the experimental treatment, and use of random assignment. In addition, experiments may include identification of a causal mechanism and control over experimental conditions.

- Random assignment of subjects to experimental and comparison groups eliminates systematic bias in group assignment. The odds of a difference between the experimental and comparison groups on the basis of chance can be calculated.

- Random assignment and random sampling both rely on a chance selection procedure, but their purposes differ. Random assignment involves placing predesignated subjects into two or more groups on the basis of chance; random sampling involves selecting subjects out of a larger population on the basis of chance. Matching cases in the experimental and comparison groups is a poor substitute for randomization because it is not possible to identify in advance all important variables to use for the match. However, matching can improve the comparability of groups when it is used to supplement randomization.

- Causal conclusions derived from experiments can be invalid due to selection bias, endogenous change, the effects of external events, cross-group contamination, or treatment misidentification. In true experiments, randomization should eliminate selection bias and bias due to endogenous change. External events, cross-group contamination, and treatment misidentification can threaten the validity of causal conclusions in both true experiments and quasi-experiments.

- Process analysis can be used in experiments to identify how the treatment had (or did not have) its effect, a matter of particular concern in field experiments. Treatment misidentification is less likely when process analysis is used.

- The generalizability of experimental results declines if the study conditions are artificial and the experimental subjects are unique. Field experiments are likely to produce more generalizable results than experiments conducted in the laboratory.

- Causal conclusions can be considered externally valid if they apply to all the subgroups in a study. When causal conclusions do not apply to all the subgroups, they are not generalizable to corresponding subgroups in the population and are not externally valid. Causal conclusions can also be considered externally invalid when they occur only under the experimental conditions.

- Nonexperimental designs use statistical controls to reduce the risk of spuriousness. A variable is controlled when it is held constant so that the association between the independent and dependent variables can be assessed without being influenced by the control variable.

- Longitudinal designs are usually preferable to cross-sectional designs for establishing the time order of effects. Longitudinal designs vary even when the same persons are measured at different times, and depend on how the population of interest is defined and how frequently follow-up measurements are taken. Fixed-sample panel designs provide the strongest test for the time order of effects, but they can be difficult to carry out successfully because of their expense and the attrition and fatigue of subjects.

- Subject deception is common in laboratory experiments and poses unique ethical issues. Researchers must weigh the potential harm to subjects and debrief subjects who have been deceived. In field experiments, a common ethical problem is selective distribution of benefits. Random assignment may be the fairest way of allocating treatment when treatment openings are insufficient for all eligible individuals and when the efficacy of the treatment is unknown.

EXERCISES

1. Read the original article reporting one of the experiments described in this chapter. Critique the article, using as your guide the article review questions presented in Appendix B. Focus on the extent to which experimental conditions were controlled and the causal mechanism was identified. Did inadequate control over conditions or inadequate identification of the causal mechanism make you feel uncertain about the causal conclusions?

2. Design a laboratory experiment to determine whether alcohol consumption leads to increased violent attitudes or actions. Describe how your design incorporates the three components of experiments. Evaluate how well your design meets the criteria for identifying causality. What are the ethical issues you face in this experiment?

3. Select a true experiment, from the *Journal of Experimental and Social Psychology*, the *Journal of Personality and Social Psychology*, or sources suggested by your instructor. Diagram the experiment using Exhibits 6.3 and 6.4 in this chapter as models. Discuss the extent to which experimental conditions were controlled and the causal mechanism was identified. How confident can you be in the causal conclusions from the study, based on review of the threats to internal validity discussed in this chapter: selection bias, endogenous change, external events, contamination, and treatment misidentification?

4. Repeat Exercise 3 with a quasi-experiment.

5. Critique the ethics of one of the articles from Exercise 3 or 4. What specific rules do you think should guide researchers' decision about subject deception and the selective distribution of benefits?

DEVELOPING A RESEARCH PROPOSAL

1. Design a laboratory experiment to text one of your hypotheses from Chapter 5 or a related hypothesis. Describe the experimental design, commenting on each component of a true experiment. Specify clearly how the independent variable will be manipulated and how the dependent variable will be measured.

2. Assume that your experiment will be conducted on campus. Formulate recruitment and randomization procedures.

3. Discuss the extent to which each source of internal validity is a problem in the study. Propose procedures like process analysis to cope with these sources of invalidity.

4. How generalizable would you expect the study's findings to be? What can be done to increase generalizability?

5. Develop appropriate procedures for the protection of human subjects in your experiment, including a consent form. Give particular attention to any aspects of the study that are likely to raise ethical concerns.

Student Study Site

The companion Web site for *The Practice of Research in Criminology and Criminal Justice*, Third Edition

http://www.sagepub.com/prccj3

Visit the Web-based Student Study Site to enhance your understanding of the chapter content and to discover additional resources that will take your learning one step further. You can enhance your understanding of the chapters by using the comprehensive study material, which includes e-flashcards, Web exercises, practice self-tests, and more. You will also find special features, such as Learning from Journal Articles, which incorporates SAGE's online journal collection.

WEB EXERCISES

1. Go to Sociosite at www.pscw.uva.nl/sociosite. Choose "Subject Areas." Choose "Crime" or "Criminology." Find an example of research that has been done using experimental methods. Explain the experiment. Choose at least five of the key terms listed at the end of this chapter that are relevant to and incorporated in the research experiment you have located on the Web. Explain how each of the five key terms you have chosen plays a role in the research example you have found on the Web.

2. Go to the Stanford Prison Experiment Web site at www.prisonexp.org. Read the information provided about the Stanford Prison Experiment.

   a. Write a brief essay on the study, including the purpose of the study, the type of experiment, the participant selection process, the problems encountered during the study, and the results.

Despite over two decades of research, the magnitude of intimate-perpetrated violence against men and women is still frequently disputed. For many reasons, including the historical stigma attached to intimate partner violence (IPV), fear of retaliation from their perpetrators, and other safety concerns, estimating incidence rates of this violence has always been a difficult task. The most enduring source of statistical information about violent crime in the United States is the Uniform Crime Report (UCR) Program compiled by the Federal Bureau of Investigation (FBI). The UCR collects information about criminal incidents of violence that are reported to the police. However, using police reports to estimate incidence rates of violence between intimates and family members is problematic for several reasons. Perhaps foremost is the fact that a large percentage of these crimes are never reported to police. Based on comparisons with national survey data, it is estimated that only 40–50 % of crimes become known to police, and the percentage may be much lower for crimes committed by intimates and other family members (Reiss & Roth 1993). In addition, except for the crime of homicide, the current UCR program does not include information on the relationship between the victim and the offender.

Due to the weaknesses of the UCR program in estimating rates of violence within families, random sample surveys of the population are now being used as a social science tool of choice for uncovering incidents of violence within families. However, as can be expected, surveys employing diverse methodologies and different definitions of violence result in tremendously diverse estimates. For example, studies show that the number of women who experience violence by an intimate partner annually range from 9.3 per 1,000 women (Bachman & Saltzman 1995) to 116 per 1,000 women (Straus & Gelles 1990). Furthermore, the methodological differences among random survey methodologies often preclude any direct comparisons.

To increase our understanding of violent victimization, the National Institute of Justice and the Centers for Disease Control and Prevention cosponsored a national telephone survey called the National Violence Against Men and Women (NVAMW) Survey, conducted by the Center for Policy Research in 1995 (Tjaden & Thoennes 2000). Respondents to the NVAMW Survey were asked about a number of health and safety issues, including physical assault they experienced as children by adult caretakers, physical assault they experienced as adults by any type of perpetrator, and forcible rape or stalking they experienced at any time in their life by any type of perpetrator. In this chapter, we will use this project along with the U.S. Justice Department's National Crime Victimization Survey (NCVS) to illustrate some key features of survey research. After an initial review of the reasons for using survey methods, we explain the major steps in questionnaire design and then consider the features of four types of surveys, highlighting the unique problems of each and suggesting possible solutions. Important ethical issues are discussed in the final section. By the chapter's end, you should be well on your way to becoming an informed consumer of survey reports and a knowledgeable developer of survey designs. In addition, you will become a more informed student of the methodological issues surrounding the measurement of violent victimization in the United States.

SURVEY RESEARCH IN THE SOCIAL SCIENCES

Survey research involves the collection of information from a sample of individuals through their responses to questions. In addition to social scientists, many newspaper editors, political pundits, and marketing gurus have turned to survey research because it is an efficient method for systematically collecting data from a broad spectrum of individuals and social settings. The results of surveys are broadcast daily on most network news programs. In fact, surveys have become such a vital part of our nation's social fabric that we cannot assess much of what we read in the newspaper or see on TV without having some understanding of this method of data collection (Converse 1984).

Modern survey research methods were developed in the early and middle years of the 20th century (Lazarsfeld & Oberschall 1965; Oberschall 1972). Beginning in the 1920s, psychologists used surveys to test the abilities of students, employees, and army recruits. For their part, sociologists conducted a few ambitious attitude surveys of the general population, such as Emory S. Bogardus's study of race relations, which had a sample of 1,725 respondents (Converse 1984). Industrial psychologists began to rely on surveys to study employee morale and other attitudes. Psychologist Paul F. Lazarsfeld studied consumers' reactions to particular products before moving into sociology and becoming one of the discipline's leading survey proponents (Converse 1984).

Political demands also stimulated the growth of survey research. Professional polling became a regular fixture after the 1936 presidential election. Professional polling firms such as the one operated by George Gallup correctly predicted that President Franklin Delano Roosevelt would win reelection, whereas the poorly designed *Literary Digest* poll led to an infamous forecast that challenger Alfred M. Landon would win (see Chapter 4). The U.S. Bureau of the Census began to supplement its 10-year census with more frequent surveys of population samples to monitor respondents' income and other economic variables. And in 1973, the U.S. Department of Justice implemented the National Crime Surveys to capture incidents of crime victimization not reported to police.

Since the early days of survey research, professional survey organizations have provided a base of support for social science researchers affiliated with universities. The development of electronic computers also aided the growth of survey research, allowing great increases in the speed and accuracy for processing and reporting data. Surveys soon became the most popular research method in the social sciences.

Attractive Features of Survey Research

Regardless of its scope, survey research owes its continuing popularity to three features: versatility, efficiency, and generalizability.

Versatility

The first and foremost reason for the popularity of survey methods is their versatility. Although a survey is not the ideal method for testing all hypotheses or learning about every social process, a well-designed survey can enhance our understanding of just about any social issue. In fact, there is hardly any topic of interest to social scientists that has not been studied at some time with survey methods. Politicians campaigning for election use surveys, as do businesses marketing a product, governments assessing community needs, agencies monitoring program effectiveness, and lawyers seeking to buttress claims of discrimination or select favorable juries.

Efficiency

Surveys also are popular because data can be collected from many people at relatively low cost and, depending on the survey design, relatively quickly. For the NVAMW Survey, Tjaden and Thoennes (2000) contracted with Schulman, Ronca, and Bucuvalas, Inc. (SRBI) to conduct the telephone survey, primarily because SRBI had conducted nationally representative sample surveys before on sensitive subject matters such as victimization. Surveys were conducted from November 1995 to May 1996 with an average cost of $30 per respondent. Large, mailed surveys are less expensive, at $10 to $15 per potential respondent, but the cost can increase greatly when intensive follow-up efforts are made. Surveys of the general population using personal interviews are much more expensive, with costs ranging from $100 per potential respondent for studies in a limited geographical area to $300 or more when lengthy travel or repeat visits are needed to connect with respondents (Floyd J. Fowler, personal communication, January 7, 1998, Center for Survey Research, University of Massachusetts–Boston; see also Dillman 1982; Groves & Kahn 1979). As you would expect, phone surveys are the quickest survey method, which accounts for their popularity in political polling.

Surveys are efficient research methods because many variables can be measured without substantially increasing the time or cost. Mailed questionnaires can include up to 10 pages of questions before respondents lose interest (and before more postage must be added). The maximum time limit for phone surveys seems to be about 45 minutes. In-person interviews can last much longer, taking more than an hour. For example, the 1991 General Social Survey (GSS) included 196 questions, many with multiple parts, and was 75 pages long.

Of course, the efficiency of the surveys can be attained only in a place with a reliable communications infrastructure (Labaw 1980:xiii–xiv). A reliable postal service, required for mail surveys, has generally been available in the United States, and phone surveys can be effective in the United States because 95% of its households have phones (Czaja & Blair 1985). Also important to efficiency are the services of the many survey organizations that provide the trained staff and the proper equipment for conducting high-quality surveys.

Generalizability

Survey methods lend themselves to probability sampling from large populations. Thus, survey research is very appealing when sample generalizability is a central research goal. In fact, survey research is often the only means available for developing a representative picture of the attitudes and characteristics of a large population.

Surveys also are the research method of choice when cross-population generalizability is a primary concern. They allow a range of social contexts and subgroups to be sampled, and the consistency of relationships can be examined across the various subgroups.

The Omnibus Survey

Most surveys are directed at a specific research question. In contrast, an **omnibus survey** covers a range of topics of interest to different social scientists. It has multiple sponsors or is designed to generate data useful to a broad segment of the social science community rather than answer one particular research question.

One of the most successful omnibus surveys is the General Social Survey (GSS) of the National Opinion Research Center at the University of Chicago. After taking a teaching position at Dartmouth College, James A. Davis, who worked at the National Opinion Research Center for 10 years, was dismayed to discover how difficult it was for faculty in small schools to obtain worthwhile survey data. Starting in 1972, the National Science Foundation agreed to fund the GSS as an annual, publicly available national survey on topics of general interest to sociologists. In 1992, the GSS changed to a biennial schedule.

Today, the GSS is administered every 2 years as a 90-minute interview to a probability sample of almost 3,000 Americans. It includes more than 500 questions about background characteristics and opinions, with an emphasis on social stratification, race relations, family issues, law and social control, and morale. It explores political views, work experiences, social ties, news sources, and views on law, health, and religion. Questions and topic areas are chosen by a board of overseers drawn from the best survey researchers.

The core of the GSS is the set of questions asked in every survey. Other questions are repeated only in alternating surveys, and some questions are added in single or multiyear supplements paid for by special grants. Since 1988, most of the questions have been asked in each survey, but many of them have been asked of only a randomly selected subset of respondents. This **split-ballot design** allows the inclusion of more questions without increasing the survey's cost. The split-ballot design also allows for experiments on the effect of question wording; different phrasings of the same question are included in the split-ballot subsets.

What is most remarkable about the GSS is its availability. The survey data sets are distributed at cost to any interested individual (only $95 for a single year's data set), and most universities obtain the GSS each year. The data are then available to faculty and students free of charge. By 1992, over 2,800 articles, scholarly papers, books, and dissertations had used GSS data. Many instructors use GSS data in methods, statistics, and other courses (Davis & Smith 1992; National Opinion Research Center 1992).

Although the NVAMW Survey is not exactly an omnibus survey, it was developed to obtain detailed information on a number of phenomena related to victimization. For example, because the Centers for Disease Control and Prevention was a co-sponsor of the survey, many questions were added to obtain detailed information on a respondent's history of health and injuries in addition to injuries resulting from victimization.

Most survey data sets sponsored by the federal government are made available for use by scholars through the Inter-University Consortium for Political and Social Research (ICPSR; more details about this are in Chapter 10). For example, all the data sets produced by the Bureau of Justice Statistics, including the NCVS, the National Prison Survey, and many others, can be obtained almost instantaneously from the ICPSR Web site (www.icpsr.umich.edu).

The deficiency of the omnibus approach is the limited depth that can be achieved in any one substantive area. In some years, the GSS avoids this problem by going into greater depth in one particular area. But the best way to get survey data about one particular topic is still the survey developed around the topic alone. The surveys we are going to highlight in this chapter were all developed to measure one topic, victimization.

QUESTIONNAIRE DEVELOPMENT AND ASSESSMENT

The **questionnaire** (or **interview schedule**, as it is often called in interview-based studies) is the central feature of the survey process. Without a well-designed questionnaire tailored to the study's purposes, survey researchers have little hope of achieving their research goals.

Questionnaire The survey instrument containing the questions for a self-administered survey.

Interview schedule The survey instrument containing the questions asked by the interviewer for an in-person interview or phone survey.

The correct design of a questionnaire varies with the specific survey method used and the other particulars of a survey project. There is no precise formula for well-designed questionnaire. Nonetheless, some key principles should guide the design of any questionnaire, and some systematic procedures should be considered for refining it.

Maintain Consistent Focus

A survey (with the exception of an omnibus survey) should be guided by a clear conception of the research problem under investigation and the population to be sampled. Does the study seek to describe some phenomenon in detail, explain some behavior, or explore some type of social relationship? Until the research objective is clearly formulated, survey design cannot begin. Throughout the process of questionnaire design, this objective should be the primary basis for making decisions about what to include and exclude, and what to emphasize or treat in a cursory fashion. Moreover, the questionnaire should be viewed as an integrated whole, in which each section and every question serves a clear purpose related to the study's objective and is a complement to other sections or questions.

Surveys often include too many irrelevant questions and fail to include questions that the researchers realize later are crucial. One way to ensure that all possibly relevant questions are asked is to use questions suggested by prior research, theory, or experience, or by experts (including participants) who are knowledgeable about the setting under investigation. Of course, even the best researcher cannot anticipate every question that could be worthwhile, nor those that are worthless.

Build on Existing Instruments

When another researcher already has designed a set of questions to measure a key concept in your study, that existing set of questions can be called a survey instrument. If evidence from previous surveys indicates that these questions provide a good measure of the concept or behaviors you are interested in, then why reinvent the wheel? To measure incidents of physical assault, Tjaden and Thoennes (2000) modified an instrument that had already been widely used in the literature: the Conflict Tactics Scale (CTS) (Straus 1979). To measure incidents of rape and sexual victimization, Tjaden and Thoennes utilized questions that had already been used by Kilpatrick et al. at the University of South Carolina in the National Women's Study (National Victim Center and the Crime Victims Research and Treatment Center 1992). As you can see in Exhibit 7.1, these questions were very specific in nature and covered all behaviors that are legally defined as rape or sexual assault in most states. In contrast, you will also see in Exhibit 7.1 that the screening questions used by the NCVS to uncover incidents of rape and sexual assault are not as specific about behavior.

EXHIBIT 7.1   Rape Screening Questions Used by the NVAMW Survey and by the NCVS Survey

National Violence Against Men and Women (NVAMW) Survey

[For females only] Regardless of how long ago it happened, has a man ever made you have sex by using force or threatening to harm you or someone close to you? Just so there is no mistake, by sex we mean putting a penis in your vagina.

[For males and females] Has anyone, male or female, ever made you have oral sex by using force or threat of harm? Just so there is no mistake, by oral sex we mean that a man or boy put his penis in your mouth or someone, male or female, penetrated your vagina or anus with their mouth or tongue.

[For males and females] Has anyone ever made you have anal sex by using force or threat of harm? Just so there is no mistake, by anal sex we mean that a man or boy put his penis in your anus.

[For males and females] Has anyone, male or female, ever put fingers or objects in your vagina or anus against your will by using force or threats?

[For males and females] Has anyone, male or female, ever attempted to make you have vaginal, oral, or anal sex against your will, but intercourse or penetration did not occur?

National Crime Victimization Survey (NCVS)

Incidents involving forced or unwanted sexual acts are often difficult to talk about. Have you been forced or coerced to engage in unwanted sexual activity by

a.   someone you didn't know before,
b.   a casual acquaintance, or
c.   someone you know well?

Respondents to the NVAMW Survey who answered yes to any of these screening questions were asked many specific questions about the nature of the victimization, including information about the offender, injuries sustained, and reports to police. Respondents to the NCVS who answered yes to their screening questions were further queried, "Do you mean forced or coerced sexual intercourse?" Only if respondents answered yes to this question were they counted as rape victims; the others were classified as other sexual assault victims. In the end, both surveys purport to measure incidents of rape in the United States, but which methodology is most appropriate? In the next section, we will provide you with some specific guidelines for writing questions and you will see that the answer to this question is not so clear-cut.

WRITING QUESTIONS

Asking people questions is the most common operation for measuring social variables, and probably the most versatile. In principle, survey questions can be a straightforward and efficient means of measuring individual characteristics, facts about events, levels of knowledge,

two surveys indicate that the NVAMW Survey uncovers about four times as many rapes as does the NCVS (Bachman 2000).

Another powerful illustration of the effect of questions on responses comes from the redesign of the NCVS itself. The NCVS was implemented in 1972 to complement what was known about crime from police reporting data from the UCR. In the 1980s, the survey's sponsor, the Bureau of Justice Statistics (BJS), redesigned the screening instrument used by the NCVS to more accurately measure incidents of violent crime, particularly those perpetrated by known offenders such as intimates and other family members (Bachman & Taylor 1994). Among the most important changes to the questionnaire were (1) additional cues to help survey respondents recall incidents, (2) questions encouraging respondents to report victimizations that they themselves may not define as crimes, (3) more direct questions on rape and sexual assault, and (4) new questions to encourage disclosure of victimizations by nonstrangers.

Exhibit 7.2 displays the old screening questions of the NCVS along with the redesigned violent crime screen questions that were implemented into a small percentage of the total NCVS sample beginning in 1989; by 1992 the entire sample was asked these questions. As you can see, not only are there more questions in the redesigned instrument, but the questions also are much more behaviorally specific than those in the old questionnaire. In fact, note that the old questionnaire did not contain a specific question about rape. Prior to the redesign, only those respondents who said they were "attacked some other way" and who self-reported that attack as a rape were classified as rape victims.

In 1994, the BJS began publishing the results of the redesigned survey. Not surprisingly to survey methodologists, rates of violent crime such as rape, intimate perpetrated assault, and other family violence increased dramatically. Exhibit 7.3 displays the rate changes for various types of violent crime against men and women. Do these dramatic increases from 1991 to 1992 actually reflect increases in victimizations? Obviously not. The increase is due primarily to the new screening questionnaire. The more specific questions increased the extent to which respondents disclosed their victimizations. As you might imagine, BJS had to educate the media and public about this methodology before these rates were released, to avoid headlines such as "The Justice Department Finds Rates of Family Violence Skyrocketing."

In addition to writing clear and meaningful questions, there are several other rules to follow and pitfalls to avoid that we will highlight in the next section.

Avoid Confusing Phrasing

Good grammar is a basic requirement for clear questions. Clearly and simply phrased questions are most likely to have the same meaning for different respondents. So be brief and to the point.

Avoid Vagueness

Virtually all questions about behavior and feelings will be more reliable if they refer to specific times or events (Turner & Martin 1984:300). Without a **reference period**, a researcher will not know how to interpret an answer. For example, the question "How often do you carry a method of self-protection such as pepper spray?" will produce answers that have no common reference period and can therefore not be reliably compared to answers

EXHIBIT 7.2 Comparison of the Redesigned and Old Screener Questions Used by the National Crime Victimization Survey

Screener questions for violent crimes
New (beginning January 1992)*

1. Has anyone attacked or threatened you in any of these ways—
   a. With any weapon, for instance, a gun or knife—
   b. With anything like a baseball bat, frying pan, scissors, or stick—
   c. By something thrown, such as a rock or bottle—
   d. Include any grabbing, punching, or choking,
   e. Any rape, attempted rape, or other type of sexual assault—
   f. Any face to face threats—
      OR
   g. Any attack or threat or use of force by anyone at all? Please mention it even if you were not certain it was a crime.
2. Incidents involving forced or unwanted sexual acts are often difficult to talk about. Have you been forced or coerced to engage in unwanted sexual activity by—
   a. Someone you didn't know before
   b. A casual acquaintance OR
   c. Someone you know well.

Old (1972–1992)

1. Did anyone take something directly from you by using force, such as by a stickup, mugging, or threat?
2. Did anyone TRY to rob you by using force or threatening to harm you?
3. Did anyone beat you up, attack you, or hit you with something, such as a rock or bottle?
4. Were you knifed, shot at, or attacked with some other weapon by anyone at all?
5. Did anyone THREATEN to beat you up or THREATEN you with a knife, gun, or some other weapon, NOT including telephone threats?
6. Did anyone TRY to attack you in some other way?

*During 1992 half of the sampled households responded to the old questionnaire, and half to the redesigned survey.

Screener questions for all types of crimes
New

1. Were you attacked or threatened OR did you have something stolen from you—
   a. At home including the porch or yard—
   b. At or near a friend's, relative's, or neighbor's home—
   c. At work or school—
   d. In place such as a storage shed or laundry room, a shopping mall, restaurant, bank or airport—
   e. While riding in any vehicle—
   f. On the street or in a parking lot—
   g. At such places as a party, theater, gym, picnic area, bowling lanes, or while fishing or hunting.
      OR
   h. Did anyone ATTEMPT to attack or attempt to steal anything belonging to you from any of these places?
2. People often don't think of incidents committed by someone they know. Did you have something stolen from you OR were you attacked or threatened by—
   a. Someone at work or school—
   b. A neighbor or friend—
   c. A relative or family member—
   d. Any other person you've met or known?
3. Did you call the police to report something that happened to YOU which you thought was a crime?
4. Did anything happen to you which you thought was a crime, but did NOT report to the police?

Old

1. Was anything stolen from you while you were away from home, for instance, at work, in a theater or restaurant, or while traveling?
2. Did you call the police to report something that happened to YOU that you thought was a crime?
3. Did anything happen to YOU that you thought was a crime, but did NOT report to the police?

EXHIBIT 7.3 Comparison of Violent Crime Rates per 1,000 Males and Females by Victim-Offender Relationship Obtained by the Redesigned Screener and the Older Screener of the NCVS

Relationship	Female Rate	Male Rate
Older Questionnaire		
Intimate	5.4	.5
Other relative	1.1	.7
Acquaintance/friend	7.6	13.0
Stranger	5.4	12.2
Redesigned Questionnaire		
Intimate	9.3	1.4
Other relative	2.8	1.2
Acquaintance/friend	12.9	17.2
Stranger	7.4	19.0

Source: R. Bachman & L. Saltzman. 1995. *Violence Against Women: Estimates from the Redesigned Survey (NCJ-154348)*, Appendix Table. Washington, DC: Bureau of Justice Statistics, U.S. Department of Justice.

To test your knowledge of writing survey questions, go to the Survey Research Interactive Exercises on the Student Study Site.

from other respondents. A more specific way to ask the question is "In the last month, how many days did you carry a method of self-protection such as pepper spray?"

In general, research shows that the longer the reference period, the greater the under-reporting of a given behavior (Cantor 1984, 1985; Kobelarcik et al. 1983). For example, one survey asked, "During the past 12 months, about how many times did you see or talk to a medical doctor?" When official records were checked, respondents forgot about 60% of their doctor visits (Goleman 1993b:C11). In general, when information about mundane or day-to-day activities is solicited, reference periods should be no longer than "in the past month." However, when rare events are being measured, such as experiences with victimizations, "the last 6 months," as utilized by the NCVS Survey, or "the past 12 months," as used by the NVAMW, are both acceptable.

Avoid Negatives and Double Negatives

Picture yourself answering the following question: "Do you disagree that juveniles should not be tried as adults if they commit murder?" It probably took a long time for you to figure out if you would actually agree or disagree with this statement, because it is written as a **double-negative question**. For example, if you think juveniles who commit murder should be tried as adults, you would actually agree with this statement. Even questions that are written with a single negative are usually difficult to answer. For example, suppose you were asked to respond to "I can't stop thinking about the terrorist attacks on 9–11" using a five-point response set of "very rarely" to "very often." A person who marks "very rarely" is actually saying "I very rarely can't stop thinking about the terrorist attacks on 9–11." Confusing, isn't it? Even the most experienced survey researchers can unintentionally make this mistake, perhaps while trying to avoid some other wording problem. For instance, in a survey commissioned by the American Jewish Committee, the Roper polling organization wrote a question about the Holocaust that was carefully worded to be neutral and value-free: "Does it seem possible or does it seem impossible to you that the Nazi extermination of the Jews never happened?" Among a representative sample of adult Americans, 22% answered it was possible the extermination never happened (Kifner 1994:A12). Many Jewish leaders and politicians were stunned, wondering how one in five Americans could be so misinformed. But a careful reading of the question reveals how confusing it is" Choosing "possible," the seemingly positive response, means that you do not believe the Holocaust happened. In fact, the Gallup organization then rephrased the question to avoid the double negative, giving a brief definition of the Holocaust and then asking, "Do you doubt that the Holocaust actually happened or not?" Only 9% responded that they doubted it happened. When a wider range of response choices was given, only 2.9% said that the Holocaust "definitely" or "probably" did not happen. This should convince you that it is best to avoid negative statements altogether. If you feel you must use a negative in a statement, be certain it stands out with boldface type or italics so that it is clear exactly what you are asking.

Avoid Double-Barreled Questions

When a question is really asking more than one question it is called a **double-barreled question**. For example, the statement, "I believe we should stop spending so much money building prisons and put it into building more schools," is really asking respondents two different questions. Some respondents may believe we should stop building so many prisons but may not want the revenue to go into building more schools. Double-barreled questions can also show up in the response categories. For example, the item below is really asking two questions and would be better stated as such:

Do you know anyone who has ever used cocaine?

____Yes ____No ____I have used cocaine

Avoid Making Either Disagreement or Agreement Disagreeable

People often tend to "agree" with a statement just to avoid seeming disagreeable. You can see the impact of this human tendency in a 1974 Michigan Survey Research Center survey that asked who was to blame for crime and lawlessness in the United States (Schuman & Presser 1981:208). When one question stated that individuals were more to blame than social conditions, 60% of the respondents agreed. But when the question was rephrased so respondents were asked, in a balanced fashion, whether individuals or social conditions were more to blame, only 46% chose individuals.

You can take several steps to reduce the likelihood of agreement bias. As a general rule, you should present both sides of attitude scales in the question itself: "In general, do you believe that *individuals* or *social conditions* are more to blame for crime and lawlessness in the United States?" (Dillman 2000:61–62). The response choices themselves should be phrased to make each one seem as socially approved, as "agreeable," as the others. You should also consider replacing a range of response alternatives that focus on the word "agree" with others. For example, "To what extent do you support or oppose the new health care plan?" (response choices range from "strongly support" to "strongly oppose") is probably a better approach than the question "To what extent do you agree or disagree with the statement: 'The new health care plan is worthy of support'?" (response choices range from "strongly agree" to "strongly disagree").

You may also gain a more realistic assessment of respondents' sentiment by adding to a question a counterargument in favor of one side to balance an argument in favor of the other side. Thus, don't just ask in an employee survey whether employees should be required to join the union; instead, ask whether employees should be required to join the union or be able to make their own decision about joining. In one survey, 10% more respondents said they favored mandatory union membership when the counterargument was left out than when it was included. It is reassuring to know, however, that this approach does not change the distribution of answers to questions about which people have very strong beliefs (Schuman & Presser 1981:186).

When an illegal or socially disapproved behavior or attitude is the focus, we have to be concerned that some respondents will be reluctant to agree that they have ever done or thought such a thing. In this situation, the goal is to write a question and response choices that make agreement seem more acceptable. For example, Dillman (2000) suggests that we ask, "Have you ever taken anything from a store without paying for it?" rather than "Have you ever shoplifted something from a store?" (p. 75). Asking about a variety of behaviors or attitudes that range from socially acceptable to socially unacceptable will also soften the impact of agreeing with those that are socially unacceptable.

Additional Guidelines for Fixed-Response Questions

Creating questions that are clear and meaningful is only half of the formula to creating a good survey instrument. The choices you provide respondents in fixed-choice questions are also important. In this section, we provide you with several rules that will help to ensure that the response choices you provide to your questions will also be clear and concise.

Response Choices Should Be Mutually Exclusive

When you want respondents to make only one choice, the fixed-response categories must not overlap. For example, if you were interested in the ways foot patrol officers spent their time while working, you might ask the following question:

On average, how much time do you spend on the job each week taking care of traffic violations?

□ Less than 1 hour
□ 1–3 hours
□ 3–6 hours
□ 6–10 hours
□ 10 hours or more

The choices provided for respondents in this question are not **mutually exclusive responses**, because they overlap. Which choice would an officer select if he or she spent an hour a week on traffic violations? Choices that are mutually exclusive would look like this:

- ☐ Less than 1 hour
- ☐ 2–3 hours
- ☐ 4–6 hours
- ☐ 7–10 hours
- ☐ 11 hours or more

Make the Response Categories Exhaustive

In addition to mutual exclusivity, fixed-response categories must also allow all respondents to select an option. Consider the same research question about foot patrol officers. Suppose we asked a question such as this:

In what activity do you spend the most time in an average week on the job?

____ Traffic violations

____ Disturbance-related issues

____ Felony arrests

____ Misdemeanor arrests

Regardless of how exhaustive we think the response categories are, there must always be an option for respondents who require another choice. Response categories can easily be made **exhaustive** if respondents are provided with a choice labeled:

_____ Other, please specify: ____

Note, however, that "Other" should be used only after you have included all options that you believe to be relevant. Otherwise, a large percentage of respondents will select the "Other" category and you will have to spend time coding their responses.

Utilize Likert-Type Response Categories

Likert-type responses generally ask respondents to indicate the extent to which they agree or disagree with statements. Why the name? Well, this format is generally believed to have been developed by Rensis Likert back in the 1930s. Likert-type response categories list choices for respondents to select their level of agreement with a statement and may look something like this:

I think "three-strikes" laws that increase penalties for individuals convicted of three or more felonies will help to decrease the crime rate.

1	2	3	4
Strongly Agree	Agree	Disagree	Strongly Disagree

Minimize Fence-Sitting and Floating

Two related problems in question writing stem from the respondent's desire to choose an acceptable or socially desirable answer. There is no uniformly correct solution to these problems, so you must carefully select an alternative.

Fence-sitters are people who see themselves as neutral in their attitudes toward a particular issue. If you are truly interested in those who do not have strong feelings on an issue, one alternative is to provide a neutral or undecided response option. The disadvantage to these options is that they may encourage some respondents to take the easy way out rather than really thinking about their feelings. They may also provide an out for respondents who do not want to reveal how they truly feel about an issue. On the other hand, not providing respondents who really have no opinion on an issue with an option such as "undecided" can be very frustrating and may encourage them to leave the item blank. Whatever you decide, it is generally a good idea to provide respondents with instructions that ask them to "select the choice in each item that most closely reflects your opinion." This should help make all respondents feel more comfortable about their answers, particularly those who only slightly feel one way or the other.

Floaters are respondents who choose a substantive answer even when they do not know anything about a particular question. For example, research has shown that a third of the public will provide an opinion on a proposed law they know nothing about if they are not provided with a "don't know" option (Schuman & Presser 1981). Of course, providing a "don't know" option has the same disadvantage as providing a neutral response option; its inclusion leads some people who have an option to take the way out and choose "don't know."

If you are really interested in informed opinions about an issue, it is best to provide detailed information about that issue when asking a question. For example, let us say we were interested in attitudes about the treatment of juvenile offenders by the criminal justice system. Suppose we asked respondents to provide their opinion to the following statement: "The Juvenile Justice Bill before Congress will help reduce crime committed by juveniles." Do you know what the Juvenile Justice Bill is? If we did not provide a "don't know" option, respondents who knew nothing about the Juvenile Justice Bill would be forced to select a response that would not be meaningful and may bias the results of the entire survey. Instead of a "don't know" option, another way to handle the problem is to provide details of the issue you are interested in. For example, you could tell respondents that one component of the bill encourages states to adjudicate all juvenile homicide offenders 13 years of age or older as adults and then ask respondents their opinion about this particular issue. If you are truly interested in the extent to which respondents have knowledge about a particular issue and you want to include a "don't know" response, it should be set apart from the other choices so that respondents do not mistake it as a neutral or undecided choice. Of course, as with all questions, there should be clear instructions about what the response options actually mean. For example, if you wanted to examine citizens' knowledge and opinion about the Independent Counsel statute, you could ask:

Instructions: For each statement, check the box that best indicates your agreement about the statement. If you do not have enough information about a statement to determine your level of agreement, leave the boxes blank and put an X on the diamond next to "Don't Know."

I think the Independent Counsel Law should remain as it is.

☐ Strongly Agree
☐ Agree
☐ Disagree
☐ Strongly Disagree
    Don't Know _____

Utilize Filter Questions

The use of filter questions is important to ensure that questions are asked only of relevant respondents. For example, if you are interested in the utilization of police services by robbery victims, you would first need to establish victimization with a **filter question**. These filter questions create **skip patterns**. For example, respondents who answer no to one question are directed to skip ahead to another question, but respondents who answer yes go on to the **contingent question** or questions. That's why these questions are sometimes called contingency questions. Skip patterns should be indicated clearly with arrows or other direction in the questionnaire, as demonstrated in Exhibit 7.4.

Combining Questions in Indexes

Measuring variables with single questions is very popular. Public opinion polls based on answers to single questions are reported frequently in newspaper articles and TV newscasts: "Do you favor or oppose U.S. policy in . . . ?" "If you had to vote today, for which candidate would you vote?" The primary problem with using a single question is that if respondents misunderstand the question or have some other problem with the phrasing, there is no way to tell. Single questions are prone to this **idiosyncratic variation**, which occurs when individuals' responses vary because of their reactions to particular words or ideas in the question. Differences in respondents' backgrounds, knowledge, and beliefs almost guarantee that they will understand the same question differently. If some respondents do not know some of the words in a question, we will not know what the respondents' answers mean—if they answer at all. If a question is too complex, respondents may focus on different parts of the question. If prior experiences or culturally biased orientations lead different groups to interpret questions differently, answers will not have a consistent meaning.

If just one question is used to measure a variable, the researcher may not realize respondents had trouble with a particular word or phrase in the questions. Although writing carefully worded questions will help reduce idiosyncratic variation, the best option is to devise multiple rather than single questions to measure concepts.

When several questions are used to measure one concept, the responses may be combined by taking the sum or average of the responses. A composite measure based on this type of sum or average is called an **index** or **scale**. The idea is that idiosyncratic variation in response to single questions will average out, so the main influence on the combined

EXHIBIT 7.4   Filter Questions and Skip Patterns

14. In the past 6 months, has anyone taken something from you by force or the threat of force?

_____ Yes (If yes, please answer questions 15 through 16)

_____ No (If no, please skip to question 17)

15. What was the approximate monetary value of the items taken?

_____ Under $50

_____ $51 to $99

_____ $100 to $299

_____ $300 to $500

_____ Over $500

16. Was the incident reported to the police?

_____ Yes

_____ No

17. How fearful are you of walking alone at night in your neighborhood?

_____ Extremely afraid

_____ Afraid

_____ Unafraid

_____ Extremely unafraid

measure will be the concept focused on by the questions. In addition, the index can be considered a more complete measure of the concept than can any one of the component questions.

Creating an index, however, is not just a matter of writing a few questions that seem to focus on one concept. Questions that seem to you to measure a common concept might seem to respondents to concern several different issues. The only way to know that a given set of questions does, in fact, form an index is to administer the questions to people similar to those you plan to study. If a common concept is being measured, people's responses to the different questions should display some consistency. Special statistics called **reliability measures** help researchers decide whether responses are consistent. Most respondent attitudes are complex and consist of many elements.

Be aware of response sets when constructing an index measuring attitudes. For example, some people tend to agree with almost everything asked of them, whereas others tend to disagree. Still others are prone to answer neutrally to everything if given the option. To decrease the likelihood of this happening, it is a good idea to make some statements favorable to a particular attitude and some unfavorable. In this way, respondents are forced to be more careful in their responses to individual items. Exhibit 7.5 displays a hypothetical set of questions designed to solicit respondents' attitudes toward police in their community.

When scoring indexes and scales made up of both favorable and unfavorable statements, you must remember to reverse code the unfavorable items. For example, marking "strongly agree" to the first item in Exhibit 7.5 should not be scored the same as a "strongly agree" response to the second item.

Due to the popularity of survey research, indexes already have been developed to measure many concepts, and some of these indexes have proved to be reliable in a range of studies. It usually is much better to use these indexes to measure concepts than to try to devise new questions to form a new index. As noted earlier in this chapter, the use of a preexisting measure both simplifies the work involved in designing a study and facilitates a comparison of findings to those obtained in previous studies.

One index available in the NCVS is designed to measure what is referred to in criminological literature as *routine activities*. To explain crime victimization, the Routine Activities

---

EXHIBIT 7.5   Items in an "Attitude Toward Police" Index

1. I think police officers are generally fair to all people regardless of their race or ethnicity.

    _____ Strongly Agree _____ Agree _____ Disagree _____ Strongly Disagree

2. Police officers are given too much freedom to stop and frisk community residents.

    _____ Strongly Agree _____ Agree _____ Disagree _____ Strongly Disagree

3. I think if someone resisted arrest, even a little, most police officers would become assaultive if they thought they could get away with it.

    _____ Strongly Agree _____ Agree _____ Disagree _____ Strongly Disagree

4. Police officers put their lives on the line every day trying to make it safe for residents of this community.

    _____ Strongly Agree _____ Agree _____ Disagree _____ Strongly Disagree

5. I think the majority of police officers have lied under oath at least once.

    _____ Strongly Agree _____ Agree _____ Disagree _____ Strongly Disagree

6. The majority of police officers are honest and fair.

    _____ Strongly Agree _____ Agree _____ Disagree _____ Strongly Disagree

Theory focuses on the circumstances in which crimes are committed rather than on the circumstances of the offender. According to Cohen and Felson (1979), each criminal act requires the convergence of three elements: likely and motivated offenders, suitable targets, and an absence of capable guardians to prevent the would-be offender from committing the crime. Thus, the routine patterns of work, play, and leisure time affect the convergence in time and place of the motivated offenders, the suitable targets, and the absence of guardians. To measure the extent to which respondents engage in routine activities away from the home, the NCVS asks three questions that are displayed in Exhibit 7.6. These questions measure different types of activities people routinely engage in away from the home and, when combined, create a better measure of an individual's activities away from the home than would a single question.

Another example of an index is that used to measure student perceptions of tolerance for substance abuse on college campuses (Core Institute 1994). An excerpt from this is shown in Exhibit 7.7. Alone, no single question would be sufficient to capture the overall tolerance of substance abuse on campus. A person's total response to these questions is likely to provide a more accurate indication of tolerance for substance abuse than would a single, general question such as "Do students on this campus feel that drinking or using drugs is OK?"

The advantages of using indexes rather than single questions to measure important concepts are very clear. Surveys often include sets of multiple-item questions, and the following are four cautions to consider:

Our presupposition that each component question is indeed measuring the same concept may be mistaken. Although we may include multiple questions in a survey to measure one concept, we may find that answers to the questions are not related to one another, and so the index cannot be created. Alternatively, we may find that answers to just a few of the questions are not related to the answers given to most of the other questions. We may therefore decide to discard these particular questions before computing the average that comprises the index.

Combining responses to specific questions can obscure important differences in meaning among the questions. Schutt, Gunston, and O'Brien's (1992) research projects on the impact of AIDS prevention education in shelters for the homeless providers an example. In this research, Schutt et al. asked a series of questions to ascertain respondents' knowledge about HIV risk factors and about methods of preventing exposure to those risk factors. After combining the responses into an overall knowledge index, they were surprised to find that the knowledge index scores were no higher in a shelter with an AIDS education program than in a shelter without such a program. Further analysis, however, showed that respondents in the shelter with an AIDS education program were more knowledgeable than the other respondents about the specific ways to prevent AIDS, which were in fact the primary focus of the program. Combining the responses to these questions with the other responses about general knowledge of HIV risk factors obscured this important finding (Schutt, Gunston, & O'Brien 1992).

The questions in an index may cluster together in subsets. All the questions may measure the intended concept, but we may conclude that the concept actually has several different aspects. A multidimensional index has then been obtained. This conclusion can in turn help

EXHIBIT 7.6 Example of an Index: NCVS Questions Used to Make Up an Index of Routine Activities Away From the Home

Before we get to the crime questions, I'd like to ask you about some of your usual activities. We have found that people with different lifestyles may be more or less likely to become victims of crime.

30. On average, during the past 6 months, that is since _____, 20 _____, how often have you gone shopping? For example, at drug, clothing, grocery, hardware, and convenience stores. (Read answer categories until respondent answers yes.)

_____ Almost every day (or more frequently)

_____ At least once a week

_____ At least once a month

_____ Less often

_____ Never

_____ Don't know

31. On average, during the past 6 months, how often have you spent the evening away from home for work, school, or entertainment? (Read answer categories until respondent answers yes.)

_____ Almost every day (or more frequently)

_____ At least once a week

_____ At least once a month

_____ Less often

_____ Never

_____ Don't know

32. On average, during the past 6 months, how often have you ridden public transportation? Do not include school buses. (Read answer categories until respondent answers yes.)

_____ Almost every day (or more frequently)

_____ At least once a week

_____ At least once a month

_____ Less often

_____ Never

_____ Don't know

EXHIBIT 7.7 Example of an Index: Excerpt from the Index of Student Tolerance of Substance Abuse

37. During the past 30 days, to what extent have you engaged in any of the following behaviors?

(mark one for each line)	0 Times	1 Time	2 Times	3–5 Times	6–9 Times	10 or More Times
a. Refused an offer of alcohol or other drugs						
b. Bragged about your alcohol or other drug use						
c. Heard someone else brag about his or her alcohol or other drug use						
d. Carried a weapon such as a gun, knife, and so on (do not count hunting situations or weapons used as part of your job)						
e. Experienced peer pressure to drink or use drugs						
f. Held a drink to have people stop bothering you about why you weren't drinking						
g. Thought a sexual partner was not attractive because he or she was drunk						
h. Told a sexual partner that he or she was not attractive because he or she was drunk						

Source: Core Institute 1994.

us refine our understanding of the original concept. For example, over two decades of work have culminated in the development of a multidimensional index of stress at the state level. Research has found that this index, the State Stress Index (SSI), is related to a number of aggressive behaviors also measured at the state level such as state rates of homicide and rape (Linsky, Bachman, & Straus 1995). The SSI is based on the life-events theory of stress, which holds that the more events such as divorce or the movement to a new community individuals must adapt to, the greater risk they have of not being able to cope. The administration and scoring of life-event measures of stressors experienced by an individual are straightforward. Respondents are asked to indicate events on the list they experienced in the recent past. The number of life events is then added to provide a score that indicates the extent to which the respondent has experienced these stressful events. The SSI is a translation of the life-events approach from the original individual level to the macro or societal level. The basic strategy was to measure the rate at which these events occurred in each state to evaluate the stressfulness of living in each state. What developed was a multidimensional index composed of 15 indicators that statistically fell into three categories: economic stressors, family stressors, and community stressors. The specific variables used to make up the SSI are presented in Exhibit 7.8.

EXHIBIT 7.8 Life-Events Indicators in the 1992 State Stress Index (SSI)

Economic Stressors

> Business failures per 1 million population, 1982
>
> Unemployment claims per 100 thousand adults age 18 and over, 1982
>
> Striking workers per 100 thousand adults age 18 and over, 1981
>
> Bankruptcy cases per 100 thousand population, 1982
>
> Mortgage foreclosures per 100 thousand population, 1982

Family Stressors

> Divorces per 1 thousand population, 1982
>
> Abortions per 100 thousand population, 1982
>
> Illegitimate births per 100 thousand population age 14 and over, 1982
>
> Infant deaths per 1 thousand live births, 1982
>
> Fetal deaths per 1 thousand live births, 1982

Community Stressors

> Disaster assistance per 100 thousand population, 1982
>
> Percentage residing in state less than 5 years, 1980
>
> New housing units per 1 thousand population, 1982
>
> New welfare recipients per 100 thousand population, 1982
>
> High school dropouts per 100 thousand population, 1982

Source: Linsky, Bachman & Straus 1995.

Sometimes particular questions are counted, or weighted, more than others in the calculation of the index. Some questions may be more central to the concept being measured than others, and so may be given greater weight in the index score. It is difficult to justify this approach without extensive testing, but some well-established indexes do involve differential weighting. Another approach to creating an index score is by giving different weights to the responses to different questions before summing or averaging the responses. Such a weighted index is also termed a *scale*. The scaling procedure might be as simple as arbitrarily counting responses to one question as worth two or three times as much as responses to another question, but most often, the weight applied to each question is determined through empirical testing. For example, based on Mooney and Lee's (1995) research on abortion law reform, the scoring procedure for a scale of support for abortion might give a "1" to agreement that abortion should be allowed "when the pregnancy resulted from rape or incest" and a "4" to agreement with the statement that abortion should be allowed "whenever a woman decided she wanted one." In other words, agreeing that abortion is allowable in any circumstances is much stronger support for abortion rights than agreeing that abortion should be allowed in the case of rape or incest.

Demographic Questions

Almost all questionnaires include a section on demographic information such as sex, age, race or ethnicity, income, and religion. For many research studies, these questions are important independent variables. For example, research has shown that all four of these factors are related to the probability of victimization. Many researchers, however, include demographic questions that are not necessary for purposes of their research. Try to avoid this, particularly for questions on income because it makes the questionnaire more intrusive than necessary. In fact, many respondents feel that questions about their income invade their privacy. If you believe income is an essential variable for your study, providing fixed responses that include a range of values to select from is less intrusive than asking respondents for specific annual incomes. This format was utilized by the NVAMW Survey, as shown in Exhibit 7.9.

Care should also be taken when writing questions about race and ethnicity. Many people are justifiably sensitive to these questions. Even the U.S. Bureau of the Census has been struggling with appropriate categories to offer respondents. In fact, the bureau still utilizes two questions, one of race and one for respondent ethnicity (Hispanic or non-Hispanic), which is obviously problematic. Most survey researchers now include questions such as the following:

Which of the following best describes your racial or ethnic background? Please check one.

- ☐ Asian
- ☐ Black or African American
- ☐ White or Caucasian
- ☐ Hispanic (may be of any race)
- ☐ Native American
- ☐ Of Mixed Race or Ethnicity
- ☐ Other (Please specify: _____)

EXHIBIT 7.9 Question on Income from the NVAMW Survey

Including income from all sources, such as work, child support, and AFDC, how much income did you personally receive in 1995 before taxes? Stop me when I get to the category that applies. Was it . . .

01	Less than $5,000
02	$5,001 to $10,000
03	$10,001 to $15,000
04	$15,001 to $20,000
05	$20,001 to $25,000
06	$25,001 to $35,000
07	$35,001 to $50,000
08	$50,001 to $80,000
09	$80,001 to $100,000
10	Over $100,001
11	(Volunteer) None
12	(Volunteer) Don't Know
13	(Volunteer) Refused

Source: Tjaden & Thoennes 2000.

Deciding which categories to include remains difficult. Some researchers still prefer to exclude the Mixed category, because they believe most respondents will identify primarily with one race.

Questions on marital status can also be tricky to compose. The traditional categories of married, single, divorced, and widowed, still used by unsavvy researchers, can be interpreted very differently by respondents. Why? Well, isn't someone currently divorced also single? And what about someone not officially divorced but separated? To avoid confusing respondents, the U.S. Bureau of the Census adopted the following response categories: Married, Separated, Widowed, Divorced, and Never Married.

Because demographic questions are usually perceived as private by respondents, some researchers place them in a section at the end of the questionnaire with an introduction reassuring respondents that the information will remain confidential. However, when the information being gathered in the rest of the questionnaire is even more sensitive, such as violence respondents may have experienced at the hands of a family member or intimate partner, some researchers opt to keep demographic questions near the beginning of the questionnaire. For example, the NVAMW Survey asked demographic questions after the first section, which asked respondents about their general perceptions of fear and safety.

Don't Forget to Pretest!

Adhering to the preceding question-writing guidelines will go a long way toward producing a useful questionnaire. However, simply asking what appear to be clear questions does not ensure that people have a consistent understanding of what you are asking. You need some external feedback, and the more of it the better.

No questionnaire should be considered ready for use until it has been **pretested**. Try answering the questionnaire yourself, and then revise it. Try it out on some colleagues or other friends, and then revise it. Then select a small sample of individuals from the population you are studying or one very similar to it, and try out the questionnaire on them. Audiotape the test interviews for later review, or for a written questionnaire, include in the pretest version some space for individuals to add comments on each key question.

The NVAMW Survey was pretested in *four* different trials on a total of 107 respondents. Yes, four! The pretests were used to assess the effectiveness of the interview introduction, identify confusion and awkwardness in question wording and response categories, and evaluate the flow and length of the interview. Once the fourth pretest established that the instrument was workable, it was then fielded to replicate samples of 500, 1,000, or 2,000 completed interviews. This additional pretesting allowed Tjaden and Thoennes (2000) to determine whether they should retain, drop, or revise particular questions.

Another important form of feedback results from simply discussing the questionnaire content with others. Persons to consult should include other researchers, key figures in the locale or organization to be surveyed (such as elected representatives, company presidents, and community leaders), and some individuals from the population to be sampled. If you find you have to explain a particular question in detail, you probably need to rewrite it. Run your list of variables and specific questions by such figures whenever you have a chance. Reviewing the relevant literature to find results obtained with similar surveys and comparable questions is also an important step to take. Of course, you should have already conducted such a review before writing your questions!

Another increasingly popular form of feedback comes from guided discussions among potential respondents, called focus groups, to check for consistent understanding of terms and to identify the range of events or experiences about which people will be asked to report. By listening and observing the focus group discussions, researchers can validate their assumptions about what level of vocabulary is appropriate and what people are going to be reporting (Fowler 1995).

Professional survey researchers have also developed a technique for evaluating questions called the **cognitive interview** (Fowler 1995). Although the specifics vary, the basic approach is to ask people test questions, then probe with follow-up questions to learn how they understood the question and what their answers mean.

Review the distribution of responses to each question, listen to the audiotapes, or read all the comments, and then code what you heard or read to identify problems in question wording or delivery. Revise any questions that respondents do not seem to interpret as you had intended or that are not working well for other reasons.

To create a good questionnaire takes several drafts. By the time you have gone through just a couple of drafts, you may not be scanning the instrument as clearly as you think. A very honest illustration of this is provided by Don Dillman, the Director of the Social and

Economic Sciences Research Center at Washington State University. His research team was about ready to mail a questionnaire with the following response categories:

What is your opinion?

Strongly Oppose	Oppose	Neither	Favor	Strongly Oppose

This Likert-type response format not only slipped by Dillman, but by his typist, a research assistant, and another person working on the project. He explains:

> By the time a would-be surveyor has reached the final draft, he or she is often scanning the questionnaire for errors but not absorbing the detail. . . . All of us were looking for other things, such as spacing, punctuation, and content. The uniform appearance of the response categories looked right to all of us. (Seltzer 1996:98)

A careful scrutiny of your questionnaire using the procedures outlined in this section will help you detect these and other problems.

ORGANIZATION OF THE QUESTIONNAIRE

Once the basic topics and specific variables for a questionnaire have been identified, they can be sorted into categories (which may become separate sections) and listed in tentative order. Throughout the question-writing process, the grouping of variables into sections and the ordering of questions within sections could be adjusted. These adjustments will in turn require changes in the specific questions, in an iterative process that leads to a polished, coherent questionnaire.

The first thing needed is a descriptive title for the questionnaire that indicates the overall topic. The title is essential because it sets the context for the entire survey. For example, the NCVS and the NVAMW Survey are both interested in measuring the magnitude of crime victimization in the United States. The NVAMW Survey, however, was presented as a survey interested in a number of personal safety-related issues, including tactics used in conflict resolution. The NCVS is named a crime survey and is a survey interested in obtaining information only about crimes. Unfortunately, some survey participants still may not view assaults by intimates and other family members as criminal acts.

Even though many of the behaviors conveyed in the screening instruments (e.g., kicking and punching) are the same for both surveys, the context in which these questions are asked must inevitably play a role in the candor respondents are willing to provide. The contextual message of crime may take precedence over the questions and may discourage respondents from reporting incidents they do not perceive as crimes. Lynch (1996) illuminates the importance of context by citing a study undertaken by BJS in 1975 when a fear-of-crime supplement was asked of a subsample of the NCVS prior to the screening interview. Reports of victimization in the subsample that received the fear of crime supplement increased significantly; this finding was attributed to a warm-up effect of the fear supplement, such that the additional questions about fear of crime stimulated recall of crime events.

In addition to the title, question order is important because this also can influence responses. Consider the issue of capital punishment. For nearly a decade after the 1976 Supreme Court case of *Gregg v. Georgia*, which made capital punishment constitutional under guided discretion statutes, proponents of the death penalty were quick to note public sentiment favoring the use of capital punishment. But refined public opinion polls reveal how this sentiment can change dramatically when respondents are given other choices. For example, a 1986 Gallup poll (Gallup 1986) revealed that 70% of those surveyed approved of the death penalty for murder. However, when respondents were given the alternative punishment of life imprisonment without the possibility of parole, the support declined to 55%. This suggests that public opinion questions that simply ask whether the person approves of capital punishment for convicted murderers are a misleading indicator of the strength of support for the death penalty. As Paternoster (1991) explains,

> These questions, which have been asked since the 1930s, have formed the basis of a folklore that Americans mandate capital punishment. This folklore has informed the attitudes of both state legislatures and Supreme Court justices who are led by these opinion polls into believing that this represents "the will of the people." What Americans may be expressing, however, is a desire for protection against dangerous criminals, not a desire for capital punishment. More sophisticated polling questions indicate that the public does not necessarily want to repay one life with another, but wants the murderer to be unable to offend again, and to ease the hardship and loss for those left behind. (P. 30)

A similar phenomenon has been observed with questions about abortion. When a sample of the general public was asked, "Do you think it should be possible for a pregnant woman to obtain a legal abortion if she is married and does not want any more children?" 58% of the respondents said yes. However, when this abortion question was preceded by one asking whether the respondent would allow abortion of a damaged fetus, only 40% said yes to the more permissive question. Clearly the first question altered the respondents' frame of reference, perhaps by making abortion as a simple way to avoid having more children seem frivolous compared to the problem of having a damaged fetus (Turner & Martin 1984:135). The point to take away from these cases is that question order is extremely important. As Schuman and Presser (1981) acknowledge:

> Both examples illustrate the potential impact of question order on the respondents' answers. This potential is greatest when two or more questions concern the same issue or closely related issues so that asking one question affects reactions to the next question. The impact of question order also tends to be greatest for general summary-type questions. (P. 23)

There is no real cure for this potential problem, although the split-ballot technique may help the problem to at least be identified when the question order is reversed on a subset of the questionnaires. What is most important is to be aware of the potential for problems due to question order and carefully evaluate the likelihood of their occurrence in any particular questionnaire. Those who survey results should mention, at least in a footnote, the order in which key questions were asked when more than one such question was used (Labaw 1980). Questionnaires should conform to several other organizational guidelines as well:

- Major topic divisions within the questionnaire should be organized in separate sections, and each section introduced with a brief statement. This helps respondents understand the organization of the questionnaire.
- Instructions should be used liberally to minimize respondent confusion. Instructions should explain how each type of question is to be answered (such as circling a number or writing a response) in a neutral way that is not likely to influence responses. This type of instruction is particularly important with groups of questions that have standard answers laid out in a matrix format. The same set of response choices appear next to each question, because many respondents do not realize that they should circle one response on each line. Instructions also should route respondents through skip patterns.
- Instructions may also be used to clarify for respondents why some information is included in the questionnaire (such as the implicit instruction to skip a set-off box of lines that are "For coding purposes only").
- The questionnaire should look attractive, be easy to complete, and have substantial open space. Resist the temptation to cram as many questions as possible onto one page. Response choices should be printed in a font or format and location different from the questions (such as in all capital letters down the middle of the page).
- Response choices should be designated by numbers to facilitate coding and data entry after the questionnaire is completed.

There were several sections in the NVAMW Survey, including sections on fear of violence and accommodation behavior, previous relationships, current partner (spouse, boyfriend or girlfriend) characteristics, sexual victimization, physical victimization, stalking victimization, violence in current relationships, and threat victimization. Exhibit 7.10 presents the introduction and a portion of the section on stalking victimization given to female respondents. Even though this questionnaire was given over the phone by professional interviewers, notice the specific instructions provided throughout.

The **cover letter** for a mailed questionnaire and the introductory statement read by interviewers in telephone or in-person interviews are critical to the survey's success. Similar to the context set by the title of the survey, the initial statement of the cover letter sets the tone for the entire questionnaire. For example, the first thing interviewers said to respondents of the NVAMW Survey was, "Hello, I'm _____ from SRBI, the national research organization. We are conducting a national survey on personal safety for the Center for Policy Research, under a grant from the federal government." Notice that even though the survey's primary purpose was to uncover incidents of victimization, it was presented to respondents as a survey interested in issues of personal safety. This was done to increase the probability of respondents disclosing incidents of victimization even if they did not perceive them as crimes. Also note that the introductory statement disclosed the researcher's affiliation and the project sponsor. In addition, the purposes of the survey should also be briefly described and a contact number of those who wish to ask questions or register complaints should be included.

EXHIBIT 7.10   Section of the NVAMW Survey on Stalking Victimizations Given to Female Respondents

SECTION H: STALKING VICTIMIZATION

H1. Now I'd like to ask you some questions about following or harassment you may have experienced on more than one occasion by strangers, friends, relatives, or even husbands and partners. Not including all bill collectors, telephone solicitors or other sales people, has anyone, male or female, ever. . . . MARK ALL THAT APPLY

    01 Followed you or spied on you?
    02 Sent you unsolicited letters or written correspondence?
    03 Made unsolicited phone calls to you?
    04 Stood outside your home, school, or workplace?
    05 Showed up at places you were even though he or she had no business being there?
    06 Left unwanted items for you to find?
    07 Tried to communicate with you in ways against your will?
    08 Vandalized your pregnancy or destroyed something you loved?
    09 (Volunteered) Don't Know   GO TO SECTION I
    10 (Volunteered) Refused   GO TO SECTION I
    11 (Volunteered) None   GO TO SECTION I

H2. If H1 = ANY OF 1–8 (RESPONDENT HAS BEEN STALKED) GO TO H3, ELSE GO TO SECTION I.

H3. Has anyone ever done any of these things to you on more than one occasion?

    1 Yes
    2 No   GO TO SECTION I
    3 (Volunteered) Don't Know   GO TO SECTION I
    4 (Volunteered) Refused   GO TO SECTION I

H4. How many different people have ever done this to you on more than one occasion?
    _____ Number of people [RANGE IS 1–97]
    98 (Volunteered) Don't Know   GO TO SECTION I
    99 (Volunteered) Refused   GO TO SECTION I

H5. Was this person or these persons. . . . MARK ALL THAT APPLY

    01 Your current spouse?
    02 An ex-spouse?
    03 A male live-in partner?
    04 A female live-in partner?
    05 A relative?
    06 Someone else you know?
    07 A stranger?
    08 (Volunteered) Don't Know
    09 (Volunteered) Refused

Source: Tjaden & Thoennes 2000.

A carefully prepared cover letter or initial statement should increase the response rate and result in more honest and complete answers to the survey questions; a poorly prepared cover letter or initial statement can have the reverse effects. There is no opportunity to clarify misunderstandings.

The cover letter or introductory statement must have the following characteristics:

Credible. The letter should establish that the research is conducted by a researcher or organization the respondent is likely to accept as a credible, unbiased authority. Research conducted by government agencies, university personnel, and recognized research organizations (such as Gallup or RAND) is usually considered credible. On the other hand, a questionnaire from an animal rights group on the topic of animal rights will probably be viewed as biased.

Personalized. The cover letter should include a personalized salutation (not just "Dear Student"), refer to the respondent in the second person ("Your participation . . ."), and close with the researcher's signature.

Interesting. The statement should interest the respondent in the contents of the questionnaire. Never make the mistake of assuming that what interests you will also interest your respondents. Try to put yourself in their shoes before composing the statement, and then test your appeal with a variety of potential respondents.

Responsible. Reassure the respondent that the information you obtain will be treated confidentially, and include a phone number the respondent can use if he or she has questions or would like a summary of the final report. Point out that the respondent's participation is completely voluntary (Dillman 1978:165–172). For example, in the NVAMW Survey, respondents were told, "I will be asking you about your personal experiences and opinions. You don't have to answer any questions you don't want to. All your answers will be treated as strictly confidential. Your participation is completely voluntary, but very important to the study. You can confirm the authenticity of the survey by calling our 800 number and asking for the Women's Safety Survey Coordinator."

SURVEY DESIGNS

The five basic survey designs are the mailed survey, group-administered survey, phone survey, in-person survey, and electronic survey. Exhibit 7.11 summarizes the typical features of the five different survey designs. Each survey has some unique advantages and disadvantages:

Manner of administration. The five survey designs differ in the manner in which the questionnaire is administered. Mailed, group, and electronic surveys are completed by the respondents themselves. During phone and in-person interviews, however, the researcher or a staff person asks the questions and records the respondent's answers.

Questionnaire structure. Survey designs also differ in the extent to which the content and order of questions are structured in advance by the researcher. Most mailed, group, phone, and electronic surveys are highly structured, fixing in advance the content and order of questions and

EXHIBIT 7.11 Typical Features of the Five Survey Designs

Design	Manner of Administration	Setting	Questionnaire Structure	Cost
Mailed survey	Self	Individual	Mostly structured	Low
Group survey	Self	Group	Mostly structured	Very low
Phone survey	Professional	Individual	Structured	Moderate
In-person interview	Professional	Individual	Structured or unstructured	High
Electronic survey	Self	Individual	Mostly structured	Very low

response choices. Some of these types of surveys, particularly mailed surveys, may include some open-ended questions (respondents write in their answers rather than checking off one of several response choices). In-person interviews are often highly structured, but they may include many questions without fixed response choices. Moreover, some interviews may proceed from an interview guide rather than a fixed set of questions. In these relatively unstructured interviews, the interviewer covers the same topics with respondents but varies questions according to the respondent's answers to previous questions. Extra questions are added as needed to clarify or explore answers to the most important questions.

Setting. Most surveys are conducted in settings where only one respondent completes the survey at a time; most mail and electronic questionnaires and phone interviews are intended for completion by only one respondent. The same is usually true of in-person interviews, although sometimes researchers interview several family members at once. On the other hand, a variant of the standard survey is a questionnaire distributed simultaneously to a group of respondents, who complete the survey while the researcher (or assistant) waits. Students in classrooms are typically the group involved, although this type of group distribution also occurs in surveys of employees and members of voluntary groups.

Cost. As mentioned earlier, in-person interviews are the most expensive type of survey. Phone interviews are much less expensive, but surveying by mail is cheaper yet. Electronic surveys are now the least expensive method because there are no interviewer costs, no mailing costs, and, for many designs, almost no costs for data entry. Of course extra staff time and expertise is required to prepare an electronic questionnaire.

Because of their different features, the five designs vary in the types of errors to which they are most prone and the situations in which they are most appropriate. The rest of this section focuses on the unique advantages and disadvantages of each design.

Mailed, Self-Administered Surveys

A **mailed (self-administered) survey** is conducted by mailing a questionnaire to respondents, who then administer the survey themselves. The central concern in this method of survey administration is maximizing the response rate. The final response rate is unlikely to be much above 80%, and almost surely will be below 70% unless procedures to maximize the response rate are precisely followed. A response rate below 60% is a disaster, and even a 70% response rate is not much more than minimally acceptable. It is hard to justify the representativeness of the sample if more than a third of those surveyed fail to respond.

Some ways to maximize the response rate (Fowler 1988:99–106; Mangione 1995:79–82; Miller 1991:144) include the following:

- Make the questionnaire attractive, with plenty of white space.
- Use contingent questions and skip patterns infrequently. When they are necessary, guide respondents visually through the pattern.
- Make individual questions clear and understandable to all the respondents. No interviewers will be on hand to clarify the meaning of the questions or probe for additional details.
- Use no more than a few open-ended questions. Respondents are likely to be put off by the idea of having to write out answers.
- Include a personalized and professional cover letter. Using an altruistic appeal (informing respondents that their response will do some good) seems to produce a response rate 7% higher than indicating that respondents will receive something for their participation.
- Have a credible research sponsor. According to one investigation, a sponsor known to respondents may increase the rate of response by as much as 17%. The next most credible sponsors are the state headquarters of an organization and then other people in a similar field. Publishing firms, college professors or students, and private associations elicit the lowest response rates.
- Write an identifying number on the questionnaire so you can determine the nonrespondents.
- A small incentive can help. Even a coupon or ticket worth $2 can be enough to increase the response rate.
- Include a stamped, self-addressed return envelope with the questionnaire.

Most important, use follow-up mailings to encourage initial nonrespondents to return a completed questionnaire. Dillman (1978) recommends a standard procedure for follow-up mailings:

1. Send a reminder postcard, thanking respondents and reminding nonrespondents, to all sample members 2 weeks after the initial mailing.

2. Send a replacement questionnaire with a new cover letter only to nonrespondents about 3 or 4 weeks after the initial mailing.

3. Send another replacement questionnaire with a new cover letter 8 weeks after the initial mailing by certified mail if possible (it's pretty expensive). If enough time and resources are available for telephone contacts or in-person visits for interviews, they will also help.

If Dillman's procedures are followed, and the guidelines for cover letters and question-naire design are followed, the response rate is almost certain to approach 70%. One review of studies using Dillman's method to survey the general population indicates that the average response to a first mailing will be about 24%; the response rate will rise to 42% after the postcard follow-up, to 50% after the first replacement questionnaire, and to 72% after a second replacement questionnaire is sent by certified mail (Dillman et al. 1974).

The response rate may be higher with particular populations surveyed on topics of interest to them, and it may be lower with surveys of populations that do not have much interest in the topic. When a survey has many nonrespondents, getting some ideas about who they are by comparing late respondents to early respondents can help to determine the likelihood of bias due to the low rate of response. If the early respondents are more educated or more interested in the topic of the questionnaire, the sample may be biased; if the respondents are not more educated or more interested than nonrespondents, the sample will be more credible.

Related to the threat of nonresponse in mailed surveys is the hazard of incomplete response. Some respondents may skip some questions or just stop answering questions at some point in the questionnaire. Fortunately, this problem does not often occur with well-designed questionnaires. Potential respondents who decide to participate in the survey will usually complete it. But there are many exceptions to this observation, since questions that are poorly written, too complex, or about sensitive personal issues simply turn off some respondents. Revising or eliminating such questions during the design phase should minimize the problem.

Group-Administered Surveys

A **group-administered survey** is completed by individual respondents assembled in a group. The response rate is not usually a concern in surveys that are distributed and collected in a group setting because most group members will participate. The difficulty with this method is that assembling a group is seldom feasible because it requires a captive audience. Individuals going about their daily activities are usually not amenable to group-administered surveys. With the exception of students, employees, members of the armed forces, and some institutionalized populations, most populations cannot be sampled in such a setting.

One issue of special concern with group-administered surveys is the possibility that respondents will feel coerced to participate and as a result will be less likely to answer questions honestly. Also, because administering a survey to a group requires approval of the group's administrators and this sponsorship is made quite obvious by the fact that the survey is conducted on the organization's premises, respondents may infer that the researcher is not at all independent of the sponsor. No complete solution to this problem exists, but it helps to make an introductory statement that emphasizes the researcher's independence and gives participants a chance to ask questions about the survey. The sponsor should also understand the need to keep a low profile and to allow the researcher both control over the data and autonomy in report writing.

Surveys by Telephone

In a **phone survey**, interviewers question respondents over the phone and then record respondents' answers. Phone interviewing has become a very popular method of conducting surveys in the United States because almost all families have phones. But two matters may undermine the validity of a phone survey: not reaching the proper sampling units and not getting enough complete responses to make the results generalizable.

Reaching Sampling Units

Today, drawing a random sample is easier than ever due to random digit dialing (Lavrakas 1987). A machine calls random phone numbers within designated exchanges, regardless of whether the numbers are published. When the machine reaches an inappropriate household (such as a business in a survey directed to the general population), the phone number is simply replaced with another.

The NVAMW Survey collected a national probability sample of 8,000 English- and Spanish-speaking women and 8,000 English- and Spanish-speaking men 18 years of age and older residing in households throughout the United States through random digit dialing (RDD). The generation of the NVAMW Survey sample is described below:

> The geographic frame for the sample was the fifty states and the District of Columbia. Within this geographic frame SRBI systematically selected household telephone bands and conducted a random digit dialing (RDD) sampling screen for gender and age eligible respondents. In households with more than one eligible respondent a designated respondent was randomly selected using the "most recent birthday method." If the interviewer encountered a language barrier, either with the person answering the phone or with the designated respondent, the interviewer thanked the person and terminated the call. If a case was designated as Spanish language, it was turned over to a Spanish-speaking interviewer.

However households are contacted, the interviewers must always ask a series of questions at the start of the survey to ensure that they are speaking to the appropriate member of the household. Exhibit 7.12 displays the beginning of the phone interview schedule utilized by the NVAMW Survey. This example shows how to distinguish appropriate and inappropriate households in a phone survey, so the interviewer is guided to the correct respondent.

Maximizing Response to Phone Surveys

Three issues require special attention in phone surveys. First, because people often are not home, multiple call-backs will be necessary for many sample members. The failure to call people back was one reason for the discrepancy between poll predictions and actual votes in the 1988 presidential race between George Bush and Michael Dukakis. Kohut (1988) found that if pollsters in one Gallup poll had stopped attempting to contact unavailable respondents after one call, a 6-percentage-point margin for Bush would have been replaced by a 2-point margin for Dukakis. Those with more money and education are more likely to be away from home, and more likely to vote Republican.

In addition, interviewers must be prepared for distractions if the respondent is interrupted by other household members. Sprinkling interesting questions throughout the questionnaire may help to maintain respondent interest. In general, rapport between the interviewer and the respondent is likely to be lower with phone surveys than in-person interviews, as respondents may tire and refuse to answer all the questions (Miller 1991:166).

EXHIBIT 7.12   Phone Interview Procedures for Respondent Designation in the NVAMW Survey

CONTROL FORM

Respondent Number:

ID   1:1-6

Interview Date:

MO   1:7-8
DAY   1:9-10
YEAR   1:11-12

Telephone Number:

PHONE   1:13-22

Interviewer Gender: 1 Male   2 Female

IGENDER   1:23

Respondent Gender: 1 Male   2 Female

RGENDER   1:24

Zip Code:

ZIP   1:25-29

FIPS Code:

FIPS   1:30-34

State Code:

01	Alabama	27	Montana
02	Alaska	28	Nebraska
03	Arizona	29	Nevada
04	Arkansas	30	New Hampshire
05	California	31	New Jersey
06	Colorado	32	New Mexico
07	Connecticut	33	New York
08	Delaware	34	North Carolina
09	District of Columbia	35	North Dakota
10	Florida	36	Ohio
11	Georgia	37	Oklahoma
12	Hawaii	38	Oregon
13	Idaho	39	Pennsylvania
14	Illinois	40	Rhode Island
15	Indiana	41	South Carolina
16	Iowa	42	South Dakota
17	Kansas	43	Tennessee
18	Kentucky	44	Texas
19	Louisiana	45	Utah
20	Maine	46	Vermont
21	Maryland	47	Virginia
22	Massachusetts	48	Washington
23	Michigan	49	West Virginia
24	Minnesota	50	Wisconsin
25	Mississippi	51	Wyoming
26	Missouri		

STATE   **1:35-36**

Region Code:

01	New England	06 ESC
02	Mid-Atlantic	07 WSC
03	ENC	08 Mountain
04	WNC	09 Pacific
05	South Atlantic	

REGION   **1:37-38**

Survey Wave:

1	Cf/m	4	Ff/m
2	Df/m	5	Af
3	Ef/m	6	Bf

WAVE   **1:39**

CF2 How many persons, age 18 and older, live in this household?
[RANGE IS 0-20, 21=Don't know; 22=Refused]

PERS18   **1:40-41**

CF3 How many of these adults are:

A Women [RANGE IS 0-99]
B Men [RANGE IS 0-99]in household
C Refused [1=Don't know/refused; O = Other]

WOMEN18   **1:42-43**
MEN18   **1:44-45**
REF18   **1:46-47**

Source: Tjaden & Thoennes 2000.

Phone surveys also must cope with difficulties due to the impersonal nature of phone contact. Visual aids cannot be used, so the interviewer must be able to verbally convey all information about response choices and skip patterns. With phone surveys, instructions for the interviewer must clarify how to ask each question, and the response choices must be short.

Careful interviewer training is essential for phone surveys. Below is a brief description of how SRBI interviewers were trained before conducting the NVAMW Survey:

> Because of the complexity of the survey, only the most experienced SRBI interviewers worked on the survey. Before fielding the survey, interviewers received specialized training on the general principles of survey research and the requirements of the study at hand. Interviewers were also trained to recognize and respond appropriately to cues that the respondent may have been concerned about being overheard. Telephone numbers of local support services (e.g., domestic violence shelters, rape crisis hotlines, child protective services) were offered to respondents who disclosed current abuse or appeared in distress.

Procedures can be standardized more effectively, quality control maintained, and processing speed maximized when phone interviewers are assisted by computers. This **computer-assisted telephone interview** has become known as **CATI**, and most large surveys are now performed in this way. There are several advantages to using CATI, but perhaps the foremost advantage is that data collection and data entry can occur concurrently. Second, the CATI system has several machine edit features that help to minimize data entry error. For example, by automatically assigning single-punch fields of appropriate width for each data item in the questionnaire, the CATI system eliminates the possibility of over-punching and the possibility of including blanks as legitimate values. Third, by programming the skip patterns into its data entry program, the CATI system ensures that the aggregated database is comprehensive and accurate.

In summary, phone surveying is the best method to use for relatively short surveys of the general population. Response rates in phone surveys tend to be very high, often above 80%, because few individuals will hang up on a polite caller or refuse to stop answering questions (at least within the first 30 minutes or so). The NVAMW Survey obtained a response rate of 72% in the female survey and 69% in the male survey.

In-Person Interviews

What is unique to the **in-person interview**, compared to the other survey designs, is the face-to-face social interaction between interviewer and respondent. If money is no object, in-person interviewing is often the best survey design.

In-person interviewing has several advantages. Response rates are higher for this survey design than with any other, when potential respondents are approached by a courteous interviewer. For example, respondents for the NCVS actually stay in the sample for 3 years. The first of these surveys is performed in person. This is one reason the NCVS obtains a very high response rate of approximately 95%. In addition, in-person interviews can be much longer than mailed or phone surveys; the questionnaire can be complex, with both open-ended and closed-ended questions and frequent branching patterns; the order in which questions are read and answered can be controlled by the interviewer; the physical and social circumstances of the interview can be monitored; and respondents' interpretations of questions can be probed and clarified.

Researchers must be alert to some special hazards due to the presence of an interviewer. Respondents should experience the interview process as a personalized interaction with an interviewer who is very interested in the respondent's experiences and opinions. At the same time, however, every respondent should have the same interview experience and be asked the same questions in the same way by the same type of person, who reacts similarly to the answers. Therein lies the researcher's challenge: to plan an interview process that will be personal, engaging, consistent, and nonreactive and to hire interviewers who can carry out the plan. Without a personalized approach, the rate of response will be lower and answers will be less thoughtful and potentially less valid. Without a consistent approach, information obtained from different respondents will not be comparable because it is less reliable and less valid.

Balancing Rapport and Control

Adherence to some basic guidelines for interacting with respondents can help interviewers maintain an appropriate balance between personalization and standardization:

- Project a professional image in the interview, that of someone who is sympathetic to the respondent but nonetheless has a job to do.
- Establish rapport at the outset by explaining what the interview is about and how it will work and by reading the consent form. Ask the respondent if he or she has any questions or concerns, and respond to these honestly and fully. Emphasize that everything the respondent says is confidential.
- During the interview, ask questions at a close but not intimate distance. Stay focused on the respondent, and be certain your posture conveys interest. Maintain eye contact, respond with appropriate facial expressions, and speak in a conversational tone of voice.
- Be sure to maintain a consistent approach; deliver each question as written and in the same tone of voice. Listen empathically, but avoid self-expression or loaded reactions.
- Repeat questions if the respondent is confused. Use nondirective probes such as "Can you tell me more about that?" for open-ended questions.

As with phone interviewing, computers can be used to increase control of the in-person interview. In a **computer-assisted personal interviewing (CAPI)** project, interviewers carry a laptop computer programmed to display the interview questions and process the responses that the interviewer types in, as well as to check that these responses fall within the allowed ranges. Interviewers seem to like CAPI, and the data obtained are at least as good quality as a non-computerized interview (Shepherd et al. 1996). **Computer-assisted self interviewing (CASI)** is also an alternative. With audio-CASI, respondents interact with a computer-administered questionnaire by using a mouse and following audio instructions

Electronic surveys can be prepared in two ways (Dillman 2000:352–354). **E-mail surveys** can be sent as messages to respondent e-mail addresses. Respondents then mark their answers in the message and send them back to the researcher. This approach is easy for researchers to develop and for respondents to use. However, this approach is cumbersome for surveys that are more than four or five pages in length. By contrast, **Web surveys** are designed on a server controlled by the researcher; respondents are then asked to visit the Web site and respond to the Web questionnaire by checking answers. This approach requires more programming by the researcher and in many cases requires more skill on the part of the respondent. However, Web surveys can be quite long, with questions that are inapplicable to a given respondent hidden from them so that the survey may actually seem much shorter than it is.

Web surveys are becoming the more popular form of Internet survey because they are so flexible. The design of the questionnaire can include many types of graphic and typographic features. Respondents can view definitions of words or instructions for answering questions by clicking on linked terms. Lengthy sets of response choices can be presented with pull-down menus. Pictures and audio segments can be added when they are useful. Because answers are recorded directly in the researcher's database, data entry errors are almost eliminated and results can be reported quickly.

The most important drawback to either Internet survey approach is the large fraction of households that are not yet connected to the Internet. For special populations with high rates of Internet use, though, the technology makes possible fast and effective surveys. Another problem researchers must try their best to avoid is creating survey formats that are so complicated that some computers cannot read them or would display them in a way that differs from what the researchers intended. Also, access to a Web survey must be limited to the sample members, perhaps by requiring use of a personal identification number (PIN) (Dillman 2000:353–401).

Computerized **interactive voice response (IVR)** systems already allow the ease of Internet surveys to be achieved with a telephone-based system. In IVR surveys, respondents receive automated calls and answer questions by pressing numbers on their touchtone phones or speaking numbers that are interpreted by computerized voice recognition software. These surveys can also record verbal responses to open-ended questions for later transcription. Although they present some difficulties when many answer choices must be used or skip patterns must be followed, IVR surveys have been used successfully with short questionnaires and when respondents are highly motivated to participate (Dillman 2000:402–411). When these conditions are not met, potential respondents may be put off by the impersonality of this computer-driven approach.

Mixed-Mode Surveys

Survey researchers increasingly are combining different survey designs. **Mixed-mode surveys** allow the strengths of one survey design to compensate for the weaknesses of another and can maximize the likelihood of securing data from different types of respondents. For example, a survey may be sent electronically to sample members who have e-mail addresses and mailed to those who do not. Alternatively, nonrespondents in a mailed survey may be interviewed in person or over the phone. As noted previously, an interviewer may use a self-administered questionnaire to present sensitive questions to a respondent.

Mixing survey designs like this makes it possible that respondents will give different answers to different questions because of the mode in which they are asked, rather than because they actually have different opinions. However, use of what Dillman (2000:232–240) calls "unimode design" reduces this possibility substantially. A unimode design uses questions and response choices that are least likely to yield different answers according to the survey mode that is used. Unimode design principles include use of the same question structures, response choices, and skip instructions across modes, as well as using a small number of response choices for each question.

A Comparison of Survey Designs

Which survey design should be used when? Group-administered surveys are similar in most respects to mailed surveys, except they require the unusual circumstance of having access to the sample in a group setting. We therefore do not need to consider this survey design by itself; what applies to mailed survey designs applies to group-administered survey designs, with the exception of sampling issues. Thus, we can focus our comparison on the four survey designs that involve the use of a questionnaire with individuals sampled from a larger population: mailed surveys, phone surveys, in-person surveys, and electronic surveys. Exhibit 7.14 summarizes their strong and weak points.

The most important consideration in comparing the advantages and disadvantages of the four survey designs is the likely response rate they will generate. Because of the great weakness of mailed surveys in this respect, they must be considered the least preferred survey design from a sampling standpoint. However, researchers may still prefer a mailed survey when they have to reach a widely dispersed population and do not have enough financial resources to hire and train an interview staff or to contract with a survey organization that already has an interview staff available in many locations.

Contracting with an established survey research organization for a phone survey is often the best alternative to a mailed survey. The persistent follow-up attempts necessary to secure an adequate response rate are much easier over the phone than in person. But the process is not simple:

> Initial telephone contact with households was attempted during hours of the day and days of the week which had the greatest probability of respondent contact (between 5:30 P.M. and 10:00 P.M. on weekdays, between 9:00 A.M. and 10:00 P.M. on Saturdays, and between 10:00 A.M. and 10:00 P.M. on Sundays). Interviewers made five attempts to ringing unanswered telephones on different days and at different times over a period of at least three weeks in order to obtain the highest possible response rate. If the interview could not be conducted or completed at the time of the initial contact, the interviewer rescheduled the interview at a time convenient to the respondent. (From a description of the NVAMW Survey)

In-person surveys are clearly preferable in terms of the possible length and complexity of the questionnaire itself, as well as the researcher's ability to monitor conditions while the questionnaire is being completed. Mailed surveys often are preferable for asking sensitive questions, although this problem can be lessened in an interview by giving respondents a

EXHIBIT 7.14   Advantages and Disadvantages of Four Survey Designs

Characteristics of Design	In-Person Survey	Mail Survey	Phone Survey	Electronic Survey
Representative Sample				
Opportunity for inclusion is known				
For completely listed populations	High	High	High	Medium
For incompletely listed populations	High	Medium	Medium	Low
Selection within sampling units is controlled (e.g., specific family members must respond)	High	Medium	High	Low
Respondents are likely to be located	Medium	High	High	Low
If samples are heterogeneous	High	Medium	High	Low
If samples are homogeneous and specialized	High	High	High	High
Questionnaire Construction and Question Design				
Allowable length of questionnaire	High	Medium	Medium	Medium
Ability to include				
Complex questions	Medium	Low	High	High
Open questions	Low	High	High	Medium
Screening questions	Low	Low	High	
Tedious, boring questions	High	High	High	High
Ability to control question sequence	Low	High	High	High
Ability to ensure questionnaire completion	Medium	High	High	High
Distortion of Answers				
Odds of avoiding social desirability bias	High	Medium	Low	High
Odds of avoiding interviewer distortion	Low	High	Medium	
Odds of avoiding contamination by others	High	Medium	Medium	Medium
Administrative Goals				
Odds of meeting personnel requirements	High	High	Low	High
Odds of implementing quickly	Low	High	Low	High
Odds of keeping costs low	High	Medium	Low	High

Source: Adapted from Dillman 1978:74–75. *Mail and Telephone Surveys: The Total Design Method.* Copyright © 1978 Don A. Dillman. Reprinted by Permission of John Wiley & Sons, Inc.

separate sheet to fill out on their own. Although interviewers may themselves distort results, either by changing the wording of questions or failing to record answers properly, this problem can be lessened by careful training, monitoring, and tape-recording the answers.

A phone survey limits the length and complexity of the questionnaire but offers the possibility of very carefully monitoring interviewers (Dillman 1978; Fowler 1988:61–73). For the NVAMW Survey, SRBI assures quality control over its interviewers in the following manner:

> Throughout the project, interviewers were silently monitored by a supervisor at least twice during each of their interviewing shifts to evaluate the manner in which interviewers were conducting the interview and the accuracy with which they were entering responses.

The advantages and disadvantages of electronic surveys must be weighed in light of the capabilities at the time that the survey is to be conducted. At this time, too many people lack Internet connections for general use of Internet surveying, and too may people who have computers lack adequate computer capacity for displaying complex Web pages.

These various points about the different survey designs lead to two general conclusions. First, in-person interviews are the strongest design and generally preferable when sufficient resources and a trained interview staff are available; telephone surveys have many of the advantages of in-person interviews at a much lower cost. Second, a decision about the best survey design for any particular study must take into account the unique features and goals of the study.

COMBINING METHODS

Conducting qualitative interviews can often enhance a research design that uses primarily quantitative survey measurement techniques. Qualitative data can provide information about the quality of standardized case records and quantitative survey measures, as well as offer some insight into the meaning of particular fixed responses.

Adding Qualitative Data

It makes sense to use official records to study the treatment of juveniles accused of illegal acts because these records document the critical decisions to arrest, to convict, or to release (Dannefer & Schutt 1982). But research based on official records can be only as good as the records themselves. In contrast to the controlled interview process in a research study, there is little guarantee that officials' acts and decisions were recorded in a careful and unbiased manner.

Case Study: Juvenile Court Records

Research on official records can be strengthened by interviewing officials who create the records or by observing them while they record information. A participant observation study of how probation officers screened cases in two New York juvenile court intake units

EXHIBIT 7.15   Researchers' and Juvenile Court Workers' Discrepant Assumptions

Researcher Assumption	Intake Worker's Assumption
• Being sent to court is a harsher sanction than diversion from court.	• Being sent to court often results in more lenient and less effective treatment.
• Screening involves judgments about individual juveniles.	• Screening centers on the juvenile's social situation.
• Official records accurately capture case facts.	• Records are manipulated to achieve the desired outcome.

Source: Needleman 1981:248–256.

shows how important such information can be (Needleman 1981). As indicated in Exhibit 7.15, Needleman (1981) found that the concepts most researchers believe they are measuring with official records differ markedly from the meaning attached to these records by probation officers. Researchers assume that sending a juvenile case to court indicates a more severe disposition than retaining a case in the intake unit, but probation officers often diverted cases from court because they thought the court would be too lenient. Researchers assume that probation officers evaluate juveniles as individuals, but in these settings, probation officers often based their decisions on juveniles' current social situation (e.g., whether they were living in a stable home), without learning anything about the individual juvenile. Perhaps most troubling for research using case records, Needleman found that probation officers decided how to handle cases first and then created an official record that appeared to justify their decisions.

ETHICAL ISSUES IN SURVEY RESEARCH

Survey research usually poses fewer ethical dilemmas than do experimental or field research designs. Potential respondents to a survey can easily decline to participate, and a cover letter or introductory statement that identifies the sponsors of and motivations for the survey gives them the information required to make this decision. The methods of data collection are quite obvious in a survey, so little is concealed from the respondents. The primary ethical issue in survey research involves protecting respondents.

Protection of Respondents

If the survey could possibly have any harmful effects for the respondents, these should be disclosed fully in the cover letter or introductory statement. The procedures used to reduce such effects should also be delineated, including how the researcher will keep interviews confidential and anonymous. In addition, surveys such as the NVAMW Survey and NCVS that attempt to measure sensitive subject matter such as rape and intimate-perpetrated assault should also have other protections in place. Johnson (1996) has stated the most important of these succinctly: "Researchers must never lose sight of the possibility that with every telephone call, respondent[s] could be living with an abusive [partner] and that their safety could be jeopardized should they learn of the content of the survey" (p. 52). In addition, by asking respondents to recall incidents of abuse and violence, there is always the possibility of causing victims serious emotional trauma. How can researchers ameliorate the negative consequences that responding to these surveys may have? What responsibility do researchers have in providing respondents safety should they need it? We believe these important questions have received far too little attention. For the most part, the NCVS has done nothing to protect the safety of its respondents or offer guidance in finding emotional support should they need it. In contrast, respondents in the NVAMW Survey were given a toll-free number they could call if they needed to suddenly hang up during the course of the interview (e.g., if they felt in danger). In addition, interviewers of the NVAMW Survey were instructed to contact an attending supervisor at the first sign a respondent was becoming upset or emotionally distraught. These supervisors were provided with a source book from the National Domestic Violence Coalition that listed rape crisis and domestic violence hotline numbers from around the country.

Confidentiality

Do any of the questions have the potential to embarrass respondents or otherwise subject them to adverse consequences such as legal sanctions? If the answer to this question is no, and it often is in surveys about general social issues, other ethical problems are unlikely. But if the questionnaire includes questions about attitudes or behaviors that are socially stigmatized or generally considered to be private or questions about actions that are illegal, the researcher must proceed carefully and ensure that respondents' rights are protected.

The first step to take with potentially troublesome questions is to consider omitting or modifying them. Researchers often include some questions in surveys just out of curiosity or out of a suspicion that the questions might prove to be important. If sensitive questions fall into this category, they probably should be omitted. There is no point in asking, "Have you ever been convicted of a felony?" if the answers are unlikely to be used in the analysis of survey results.

Many surveys do include some essential questions that might prove damaging to the subjects if their answers were disclosed, particularly surveys interested in delinquent or criminal offending behavior. To prevent any possibility of harm to subjects due to disclosure of such information, it is critical to preserve subject confidentiality. No one other than research personnel should have access to information that could be used to link respondents to their responses, and even that access should be limited to what is necessary for specific research purposes. Only numbers should be used to identify respondents on their questionnaires, and the researcher should keep the names that correspond to these numbers in a safe, private location, unavailable to staff and others who might otherwise come across them. Follow-up mailings or contact attempts that require linking the ID numbers with names and addresses should be carried out by trustworthy assistants under close supervision.

Only if no identifying information about respondents is obtained can surveys provide true **anonymity** to respondents. In this way, no identifying information is ever recorded to link respondents with their responses. However, the main problem with anonymous surveys is that they preclude follow-up attempts to encourage participation by initial non-respondents, and they prevent panel designs, which measure change through repeated surveys of the same individuals. In-person surveys rarely can be anonymous because an interviewer must in almost all cases know the name and address of the interviewee. However, phone surveys that are meant only to sample opinion at one point in time, as in political polls, can safely be completely anonymous. When no follow-up is desired, group-administered surveys also can be anonymous. To provide anonymity in a mail survey, the researcher should omit identifying codes from the questionnaire but could include a self-addressed, stamped postcard so the respondent can notify the researcher that the questionnaire has been returned, without being linked to the questionnaire itself (Mangione 1995:69).

Any survey can allow anonymous responses to a subset of particularly sensitive questions. A tear-off sheet containing these questions and a separate return envelope, without identifying information, can be included with a mailed survey. In an in-person interview, this special section can be left with the respondent to be completed later and returned by mail. Of course, a response obtained in this way cannot be linked with the response of the same subject to the same question in some later follow-up survey. But if it increases the response rate, this method can provide more valid results for the initial survey.

CONCLUSION

Survey research is an exceptionally efficient and productive method for investigating a wide array of social research questions. In addition to the potential benefits for social science, considerations of time and expense frequently make a survey the preferred data-collection method. One or more of the four survey designs reviewed in this chapter can be applied to almost any research question. It is no wonder that surveys have become the most popular research method in sociology and that they frequently influence discussion and planning about important social and political questions.

The relative ease of conducting at least some types of survey research leads many people to imagine that no particular training or systematic procedures are required. Nothing could be further from the truth. As a result of this widespread misconception, you will encounter a great many nearly worthless survey results. You must be prepared to carefully examine the procedures used in any survey before accepting its finding as credible. And if you decide to conduct a survey, you must be prepared to invest the time and effort required to follow proper procedures.

KEY TERMS

Anonymity
Cognitive interview
Computer-assisted personal
    interviewing (CAPI)
Computer-assisted self
    interviewing (CASI)
Contingent question
Cover letter
Double-barreled question
Double-negative question
Electronic survey
E-mail survey
Exhaustive responses
Fence-sitter
Filter question
Fixed-choice question
Floater
Group-administered survey
Idiosyncratic variation
Index

In-person interview
Interactive voice response (IVR)
Interpretive question
Interview schedule
Likert-type responses
Mailed (self-administered) survey
Mixed-mode survey
Mutually exclusive responses
Omnibus survey
Open-ended question
Phone survey
Pretest
Questionnaire
Reference period
Reliability measures
Scale
Skip pattern
Split-ballot design
Survey research
Web survey

HIGHLIGHTS

- Surveys are the most popular form of social research because of their versatility, efficiency, and generalizability. Many survey data sets, such as the General Social Survey (GSS), are available for social scientists to use in teaching and research.

- Surveys can fail to produce useful results due to problems in sampling, measurement, and overall survey design. Political polling can produce inconsistent results because of rapid changes in popular sentiment.

- A survey questionnaire or interview schedule should be designed as an integrated whole, with each question and section serving some clear purpose and complementing the others.

- Questions must be worded carefully to avoid confusing the respondents or encouraging a less-than-honest response. Inclusion of "don't know" choices and neutral responses may help, but the presence of such options also affects the distribution of answers. Open-ended questions can be used to determine the meaning that respondents attach to their answers. Answers to any survey questions may be affected by the questions that precede them in a questionnaire or interview schedule.

- Every questionnaire and interview schedule should be pretested on a small sample that is like the sample to be surveyed.

- The cover letter for a mailed questionnaire and the introductory statement for an interview should be credible, personalized, interesting, and responsible.

- Response rates in mailed surveys are typically well below 70% unless multiple mailings are made to nonrespondents and the questionnaire and cover letter are attractive, interesting, and fully planned. Response rates for group-administered surveys are usually much higher.

- Phone interviews using random digit dialing allow fast turnaround and efficient sampling. Multiple call-backs may be required, but once the people are contacted, most can be interviewed by phone for 30 to 45 minutes.

- In-person interviews have several advantages over other types of surveys: They allow longer and more complex interview schedules, monitoring of the conditions when the questions are answered, probing for respondents' understanding of the questions, and high response rates.

- Electronic surveys may be e-mailed or posted on the Web. Interactive voice-response systems using the telephone are another option. At this time, use of the Internet is not sufficiently widespread to allow e-mail or Web surveys of the general population, but these approaches can be fast and efficient for populations with high rates of computer use.

- Mixed-mode surveys allow the strengths of one survey design to compensate for the weaknesses of another. However, questions and procedures must be designed carefully, using "unimode design" principles, to reduce the possibility that responses to the same question will vary as a result of the mode of delivery.

- Most survey research poses few ethical problems because respondents are able to decline to participate. This option should be stated clearly in the cover letter or introductory statement. Special care must be taken when questionnaires are administered in group settings (to captive audiences) and when sensitive personal questions are to be asked; subject confidentiality should always be preserved.

EXERCISES

1. Read the original article reporting one of the surveys described in this book (check the text of the chapters for ideas). Critique the article using the questions presented in Appendix B as your guide. Focus particular attention on sampling, measurement, and survey design.

2. Write 8 to 10 questions for a one-page questionnaire on fear of crime among students. Include some questions to measure characteristics (such as income or year in school) that might help to explain the attitudes. Make all but one of your questions closed-ended.

3. By interviewing two students, conduct a preliminary pretest of the questionnaire you wrote for Exercise 2. Follow up the closed-ended questions with open-ended questions that ask the students what they meant by each response or what came to mind when they were asked each question. Take account of the answers when you revise your questions.

4. Make any necessary revisions to the questionnaire you wrote in Exercise 2. Write a cover letter that presumes the survey will be administered to students in a class at your school. Submit the questionnaire and cover letter to your instructor for comment and evaluation.

DEVELOPING A RESEARCH PROPOSAL

1. Write 10 questions for a one-page questionnaire that concerns your proposed research question. Your questions should operationalize at least three of the variables on which you have focused, including at least one independent and one dependent variable (you may have multiple questions to measure some variables). Make all but one of your

questions closed-ended. If you completed the "research proposal" exercises in Chapter 3, you can select your questions from the ones you developed for those exercises.

2. Conduct a preliminary pretest of the questionnaire by conducting cognitive interviews with two students or other persons like those to whom the survey is directed. Follow up the closed-ended questions with open-ended probes that ask the students what they meant by each response or what came to mind when they were asked each question. Take account of the feedback you receive when you revise your questions.

3. Polish up the organization and layout of the questionnaire, following the guidelines in this chapter. Prepare a rationale for the order of questions in your questionnaire. Write a cover letter directed to the appropriate population that contains appropriate statements about research ethics (human subjects issues).

Student Study Site

The companion Web site for *The Practice of Research in Criminology and Criminal Justice*, Third Edition

http://www.sagepub.com/prccj3

Visit the Web-based Student Study Site to enhance your understanding of the chapter content and to discover additional resources that will take your learning one step further. You can enhance your understanding of the chapters by using the comprehensive study material, which includes e-flashcards, Web exercises, practice self-tests, and more. You will also find special features, such as Learning from Journal Articles, which incorporates SAGE's online journal collection.

WEB EXERCISES

1. Go to the Social Information Gateway (SOSIG) at http://sosig.esrc.bris.ac.uk. Search SOSIG for electronic journal articles that use surveys to collect information on crime, criminal behavior, or criminal victimization. Find at least five articles and briefly describe each.

2. Go to the Council of European Social Science Data Archives (CESSDA) at www.nsd. uib.no/cessda/index.html. Go to "The Map." View the type of data provided by various European data banks. Browse the publications list or conduct a search at the site for at least three countries for studies that use surveys to collect information on crime, criminal behavior, or criminal victimization. What type of information did you find? If possible, do a cross-national comparison between the studies reviewed in Exercise 1 and the studies found in this exercise.

3. Who does survey research and how do they do it? These questions can be answered through careful inspection of ongoing surveys and the organizations that administer them at http://www.ciser.cornell.edu/info/polls.shtml. Spend some time reading about of the different survey research organizations, and write a brief summary of the types of research they conduct, the projects in which they are involved, and the resources they offer on their Web sites. What are the distinctive features of different survey research organizations?

4. Go to the Research Triangle Institute site at http://www.rti.org. Click on "Tools and Methods," then "Surveys," and then "Survey Design and Development." Read about their methods for computer-assisted interviewing (under "Survey Methods") and their cognitive

In this chapter you will learn how qualitative methods were used to illuminate the relationships, both individually and collectively, that gang members have with other social institutions (Decker & Van Winkle 1996). You will also get an inside look at community policing in action (Miller 1999). Throughout the chapter, you will learn, from a variety of other examples, that some of our greatest insights into social processes can result from what appear to be very ordinary activities: observing, participating, listening, and talking.

But you will also learn that qualitative research is much more than just doing what comes naturally in social situations. Qualitative researchers must keenly observe respondents, sensitively plan their participation, systematically take notes, and strategically question respondents. They must also prepare to spend more time and invest more of their whole selves than often occurs with experiments or surveys. Moreover, if we are to have any confidence in the validity of a qualitative study's conclusions, each element of its design must be reviewed as carefully as the elements of an experiment or survey.

The chapter begins with an overview of the major features of qualitative research, as reflected in Venkatesh's (1997) study of Blackstone (the fictitious name given to the residential area). The next section discusses the various approaches to participant observation research, which is the most distinctive qualitative method, and reviews the stages of research using participant observation. We then review, in some detail, the issues involved in intensive interviewing, before briefly explaining focus groups, an increasingly popular qualitative method. The last two sections cover issues that are of concern in any type of qualitative research project: analyzing the data collected and making ethical decisions. By the chapter's end, you should appreciate the hard work required to translate "doing what comes naturally" into systematic research, be able to recognize strong and weak points in qualitative studies, and be ready to do some research yourself.

FUNDAMENTALS OF QUALITATIVE METHODS

Qualitative methods can often be used to enrich experiments and surveys, and refer to three distinctive research designs: **participant observation**, **intensive interviewing**, and **focus groups**. Participant observation and intensive interviewing are often used in the same project; focus groups combine some elements of these two approaches into a unique data-collection strategy.

Participant observation A qualitative method for gathering data that involves developing a sustained relationship with people while they go about their normal activities.

Intensive interviewing A qualitative method that involves open-ended, relatively unstructured questioning in which the interviewer seeks in-depth information on the interviewee's feelings, experiences, and perceptions (Lofland & Lofland 1984:12).

Focus groups A qualitative method that involves unstructured group interviews in which the focus group leader actively encourages discussion among participants on the topics of interest.

Although these three qualitative designs differ in many respects, they share several features that distinguish them from experimental and survey research designs (Denzin & Lincoln 1994; Maxwell 1996; Wolcott 1995).

Collection primarily of qualitative rather than quantitative data. Any research design may collect both qualitative and quantitative data, but qualitative methods emphasize observations about natural behavior and artifacts that capture social life as it is experienced by the participants rather than in categories predetermined by the researcher.

Exploratory research questions, with a commitment to inductive reasoning. Qualitative researchers typically begin their projects seeking not to test preformulated hypotheses but to discover what people think and how and why they act in certain social settings. Only after many observations do qualitative researchers try to develop general principles to account for their observations.

A focus on previously unstudied processes and unanticipated phenomena. Previously unstudied attitudes and actions cannot adequately be understood with a structured set of questions or within a highly controlled experiment. Therefore, qualitative methods have their greatest appeal when we need to explore new issues, investigate hard-to-study groups, or determine the meaning people give to their lives and actions.

An orientation to social context, to the interconnections between social phenomena rather than to their discrete features. The context of concern may be a program, an organization, a case study, or a broader social context. For example, in Venkatesh's (1997) analysis of the social space occupied by a street gang, he observed:

> The result of their [Saint's street gang] corporatization was the emergence of a novel social space in Blackstone, that is, a new orientation to local geography in which the symbolic distinctions of local street gangs challenged the building-centered distinctions that had previously underwritten the power of the Councils. (P. 7)

A focus on human subjectivity, on the meanings that participants attach to events and people give to their lives. "Through life stories, people account for their lives. . . . The themes people create are the means by which they interpret and evaluate their life experiences and attempt to integrate these experiences to form a self-concept" (Kaufman 1986:24–25).

A focus on the events leading up to a particular event or outcome instead of general causal explanations. With its focus on particular actors and situations and the processes that connect them, qualitative research tends to identify causes of particular events embedded within an unfolding, interconnected action sequence (Maxwell 1996:20–21). The language of variables and hypotheses appears only rarely in the qualitative literature.

Reflexive research design. The design develops as the research progresses:

> Each component of the design may need to be reconsidered or modified in response to new developments or to changes in some other component. . . .
> The activities of collecting and analyzing data, developing and modifying theory, elaborating or refocusing the research questions, and identifying and eliminating

validity threats are usually all going on more or less simultaneously, each influencing all of the others. (Maxwell 1996:2–3)

Sensitivity to the subjective role of the researcher. Little pretense is made of achieving an objective perspective on social phenomena.

Miller and Crabtree (1999a) capture the entire process of qualitative research in a simple diagram (see Exhibit 8.1). In this diagram, qualitative research begins with the qualitative researcher reflecting on the setting and her relation to it and interpretations of it. The researcher then describes the goals and means for the research. This description is followed by *sampling* and *collecting* data, *describing* the data, and *organizing* those data. Thus, the *gathering process* and the *analysis process* proceed together, with repeated description and analysis of data as they are collected. As the data are organized, *connections* are identified between different data segments, and efforts are made to *corroborate* the credibility of these connections. This *interpretive process* begins to emerge in a written account that represents what has been done and how the data have been interpreted. Each of these steps in the research process informs the others and is repeated throughout the research process.

EXHIBIT 8.1   Qualitative Research Process

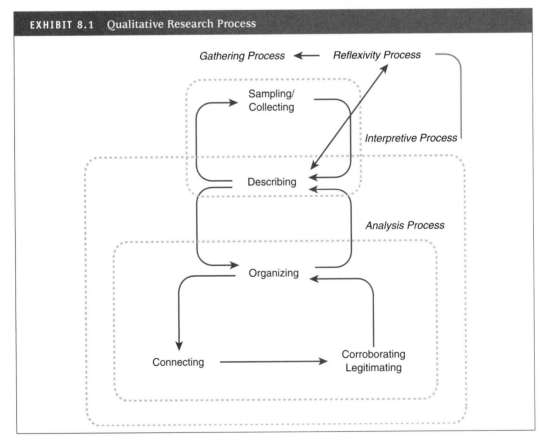

Source: Adapted from Miller & Crabtree 1999a:16.

Origins of Qualitative Research

Anthropologists and sociologists laid the foundation for modern qualitative methods while doing **field research** in the early decades of the twentieth century. Dissatisfied with studies of native peoples that relied on second-hand accounts and inspection of artifacts, anthropologists Franz Boas and Bronislaw Malinowski went to live in or near the communities they studied. Boas visited Native American villages in the American Northwest; Malinowski lived among New Guinea natives. Neither truly participated in the ongoing social life of those they studied (Boas collected artifacts and original texts, and Malinowski reputedly lived as something of a noble among the natives he studied), but both helped to establish the value of intimate familiarity with the community of interest and thus laid the basis for modern anthropology (Emerson 1983:2–5).

Many of sociology's field research pioneers were former social workers and reformers. Some brought their missionary concern with the spread of civic virtue among new immigrants to the Department of Sociology and Anthropology at the University of Chicago. Their successors continued to focus on sources of community cohesion and urban strain but came to view the city as a social science laboratory rather than as a focus for reform. They adopted the fieldwork methods of anthropology for studying the "natural areas" of the city and the social life of small towns (Vidich & Lyman 1994). By the 1930s, 1940s, and 1950s, qualitative researchers were emphasizing the value of direct participation in community life and sharing in subjects' perceptions and interpretations of events (Emerson 1983:6–13).

Case Study: Life in the Gang

The use of fieldwork techniques to study gangs has a long tradition in a variety of cities, including Thrasher's (1927) classic study of gangs in Chicago, and others, including Whyte (1943), Hagedorn (1988), Vigil (1988), Padilla (1992), Sanchez-Jankowski (1991), and Moore (1978, 1991), who spent over two decades studying the "home-boys" of Hispanic barrios all over the country. All these researchers employed a field-work approach to the study of gangs rather than the more structured approaches offered by quantitative methods.

You can get a better feel for qualitative methods by reading the following excerpts from Decker and Van Winkle's (1996) book about gangs, *Life in the Gang: Family, Friends, and Violence*, and by reasoning inductively from their observations. See if you can induce from these particulars some of the general features of field research. Ask yourself, "What were the research questions?" "How were the issues of generalizability, measurement, and causation approached?" "How did social factors influence the research?"

One of the first issues Decker and Van Winkle (1996) were challenged with was precisely defining a gang. After all, the term gang could refer to many groups of youth, including high school Debate Society or the Young Republicans. After reviewing the literature, Decker and Van Winkle developed a working definition of a gang as an "age-graded peer group that exhibits some permanence, engages in criminal activity, and has some symbolic representation of membership" (p. 31). To operationalize who was a gang member, they relied on self-identification. "Are you claiming. . . ." was a key screening question that was also verified, as often as possible, with other gang members.

There were several questions that Decker and Van Winkle (1996) were interested in:

Our study revolved around a number of activities, both gang and nongang related, that our subjects were likely to engage in. First, we were interested in motivations to join gangs, the process of joining the gang, the symbols of gang membership, the strength of associational ties, the structure or hierarchy within the gang, motivations to stay (or leave) the gang, and how this generation of St. Louis gangs began. The second set of issues concerned the activities gang members engaged in. These included such things as turf protection, drug sales and use, and violence, as well as conventional activities. An accurate picture of gang members must portray both the nature of their gang involvement and the legal status of their activities. . . . A unique feature of our work is its focus on families. There has been little research examining specifically the links between gang members and their family members. For this reason, we have separated the family from our analysis of other social institutions and devote special attention to this relationship. (Pp. 54–55)

With these research questions in mind, Decker and Van Winkle (1996) explain why they chose a fieldwork approach: "A single premise guided our study; the best information about gangs and gang activity would come from gang members contacted directly in the field" (p. 27). As stated earlier, Decker and Van Winkle combined two methods of qualitative data collection. With the help of a field ethnographer who spent the majority of each day "on the streets," direct observation was conducted along with the intensive interviewing conducted by Decker and Van Winkle.

The data in Decker and Van Winkle's (1996) study were obtained from the observations in the field and from the intensive interviews. As they state,

Learning about gangs and gang members can be best accomplished by hearing the gang member's story directly from the individuals involved. . . . We went to great lengths to ensure that each person we interviewed felt they had received the opportunity to "tell their story in their own words." (P. 45)

Because they did not rely on structured questionnaires with fixed-response formats, their data are primarily qualitative rather than quantitative.

As for their method, it was inductive. First they gathered data. Then, as data collection continued, they figured out how to interpret the data, how to make sense of the social situations they were studying. Their analytic categories ultimately came not from social theory but from the categories by which the gang members themselves described one another and their activities and how they made sense of their social world.

To summarize, Decker and Van Winkle's (1996) research began with exploratory questions and proceeded inductively throughout, developing general concepts to make sense of specific observations. Although the researchers were not gang members themselves, with observational data collected on the streets and transcripts from intensive interviews, Decker and Van Winkle were able to share many gang members' experiences and perspectives. They provided the field of criminology with in-depth descriptions and idiographic connections of sequences of events that could not have been obtained through other methodologies. They successfully used field research to explore human experiences in depth, carefully analyzing the social contexts in which the experiences occurred. As you will see, like Decker and Van Winkle's work, the goal of much qualitative research is to create a **thick description** of the setting being studied, a description that provides a sense of what it is like to experience that setting or group from the standpoint of the natural actors in that setting (Geertz 1973).

PARTICIPANT OBSERVATION

Other researchers have utilized a more direct observational strategy for studying gangs. For example, to illuminate the nuances and complexities of the role of a street gang in community social life, Venkatesh (1997) conducted intensive participant observation in Blackstone, a midsize public housing development located in a poor ghetto of a large midwestern city. As Venkatesh describes, "Having befriended these gang members, I moved into their world, accompanying them into Blackstone and other spaces where they were actively involved in illicit economic activities, member recruitment, and the general expansion of their street-based organization" (p. 4). As this quote eloquently depicts, participant observation, called fieldwork in anthropology, is a method of studying natural social processes as they happen (in the field rather than in the laboratory) and leaving them relatively undisturbed. It is the seminal field research method, a means for seeing the social world as the research subjects see it, in its totality, and for understanding subjects' interpretations of that world (Wolcott 1995:66). By observing people and interacting with them in the course of their normal activities, participant observers seek to avoid the artificiality of experimental designs and the unnatural structured questioning of survey research (Koegel 1987:8).

The term *participant observer* actually represents a continuum of roles (see Exhibit 8.2), ranging from being a complete observer who does not participate in group activities and is publicly defined as a researcher, to being a covert participant who acts just like other group members and does not disclose his or her research role. Many field researchers develop a role between these extremes, publicly acknowledging being a researcher but nonetheless participating in group activities. In some settings, it also is possible to observe covertly without acknowledging being a researcher or participating.

Choosing a Role

The first concern of all participant observers is to decide what balance to strike between observing and participating and whether to reveal their role as researchers. These decisions must take into account the specifics of the social situation being studied, the researcher's own background and personality, the larger sociopolitical context, and ethical concerns. The balance of participating and observing that is most appropriate also changes during most projects, often many times. And the researcher's ability to maintain either a covert or an overt role will be challenged many times.

Complete Observation

Miller (1999) adopted the role of a complete observer when she conducted research on community policing. Community policing, as most of you probably know, is an approach

EXHIBIT 8.2   The Observational Continuum

To study a political activist group...

You could take the role of complete observer:

You could take the role of participant and observer:

You could take the role of covert participant:

to policing that emphasizes building closer ties between police and members of the community. Miller was particularly interested in how gender affected the attitudes and behavior of neighborhood police officers (NPOs):

> I was curious as to whether the interpersonal dynamics I observed with Officer Terry [a female officer] would be like those when a male NPO was involved. I wanted to delve into the heads and hearts of the NPOs, to see for myself what worked in community policing and to see what they felt did not. I wanted to examine how such a paradigm shift in the theory and practice of policing would affect the officers who desire street action, and how they would assess their new "walk and talk" colleagues. (P. x)

In **complete observation**, researchers try to see things as they happen, without disrupting the participants. Along with intensive interviews with police officers, Miller (1999) also observed police officers on their daily shifts:

> Both neighborhood and patrol officers' shifts were observed, either on foot with neighborhood officers, or in squad cars with patrol officers. This component of the project also permitted gathering some observational information about citizens' reactions to police delivery of services. . . . Typically, we tried to work the same shifts as the neighborhood police officers, and we shadowed the NPO and each corresponding patrol officer during the same shift. Eight-hour shifts were evenly divided into four-hour blocks of walking in the neighborhood with the neighborhood officer and four-hour blocks of riding in the squad car with the patrol officer assigned to the same neighborhood. Shadowing both permitted a cross-check of how neighborhood officers perceived the role of patrol officers and of how patrol officers saw their role in conjunction with, or opposition to, the neighborhood policing concept. (Pp. 232–233)

As clearly depicted in this quote, the "shadowing" is visible. Thus, the researcher's very presence as an observer alters the social situation being observed. It is not natural in most social situations to have an observer present, one who will record at some point her or his observations for research and publication purposes. The observer thus sees what individuals do when they are being observed, which is not necessarily what they would do without an observer. This is called a reactive effect, and the extent to which it can be a problem varies with the situation. In Miller's (1999) study, her long tenure as an observer made her presence commonplace, thereby serving to decrease the problem of **reactive effects**. She states,

> Since I had spent so many hours over eighteen months with the Jackson City Police Department [fictional name], I had grown to be a familiar face; this, I believe, decreased respondents' tendencies toward social desirability. Officers took my presence for granted in the briefing room, the hallways, the interview rooms, and in the field, including me in jokes and informal conversation in the coffee shop. (P. 235)

In general, in social settings involving many people, where observing while standing or sitting does not attract attention, the complete observer is unlikely to have much effect on the social processes. On the other hand, when the social setting involves few people and observing is unlike the usual activities in the setting, or when the observer differs in obvious respects from the participants, the complete observer is more likely to have an impact.

Participation and Observation

Most field researchers adopt a role that involves some active participation in the setting. Usually they inform at least some group members of their research interests, but then they participate in enough group activities to develop rapport with members and to gain a direct sense of what group members experience. This is not an easy balancing act:

> the key to participant observation as a fieldwork strategy is to take seriously the challenge it poses to participate more, and to play the role of the aloof observer less. Do not think of yourself as someone who needs to wear a white lab coat and carry a clipboard to learn about how humans go about their everyday lives. (Wolcott 1995:100)

In his classic study of corner gangs and other social organizations in the poor Boston community he called Cornerville, Whyte (1943) spent a large part of nearly 4 years trying to be accepted by the community and seen as a good fellow. He describes his efforts:

> My aim was to gain an intimate view of Cornerville life. My first problem, therefore, was to establish myself as a participant in the society so that I would have a position from which to observe. I began by going to live in Cornerville, finding a room with an Italian family. . . . It was not enough simply to make the acquaintance of various groups of people. The sort of information that I sought required that I establish intimate social relations, and that presented special problems. Since illegal activities are prevalent in Cornerville, every newcomer is under suspicion. . . . I put in a great deal of time simply hanging around with them [the men] and participating in their various activities. This active participation gave me something in common with them so that we had other things to talk about besides the weather. It broke down the social barriers and made it possible for me to be taken into the intimate life of the group. (Pp. v–vii)

Because of the great deal of time he spent with each gang and social organization he was studying, Whyte (1943) became accepted into each group and the community. The result was his famous book, *Street Corner Society* (1943). Sudhir Alladi Ventatesh's (2000) book, *American Project* about the relationship between gangs and a public housing development will almost certainly become a classic as well. In it, he describes the evolution of his research methodology from structured interviews to participant observation:

> They read my survey instrument, informed me that I was "not going to learn shit by asking these questions," and said I would need to "hang out with them" if I really wanted to understand the experiences of African-American youth in the city. Over the next few months, I met with many of them informally to play racquetball, drink beer on the shores of Lake Michigan, attend their parties, and eat dinner with their families. . . . Over an eighteen-month period, I logged notes on the activities of their gang, called the Black Kings. But it was the gang's relationship with other people in the housing development that piqued my interest. Gang members were also schoolchildren, nephews, churchgoers, fathers, husbands, and so on. They were "gang members" at certain times and in certain contexts, such as narcotics trafficking and meetings in open park space, but most of the time their lives were characterized by involvement with work, family, school, and peers. (Pp. xiv)

Participating and observing have two clear ethical advantages. Because group members know the researcher's real role in the group, they can choose to keep some information or attitudes hidden. By the same token, the researcher can decline to participate in unethical or dangerous activities without fear of exposing his or her identity.

Most field researchers who opt for disclosure get the feeling that, after they have become known and at least somewhat trusted figures in the group, their presence does not have a palpable effect on members' actions. The major influences on individual actions and attitudes are past experiences, personality, group structure, and so on, and these continue to exert their influence even when an outside observer is present. The participant observer can presumably be ethical about identity disclosure and still observe the natural social world. Of course, the argument is less persuasive when the behavior to be observed is illegal or stigmatized, giving participants reason to fear the consequences of disclosure to any outsider. In practice it can be difficult to maintain a fully open research role even in a setting without these special characteristics.

Even when researchers maintain a public identity as researchers, the ethical dilemmas arising from participation in group activities do not go away. In fact, researchers may have to prove themselves to group members by joining in some of their questionable activities. For example, police officers gave Van Maanen (1982) a nonstandard and technically prohibited pistol to carry on police patrols. Pepinsky (1980) witnessed police harassment of a citizen but did not intervene when the citizen was arrested. Trying to strengthen his ties with a local political figure in Cornerville, Whyte (1943) illegally voted multiple times in a local election.

Experienced participant observers try to lessen some of the problems of identity disclosure by evaluating both their effect on others in the setting and the effect of others on the observers. The observers must write about these effects throughout the time they are in the field and as they analyze their data. While in the field they must preserve some regular time when they can concentrate on their research and schedule occasional meetings with other researchers to review the fieldwork. Participant observers modify their role as circumstances seem to require, perhaps not always disclosing their research role at casual social gatherings or group outings but always informing new members of their role.

Covert Participation

To lessen the potential for reactive effects and to gain entry to otherwise inaccessible settings, some field researchers have adopted the role of covert participant. By doing so they keep their research secret and do their best to act like other participants in a social setting or group. **Covert participation** is also known as *complete participation*. Laud Humphreys (1970) served as a "watch queen" so that he could learn about men engaging in homosexual acts in a public restroom. Randall Alfred (1976) joined a group of Satanists to investigate group members and their interaction. Goffman (1961) worked as a state hospital assistant while studying the treatment of psychiatric patients.

Although the role of covert participant lessens some of the reactive effects encountered by the complete observer, covert participants confront other problems. The following are a few examples:

- *Covert participants cannot openly take notes or use any obvious recording devices.* They must write up notes based solely on memory and must do so at times when it is natural for them to be away from group members.

- *Covert participants cannot ask questions that will arouse suspicion.* Thus, they often have trouble clarifying the meaning of other participants' attitudes or actions.
- *The role of covert participation is difficult to play successfully.* Covert participants will not know how regular participants act in every situation in which the researchers find themselves. Regular participants enter the observed situation with social backgrounds and goals different from the researchers, whose spontaneous reactions to every event are unlikely to be consistent with those of the regular participants. Suspicion that researchers are not "one of us" may then have reactive effects, obviating the value of complete participation (Erikson 1967). In his study of the Satanists, for example, Alfred (1976) pretended to be a regular group participant until he completed his research, at which time he informed the group leader of his covert role. Rather than act surprised, the leader told Alfred that he had long considered Alfred to be strange, not like the others. We will never be certain how Alfred's observations were affected.
- *Covert participants must keep up the act at all times while in the setting under study.* Researchers may experience enormous psychological strain, particularly in situations where they are expected to choose sides in intragroup conflict or to participate in criminal or other acts. Of course, some covert observers may become so wrapped up in their role that they adopt not just the mannerisms but also the perspectives and goals of the regular participants; they "go native." At this point, they abandon research goals and cease to critically evaluate their observations.

Ethical issues have been at the forefront of debate over the strategy of covert participation. Erikson (1967) argues that covert participation is by its very nature unethical and should not be allowed except in public settings. Covert researchers cannot anticipate the unintended consequences of their actions for research subjects, Erikson points out. If others suspect the researcher's identity or if the researcher contributes to, or impedes, group action, these consequences can be adverse. In addition, other social research is harmed when covert research is disclosed, either during the research or upon its publication, because distrust of social scientists increases and access to research opportunities may decrease.

But a total ban on covert participation would "kill many a project stone dead" (Punch 1994:90). Studies of unusual religious or sexual practices and institutional malpractice would rarely be possible. "The crux of the matter is that some deception, passive or active, enables you to get at data not obtainable by other means" (Punch 1994:91). Therefore, some field researchers argue that covert participation is legitimate in some settings. If the researcher maintains the confidentiality of others, keeps commitments to others, and does not directly lie to others, some degree of deception may be justified in exchange for the knowledge gained (Punch 1994:90).

Entering the Field

Entering the field, the setting under investigation, is a critical stage in a participant observation project because it can shape many subsequent experiences. Some background work is necessary before entering the field, at least enough to develop a clear understanding of what the research questions are likely to be and to review one's personal stance toward the people and problems likely to be encountered. With participant observation, researchers must also learn in advance about the participants' dress and their typical activities to avoid being caught completely unprepared.

Entering the field even required Whyte (1943) to learn a new language:

Since the mother and father of the family spoke no English, I began studying Italian. Conversations with them and practice with the Linguaphone enabled me to learn enough to talk fairly fluently with the older generation. As I became largely concerned with the second-generation men, who conducted their activities in English, Italian was not essential to me; but the fact that I made the effort to learn the language was important, since it gave the impression that I had a sincere and sympathetic interest in Cornerville people. (P. v)

Many field researchers avoid systematic study and extensive reading about a setting for fear that it will bias their first impressions, but entering without a sense of the social norms can lead to disaster. Whyte (1943) came close to such disaster when he despaired about not making social contacts in Cornerville and decided to try an unconventional entry approach (unconventional for a field researcher, that is). He describes what happened when he went to a hotel bar in search of women to talk with:

I looked around me again and now noticed a threesome: one man and two women. It occurred to me that here was a maldistribution of females which I might be able to rectify. I approached the group and opened with something like this: "Pardon me. Would you mind if I joined you?" There was a moment of silence while the man stared at me. He then offered to throw me downstairs. I assured him that this would not be necessary and demonstrated as much by walking right out of there without any assistance. (P. 289)

Developing trust with at least one member of the research setting is a necessity in qualitative research. Such a person can become a valuable informant throughout the project, and most participant observers make a point of developing trust with at least one informant in a group under study. The entry gambit that finally worked for Whyte (1943) was to rely on a local community leader for introductions. A helpful social worker at the local settlement house introduced Whyte to "Doc," who agreed to help:

Well, any nights you want to see anything, I'll take you around. I can take you to the joints—gambling joints—I can take you around to the street corners. Just remember that you're my friend. That's all they need to know [so they won't bother you]. (P. 291)

Miller (1999) gained access to the police department she studied through a chief of police who was extremely open to research. She also had two friends on the police force at the time of her study:

Both of my friends were well-liked on the force and had great credibility with their colleagues. They vouched for me to others, tracked down retired officers for me to interview, helped with scheduling, answered my questions, and provided clarification and other assistance as the need arose. . . . Whenever I encountered scheduling snafus or any reluctance by a police offer to schedule an interview or a walk-along, my friends on the force would make a call and easily arrange the time I needed with other officers. (P. 230)

In short, field researchers must be very sensitive to the impression they make and the ties they establish when entering the field. This state of research lays the groundwork for collecting data from people who have different perspectives and for developing relationships that the researcher can use to surmount the problems that inevitably arise in the field.

Developing and Maintaining Relationships

Researchers must be careful to manage their relationships in the research setting so they can continue to observe and interview diverse members of the social setting throughout the long period typical of participant observation (Maxwell 1996:66). Every action the researcher takes can develop or undermine this relationship. As Decker and Van Winkle (1996) describe,

> There are a number of groups and individuals with whom field relationships must be maintained. Doing so effectively often involves balancing the competing demands of confidentiality, trust, and danger that emerge in a field study of individuals actively engaged in offending. (P. 45)

Maintaining trust is the cornerstone to successful research engagement, as Decker and Van Winkle (1996) further elaborate:

> We were able to maintain good field relations with our subjects by strictly observing our own commitment to the confidentiality of their statements. Since we interviewed many individuals from the same gang, it was often the case that one member would want to know what an earlier participant had told us. We refused to honor such inquiries, reminding them that the same confidentiality that applied to their own answers also covered those of their fellow gang members. We received numerous requests from gang members to sit in on the interview of a fellow member. These requests were declined a well. The strict confidentiality we were committed to was respected by our subjects, and appeared to enhance our own credibility as "solid" in their eyes. (P. 46)

Whyte (1943) used what was in retrospect a sophisticated two-part strategy to develop and maintain relationships with the Cornerville street-corner men. The first part of Whyte's strategy was to maintain good relations with Doc and, through Doc, stay on good terms with the others. The less obvious part of Whyte's strategy was a consequence of his decision to move into Cornerville, a move he decided was necessary to really understand and be accepted in the community. The room he rented in a local family's home became his base of operations. In some respects, this family became an important dimension of Whyte's immersion in the community. But he also recognized that he needed a place to unwind after his days of constant alertness in the field, so he made a conscious decision not to include the family as an object of study. Living in this family's home became a means for Whyte to maintain standing as a community insider without becoming totally immersed in the demands of research (pp. 294–297).

Experienced participant observers (Whyte 1943:300–306; Wolcott 1995:91–95) have developed some sound advice for others seeking to maintain relationships in the field:

- *Develop a plausible (and honest) explanation for yourself and your study.*
- *Maintain the support of key individuals in groups or organizations under study.*
- *Don't be too aggressive in questioning others (e.g., don't violate implicit norms that preclude discussion of illegal activity with outsiders).* Being a researcher requires that you not simultaneously try to be the guardian of law and order.
- *Don't fake social similarity with your subjects.* Taking a friendly interest in them should be an adequate basis for developing trust.
- *Avoid giving and receiving monetary or other tangible gifts, but do not violate norms of reciprocity.* Living with other people, taking others' time for conversations, and going out for a social evening all create expectations and incur social obligations. You cannot be an active participant without occasionally helping others. But you will lose your ability to function as a researcher if you are seen as someone who gives away money or other favors. Such small forms of assistance as an occasional ride to the store or advice on applying to college may strike the right balance.
- *Be prepared for special difficulties and tensions if multiple groups are involved.* It is hard to avoid taking sides or being used in situations of intergroup conflict.

Jody Miller (2000) describes her efforts to develop trust with the female gang members she interviewed for her book *One of the Guys*:

First, my research approach proved useful for establishing rapport. The survey began with relatively innocuous questions (demographics, living arrangements, attitudes toward school) and slowly made the transition from these to more sensitive questions about gang involvement, delinquency, and victimization. In addition, completing the survey interview first allowed me to establish a relationship with each young woman, so that when we completed the in-depth interview, there was a preexisting level of familiarity between us. . . . In addition, I worked to develop trust in the young women I interviewed through my efforts to protect their confidentiality. (Pp. 29–30).

Sampling People and Events

Decisions to study one setting or several settings and to pay attention to specific people and events will shape field researchers' ability to generalize about what they have found as well as the confidence that others can place in the results of their study. Limiting a particular study to a single setting allows a more intensive portrait of actors and activities in that setting, but also makes generalization of the findings questionable.

We may be reassured by information indicating that a typical case was selected for study or that the case selected was appropriate in some way for the research question. We also must keep in mind that many of the most insightful participant observation studies were conducted in only one setting and draw their credibility precisely from the researcher's thorough understanding of that setting. Nonetheless, studying more than one case or setting almost always strengthens the causal conclusions and makes the findings more generalizable (King, Keohane, & Verba 1994).

Decker and Van Winkle (1996) utilized the technique of snowball sampling. In addition, they chose to contact gang members directly, without the intervention of social service or criminal justice agencies, for several reasons, including their concern that they would be identified with law enforcement. To make their findings more generalizable, they interviewed members of several different gangs. Specifically, the snowball began with an earlier fieldwork project on active residential burglars (Wright & Decker 1994). The young members from this sample, along with contacts the field ethnographer had with several active street criminals, started the referral process. The initial interviewees then nominated other gang members as potential interview subjects. Because they wanted to interview members from several gangs, they had to restart the snowball sampling procedure many times to gain access to a large number of gangs. One problem, of course, was validating whether individuals claiming to be gang members, so-called "wannabes," actually were legitimate members. Over 500 contacts were made before the final sample of 99 was complete.

Theoretical sampling is a systematic approach to sampling in participant observational research (Glaser & Strauss 1967). Decker and Van Winkle (1996) used this technique to ensure that various subgroups such as race, sex, or type of gang were represented within their sample. When field researchers discover in an investigation that particular processes seem to be important, implying that certain comparisons should be made or that similar instances should be checked, the researchers then choose new settings or individuals to study as well, as diagrammed in Exhibit 8.3 (Ragin 1994:98–101). Based on the existing literature and anecdotal knowledge, Decker and Van Winkle (1996) knew that not all gang members were young minority-group males. They describe their strategy to obtain a full range of gang members as follows:

> We aggressively pursued leads for female gangs and gang members as well as opportunities to locate older and nonblack gang members. These leads were more difficult to find and often caused us to miss chances to interview other gang members. Despite these "missed opportunities," our sample is strengthened in that it more accurately represents the diverse nature of gangs and gang members in St. Louis. (P. 43)

The resulting sample of gang members in Decker and Van Winkle's (1996) study represented 29 different gangs; 16 were affiliated with the Crips and 13 were affiliated with the Bloods. Thus, Decker and Van Winkle's ability to draw from different gangs in developing conclusions gives us greater confidence in their studies' generalizability.

You already learned in Chapter 4 about nonprobability sampling methods, which can also be used to develop a more representative range of opinions and events in a field setting. Purposive sampling, which is a type of theoretical sampling, can be used to identify opinion leaders and representatives of different roles. With snowball sampling, field researchers learn from participants about who represents different subgroups in a setting.

EXHIBIT 8.3   Theoretical Sampling

Original cases interviewed in a study of cocaine users:

Realization: Some cocaine users are businesspeople.
Add businesspeople to sample:

Realization: Sample is low on women.
Add women to sample:

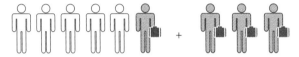

Realization: Some female cocaine users are mothers of young children.
Add mothers to sample:

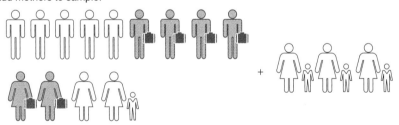

Quota sampling also may be employed to ensure the representation of particular categories of participants. Using some type of intentional sampling strategy within a particular setting can allow tests of some hypotheses that would otherwise have to wait until comparative data could be collected from several settings (King, Keohane, & Verba 1994).

When field studies do not require ongoing, intensive involvement by researchers in the setting, the **experience sampling method** (ESM) can be used. The experiences, thoughts, and feelings of a number of people are randomly sampled as they go about their daily activities. Participants in an ESM study carry an electronic pager and fill out reports when they are beeped. For example, 107 adults carried pagers in Kubey's (1990) ESM study of television habits and family quality of life. Participants' reports indicated that heavy TV viewers were less active during non-TV family activities, although heavy TV viewers also spent more time with their families. They felt as positively toward other family members as did those who watched less TV. Although ESM is a powerful tool for field research, it is still limited by the need to recruit people to carry pagers. Ultimately, the generalizability of ESM findings relies on the representativeness, and reliability, of the persons who cooperate in the research.

Taking Notes

Written **field notes** are the primary means of recording participant observation data (Emerson, Fretz, & Shaw 1995). Of course, written no longer means handwritten; many field researchers jot down partial notes while observing and then retreat to their computers to write up more complete notes on a daily basis. The computerized text can then be inspected and organized after it is printed out, or it can be marked up and organized for analysis using one of several computer programs designed especially for the task.

It is almost always a mistake to try to take comprehensive notes while engaged in the field; the process of writing extensively is just too disruptive. The usual procedure (see Exhibit 8.4) is to jot down brief notes about the highlights of the observation period. These brief notes, called **jottings**, then serve as memory joggers when writing the actual field notes at a later time. With the aid of the brief notes and some practice, researchers usually remember a great deal of what happened, as long as the comprehensive field notes are written within the next 24 hours, that night or upon arising the next day.

EXHIBIT 8.4   The Note-Taking Process

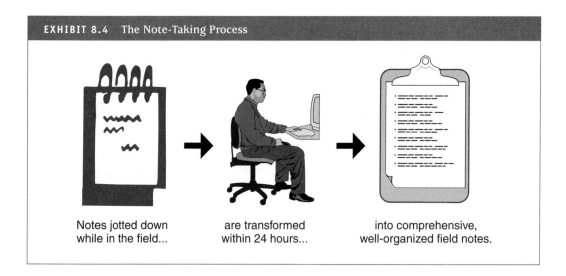

Notes jotted down while in the field... are transformed within 24 hours... into comprehensive, well-organized field notes.

In her study of community policing, Miller (1999) describes how her research team monitored what they observed and heard on ride-alongs and walk-alongs:

Before beginning, the researchers were trained to follow Lofland and Lofland's fieldwork steps (1995:89–98): during the period of observation, take notes to aid memory and to let respondents know that they are being taken seriously; convert these to full fieldnotes at the end of each shift to minimize the time between observation and writing so that crucial material is not lost; write up observations fully before the next trip to the field; and, when additional information is recalled, add it to the written notes. For the research team, fieldnotes were a "running description of events, people, things heard and overheard, conversations among people, conversations with people. Each new physical setting and person encountered merit[ed] a description" (Lofland & Lofland 1995:93). Investigators distinguished between the respondents' verbatim accounts and their own paraphrasing and general recall. The researchers also recorded their private emotional responses, based on Lofland and Lofland's admonition (1995:95) that their "emotional experience, even if not shared by others in the setting, may still suggest important analytical leads."

Usually, writing up notes takes as long as making the observations. Field notes must be as complete, detailed, and true to what was observed and heard as possible. Quotes should be clearly distinguished from the researcher's observations and phrased in the local vernacular; pauses and interruptions should be indicated. The surrounding context should receive as much attention as possible, and a map of the setting always should be included, with indications of where individuals were at different times.

Careful note-taking yields a big payoff for the researcher. On page after page, field notes will suggest new concepts, causal connections, and theoretical propositions. Social processes and settings can be described in rich detail, with ample illustrations. Exhibit 8.5, for example, contains field notes recorded by Miller (1999) for her study of community police officers. The notes include observations of the setting, the questions Miller asked and the answers she received, and her analytic thoughts about one of the police officers. What can be learned from just this one page of field notes? You can vividly visualize the neighborhood patrolled on the evening described. Key concepts and phrases are identified in the notes, such as "the hole," which refers to the worst apartment houses in the city ("absent landlords, repairs unattended"), and "jackets," which are tiny bags used to sell crack. The notes depict the nature and tone of the interactions between the patrol officer and several residents, with the officer knowing most of the residents by name and asking about other family members. From such notes, researchers can develop a theoretical framework for understanding the setting and a set of concepts and questions to inform subsequent observations.

Complete field notes must provide more than only a record of what was observed or heard. Notes also should include descriptions of the methodology: where researchers were standing while observing, how they chose people for conversation or observation, and what counts of people or events they made and why. Sprinkled throughout the notes also should be a record of the researchers' feelings and thoughts while observing: when they were disgusted by some statement or act, when they felt threatened or intimidated, why their attention shifted from one group to another. Notes like these provide a foundation for later review of the likelihood of bias or inattention to some salient features of the situation.

EXHIBIT 8.5   Field Notes From Susan Miller's (1999) Study of Community Policing

(7–6–95, Officer P, NBH #4)

Once the sergeant arrived, we walked around in groups for high visibility in the 1900 block, "the hole" (worst apartment houses: absent landlords, repairs unattended, no background check for tenants). Planned to search some of the basements for drugs. Several officers commented that "Linc" was seen hanging around the hole more frequently than usual—could indicate increased drug activity. We drove towards the hole. When we got within a block, Officer P. slowed down and asked over radio if everybody was ready. In we went. We parked by the other police mini-station in case anyone wanted sodas later. P. noticed that the windows had been broken, with rocks around. A white woman (early 30s) from the apt across the street came over and said it happened Wed night, and she had already reported it to the other neighborhood officers. Unclear if break happened after incidents on Wed or even if they were related. P. felt they were related since they picked police windows to break. She was very friendly and cooperative, as were her kid (white 12 years old) and her boyfriend (black). We started to walk up to the building on the way 1900 block. Kids came up to us to say hi to P. adults (mostly black) asked P. how he was doing and how things were since the mini-riot last night. Residents seemed to know him, and they answered his questions with ease, some volunteering more information than others. P. knew many by name and often asked about other family members. Easy rapport. At each building, P. would knock on the door of one of the ground floor apartments to ask for a basement key. Women always answered and were friendly and cooperative. Before P. got to the door with a key, one of the private security guards had already forced it open with a knife. The women, upon seeing this, complained to P. that everyone got in that way. (All women are African American unless otherwise noted.) One woman ran after P. asking him to get the landlord to fix the lock on the basement door. P. whipped out his cell phone and dialed right there in front of her (she was happy about that). In the basements, officers poked at the insulation where the walls met ceilings. Last week, they found a gun hidden there, and often find stashes of drugs. Our search found two different baggies full of "jackets," which is a way to sell crack (tiny bags) and a bag of pot. Corners of baggies used to sell crack in $20 quantities on the street. P. documented all of this to show the absent landlord. Back outside (around 8:10), several officers and squad cars surrounding a bunch of on-lookers as well as 3 young males (2 black, 1 white, in 20s) being frisked. They were spotted going from their car into an apartment where they don't live—a known apt where people went to do heroin. One of the bystanders (black, mid 20s) took P. aside to talk about getting some of his personal items back from jail (P. had visited him earlier in jail to get his version of the mini-riot). Search revealed a rock of cocaine.

Managing the Personal Dimensions

Our overview of participant observation is not complete without considering its personal dimensions. Because field researchers become a part of the social situation they are studying, they cannot help but be affected on a personal, emotional level. At the same time, those being studied react to researchers not just as researchers but as personal acquaintances, often as friends, sometimes as personal rivals. Managing and learning from this personal side of field research is an important part of any project.

The impact of personal issues varies with the depth of researchers' involvement in the setting. The more involved researchers are in multiple aspects of the ongoing social situation, the more important personal issues become and the greater the risk of "going native." Even when researchers acknowledge their role, "increased contact brings sympathy, and sympathy in its turn dulls the edge of criticism" (Fenno 1978:277). To study the social life of "corner boys," however, Whyte (1943) could not stay so disengaged. Recall that he moved into an apartment with a Cornerville family and lived for about 4 years in the community he was investigating:

> The researcher, like his informants, is a social animal. He has a role to play, and he has his own personality needs that must be met in some degree if he is to function successfully. Where the researcher operates out of a university, just going into the field for a few hours at a time, he can keep his personal social life separate from field activity. His problem of role is not quite so complicated. If, on the other hand, the researcher is living for an extended period in the community he is studying, his personal life is inextricably mixed with his research. (P. 279)

The correspondence between researchers' social attributes—age, sex, race, and so on—and those of their subjects also shapes personal relationships, as Miller (1999) noted:

> In the face-to-face interviews with neighborhood police officers it was my sense that being a woman facilitated the conversation. In fact, other investigators who have considered how the researcher's gender could impede or enhance rapport with respondents have found that women interviewing men may facilitate the subjects' ability to talk openly about their feelings. Men may be more comfortable speaking of intimate topics with women than with other men (Williams & Heikes 1993:281). . . . It made sense to the police that a female researcher would ask them about gender issues and that, as a criminologist, I would ask these questions in the context of community policing. Thus, I was able to examine gendered behavior and assumptions of masculinity and femininity within community policing with greater ease. (P. 232)

There is no formula for successfully managing the personal dimension of field research. It is much more art than science and flows more from the researcher's own personality and natural approach to other people than from formal training. Novice field researchers often neglect to consider how they will manage personal relationships when they plan and carry out their projects. Then suddenly, they find themselves doing something they do not believe they should, just to stay in the good graces of research subjects, or juggling the emotions resulting from conflict within the group. As Whyte (1943) noted,

> The field worker cannot afford to think only of learning to live with others in the field. He has to continue living with himself. If the participant observer finds himself engaging in behavior that he has learned to think of as immortal, then he is likely to begin to wonder what sort of a person he is after all. Unless the field worker can carry with him a reasonably consistent picture of himself, he is likely to run into difficulties. (P. 317)

These issues are even more salient when researchers place themselves in potentially dangerous situations. As Decker and Van Winkle (1996) explain,

> In part, gang members were of interest to us because of their involvement in violence. Because of this, we took steps to insure our own safety. One of the guiding principles was to limit the number of people being separately interviewed at the same time and location. In addition, we steadfastly avoided interviewing members of rival gangs at the same time. The field ethnographer carried a portable phone with him at all times, to insure that he could check in with us and we with him. Despite our best efforts, there were occasions when these precautions did not work. The field ethnographer witnessed several drive-by shootings while on the way to pick up interview subjects, and on one occasion, he saw three of our subjects shot while waiting to be picked up for an interview. . . . Not all exposure to risk of physical danger comes through such obvious means; however, during one interview, when asked whether he owned any guns, a gang member reached into his coat pocket and pulled out a .32 caliber pistol. We assured him that we would have taken his word for it. (P. 46)

If you plan a field research project, there are some general guidelines to follow:

- Take the time to consider how you want to relate to your potential subjects as people.
- Speculate about what personal problems might arise and how you will respond to them.
- Keep in touch with other researchers and personal friends outside the research setting.
- Maintain standards of conduct that make you comfortable as a person and that respect the integrity of your subjects. (Whyte 1943:300–317)

When you evaluate participant observers' reports, pay attention to how they defined their role in the setting and dealt with personal problems. Do not place too much confidence in such research unless the report provides this information.

SYSTEMATIC OBSERVATION

Observations can be made in a more systematic, quantitative design that allows systematic comparisons and more confident generalizations. A researcher using systematic observation develops a standard form on which to record variation within the observed setting in terms of variables of interest. Such variables might include the frequency of some behavior(s), the particular people observed, the weather or other environmental conditions, and the number and state of repair of physical structures. In some systematic observation studies, records will be obtained from a random sample of places or times.

Case Study: Systematic Observation of Public Spaces

You first learned about Robert Sampson and Stephen Raudenbush's (1999) study of disorder and crime in urban neighborhoods in Chapter 5. In this section, we'll elaborate on their use of the method of systematic social observation of public spaces to learn about these neighborhoods. A **systematic observational** strategy increases the reliability of observational data by using explicit rules that standardize coding practices across observers (Reiss 1971b). It is a method particularly well suited to overcoming one of the limitations of survey research on crime and disorder: Residents who are fearful of crime perceive more neighborhood disorder than do residents who are less fearful, even though both are observing the same neighborhood (Sampson & Raudenbush 1999:606).

This ambitious multiple methods investigation combined observational research, survey research, and archival research. The observational component involved a stratified probability (random) sample of 196 Chicago census tracts. A specially equipped sport utility vehicle was driven down each street in these tracts at the rate of 5 miles per hour. Two video recorders taped the blocks on both sides of the street, while two observers peered out the vehicle's windows and recorded their observations in logs. The result was an observational record of 23,816 face blocks (the block on one side of the street is a face block). The observers recorded in their logs codes that indicated land use, traffic, physical conditions, and evidence of physical disorder (see Exhibit 8.6). The videotapes were sampled and then coded for 126 variables, including housing characteristics, businesses, and social interactions. Physical disorder was measured by counting such features as cigarettes or cigars in the street, garbage, empty beer bottles, graffiti, condoms, and syringes. Indicators of social disorder included adults loitering, drinking alcohol in public, fighting, and selling drugs. To check for reliability, a different set of coders recoded the videos for 10% of the blocks. The repeat codes achieved 98% agreement with the original codes.

Sampson and Raudenbush (1999) also measured crime levels with data from police records, census tract socioeconomic characteristics with census data, and resident attitudes and behavior with a survey. As you learned in Chapter 5, the combination of data from these sources allowed a test of the relative impact on the crime rate of informal social control efforts by residents and of the appearance of social and physical disorder.

This study illustrates both the value of multiple methods and the technique of recording observations in a form from which quantitative data can be obtained. The systematic observations give us much greater confidence in the measurement of relative neighborhood disorder than we would have in unstructured descriptive reports or in responses of residents to survey questions. However, for some purposes, it might be more important to know how disordered the neighborhood is in the eyes of the residents, so interviews might be preferred or perhaps participant observation reports on "what it is really like."

INTENSIVE INTERVIEWING

Asking questions is part of almost all participant observation (Wolcott 1995:102–105). Many qualitative researchers employ intensive interviewing exclusively, without systematic observation of respondents in their natural setting. Unlike the more structured interviewing that

EXHIBIT 8.6 Neighborhood Disorder Indicators Used in Systematic Observation Log

Variable	Category	Frequency
Physical Disorder		
Cigarettes, cigars on street or gutter	no yes	6815 16758
Garbage, litter on street or sidewalk	no yes	11680 11925
Empty beer bottles visible in street	no yes	17653 5870
Tagging graffiti	no yes	12859 2252
Graffiti painted over	no yes	13390 1721
Gang graffiti	no yes	14138 973
Abandoned cars	no yes	22782 806
Condoms on sidewalk	no yes	23331 231
Needles or syringes on sidewalk	no yes	23392 173
Political message graffiti	no yes	15097 14
Social Disorder		
Adults loitering or congregating	no yes	14250 861
People drinking alcohol	no yes	15075 36
Peer group, gang indicators present	no yes	15091 20
People intoxicated	no yes	15093 18
Adults fighting or hostilely arguing	no yes	15099 12
Prostitutes on street	no yes	15100 11
People selling drugs	no yes	15099 12

Source: Raudenbush & Sampson 1999:15.

may be used in survey research (discussed in Chapter 7), intensive interviewing relies on open-ended questions. Qualitative researchers do not presume to know the range of answers that respondents might give and they seek to hear these answers in the respondents' own words. Rather than asking standard questions in a fixed order, intensive interviewers allow the specific content and order of questions to vary from one interviewee to another.

What distinguishes intensive interviewing from more structured forms of questioning is consistency and thoroughness. The goal is to develop a comprehensive picture of the interviewees' background, attitudes, and actions, in their own terms; to "listen to people as they describe how they understand the worlds in which they live and work" (Rubin & Rubin 1995:3). For example, even though Decker and Van Winkle (1996) had an interview guide, they encouraged elaboration on the part of their respondents and "went to great lengths to insure that each person we interviewed felt they had received the opportunity to tell their story in their own words" (p. 45).

Intensive interview studies do not directly reveal the social context in which action is taken and opinions are formed. Similar to participant observation studies, intensive interviewing engages researchers more actively with their subjects than does standard survey research. The researchers must listen to lengthy explanations, ask follow-up questions tailored to the preceding answers, and seek to learn about interrelated belief systems or personal approaches to things rather than measure a limited set of variables. As a result, intensive interviews are often much longer than standardized interviews, sometimes as long as 15 hours, conducted in several different sessions (Kaufman 1986:22).

The intensive interview becomes more like a conversation between partners than between a researcher and a subject. Intensive interviewers actively try to probe understandings and engage interviewees in a dialogue about the intended meaning of their comments. The interview typically follows a preplanned outline of topics, which often are asked of selected group members or other participants in a reasonably consistent manner. Some projects may use relatively structured interviews, particularly when the focus is on developing knowledge about prior events or some narrowly defined topic. But more exploratory projects, particularly those aimed at learning the interviewees' interpretations of the world, may let each interview flow in a unique direction in response to the interviewee's experiences and interests (Kvale 1996:3–5; Rubin & Rubin 1995:6; Wolcott 1995:113–114). In either case, qualitative interviewers must nimbly adapt throughout the interview, paying attention to nonverbal cues, expressions with symbolic value, and the ebb and flow of the interviewee's feelings and interests. "You have to be free to follow your data where they lead" (Rubin & Rubin 1995:64).

Random selection is rarely used to select respondents for intensive interviews, but the selection method still must be carefully considered. If interviewees are selected in a haphazard manner, as by speaking just to those who happen to be available at the time that the researcher is on site, the interviews are likely to be of less value than when a more purposive selection strategy is used. Researchers should try to select interviewees who are knowledgeable about the subject of the interview, who are open to talking, and who represent the range of perspectives (Rubin & Rubin 1995:65–92). Selection of new interviewees should continue, if possible, at least until the **saturation point** is reached, the point when new interviews seem to yield little additional information (see Exhibit 8.7). As new issues are uncovered, additional interviewees may be selected to represent different opinions about these issues.

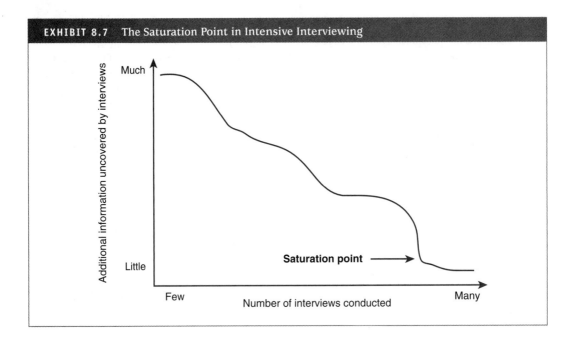

EXHIBIT 8.7 The Saturation Point in Intensive Interviewing

A recent book by Fleury-Steiner (2003) that examines the thoughts and emotions of jurors in death penalty cases is an excellent illustration of the tremendous insights that can be uncovered through intensive interviewing. In *Jurors' Stories of Death*, Fleury-Steiner reports on his work with the Capital Jury Project (CJP), which was a national study of the experiences of citizens who served as jurors on death penalty cases. To encourage respondents to tell stories about their experiences, the CJP survey explicitly asked jurors to tell interviewers about important moments during the trial and deliberations, and their impressions of the defendant. Fleury-Steiner states, "The goal of these questions was to facilitate jurors to construct their responses in their own ways. . . . Given the leeway to answer as they saw fit, in many instances jurors' stories emerged when I least expected them to" (p. 44).

The inductive analytic process of generating theory and making conclusions based on intensive interview narratives is often a time-consuming process. Fleury-Steiner (2003) explains,

> Through numerous rounds of retranscribing and revising, I was able to clarify my interpretations of jurors' stories. . . . Expanding beyond a literal interpretation of what was on the page, I began to notice consistencies in the way jurors made sense in their stories, including the taken for granted normative grammars of both speaker and listener. By privileging the "telling" of jurors' stories, I was able to make subsequent analytical interpretation. Indeed, the more I returned to the data, the more I began to connect the particularities of jurors' stories to broader sociohistorical, cultural, and institutional interpretations of identity, morality, and punishment. (P. 47)

Establishing and Maintaining a Partnership

Because intensive interviewing does not engage researchers as participants in subjects' daily affairs, the problems of entering the field are much reduced. However, the logistics of arranging long periods for personal interviews can still be fairly complicated. It is important to establish rapport with subjects by considering in advance how they will react to the interview arrangement and by developing an approach that does not violate their standards for social behavior. Interviewees should be treated with respect, as knowledgeable partners whose time is valued. (In other words, avoid coming late for appointments.) A commitment to confidentiality should be stated and honored (Rubin & Rubin 1995).

But the intensive interviewer's relationship with the interviewee is not an equal partnership, for the researcher seeks to gain certain types of information and strategizes throughout to maintain an appropriate relationship (Kvale 1996:6). In the first few minutes of the interview, the goal is to show interest in the interviewee and to clearly explain the purpose of the interview (Kvale 1996:128). During the interview, the interviewer should maintain an appropriate distance from the interviewee, one that does not violate cultural norms; the interviewer should maintain eye contact and not engage in distracting behavior. An appropriate pace is also important; pause to allow the interviewee to reflect, elaborate, and generally not feel rushed (Gordon 1992). When an interview covers emotional or otherwise stressful topics, at the end the interviewer should give the interviewee an opportunity to unwind (Rubin & Rubin 1995:138).

Asking Questions and Recording Answers

Intensive interviewers must plan their main questions around an outline of the interview topic. The questions should generally be short and to the point. More details can then be elicited through nondirective probes (such as "Can you tell me more about that?"), and follow-up questions can be tailored to answers to the main questions. Interviewers should strategize throughout an interview about how best to achieve their objectives while taking into account interviewees' answers.

Decker and Van Winkle's (1996) interview narrative illustrates this well:

Nearly half of the gang members identified leaders as persons who could provide material advantage, thus ascribing a functional character to leadership within the gang. Since half of our sample was in their early teens, someone with the ability to procure cars, drugs, guns, or alcohol could play a valuable role in the gang. Consequently, it was no surprise to find that over half of gang members identified leaders as persons who could "deliver." Because of the situational nature of leadership, persons moved in and out of this role. This was especially true in the case of being able to provide drugs in large quantities for street sales:

Q: Does someone have more juice in the gang?

A: Yeah, you always got someone that got more juice.

Q: What is the type of person who usually has more juice?

The very characteristics that make qualitative research techniques so appealing restrict their use to a limited set of research problems. It is not possible to draw representative samples for study using participant observation, and for this reason the generalizability of any particular field study's results cannot really be known. Only the accumulation of findings from numerous qualitative studies permits confident generalization, but here again the time and effort required to collect and analyze the data make it unlikely that many particular field research studies will be replicated.

Even if qualitative researchers made an effort to replicate key studies, their notion of developing and grounding explanations inductively in the observations made in a particular setting would hamper comparison of findings. Measurement reliability is thereby hindered, as are systematic tests for the validity of key indicators and formal tests for causal connections.

In the final analysis, qualitative research involves a mode of thinking and investigating different from that used in experimental and survey research. Qualitative research is inductive and idiographic; experiments and surveys tend to be conducted in a deductive, quantitative framework. Both approaches can help social scientists learn about the social world; the proficient researcher must be ready to use either. Qualitative data are often supplemented with many quantitative characteristics or activities. And as you have already seen, quantitative data are often enriched with written comments and observations, and focus groups have become a common tool of survey researchers seeking to develop their questionnaires. Thus, the distinction between qualitative and quantitative research techniques is not always clear-cut.

KEY TERMS

Complete observation
Covert (complete) participation
Experience sampling method (ESM)
Field notes
Field research
Focus group
Intensive interviewing

Participant observation
Qualitative methods
Reactive effect
Saturation point
Tacit knowledge
Theoretical sampling

HIGHLIGHTS

- Qualitative methods are most useful in exploring new issues, investigating hard-to-study groups, and determining the meaning people give to their lives and actions. In addition, most social research projects can be improved in some respects by taking advantage of qualitative techniques.

- Qualitative researchers tend to develop ideas inductively, try to understand the social context and sequential nature of attitudes and actions, and explore the subjective meanings that participants attach to events. They rely primarily on participant observation, intensive interviewing, and in recent years, focus groups.

- Participant observers may adopt one of several roles for a particular research project. Each role represents a different balance between observing and participating, which may or may not include public acknowledgment of the researcher's real identity. Many field researchers prefer a moderate role, participating as well as observing in a group but publicly acknowledging the researcher role. Such a role avoids the ethical issues posed by covert participation while still allowing the customary insights into the social world derived from directly participating in it. The role that the participant observer chooses should be based on an evaluation of the problems likely to arise from reactive effects, the ethical dilemmas of covert observation, and the consequences of identity disclosure in the particular setting.

- Field researchers must develop strategies for entering the field, developing and maintaining relations in the field, sampling, and recording and analyzing data. Sampling techniques commonly used in field research include theoretical sampling, purposive sampling, snowball sampling, quota sampling, and in special circumstances, random selection with the experience sampling method.

- Recording and analyzing notes is a crucial step in field research. Detailed notes should be recorded and analyzed daily to refine methods and to develop concepts, indicators, and models of the social system observed.

- Intensive interviews involve open-ended questions and follow-up probes, with specific question content and order varying from one interview can supplement participant observation data.

- Focus groups combine elements of participant observation and intensive interviewing. They can increase the validity of attitude measurement by revealing what people say when presenting their opinions in a group context.

- The four main ethical issues in field research concern voluntary participation, subject well-being, identity disclosure, and confidentiality.

Ethical issue

EXERCISES

1. Review the experiments and surveys described in previous chapters. Choose one and propose a field research design that would focus on the same research question but with participant observation techniques in a local setting. Propose the role that you would play in the setting, along the participant observation continuum, and explain why you would favor this role. Describe the stages of your field research study, including your plans for entering the field, developing and maintaining relationships, sampling, and recording and analyzing data. Then discuss what you would expect your study to add to the findings resulting from the study described in the book.

2. Explore a qualitative project using the software HyperRESEARCH (see Student Study Site) and your own data. Conduct a brief observational study in a public location on campus where students congregate. A cafeteria, a building lobby, or a lounge would be ideal. You can sit and observe, taking occasional notes unobtrusively, without violating any expectations of privacy. Observe for 30 minutes. Write up field notes, being sure to include a description of the setting and a commentary on your own behavior and your reactions to what you observed. Then load the demonstration copy of HyperRESEARCH, and make your own project.

3. Develop an interview guide that focuses on a research question addressed in one of the studies in this book. Using this guide, conduct an intensive interview with one person who is involved with the topic in some way. Take only brief notes during the interview, and

then write as complete a record of the interviews as you can immediately afterward. Turn in an evaluation of your performance as an interviewer and note-taker, together with your notes.

4. Read about focus groups in one of the references cited in this chapter and then devise a plan for using a focus group to explore and explain student perspectives about crime on campus. How would you recruit students for the group? What types of students would you try to include? How would you introduce the topic and the method to the group? What questions would you ask? What problems would you anticipate (e.g., discord between focus group members or digressions from the chosen topic)? How would you respond to these problems?

5. Read and summarize one of the qualitative studies discussed in this chapter or another classic study recommended by your instructor. Review and critique the study using the article review questions presented in Appendix B. What questions are answered by the study? What questions are raised for further investigation?

6. The April 1992 issue of the *Journal of Contemporary Ethnography* is devoted to a series of essays reevaluating Whyte's (1943) classic field study, *Street Corner Society*. A social scientist interviewed some of the people described in Whyte's book and concluded that the researcher had made methodological and ethical errors. Whyte and others offer able rejoinders and further commentary. Reading the entire issue of this journal will improve your appreciation of the issues that field researchers confront.

7. Find the Qualitative Research lesson in the interactive exercises on the Student Study Site http://www.sagepub.com/prccj3. Answer the questions in this lesson in order to review the types of ethical issues that can arise in the course of participant observation research.

DEVELOPING A RESEARCH PROPOSAL

Add a qualitative component to your proposed study. You can choose to do this with a participant observation project or intensive interviewing. Pick the method that seems most likely to help answer the research question for the overall survey project.

1. For a participant observation component, propose an observational plan that would complement the overall survey project. Present in your proposal the following information about your plan: (1) choose a site and justify its selection in terms of its likely value for the research, (2) choose a role along the participation-observation continuum and justify your choice, (3) describe access procedures and note any likely problems, (4) discuss how you will develop and maintain relations in the site, (5) review any sampling issues, and (6) present an overview of the way in which you will analyze the data you collect.

2. For an intensive interview component, propose a focus for the intensive interviews that you believe will add the most to findings from the survey project. Present in your proposal the following information about your plan: (1) present and justify a method for selecting individuals to interview, (2) write out three introductory biographical questions and five "grand tour" questions for your interview schedule, (3) list at least six different probes you may use, (4) present and justify at least two follow-up questions for one of your grand tour questions, and (5) explain what you expect this intensive interview component to add to your overall survey project.

Student Study Site

The companion Web site for *The Practice of Research in Criminology and Criminal Jus*
Edition
  http://www.sagepub.com/prccj3
  Visit the Web-based Student Study Site to enhance your understanding of the chapter content and to discover additional resources that will take your learning one step further. You can enhance your understanding of the chapters by using the comprehensive study material, which includes e-flashcards, Web exercises, practice self-tests, and more. You will also find special features, such as Learning from Journal Articles, which incorporates SAGE's online journal collection.

WEB EXERCISES

1. Go to the *Annual Review of Sociology*'s Web site by following the publication link at http://soc.AnnualReviews.org. Search for articles that use field research as the primary method of gathering data on gangs or delinquency. Find at least three articles and report on the specific method of field research used in each.

2. Search the Web for information on focus groups (previous, upcoming, or ongoing) involving victims, offenders, fear of crime, crime prevention, or another criminological topic. List the Web sites you found, and write a paragraph about the purpose of each focus group and the sample involved. How might these focus groups be used to influence public policy?

ETHICS EXERCISES

1. Covert participation may be the only way for researchers to observe the inner workings of some criminal or other deviant groups, but this strategy is likely to result in the researcher witnessing, and perhaps being asked to participate in, illegal acts. Do you think that covert participation is ever ethical? If so, under what conditions? Can the standards of "no harm to subjects," "identity disclosure," and "voluntary" participation be maintained in covert research?

2. A *New York Times* reporter (Wines 2006) recently talked about the dilemma many reporters have: whether or not to provide monetary or other compensation like food or medical supplies to people they interview for a story. In journalism, paying for information is a "cardinal sin" because journalists are "indoctrinated with the notion that they are observers." They are trained to report on situations, but not to influence a situation. This is what many scholars believe a researcher's role should be. Nevertheless, as we learned in this chapter, it is common in research to offer small gratuities for information and interviews. However, does paying for information unduly influence the truthfulness of the information being sought? Do you believe some people will say anything to earn money? What are your thoughts on paying for information? What if you were investigating the problems faced by families living below the poverty level and during an interview, you noticed that the family refrigerator and cupboards were empty and the baby was crying from hunger? What is the ethical reaction? If you believe the most ethical response would be to provide food or money for food, is it fair that there is another family next door in the same condition that did not happen to be on your interview list? How should gratuities be handled? Write a paragraph on how gratuities should be handled.

SPSS EXERCISES

The YOUTH.POR data set includes some questions on opinions regarding friends' attitudes toward delinquent acts and the extent to which getting caught for committing a crime would negatively affect the respondent's life.

1. Describe the opinions about friends' attitudes and personal misfortune based on the frequencies for these variables (V77; V79; V109; V119).

2. What explanation can you develop (inductively) for these attitudes? Do you believe that either friends' attitudes toward delinquent acts or getting caught for committing a crime would influence behavior? Explain.

3. Propose a participant observation, a focus group, or an intensive interview study to explore these attitudes further. Identify the sample for the study, and describe how you would carry out your observations, focus groups, or interviews.

Qualitative Data Analysis

I don't think most girls would go out there and kill somebody. It just depends on how crazy you are and how much you hate that person. But I don't really think, I don't think they would do it as much as the boys would do. I wouldn't. I wouldn't go out there and kill somebody just 'cause they wearin' that color. I wouldn't do that. I might beat 'em up or get me, I might get beat up. But I would never go out to that certain extent to kill 'em.

—Miller 2000

The statement above was made by a young female gang member. This statement, along with several other narratives about the reality of violence in gang life for female gang members, led Miller (2000) to conclude that female gang members were less likely to resort to serious violence compared to their male counterparts. Narratives such as these often represent the type of data that are analyzed in qualitative data analysis. The first difference, then, between qualitative and quantitative data analysis is that the data to be analyzed are text, rather than numbers, at least when the analysis first begins.

In this chapter, we will present the features that most qualitative data analyses share, and illustrate these features with research on several topics including youth victimization and community oriented policing. You will quickly learn that there is no one way to analyze textual data. To quote Michael Quinn Patton (2002),

> Qualitative analysis transforms data into findings. No formula exists for that transformation. Guidance, yes. But no recipe. Direction can and will be offered, but the final destination remains unique for each inquirer, known only when—and if—arrived at. (P. 432)

We will discuss some of the different types of qualitative data analysis before focusing on computer programs for qualitative data analysis; you will see that these increasingly popular programs are blurring the distinctions between quantitative and qualitative approaches to textual analysis.

FEATURES OF QUALITATIVE DATA ANALYSIS

The distinctive features of qualitative data collection methods that you studied in Chapter 8 are also reflected in the methods used to analyze those data. The focus on text, on qualitative data rather than on numbers, is the most important feature of qualitative analysis. The "text" that qualitative researchers analyze is most often transcripts of interviews or notes from participant observation sessions, but text can also refer to pictures or other images that the researcher examines.

What can the qualitative data analyst learn from a "text"? Here qualitative analysts may have two different goals. Some view analysis of a text as a way to understand what participants "really" thought, felt, or did in some situation or at some point in time. The text becomes a way to get "behind the numbers" that are recorded in a quantitative analysis to see the richness of real social experience. Other qualitative researchers have adopted a **hermeneutic perspective** on texts, that is, a perspective that views a text as an interpretation that can never be judged true or false. The text is only one possible interpretation among many (Patton 2002:114).

From a hermeneutic perspective, the meaning of a text is negotiated among a community of interpreters, and to the extent that some agreement is reached about meaning at a particular time and place, that meaning can only be based on consensual community validation. A researcher is constructing a "reality" with her interpretations of a text provided by the subjects of research; other researchers, with different backgrounds, could come to markedly different conclusions.

You can see in this discussion about text that qualitative and quantitative data analyses also differ in the priority given to the prior views of the researcher and to those of the subjects of the research. Qualitative data analysts seek to describe their textual data in ways that capture the setting or people who produced this text on their own terms rather than in terms of predefined measures and hypotheses. This means that qualitative data analysis tends to be inductive; the analyst identifies important categories in the data, as well as patterns and relationships, through a process of discovery. There are often no predefined measures or hypotheses. Anthropologists term this an **emic focus**, which means representing the setting in terms of the participants, rather than an **etic focus**, in which the setting and its participants are represented in terms that the researcher brings to the study.

Good qualitative data analyses also are distinguished by their focus on the interrelated aspects of the setting, group, or person under investigation—the case—rather than breaking the whole into separate parts. The whole is always understood to be greater than the sum of its parts, and so the social context of events, thoughts, and actions becomes essential for interpretation. Within this framework, it does not really make sense to focus on two variables out of an interacting set of influences and test the relationship between just those two.

Qualitative data analysis is an iterative and reflexive process that begins as data are being collected rather than after data collection has ceased (Stake 1995). Next to his field notes or interview transcripts, the qualitative analyst jots down ideas about the meaning of the text and how it might relate to other issues. This process of reading through the data and interpreting them continues throughout the project. The analyst adjusts the data collection process itself when it begins to appear that additional concepts need to be investigated or new relationships explored. This process is termed **progressive focusing** (Parlett & Hamilton 1976).

> We emphasize placing an interpreter in the field to observe the workings of the case, one who records objectively what is happening but simultaneously examines its meaning and redirects observation to refine or substantiate those meanings. Initial research questions may be modified or even replaced in mid-study by the case researcher. The aim is to thoroughly understand [the case]. If early questions are not working, if new issues become apparent, the design is changed. (Stake 1995:9)

Progressive focusing the process by which a qualitative analyst interacts with the data and gradually refines his or her focus.

We want to reiterate the narrative from Venkatesh's (2000) study of gang life in a Chicago public housing project because it vividly illustrates how progressive focusing affects the entire research process as well:

> [The African-American youth] read my survey instrument, informed me that "I was not going to learn shit by asking these questions," and said I would need to "hang out with them" if I really wanted to understand the experiences of African-American youth in the city. Over the next few months, I met with many of them informally to play racquetball, drink beer on the shores of Lake Michigan, attend their parties, and eat dinner with the families. . . . Their views of life, getting ahead in American, the status of blacks, and "gangland" challenged some of my preconceived notions about these topics. (P. xiv)

Carrying out this process successfully is more likely if the analyst reviews a few basic guidelines when he or she starts the process of analyzing qualitative data (Miller & Crabtree 1999b:142–143):

- Know yourself, your biases, and preconceptions.
- Know your question.
- Seek creative abundance. Consult others and keep looking for alternative interpretations.
- Be flexible.
- Exhaust the data. Try to account for all the data in the texts, then publicly acknowledge the unexplained and remember the next principle.
- Celebrate anomalies. They are the windows to insight.
- Get critical feedback. The solo analyst is a great danger to self and others.
- Be explicit. Share the details with yourself, your team members, and your audiences.

Qualitative Data Analysis as an Art

If you find yourself longing for the certainty of predefined measures and deductively derived hypotheses, you are beginning to understand the difference between setting out to analyze data quantitatively and planning to do so with a qualitative approach in mind. Or maybe you are now appreciating better the contrast between the positivist and interpretivist research philosophies that were summarized in Chapter 2. When it comes right down to it, the process of qualitative data analysis is even described by some as involving as much "art" as science, as a "dance," in the words of Miller and Crabtree (1999b):

> Interpretation is a complex and dynamic craft, with as much creative artistry as technical exactitude, and it requires an abundance of patient plodding, fortitude, and discipline. There are many changing rhythms; multiple steps; moments of jubilation, revelation, and exasperation. . . . The dance of interpretation is a dance for two, but those two are often multiple and frequently changing, and there is always an audience, even if it is not always visible. Two dancers are the interpreters and the texts. (Pp. 138–139)

The "dance" of qualitative data analysis is represented in Exhibit 9.1, which captures the alternation between immersion in the text to identify meanings and editing the text to create categories and codes. The process involves three different modes of reading the text:

1. When the researcher reads the text literally (L, in Exhibit 9.1), she is focused on its literal content and form, so the text "leads" the dance.

2. When the researcher reads the text reflexively (R), she focuses on how her own orientation shapes her interpretations and focus. Now, the researcher leads the dance.

3. When the researcher reads the text interpretively (I), she tries to construct her own interpretation of what the text means.

EXHIBIT 9.1   Dance of Qualitative Analysis

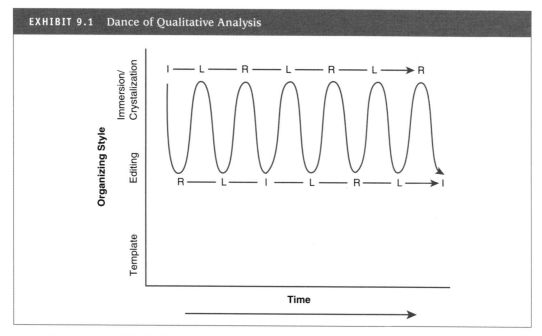

Source: Miller & Crabtree 1999b:139, Figure 7.1. Based on Addison 1999.

In this artful way, analyzing text involves both inductive and deductive processes: The researcher generates concepts and linkages between them based on reading the text and also checks the text to see whether her concepts and interpretations are reflected in it.

Qualitative Compared to Quantitative Data Analysis

With these points in mind, let us review the ways in which qualitative data analysis differs from quantitative analysis (Denzin & Lincoln 2000a:8–10; Patton 2002:13–14).

- A focus on meanings rather than on quantifiable phenomena
- Collection of many data on a few cases rather than few data on many cases
- Study in depth and detail, without predetermined categories or directions, rather than emphasis on analyses and categories determined in advance
- Conception of the researcher as an "instrument," rather than as the designer of objective instruments to measure particular variables
- Sensitivity to context rather than seeking universal generalizations
- Attention to the impact of the researcher's and others' values on the course of the analysis rather than presuming the possibility of value-free inquiry
- A goal of rich descriptions of the world rather than measurement of specific variables

You will also want to keep in mind features of qualitative data analysis that are shared with those of quantitative data analysis. Both qualitative and quantitative data analysis can involve making distinctions about textual data. You also know that textual data can be transposed to quantitative data through a process of categorization and counting. Some qualitative analysts also share with quantitative researchers a positivist goal of describing better the world as it "really" is, but others have adopted a postmodern goal of trying to understand how different people see and make sense of the world, without believing that there is any "correct" description.

TECHNIQUES OF QUALITATIVE DATA ANALYSIS

Exhibit 9.2 outlines the different techniques that are shared by most approaches to qualitative data analysis:

1. Documentation of the data and the process of data collection
2. Organization or categorization of the data into concepts
3. Connection of the data to show how one concept may influence another
4. Corroboration or legitimization, by evaluating alternative explanations, disconfirming evidence, and searching for negative cases
5. Representing the account (reporting the findings)

The analysis of qualitative research notes begins in the field, at the time of observation, interviewing, or both, as the researcher identifies problems and concepts that appear likely to help in understanding the situation. Simply reading the notes or transcripts is an important step in the analytic process. Researchers should make frequent notes in the margins to identify important statements and to propose ways of coding the data: "husband/wife conflict," perhaps, or "tension reduction strategy."

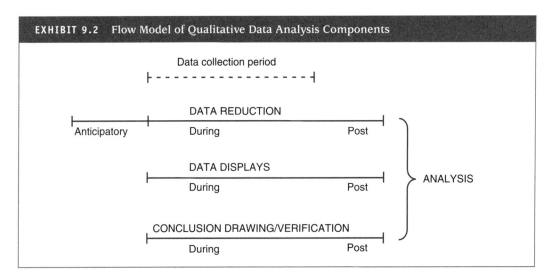

EXHIBIT 9.2   Flow Model of Qualitative Data Analysis Components

Source: Miles & Huberman 1994:10, Figure 1.3. Used with permission.

An interim stage may consist of listing the concepts reflected in the notes and diagramming the relationships among concepts (Maxwell 1996:78–81). In a large project, weekly team meetings are an important part of this process. Miller (1999) described this process in her study of neighborhood police officers. Miller's research team met both to go over their field notes and to resolve points of confusion, as well as to dialogue with other skilled researchers who helped to identify emerging concepts:

> The fieldwork team met weekly to talk about situations that were unclear and to troubleshoot any problems. We also made use of peer-debriefing techniques. Here, multiple colleagues, who were familiar with qualitative data analysis but not involved in our research, participated in preliminary analysis of our findings. (P. 233)

This process continues throughout the project and should assist in refining concepts during the report-writing phase, long after data collection has ceased. Let us examine each of the stages of qualitative research in more detail.

Documentation

The data for a qualitative study most often are notes jotted down in the field or during an interview—from which the original comments, observations, and feelings are reconstructed—or text transcribed from audiotapes. "The basic data are these observations and conversations, the actual words of people reproduced to the best of my ability from the field notes" (Diamond 1992:7). What to do with all this material? Many field research projects have slowed to a halt because a novice researcher becomes overwhelmed by the quantity of information that has been collected. A one-hour interview can generate 20 to 25 pages of single-spaced text (Kvale 1996:169). Analysis is less daunting, however, if the researcher maintains a disciplined transcription schedule.

> Usually, I wrote these notes immediately after spending time in the setting or the next day. Through the exercise of writing up my field notes, with attention to "who" the speakers and actors were, I became aware of the nature of certain social relationships and their positional arrangements within the peer group. (Anderson 2003:38)

You can see the analysis already emerging from this simple process of taking notes.

The first formal analytical step is documentation. The various contacts, interviews, written documents, and whatever it is that preserves a record of what happened all need to be saved and listed. Documentation is critical to qualitative research for several reasons: It is essential for keeping track of what will be a rapidly growing volume of notes, tapes, and documents; it provides a way of developing an outline for the analytic process; and it encourages ongoing conceptualizing and strategizing about the text.

An excellent example of a documentation guide is provided by Tammy Anderson (forthcoming), who conducted a large study of drug use and victimization in nightclub events (e.g., raves, hip-hop, and EDM events) in Philadelphia. When she and her team entered an event for direct observation, they followed a guideline form, which cued the researcher to obtain information about all pertinent data points (see Exhibit 9.3).

EXHIBIT 9.3   Example of a Direct Observation Form for Nightclub Observation.

1. Site:

2. Period of observation:

3. Type of event:

4. Description of clubber demographics:

5. Description of conversations with clubbers and staff:

Description of response to ethnographer's presence, conversations:

Social organization of event	Physical layout (chill areas, dance floor, dj box, exits, bars)	Utilization of area by clubbers	Clubbers' interactions within areas	Entertainment personnel
Social organization of event II	Staffing patterns	Roles and Behaviors	Interaction with clubbers by security, managers, bartenders	Entertainment personnel
Club's cultural ethos	Vibe	Music	Norms	Identity markers or props
Outside support agencies	Public safety	Law enforcement	Medical personnel	
Behaviors at event	Type	Frequency	Impact	
Clubbers' typologies	Dress and props	Status indicators	Clubbing motives	
Drug or alcohol consumption	Clubbers	Staff	Consequences	
Victimization	Observed	Rumored	Victim or offender	Consequences (e.g., clubber and staff reaction)
Personal reflections				

Source: Anderson, forthcoming.

Conceptualization, Coding, and Categorizing

Identifying and refining important concepts is a key part of the iterative process of qualitative research. Sometimes, conceptualizing begins with a simple observation that is interpreted directly, "pulled apart," and then put back together more meaningfully. Stake (1995) provides an example:

To test your understanding of qualitative data analysis techniques, go to the Qualitative Data Analysis interactive exercises on the Student Study Site.

> When Adam ran a pushbroom into the feet of the children nearby, I jumped to conclusions about his interactions with other children: aggressive, teasing, arresting. Of course, just a few minutes earlier I had seen him block the children climbing the steps in a similar moment of smiling bombast. So I was aggregating, and testing my unrealized hypotheses about what kind of kid he was, not postponing my interpreting. . . . My disposition was to keep my eyes on him. (P. 74)

The focus in this conceptualization "on the fly" is to provide a detailed description of what was observed and a sense of why that was important.

More often, analytic insights are tested against new observations, the initial statement of problems and concepts is refined, the researcher then collects more data, interacts with the data again, and the process continues. Elijah Anderson (2003) recounts how his conceptualization of social stratification at Jelly's Bar developed over a long period of time:

> I could see the social pyramid, how certain guys would group themselves and say in effect, "I'm here and you're there." I made sense of these crowds [initially] as the "respectables," the "non-respectables," and the "near-respectables." . . . Inside, such non-respectables might sit on the crates, but if a respectable came along and wanted to sit there, the lower status person would have to move. (Pp. 18–19)

But this initial conceptualization changed with experience, as Anderson (2003) realized that the participants themselves used other terms to differentiate social status: "winehead," "hoodlum," and "regular" (p. 28). What did they mean by these terms? "The 'regulars' basically valued 'decency.' They associated decency with conventionality but also with 'working for a living,' or having a 'visible means of support'" (p. 29). In this way, Anderson progressively refined his concept as he gained experience in the setting.

Miller (2000) provides another excellent illustration of this iterative process of conceptualization in her study of girls in gangs:

> I paid close attention to and took seriously respondents' reactions to themes raised in interviews, particularly instances in which they "talked back" by labeling a topic irrelevant, pointing out what they saw as misinterpretations on my part, or offering corrections. In my research, the women talked back the most in response to my efforts to get them to articulate how gender inequality shaped their experiences in the gang. Despite stories they told to the contrary, many maintained a strong belief in their equality within the gang. Consequently, I developed an entire theoretical discussion around the contradictory operation of gender within the subject. As the research progressed, I also took emerging themes back to respondents in subsequent interviews to see if they felt I had gotten it right. In addition to conveying that I was interested in their perspectives and experiences, this process also proved useful for further refining my analyses. (P. 30)

The process described in this quote illustrates the reflexive nature of qualitative data collection and analysis. In qualitative research, the collection of data and their analysis are not typically mutually exclusive activities. This excerpt shows how the researcher first was alerted to a concept by observations in the field, then refined his understanding of this concept by investigating its meaning. By observing the concept's frequency of use, he came to realize its importance. Then he incorporated the concept into an explanatory model of student-patient relationships.

Examining Relationships and Displaying Data

Examining relationships is the centerpiece of the analytic process, because it allows the researcher to move from simple description of the people and settings to explanations of why things happened as they did with those people in that setting. The process of examining relationships can be captured in a matrix that shows how different concepts are connected, or perhaps what causes are linked with what effects.

Exhibit 9.4 displays a matrix used in evaluation research to capture the relationship between the extent to which stakeholders in a new program had something important at stake in the program and the researcher's estimate of their favorability toward the program. Each cell of the matrix was to be filled in with a summary of an illustrative case study. In other matrix analyses, quotes might be included in the cells to represent the opinions of these different stakeholders, or the number of cases of each type might appear in the cells. The possibilities are almost endless. Keeping this approach in mind will generate many fruitful ideas for structuring a qualitative data analysis.

The simple relationships that are identified with a matrix like that shown in Exhibit 9.4 can be examined and then extended to create a more complex causal model. Such a model represents the multiple relationships among the constructs identified in a qualitative analysis as important for explaining some outcome. A great deal of analysis must precede the construction of such a model, with careful attention to identification of important variables and the evidence that suggests connections between them.

EXHIBIT 9.4 Coding Form for Relationships: Stakeholders' Stakes

Estimate of Various Stakeholders' Inclination Toward the Program		
How high are the stakes for various primary stakeholders?	*Favorable*	*Neutral or Unknown*	*Antagonistic*
High			
Moderate			
Low			

Source: Patton 2002:472.

Note: Construct illustrative case studies for each cell based on fieldwork.

Exhibit 9.5 provides an excellent example of a causal model developed by Baskin and Sommers (1998) to explain the desistance process for the sample of violent female offenders they interviewed in the state of New York. They described the process for the women who made it out of their lives of crime as follows:

> Desistance is a process as complex and lengthy as the process of initial involvement. It was interesting to find that some of the key concepts in initiation of deviance—social bonding, differential association, deterrence, age—were equally important in the process of desistance. We see the aging offender take the threat of punishment seriously, reestablish links with conventional society and sever associations with subcultural street elements. We found, too, that the decision to give up crime was triggered by a shock of some sort that was followed by a period of crisis. Anxious and dissatisfied, the women took stock of their lives and criminal activity. They arrived at a point at which the deviant way of life seemed senseless. (P. 139)

EXHIBIT 9.5   The Desistance Process for Violent Female Offenders (Baskin & Sommers 1998)

Stage 1: Problems Associated With Criminal Participation	
Socially Disjunctive Experiences Hitting rock bottom Fear of death Tiredness Illness	*Delayed Deterrence* Increased probability of punishment Increased difficulty in "doing time" Increased severity of sanctions Increasing fear
Assessment Reappraisal of life and goals Psychic change	
Decision Decision to quit or initial attempts at desistance Continuing possibility of criminal participation	
Stage 2: Restructuring of Self	
Public pronouncement of decision to end criminal participation Claim to a new identity	
Stage 3: Maintenance of the Decision to Stop	
Ability to successfully renegotiate identity Support of significant others Integration into new social networks Ties to conventional roles Stabilization of new social identity	

Source: Baskin & Sommers 1998:141.

Authenticating Conclusions

No set standards exist for evaluating the validity or "authenticity" of conclusions in a qualitative study, but the need to consider carefully the evidence and methods on which conclusions are based is just as great as with other types of research. Individual items of information can be assessed in terms of at least three criteria (Becker 1958:654–656):

- *How credible was the informant?* Were statements made by someone with whom the researcher had a relationship of trust or by someone the researcher had just met? Did the informant have reason to lie? If the statements do not seem to be trustworthy as indicators of actual events, can they at least be used to help understand the informant's perspective?
- *Were statements made in response to the researcher's questions, or were they spontaneous?* Spontaneous statements are more likely to indicate what would have been said had the researcher not been present.
- *How does the presence or absence of the researcher or the researcher's informant influence the actions and statements of other group members?* Reactivity to being observed can never be ruled out as a possible explanation for some directly observed social phenomenon. However, if the researcher carefully compares what the informant says goes on when the researcher is not present, what the researcher observes directly, and what other group members say about their normal practices, the extent of reactivity can be assessed to some extent.

A qualitative researcher's conclusions should also be assessed by her ability to provide a credible explanation for some aspect of social life. That explanation should capture group members' **tacit knowledge** of the social processes that were observed, not just their verbal statements about these processes. Tacit knowledge, "the largely unarticulated, contextual understanding that is often manifested in nods, silences, humor, and naughty nuances " is reflected in participants' actions as well as their words and in what they fail to state but nonetheless feel deeply and even take for granted (Altheide & Johnson 1994:492–493). These features are evident in Whyte's (1955) analysis of Cornerville social patterns:

> The corner-gang structure arises out of the habitual association of the members over a long period of time. The nuclei of most gangs can be traced back to early boyhood. . . . Home plays a very small role in the group activities of the corner boy. . . . The life of the corner boy proceeds along regular and narrowly circumscribed channels. . . . Out of [social interaction within the group] arises a system of mutual obligations which is fundamental to group cohesion. . . . The code of the corner boy requires him to help his friends when he can and to refrain from doing anything to harm them. When life in the group runs smoothly, the obligations binding members to one another are not explicitly recognized. (Pp. 255–257)

Comparing conclusions from a qualitative research project to those obtained by other researchers conducting similar projects can also increase confidence in their authenticity. Susan Miller's (1999) study of neighborhood police officers (NPOs) found striking parallels in the ways they defined their masculinity to processes reported in research about males in nursing and other traditionally female jobs:

> In part, male NPOs construct an exaggerated masculinity so that they are not seen as feminine as they carry out the social-work functions of policing. Related to this is the almost defiant expression of heterosexuality, so that the men's sexual orientation can never truly be doubted even if their gender roles are contested. Male patrol officers' language—such as their use of terms like "pansy police" to connote neighborhood police officers—served to affirm their own heterosexuality. . . . In addition, the male officers, but not the women, deliberately wove their heterosexual status into conversations, explicitly mentioning their female domestic partner or spouse and their children. This finding is consistent with research conducted in the occupational field. The studies reveal that men in female-dominated occupations, such as teachers, librarians, and pediatricians, over-reference their heterosexual status to ensure that others will not think they are gay. (P. 222)

Reflexivity

Confidence in the conclusions from a field research study is also strengthened by an honest and informative account about how the researcher interacted with subjects in the field, what problems he or she encountered, and how these problems were or were not resolved. Such a "natural history" of the development of the evidence enables others to evaluate the findings. Such an account is important first and foremost because of the evolving and variable nature of field research: To an important extent, the researcher "makes up" the method in the context of a particular investigation rather than applying standard procedures that are specified before the investigation begins.

Qualitative data analysts, more often than quantitative researchers, display real sensitivity to how a social situation or process is interpreted from a particular background and set of values and not simply based on the situation itself (Altheide & Johnson 1994). Researchers are only human, after all, and must rely on their own senses and process all information through their own minds. By reporting how and why they think they did what they did, they can help others determine whether, or how, the researchers' perspectives influenced their conclusions. "There should be clear 'tracks' indicating the attempt [to show the hand of the ethnographer] has been made" (Altheide & Johnson 1994:493).

Anderson's (2003) memoir about the Jelly's Bar research illustrates the type of "tracks" that an ethnographer makes as well as how he can describe those tracks. He acknowledges that his tracks began as a child:

> While growing up in the segregated black community of South Bend, from an early age, I was curious about the goings on in the neighborhood, but particularly streets, and more particularly, the corner taverns that my uncles and my dad would go to hang out and drink in. . . . Hence, my selection of Jelly's as a field setting was a matter of my background, intuition, reason, and with a little bit of luck. (Pp. 1–2)

After starting to observe at Jelly's, Anderson's (2003) "tracks" led to Herman:

> After spending a couple of weeks at Jelly's, I met Herman and I felt that our meeting marked a big achievement. We would come to know each other well . . . something of an informal leader at Jelly's. . . . We were becoming friends. . . . He seemed to genuinely like me, and he was one person I could feel comfortable with. (P. 4)

So we learn that Anderson's (2003) observations were to be shaped, in part, by Herman's perspective, but we also find out that Anderson maintained some engagement with fellow students. This contact outside the bar helped to shape his analysis: "By relating my experiences to my fellow students, I began to develop a coherent perspective or a 'story' of the place which complemented the accounts that I had detailed in my accumulating field notes" (p. 6).

In this way, the outcome of Anderson's (2003) analysis of qualitative data resulted, in part, from the way in which he "played his role" as a researcher and participant, not just from the setting itself.

ALTERNATIVES IN QUALITATIVE DATA ANALYSIS

The qualitative data analyst can choose from many interesting alternative approaches. Of course, the research question under investigation should shape the selection of an analytic approach, but the researcher's preferences and experiences inevitably also will have an important influence on the method chosen. The alternative approaches we present here (ethnography, ethnomethodology, qualitative comparative analysis, narrative analysis, conversation analysis, case-oriented understanding, and grounded theory) give you a good sense of the different possibilities (Patton 2002).

Ethnography

Ethnography is the study of a culture or cultures that a group of people share (Van Maanen 1995:4). As a method, it usually is meant to refer to the process of participant observation by a single investigator who immerses himself or herself in the group for a long period of time (often one or more years). Ethnographic research can also be called "naturalistic," because it seeks to describe and understand the natural social world as it really is, in all its richness and detail. As you learned in Chapter 8, anthropological field research has traditionally been ethnographic, and much sociological fieldwork shares these same characteristics. But there are no particular methodological techniques associated with ethnography, other than just "being there." The analytic process relies on the thoroughness and insight of the researcher to "tell it like it is" in the setting, as he or she experienced it.

Code of the Street, Anderson's (1999) award-winning study of Philadelphia's inner city, captures the flavor of this approach:

> My primary aim in this work is to render ethnographically the social and cultural dynamics of the interpersonal violence that is currently undermining the quality

of life of too many urban neighborhoods. . . . How do the people of the setting perceive their situation? What assumptions do they bring to their decision making? (Pp. 10–11)

The methods of investigation are described in the book's preface: participant observation, including direct observation and in-depth interviews, impressionistic materials drawn from various social settings around the city, and interviews with a wide variety of people. Like most traditional ethnographers, Anderson (1999) describes his concern with being "as objective as possible" and using his training as other ethnographers do, "to look for and to recognize underlying assumptions, their own and those of their subjects, and to try to override the former and uncover the latter" (p. 11).

From analysis of the data obtained in these ways, a rich description emerges of life in the inner city. Although we often do not "hear" the residents speak, we feel the community's pain in Anderson's (1999) description of "the aftermath of death":

> When a young life is cut down, almost everyone goes into mourning. The first thing that happens is that a crowd gathers about the site of the shooting or the incident. The police then arrive, drawing more of a crowd. Since such a death often occurs close to the victim's house, his mother or his close relatives and friends may be on the scene of the killing. When they arrive, the women and girls often wail and moan, crying out their grief for all to hear, while the young men simply look on, in studied silence. . . . Soon the ambulance arrives. (P. 138)

Anderson (1999) uses these descriptions as a foundation on which he develops the key concepts in his analysis, such as "code of the street":

> The "code of the street" is not the goal or product of any individual's action but is the fabric of everyday life, a vivid and pressing milieu within which all local residents must shape their personal routines, income strategies, and orientations to schooling, as well as their mating, parenting, and neighbor relations. (P. 326)

Anderson's (2003) report on his Jelly's Bar study illustrates how an ethnographic analysis deepened as he became more socially integrated into the Jelly's Bar group. He thus became more successful at "blending the local knowledge one has learned with what we already know sociologically about such settings" (p. 39):

> I engaged the denizens of the corner and wrote detailed field notes about my experiences, and from time to time looked for patterns and relationships in my notes. In this way, an understanding of the setting came to me in time, especially as I participated more fully in the life of the corner and wrote my field notes about my experiences; as my notes accumulated, and as I reviewed them occasionally and supplemented them with conceptual memos to myself, their meanings became more clear, while even more questions emerged. (P. 15)

This rich ethnographic tradition is being abandoned by some qualitative data analysts, however. Many have become skeptical of the ability of social scientists to perceive the social world in a way that is not distorted by their own subjective biases or to a receive impressions from the actors in that social world that are not altered by the fact of being studied (Van Maanen 2002). As a result, both specific techniques and alternative approaches to qualitative data analysis have proliferated. The next sections introduce several of these alternative approaches.

Ethnomethodology

Ethnomethodology focuses on the way that participants construct the social world in which they live, how they "create reality," rather than on describing the social world itself. In fact, ethnomethodologists do not necessarily believe that we can find an objective reality; it is the way that participants come to create and sustain a sense of "reality" that is of interest. In the words of Gubrium and Holstein (1997), in ethnomethodology, as compared to the naturalistic orientation of ethnography,

> The focus shifts from the scenic features of everyday life onto the ways through which the world comes to be experienced as real, concrete, factual, and "out there." An interest in members' methods of constituting their world supersedes the naturalistic project of describing members' worlds as they know them. (P. 41)

Unlike the ethnographic analyst, who seeks to describe the social world as the participants see it, the ethnomethodological analyst seeks to maintain some distance from that world. The ethnomethologist views a "code" of conduct like that described by Anderson (2003) not as a description of a real normative force that constrains social action, but as the way that people in the setting create a sense of order and social structure (Gubrium & Holstein 1997:44–45). The ethnomethodologist focuses on how reality is constructed, not on what it *is*.

Qualitative Comparative Analysis

In previous chapters, we have discussed many studies that have attempted to explain violent crime. Another novel approach to the study of violence, specifically of homicide, was recently undertaken by Miethe, Regoeczi, and Drass (2004). In their book *Rethinking Homicide*, instead of focusing on the victim or the offender, they analyzed the homicide situation as the unit of analysis, exploring the structure and process underlying the lethal outcome. They state, "Homicide situations are defined by the nexus of offender, victim, and offense elements in time and space. It is the combination of these elements, not their operation in isolation, which provides the context for lethal violence" (p. xvii).

To analyze the homicide narratives from cases from 4 large cities in the U.S., Miethe, Regoeczi, and Drass (2004) used qualitative comparative analysis (QCA). They describe why QCA was the most appropriate methodology for their study:

> As an analytical method for case comparisons, QCA considers cases holistically, as complex configurations of attributes measured by a set of variables. It also assumes that events or outcomes are produced by variables acting together in combination. Under this specification, the effect of any particular variable may be different from

one case to another, depending upon the values of other attributes of a case. QCA places importance on the total context of elements to assess whether the same outcome can be produced by different combinations of variables. (P. 50)

It is beyond the scope of this book to detail the process of QCA here. Readers who want more information should seek other sources (Amenta & Poulsen 1994). QCA actually combines elements of quantitative and qualitative analysis, and as such, is a hybrid of sorts. For example, similar to a statistical analysis, QCA requires the specification of a model relating a set of independent variables to an outcome variable. But it examines the data holistically by developing typologies across a complex arrangement of cases and variables.

After their QCA analysis of over 400,000 homicides, Miethe, Regoeczi, and Drass (2004) concluded that most homicides occur in situational contexts that have changed little over the past 3 decades. They found expressive homicides involving disputes and arguments to be more prevalent than instrumental homicides with males being the primary perpetrators of both, but particularly of instrumental homicides. Contrary to previous research that characterized confrontational homicides to involve public disputes among marginalized males, Miethe, Regoeczi, and Drass also found that these "character contests" did most often involve males, but they also involved female offenders, occurred in both public and private settings, and were not restricted to minority group members. Like other research, they found that these conflicts often involved what many would call "trivial altercations" where alcohol was involved and where the victim often provoked the offender in some way (e.g., with an insult). This was particularly true for youth homicides wherein many contained elements of honor contests and issues of respect.

Narrative Analysis

Narrative "displays the goals and intentions of human actors; it makes individuals, cultures, societies, and historical epochs comprehensible as wholes" (Richardson 1995:200). **Narrative analysis** focuses on "the story itself" and seeks to preserve the integrity of personal biographies or a series of events that cannot adequately be understood in terms of their discrete elements (Riessman 2002:218). The coding for a narrative analysis is typically of the narratives as a whole, rather than of the different elements within them. The coding strategy revolves around reading the stories and classifying them into general patterns.

For example, Morrill et al. (2000:534) read through 254 conflict narratives written by the ninth graders they studied and found four different types of stories:

1. *Action tales*, in which the author represents himself or herself and others as acting within the parameters of taken-for-granted assumptions about what is expected for particular roles among peers.

2. *Expressive tales*, in which the author focuses on strong, negative emotional responses to someone who has wronged him or her.

3. *Moral tales*, in which the author recounts explicit norms that shaped his or her behavior in the story and influenced the behavior of others.

4. *Rational tales*, in which the author represents him- or herself as a rational decision maker navigating through the events of the story.

In addition to these dominant distinctions, Morrill et al. (2000:534–535) also distinguished the stories in terms of four stylistic dimensions: plot structure (such as whether the story unfolds sequentially), dramatic tension (how the central conflict is represented), dramatic resolution (how the central conflict is resolved), and predominant outcomes (how the story ends). Coding reliability was checked through a discussion by the two primary coders, who found that their classifications agreed for a large percentage of the stories.

According to Morrill et al. (2000), action tales:

> unfold in matter-of-fact tones kindled by dramatic tensions that begin with a disruption of the quotidian order of everyday routines. A shove, a bump, a look . . . triggers a response. . . . Authors of action tales typically organize their plots as linear streams of events as they move briskly through the story's scenes. . . . This story's dramatic tension finally resolves through physical fighting, but . . . only after an attempted conciliation. (P. 536)

You can contrast that "action tale" with the following narrative, which Morrill et al. (2000) classify as a "moral tale," in which the students explicitly tell about their moral reasoning, often referring to how normative commitments shape their decisionmaking":

> I . . . got into a fight because I wasn't allowed into the basketball game. I was being harassed by the captains that wouldn't pick me and also many of the players. The same type of things had happened almost every day where they called me bad words so I decided to teach the ring leader a lesson. I've never been in a fight before but I realized that sometimes you have to make a stand against the people that constantly hurt you, especially emotionally. I hit him in the face a couple of times and I got respect I finally deserved. (Pp. 545–546)

Morrill et al. (2000:553) summarize their classification of the youth narratives in a simple table that highlights the frequency of each type of narrative and the characteristics associated with each of them (see Exhibit 9.6). How does such an analysis contribute to our understanding of youth violence? Morrill et al. first emphasize that their narratives "suggest that consciousness of conflict among youths—like that among adults—is not a singular entity, but comprises a rich and diverse range of perspectives" (p. 551).

Theorizing inductively, Morrill et al. (2000:553–554) then attempt to explain why action tales were much more common than the more adult-oriented normative, rational, or emotionally expressive tales. One possibility is Gilligan's (1988) theory of moral development, which suggests that younger students are likely to limit themselves to the simpler action tales that "concentrate on taken-for-granted assumptions of their peer and wider cultures, rather than on more self consciously reflective interpretation and evaluation" (Morrill et al. 2000:554). More generally, Morrill et al. argue, "We can begin to think of the building blocks of cultures as different narrative styles in which various aspects of reality are accentuated, constituted, or challenged, just as others are deemphasized or silenced" (p. 556).

In this way, Morrill et al.'s (2000) narrative analysis allowed an understanding of youth conflict to emerge from the youths' own stories while also informing our understanding of broader social theories and processes.

EXHIBIT 9.6 Summary Comparison of Youth Narratives[a]				
Representation of	Action Tales (N = 144)	Moral Tales (N = 51)	Expressive Tales (N = 35)	Rational Tales (N = 24)
Bases of everyday conflict	Disruption of everyday routines and expectations	Normative violation	Emotional provocation	Goal obstruction
Decision making	Intuitive	Principled stand	Sensual	Calculative choice
Conflict handling	Confrontational	Ritualistic	Cathartic	Deliberative
Physical violence[b]	In 44% (N = 67)	In 27% (N = 16)	In 49% (N = 20)	In 29% (N = 7)
Adults in youth conflict control	Invisible or background	Sources of rules	Agents of repression	Institutions of social control

Source: Morrill et al. 2000:551, Table 1. Copyright 2000. Reprinted with permission of Blackwell Publishing.

a. Total N = 254.

b. Percentages based on the number of stories in each category.

Conversation Analysis

Conversation analysis is a specific qualitative method for analyzing ordinary conversation. Unlike narrative analysis, it focuses on the sequence and details of conversational interaction, rather than on the "stories" that people are telling. Like ethnomethodology, from which it developed, conversation analysis focuses on how reality is constructed, rather than on what it *is*. Three premises guide conversation analysis (Gubrium & Holstein 2000:492):

1. Interaction is sequentially organized, and talk can be analyzed in terms of the process of social interaction rather than in terms of motives or social status.
2. Talk, as a process of social interaction, is contextually oriented; it is both shaped by interaction and creates the social context of that interaction.
3. These processes are involved in all social interaction, so no interactive details are irrelevant to understanding it.

Consider these premises as you read the following dialogue between British researcher Ann Phoenix (2003) and a boy she called "Thomas" in her study of notions of masculinity, bullying, and academic performance among 11- to 14-year-old boys in 12 London schools.

Thomas: It's your attitude, but some people are bullied for no reason whatsoever just because other people are jealous of them. . . .

Q: How do they get bullied?

Thomas: There's a boy in our year called James, and he's really clever and he's basically got no friends, and that's really sad because. . . . He gets top marks in every test and everyone hates him. I mean, I like him. (P. 235)

Phoenix (2003) notes that here,

> Thomas dealt with the dilemma that arose from attempting to present himself as both a boy and sympathetic to school achievement. He . . . distanced himself from . . . being one of those who bullies a boy just because they are jealous of his academic attainments . . . [and] constructed for himself the position of being kind and morally responsible. (P. 235)

Do you see how Thomas's presentation of himself reflected his interchange with the researcher, as she probed his orientation? Do you imagine that his talk would have been quite different if this conversation had been with other boys?

Case-Oriented Understanding

A **case-oriented understanding** attempts to understand a phenomenon from the standpoint of the participants. The case-oriented understanding method reflects an interpretive research philosophy that is not geared to identifying causes but provides a different way to explain social phenomena. For example, Fischer and Wertz (2002) constructed such an explanation of the effect of being criminally victimized. They first recounted crime victims' stories and then identified common themes in these stories.

Their explanation began with a description of what they termed the process of "living routinely" before the crime: "He/she . . . feels that the defended against crime could never happen to him/her. I said, 'nah, you've go to be kidding.'"

In a second stage, "Being Disrupted," the victim copes with the discovered crime and fears worse outcomes: "You imagine the worst when it's happening. . . . I just kept thinking my baby's upstairs." In a later stage, "Reintegrating," the victim begins to assimilate the violation by taking some protective action: "But I clean out my purse now since then and I leave very little of that kind of stuff in there."

Finally, when the victim is "Going On," he or she reflects on the changes the crime produced: "I don't think it made me stronger. It made me smarter."

You can see how Fischer and Wertz (2002:288–290) constructed an explanation of the effect of crime on its victims through this analysis of the process of responding to the experience. This effort to "understand" what happened in these cases gives us a much better sense of why things happened as they did.

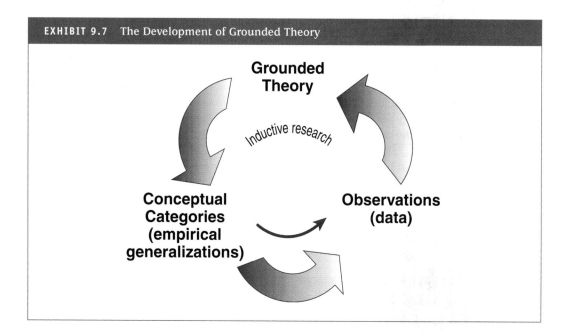

EXHIBIT 9.7   The Development of Grounded Theory

Grounded Theory

Theory development occurs continually in qualitative data analysis (Coffey & Atkinson 1996:23). The goal of many qualitative researchers is to create **grounded theory**, that is, to build up inductively a systematic theory that is "grounded" in, or based on, the observations. The observations are summarized into conceptual categories, which are tested directly in the research setting with more observations. Over time, as the conceptual categories are refined and linked, a theory evolves (Glaser & Strauss 1967; Huberman & Miles 1994:436). Exhibit 9.7 illustrates this process.

As observation, interviewing, and reflection continue, researchers refine their definitions of problems and concepts and select indicators. They can then check the frequency and distribution of phenomena: How many people made a particular type of comment? How often did social interaction lead to arguments? Social system models may then be developed, which specify the relationships among different phenomena. These models are modified as researchers gain experience in the setting. For the final analysis, the researchers check their models carefully against their notes and make a concerted attempt to discover negative evidence that might suggest the model is incorrect.

VISUAL SOCIOLOGY

For about 150 years, people have been creating a record of the social world with photography. This creates the possibility of "observing" the social world through photographs and films and of interpreting the resulting images as a "text." It is no surprise that visual sociology has been developed as a method both to learn how others "see" the social world and to create images of it for further study. As in the analysis of written text, however, the visual sociologist must be sensitive to the way in which a photograph or film "constructs" the reality that it depicts.

An analysis by Margolis (2004) of photographic representations of American Indian boarding schools gives you an idea of the value of analysis of photographs (see Exhibit 9.8). On the left is a picture taken in 1886 of Chiricahua Apaches who had just arrived at the Carlisle Indian School in Carlisle, Pennsylvania. The school was run by a Captain Richard Pratt, who, like many Americans in that period, felt tribal societies were communistic, indolent, dirty, and ignorant, while Western civilization was industrious and individualistic. So Captain Pratt set out to acculturate American Indians to the dominant culture. The second picture shows the result: the same group of Apaches looking like European, not Native, Americans, dressed in standard uniforms, with standard haircuts, and with more standard posture.

EXHIBIT 9.8   Pictures of Chiricahua Apache Children Before and After Starting Carlisle Indian School, Carlisle, Pennsylvania, 1886

Chiricahua Apache Children Upon Arrival at Carlisle School.

Chiricahua Apache Children After 4 Months at Carlisle School.

Source: Margolis 2004:78.

Many other pictures display the same type of transformation. Are these pictures each "worth a thousand words"? They capture the ideology of the school management, but we can be less certain that they document accurately the "before and after" status of the students.

Captain Pratt "consciously used photography to represent the boarding school mission as successful" (Margolis 2004:79). While he clearly tried to ensure a high degree of conformity, there were accusations that the contrasting images were exaggerated to overemphasize the change (Margolis 2004:78). Reality was being constructed, not just depicted, in these photographs.

Visual sociology will certainly become an increasingly important aspect of qualitative analyses of social settings and the people in them. The result will be richer descriptions of the social world, but whether you examine or also produce pictures for such analyses, remember Darren Newbury's (2005) reminder to readers of his journal, *Visual Studies*, that "images cannot be simply taken of the world, but have to be made within it" (p. 1).

Pictures, like other "text," are, in part, a social construction.

COMPUTER-ASSISTED QUALITATIVE DATA ANALYSIS

The analysis process can be enhanced in various ways by using a computer. Programs designed for qualitative data can speed up the analysis process, make it easier for researchers to experiment with different codes, test different hypotheses about relationships, and facilitate diagrams of emerging theories and preparation of research reports (Coffey & Atkinson 1996; Richards & Richards 1994). The steps involved in **computer-assisted qualitative data analysis** parallel those used traditionally to analyze such text as notes, documents, or interview transcripts: preparation, coding, analysis, and reporting. We use two of the most popular programs to illustrate these steps: HyperRESEARCH and QSR NVivo. (See the Student Study Site, http:www.sagepub.com/prccj3, for an extended introduction to HyperRESEARCH. The software itself and the HyperRESEARCH tutorials are found on the study site as well.)

Text preparation begins with typing or scanning text in a word processor or, with NVivo, directly into the program's rich text editor. NVivo will create or import a rich text file. HyperRESEARCH requires that your text be saved as a text file (as "ASCII" in most word processors) before you transfer it into the analysis program. HyperRESEARCH expects your text data to be stored in separate files corresponding to each unique case, such as an interview with one subject.

Coding the text involves categorizing particular text segments. This is the foundation of much qualitative analysis. Either program allows you to assign a code to any segment of text (in NVivo, you drag through the characters to select them; in HyperRESEARCH, you click on the first and last words to select text). You can make up codes as you go through a document and also assign codes that you have already developed to text segments. Exhibit 9.9 shows the screens that appear in the two programs at the coding stage, when a particular text segment is being labeled. You can also have the programs "autocode" text by identifying a word or phrase that should always receive the same code, or, in NVivo, by coding each section identified by the style of the rich text document, for example, each question or speaker (of course, you should check carefully the results of autocoding). Both programs also let you examine the coded text "in context," embedded in its place in the original document.

In qualitative data analysis, coding is not a one-time-only or one-code-only procedure. Both HyperRESEARCH and NVivo allow you to be inductive and holistic in your coding: You can revise codes as you go along, assign multiple codes to text segments, and link your own comments ("memos") to text segments. In NVivo, you can work "live" with the coded text to alter coding or create new, more subtle categories. You can also place hyperlinks to other documents in the project or any multimedia files outside it.

EXHIBIT 9.9a HyperRESEARCH Coding Stage

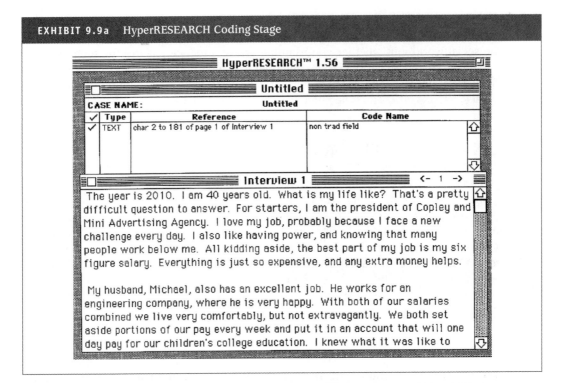

EXHIBIT 9.9b NVivo Coding Stage

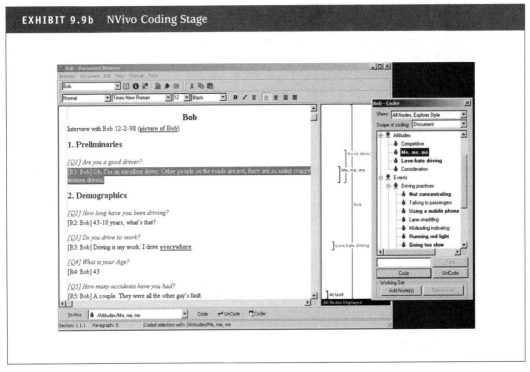

EXHIBIT 9.10   A Free-Form Model in NVivo

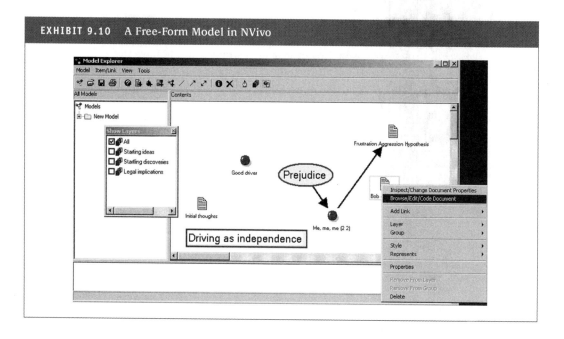

Analysis focuses on reviewing cases or text segments with similar codes and examining relationships among different codes. You may decide to combine codes into larger concepts. You may specify additional codes to capture more fully the variation among cases. You can test hypotheses about relationships among codes. NVivo allows development of an indexing system to facilitate thinking about the relationships among concepts and the overarching structure of these relationships. It will also allow you to draw more free-form models (see Exhibit 9.10). In HyperRESEARCH, you can specify combinations of codes that identify cases that you want to examine.

Reports from both programs can include text to illustrate the cases, codes, and relationships that you specify. You can also generate counts of code frequencies and then import these counts into a statistical program for quantitative analysis. However, the many types of analyses and reports that can be developed with qualitative analysis software do not lessen the need for a careful evaluation of the quality of the data on which conclusions are based.

In reality, using a qualitative data analysis computer program is not always as straightforward as it appears. Scott Decker and Barrik Van Winkle (1996) describe the difficulty they faced in using a computer program to identify instances of the concept of "drug sales":

The software we used is essentially a text retrieval package. . . . One of the dilemmas faced in the use of such software is whether to employ a coding scheme within the interviews or simply to leave them as unmarked text. We chose the first alternative, embedding conceptual tags at the appropriate points in the text. An example illustrates this process. One of the activities we were concerned with was drug sales. Our first chore (after a thorough reading of all the transcripts) was to use the software to "isolate" all of the transcript sections dealing with drug sales. One way to do this would be to search the transcripts for every instance in which the word "drugs" was used. However, such a strategy would have the disadvantages of providing information of too general a character while often missing important statements about drugs. Searching on the word "drugs" would have produced a file including every time the word was used, whether it was in reference to drug sales, drug use, or drug availability, clearly more information than we were interested in. However, such a search would have failed to find all of the slang used to refer to drugs ("boy" for heroin, "Casper" for crack cocaine) as well as the more common descriptions of drugs, especially rock or crack cocaine. (Pp. 53–54)

Decker and Van Winkle solved this problem by parenthetically inserting conceptual tags in the text whenever talk of drug sales was found. This process allowed them to examine all the statements made by gang members about a single concept (drug sales). As you can imagine, however, this still left the researchers with many pages of transcripts material to analyze.

ETHICS IN QUALITATIVE DATA ANALYSIS

The qualitative data analyst is never far from ethical issues and dilemmas. Throughout the analytic process, the analyst must consider how the findings will be used and how participants in the setting will react. Miles and Huberman (1994:293–295) suggest several specific questions that are of particular importance during the process of data analysis:

"*Privacy, confidentiality, and anonymity*. In what ways will the study intrude, come closer to people than they want? How will information be guarded? How identifiable are the individuals and organizations studied?" We have considered this issue already in the context of qualitative data collection, but it also must be a concern during the process of analysis. It can be difficult to present a rich description in a case study while at the same time not identifying the setting. It can be easy for participants in the study to identify each other in a qualitative description, even if outsiders cannot. Qualitative researchers should negotiate with participants early in the study the approach that will be taken to protecting privacy and to maintaining confidentiality. Selected participants should also be asked to review reports or other products before their public release in order to gauge the extent to which they feel privacy has been appropriately preserved.

"*Intervention and advocacy*. What do I do when I see harmful, illegal, or wrongful behavior on the part of others during a study? Should I speak for anyone's interests besides my own? If so, whose interests do I advocate?" Maintaining what is called "guilty knowledge" may force the researcher to suppress some parts of the analysis so as not to disclose the wrongful behavior, but presenting "what really happened" in a report may prevent ongoing access and violate understandings with participants.

"Research integrity and quality. Is my study being conducted carefully, thoughtfully, and correctly in terms of some reasonable set of standards?" Real analyses have real consequences, so you owe it to yourself and those you study to adhere strictly to the analysis methods that you believe will produce authentic, valid conclusions.

"Ownership of data and conclusions. Who owns my field notes and analyses: I, my organization, my funders? And once my reports are written, who controls their diffusion?" Of course, these concerns arise in any social research project, but the intimate involvement of the qualitative researcher with participants in the setting studied makes conflicts of interest between different stakeholders much more difficult to resolve. Working through the issues as they arise is essential.

"Use and misuse of results. Do I have an obligation to help my findings be used appropriately? What if they are used harmfully or wrongly?" It is prudent to develop understandings early in the project with all major stakeholders that specify what actions will be taken in order to encourage appropriate use of project results and to respond to what is considered misuse of these results.

CONCLUSION

The variety of approaches to qualitative data analysis makes it difficult to provide a consistent set of criteria for interpreting their quality. Denzin's (2002:362–363) "interpretive criteria" are a good place to start. Denzin suggests that at the conclusion of their analyses, qualitative data analysts ask the following questions about the materials they have produced. Reviewing several of them will serve as a fitting summary for your understanding of the qualitative analysis process.

- *Do they illuminate the phenomenon as lived experience?* In other words, do the materials bring the setting alive in terms of the people in that setting?
- *Are they based on thickly contextualized materials?* We should expect thick descriptions that encompass the social setting studied.
- *Are they historically and relationally grounded?* There must be a sense of the passage of time between events and the presence of relationships between social actors.
- *Are they processual and interactional?* The researcher must have described the research process and his or her interactions within the setting.
- *Do they engulf what is known about the phenomenon?* This includes situating the analysis in the context of prior research and also acknowledging the researcher's own orientation upon first starting the investigation.

When an analysis of qualitative data is judged as successful in terms of these criteria, we can conclude that the goal of "authenticity" has been achieved.

As a research methodologist, you must be ready to use both types of techniques, evaluate research findings in terms of both sets of criteria, and mix and match the methods as required by the research problem to be investigated and the setting in which it is to be studied.

KEY TERMS

Case-oriented understanding
Computer-assisted qualitative data analysis
Emic focus
Ethnography
Etic focus
Grounded theory

Matrix
Narrative analysis
Progressive focusing
Qualitative comparative analysis (QCA)
Tacit knowledge

HIGHLIGHTS

- Qualitative data analysts are guided by an emic focus of representing persons in the setting on their own terms rather than by an etic focus on the researcher's terms.
- Case studies use thick description and other qualitative techniques to provide a holistic picture of a setting or group.
- Ethnographers attempt to understand the culture of a group.
- Narrative analysis attempts to understand a life or a series of events as they unfolded, in a meaningful progression.
- Grounded theory connotes a general explanation that develops in interaction with the data and is continually tested and refined as data collection continues.
- Special computer software can be used for the analysis of qualitative, textual, and pictorial data. Users can record their notes, categorize observations, specify links between categories, and count occurrences.

EXERCISES

1. List the primary components of qualitative data analysis strategies. Compare and contrast each of these components with those relevant to quantitative data analysis.

2. Read the complete text of one of the qualitative studies presented in this chapter and evaluate its conclusions for authenticity, using the criteria in this chapter.

3. Attend a sports game as an ethnographer. Write up your analysis and circulate it for criticism.

4. Write a narrative in class about your first date, car, college course, or something else that you and your classmates agree on. Then collect all the narratives and analyze them in a "committee of the whole." Follow the general procedures discussed in the example of narrative analysis in this chapter.

5. Review one of the articles on the book Study Site (http://www.sagepub.com/prccj3) that used qualitative methods. Describe the data that were collected, and identify the steps used in the analysis. That type of qualitative data analysis was this? If it is not one of the methods presented in this chapter, describe its similarities to and differences from one of these methods. How confident are you in the conclusions, given the methods of analysis used?

DEVELOPING A RESEARCH PROPOSAL

1. Which qualitative data analysis alternative is most appropriate for the qualitative data you proposed to collect for your project? Using the approach, develop a strategy for using the techniques of qualitative data analysis to analyze your textual data.

Student Study Site

The companion Web site for *The Practice of Research in Criminology and Criminal Justice*, Third Edition

http://www.sagepub.com/prccj3

Visit the Web-based Student Study Site to enhance your understanding of the chapter content and to discover additional resources that will take your learning one step further. You can enhance your understanding of the chapters by using the comprehensive study material, which includes e-flashcards, Web exercises, practice self-tests, and more. You will also find special features, such as Learning from Journal Articles, which incorporates SAGE's online journal collection.

WEB EXERCISES

1. The Qualitative Report is an online journal about qualitative research. Inspect the table of contents for a recent issue at http://www.nova.edu/ssss/QR/index.html. Read one of the articles and write a brief article review.

2. Be a qualitative explorer! Go to the list of qualitative research Web sites and see what you can find that enriches your understanding of qualitative research (http:/www.qualitative research.uga.edu/ QualPage/). Be careful to avoid textual data overload.

ETHICS EXERCISES

1. Ethnographers and other qualitative researchers often seek to develop "thick [rich] descriptions" of social settings, but the details that are required for "thickness" can make the identity of a social setting and even of participants obvious to insiders in that setting. What requirements should be imposed on qualitative research reports to prevent identity disclosure? Changing names? Altering descriptions of people or events? Do you think it is OK to identify a community? An organization? A small group? A very well known leader?

2. Some qualitative researchers recommend sharing findings and interpretations with research subjects prior to publication, and then making changes that are suggested by the subjects. What advice would you give to qualitative researchers about engaging with research subjects during or after the process of data analysis? Would you include the reactions of research subjects to your findings in your research report? Would you alter your conclusions based on feedback from research subjects? Under what conditions?

Analyzing Content: Historical, Secondary, and Content Analysis, and Crime Mapping

The research methods we have examined so far have relied on researchers collecting the data or information themselves. Increasingly, however, those interested in criminological research questions are relying on data previously collected by other investigators (Riedel 2000). As we noted in Chapter 1, this is referred to as secondary data analysis. Secondary data analysis is simply the act of collecting or analyzing data that was originally collected for another purpose (Riedel 2000). Thus, if a researcher goes to a police department and personally compiles information from police reports to examine a research question, she is still engaging in secondary data analysis because the police records were originally collected for another purpose.

Secondary data analysis has a long history. Since the latter part of the 17th century, people have been monitoring the state of their localities by examining rates of population, mortality, marriage, disease, climate, and crime. Adolphe Quételet, an ambitious Belgian mathematician, was one of the first to show that the annual number of murders reported in France from 1826 to 1831 was relatively constant and, further, that the proportion of murders committed with guns, swords, knives, stones, kicks and punches, and strangulation was also relatively constant. He concluded that although we may not know who will kill whom by what means, we do know, with a high degree of probability, that a certain number of murders of a certain type will happen every year in France (Menand 2001). This was one of the first attempts to apply the methods of science to social phenomena. You are also probably familiar with Émile Durkheim's ([1951] 1987) use of official statistics on suicide rates in different areas to examine the relationship between religion and suicide.

In this chapter, we will tell you about a number of data sets including surveys and official records that are publicly available for research purposes. Then we examine several research methods that rely on secondary data, including historical events research, cross-cultural research, content analysis, and crime mapping. Because using data originally gathered for other purposes poses unique concerns for a researcher, we spend the latter part of the chapter highlighting these methodological issues.

ANALYZING SECONDARY DATA

In general, there are four major types of secondary data: surveys, official statistics, official records, and other historical documents. Although a data set can be obtained by an agreement between two or more researchers, many researchers obtain data through the Inter-University Consortium for Political and Social Research (ICPSR) (www.icpsr.umich.edu). Data stored at ICPSR primarily include surveys, official records, and official statistics. ICPSR stores data and information for nearly 5,000 sources and studies, including those conducted independently and those conducted by the U.S. government. Riedel (2000) has documented the majority of data sets that are available from the ICPSR and that are appropriate for crime research, including the following:

Census enumerations: historical and contemporary population characteristics. The most well-known data sets within this category are the surveys conducted every decade by the Bureau of the Census. Linking information from this data set (e.g., neighborhood characteristics including such things as poverty and residential mobility) to crime data at the same level (e.g., census block, county) has provided researchers with a rich source of data to test theories of crime.

The National Archive of Criminal Justice Data (NACJD). The Bureau of Justice Statistics and National Institute of Justice co-sponsored NACJD, which provides more than 600 criminal justice data collections to the public: A sample of these data sets include:

- Capital Punishment in the United States
- Expenditure and Employment Data for the Criminal Justice System
- Gang Involvement in Rock Cocaine Trafficking in Los Angeles, 1984–1985
- Criminal Careers and Crime Control in Massachusetts
- Longitudinal Research Design, Phase I, 1940–1965.
- Changing Patterns of Drug Use and Criminality Among Crack Cocaine Users In New York City: Criminal Histories and CJ Processing, 1983–1984, 1986
- The National Crime Victimization Survey, ongoing
- National Jail Census
- National Judicial Reporting Program
- National Survey of Jails
- Survey of Adults on Probation
- Survey of Inmates of Federal Correctional Facilities
- Survey of Inmates of Local Jails
- Survey of Inmates of State Correctional Facilities
- FBI Uniform Crime Reporting Program (UCR) data, including the Supplementary Homicide Reports (SHR)

Social indicators and behavior. There are a series of annual surveys under this heading including the General Social Survey, which has been conducted annually by the National Opinion Research Center since 1972. In addition, Monitoring the Future: A Continuing Study of the Lifestyles and Values of Youth is a survey of a nationally representative sample of high school seniors that asks them many things including self-reports of drug and alcohol use and their attitudes toward a number of issues. The National Youth Survey Series (1976–1980 and 1983) is another survey available at ICPSR that examines factors related to delinquency.

Case Study: Gender and Offending

As you can see, the research possibilities are almost limitless with the wealth of data already made available to researchers interested in issues of criminology and criminal justice. For example, Heimer and De Coster (1999) used the National Youth Survey to examine the mechanisms explaining variation in violent delinquency between male and female youth. Generally, studies of self-reported violent delinquency find gender ratios (self-reports of delinquency of males to females) ranging from approximately 1.1 to 5.3, depending on the specific aggressive offense being measured. These ratios show that although males generally commit more violent acts than females, young females do engage in violence.

One of the primary mechanisms related to violent offending that Heimer and De Coster (1999) examined was "cultural definitions of violence." Specifically, according to differential association theory, interaction with others and social structural context are important because they shape the learning of violent definitions, which in turn affects the likelihood that youths engage in violent delinquency. Heimer and De Coster believed that one factor affecting the violent offending disparities between young males and females was that, on average, boys tend to acquire more violent definitions than girls. In addition, the "gender definitions" held by boys and girls are different; femininity often is equated with a high capacity for nurturance and a tendency toward passivity rather than aggressiveness. In contrast, masculinity tends to be equated with competitiveness, independence, rationality, and strength. Heimer and De Coster tested many hypotheses including the following:

> Hypothesis: Higher levels of violent definitions among boys than girls will explain part of the gender gap in violent offending.

> Hypothesis: Learning traditional gender definitions will reduce the chances of violent delinquency among females and increase the chances of violent delinquency among males.

To operationalize the concepts of violent definitions and gender definitions, Heimer and De Coster (1999) used several questions in the National Youth Survey (NYS), which are displayed in Exhibit 10.1 along with the questions used to uncover incidents of violent offending. After statistically controlling for other important factors proven to be related to violent offending, such as socioeconomic status, age, prior violent delinquency, association with aggressive friends, and attachments to family, the results of the analyses of the NYS data provided support for both hypotheses. Heimer and De Coster concluded,

> In sum, girls are less violent than boys mainly because they are influenced more strongly by bonds to family, learn fewer violent definitions, and are taught that violence is inconsistent with the meaning of being female. These mechanisms appear to be so effective among girls, that direct, overt controls like supervision and coercive discipline contribute little to the explanation of variation in female violence, while they are important for explaining variation in male violence. (P. 303)

Others have used existing data to examine factors related to the gender differences in juvenile delinquency. For example, Piquero et al. (2005) also investigated the difference in peer influences on vandalism and shoplifting among a sample of male and female 10th-graders from data that were originally collected in 1981. Among other things, they concluded that the presence of delinquent peers increased these delinquent behaviors for males, but did not increase the likelihood of females engaging in these activities.

HISTORICAL EVENTS RESEARCH

The central insight behind historical methods is that we can improve our understanding of social processes when we make comparisons to other times and places. Although there are no rules for determining how far in the past the focus of research must be in order for it to be considered historical, in practice, research tends to be considered historical when it focuses on a period prior to the experience of most of those conducting research (Abbott 1994).

EXHIBIT 10.1   Questions From the National Youth Survey Used to Measure Violent Definitions, Gender Definitions, and Violent Delinquent Offending

Violent Delinquency

The below items were coded 1 = never; 2 = once or twice a year; 3 = once or twice every 2–3 months; 4 = once a month; 5 = once every 2–3 weeks; 6 = once a week; 7 = 2–3 times a week; 8 = once a day; 9 = 2–3 times a day.

How may times in the past year have you

  a) carried a hidden weapon other than a plain pocket knife?

  b) attacked someone with the idea of seriously hurting or killing them?

  c) been involved in gang fights?

  d) hit or threatened to hit a teacher or other adult at school?

  e) hit or threatened to hit your parents?

  f) hit or threatened to hit other students?

  g) had or tried to have sexual relations with someone against their will?

  h) used force (strong-arm methods) to get money or things from other students?

  i) used force (strong-arm methods) to get money or things from a teacher or adult at school?

  j) used force (strong-arm methods) to get money or things from other people (not teachers or students)?

Violent Definitions

The following questions were coded 1 = strongly disagree; 2 = disagree; 3 = neither agree nor disagree; 4 = agree; 5 = strongly agree.

  a) In order to gain respect from your friends, it is sometimes necessary to beat up on other kids.

  b) It is alright to beat up another person if he or she called you a dirty name.

  c) It is alright to beat up another person if he or she started the fight.

  d) Hitting another person is an acceptable way to get him or her to do what you want.

Gender Definitions

  a) In general, the father should have greater authority than the mother in the bringing up of children.

  b) Women with children should not work outside the home unless there is no one else to support the family.

  c) In a marriage, it is the woman's responsibility to care for any children and take care of the home.

  d) Women are too emotional to solve problems well.

  e) Women are physically and emotionally weaker than men and therefore need male protection and support.

Source: Heimer & De Coster 1999.

To test your understanding of secondary techniques, go to the Content Analysis and Other Secondary Techniques Interactive Exercises on the Student Study Site.

Historical research in the social sciences differs from traditional historical research because researchers seek to develop general theoretical explanations of historical events and processes instead of just detailed, "fact-centered" descriptions of them (Monkkonen 1994:8). Social scientists do not ignore the details of historical events; rather, they "unravel" unique events to identify general patterns (Abrams 1982:200). However, as in traditional history, the focus on the past presents special methodological challenges. Documents and other evidence may have been lost or damaged and what evidence there is may represent a sample biased toward those who were more newsworthy figures or who were more prone to writing. The feelings of individuals involved in past events may be hard, if not impossible, to reconstruct. Nonetheless, in many situations the historical record may support very systematic research on what occurred in the past.

When research on past events does not follow processes for a long period of time, when it is basically cross-sectional, it is usually referred to as **historical events research**. Investigations of past events may be motivated by the belief that they had a critical impact on subsequent developments or because they provide opportunities for testing the implications of a general theory (Kohn 1987).

Case Study: The Martinsville Seven

In *The Martinsville Seven: Race, Rape, and Capital Punishment,* Eric Rise (1995) offers an in-depth example of historical events research. On January 8, 1949, a 32-year-old white woman in Martinsville, Virginia, accused seven young black men of violently raping her. These men became known as the "Martinsville Seven." Within two days of the allegation, state and local police had arrested and obtained confessions from each of the suspects. In a rapid succession of brief trials held over the course of 11 days, six separate juries convicted the defendants of rape and sentenced each to death. During the first week of February 1951, each of the "Martinsville Seven" died in the electric chair of the Virginia State Penitentiary. At a time when African Americans were beginning to assert their civil rights vigorously, the executions provided a stark reminder of the harsh treatment reserved for blacks who violated southern racial codes.

In his analysis of the case, Rise (1995) examined a wealth of historical documents, including

1. Legal papers that the lawyers involved in the case made available to such archives as the Virginia State Library and Archives, the Archives and Rare Books Division of the Schomburg Center for Research in Black Culture at the New York Public Library, the Truman Library, the Margaret I. King Library at the University of Kentucky, and the United States National Archive

2. Official transcripts of the cases from the Clerk's Office in Martinsville, Virginia

3. Briefs and petitions to the Virginia State Supreme Court

4. Related case law from the Supreme Court

5. Annual reports from the Virginia Department of Corrections 1946–1951

6. Newspaper accounts of the event from 16 local and national newspapers including the *Martinsville Bulletin*, *The New York Times*, and *The Washington Post*.

The large number of defendants, the rapid pace of the trials, and the specter of multiple executions attracted the attention of two national organizations, the National Association for the Advancement of Colored People (NAACP) and the Civil Rights Congress (CRC). In addition to the discriminatory nature of the jury's harsh sentences, Rise (1995) also found that another factor that obstructed justice in this case was the escalating tension and antagonisms between these two organizations. Ironically, both were trying to obtain justice for the Martinsville Seven.

The CRC was a radical communist organization, and during this time of domestic tension and extreme anticommunist sentiments, most civil rights organizations, including the NAACP, did not want to appear aligned with them. Rise's (1995) analysis of historical documents and correspondence revealed extreme conflict between the two organizations while they were both vying for top billing in representing the Martinsville Seven. The inherent goal of the CRC, of course, was not so much to seek justice for the seven men, but to "reveal the capitalist conspiracy of racial and economic oppression in America. The divergent manner in which each organization proposed to handle the cases highlighted fundamental disagreements over strategy, ingrained suspicions of motive, and fierce competition for membership" (p. 68). Because cooperation was never established through the appeals process, both organizations continued to pursue the cases in a manner consistent with their own ideological and institutional interests. The result, of course, was deadly. As Rise notes, "Never before had a state executed so many men for a single rape incident, nor had a lynching of that magnitude for men accused of rape ever been reported" (p. 3).

Rise's (1995) analysis revealed the critical effect that this historical event had on subsequent developments. The revelation that capital sentencing disparities existed between white and African American defendants during the appeals process of the Martinsville Seven was significant for modern jurisprudence. Specifically, the Martinsville case was the first time attorneys presented in court equal protection arguments that challenged the racial disparity of death sentences for rape. Subsequent to the Martinsville case, civil rights lawyers have not only used the equal protection strategy in the pursuit of equal justice for African Americans, but they also have used it as evidence for the abolition of the death penalty.

COMPARATIVE METHODS

The limitations of examining data from a single location have encouraged many social scientists to turn to comparisons among many geographical entities. Although comparative methods are often associated with cross-national comparisons, research examining smaller aggregates such as states and cities can also be subsumed under the comparative research umbrella. **Comparative research** methods allow for a broader vision about social relations than is possible with cross-sectional research limited to one location.

Case Study: The Comparative Crime Data File

One of the largest comparative research projects undertaken in criminology was the development of the Comparative Crime Data File (CCDF), which was created by Archer and

Gartner (1984). Archer and Gartner articulated the need for comparative research in the field succinctly:

> The need for cross-national comparisons seems particularly acute for research on crime and violence since national differences on these phenomena are of a remarkable magnitude. In some societies, homicide is an hourly, highly visible, and therefore somewhat unexceptional cause of death. In other nations, homicides are so infrequent that, when they do occur, they receive national attention and lasting notoriety. (P. 4)

The CCDF continues to be updated, but originally contained crime and violence data from 110 nations and 44 major international cities covering the period from approximately 1900 to 1970.

Without investigating the patterns and correlates of crime and violence across different societies, Archer and Gartner (1984) contend that criminological research has five major problems:

- *Generalization.* It has been impossible to test the generality of a finding based on single-society research by means of replication in a sample of several societies.
- *Controlled comparison.* The absence of a sufficient number of cases (e.g., nations or cities) has hindered rigorous comparisons between those cases affected by some social change and control cases unaffected by the change.
- *Causal inference.* With longitudinal data unavailable, researchers have not been able to satisfy one of the classic requirements for making causal inferences, the correct temporal relationship among the variables under study.
- *Mediation and intervening variables.* Without a reasonably large sample of nations, it is impossible to discover whether certain variables may mediate the effects of a social change. Without a large sample of societies, a general pattern that explains or orders these different outcomes will never be seen.
- *Methodological uncertainty.* Without an archive of broadly comparative and longitudinal crime data, some key methodological issues have been largely uninvestigable. For example, it would not be possible to assess the reliability of different crime indicators like the number of offenses known or the number of arrests using data from a number of societies.

In their work, Archer and Gartner (1984) examined many research questions using the CCDF. One of these questions is related to the idea that war might increase the level of post-war homicide within nations involved in wars. There are several theoretical models that speculate about the possible effects of wars on postwar violence within a nation. For example, the social solidarity model posits a wartime *decrease* in violence because of the increase in social solidarity among a nation's citizenry. At a more individual level, the violent veteran model predicts that postwar levels of violence within a nation will *increase* as a result of the violent acts of returning war veterans. At a societal level, the legitimization of violence model postulates that during a war, a society reverses its prohibitions against killing and instead honors acts of violence that would be regarded as murderous in peacetime. This social approval or legitimation of violence, this model predicts, may produce a lasting reduction of inhibitions against taking human life, even after the war, thereby increasing levels of violence within nations.

To examine the effects of war on postwar violence, Archer and Gartner (1984) compared national rates of homicide before and after many wars, both small and large, including the two World Wars. Exhibit 10.2 displays the increase or decrease of homicide rates in combatant nations and a sample of control nations that were not involved in World War I and World War II. The researchers also controlled for a number of other factors related to war, including the number of combat deaths in war, whether nations were victorious or defeated, and whether the nation's postwar economics were improved or worsened, and for several types of homicide offenses committed by both men and women. The researchers found that most combatant nations in their analysis experienced substantial postwar increases in their rates of homicide after both small and large wars. Archer and Gartner concluded,

> The one model that appears to be fully consistent with the evidence is the legitimation of violence model, which suggests that the presence of authorized or sanctioned killing during war has a residual effect on the level of homicide in peacetime society. (P. 96)

CONTENT ANALYSIS

Do media accounts of crime, such as newspaper and television news coverage, accurately portray the true nature of crime? To answer this question, using the methodologies already discussed in this book may not be so helpful. Content analysis, however, would. Content analysis is "the systematic, objective, quantitative analysis of message characteristics" (Neuendorf 2002:1). Using this method, we can learn a great deal about popular culture and many other issues through studying the characteristics of messages delivered through the mass media and other sources.

The goal of a content analysis is to develop inferences from text (Weber 1990). You can think of a content analysis as a "survey" of some documents or other records of prior communication. In fact, a content analysis is a survey designed with fixed-choice responses so that it produces quantitative data that can be analyzed statistically. This method was first applied to the study of newspaper and film content and then developed systematically for the analysis of Nazi propaganda broadcasts in World War II, but it can also be used to study historical documents, records of speeches, and other "voices from the past" (Neuendorf 2002:31–37).

Content analysis bears some similarities to qualitative data analysis, because it involves coding and categorizing text and identifying relationships among constructs identified in the text. However, since it usually is conceived as a quantitative procedure, content analysis overlaps with qualitative data analysis only at the margins, the points where qualitative analysis takes on quantitative features or where content analysis focuses on qualitative features of the text. Content analysis also bears some similarities to secondary data analysis, because it involves taking data, text, that already exists and subjecting it to a new form of "analysis," but unlike secondary analysis of previously collected quantitative data, content analysis also involves sampling and measurement of primary data. Unlike the secondary

EXHIBIT 10.2   Homicide Rate Changes in Combatant and Control Nations After World War I and World War II

Combatant Nations

Rates Decreased	Rates Remained Unchanged	Rates Increased
Australia (I)	England (I)	Belgium
Canada (I)	France (I)	Bulgaria
Hungary (I)	S. Africa (I)	Germany
Finland (II)	Canada (II)	Italy (I)
N. Ireland (II)		Japan (I)
U.S. (II)		Portugal (I)
		Scotland (I)
		U.S. (I)
		Australia (II)
		Denmark (II)
		England (II)
		France (II)
		Italy (II)
		Japan (II)
		Netherlands (II)
		New Zealand (II)
		Norway (II)
		Scotland (II)
		S. Africa (II)

Control Nations

Rates Decreased	Rates Remained Unchanged	Rates Increased
Norway (I)	Ceylon (I)	Finland (I)
Ceylon (II)	Chile (1)	Thailand (I)
Chile (II)	Netherlands (I)	Colombia (II)
El Salvador (II)		Sweden (I)
Ireland (II)		Turkey (II)
Switzerland (II)		
Thailand (II)		

Source: Archer & Gartner 1984, Table 4.1.

analyst of qualitative data, the content analyst, by definition, applies primarily quantitative techniques to the coding and analysis of her data. These distinctions become even fuzzier, however, when we recognize that content analysis techniques can be used with all forms of messages, including visual images, sounds, and interaction patterns, as well as written text (Neuendorf 2002:24–25).

Identifying a Population of Documents or Other Textual Sources

To test your understanding of content analysis, go to the Content Analysis and Other Secondary Techniques Interactive Exercises on the Student Study Site.

The population of documents that is selected for analysis should be appropriate to the research question of interest. The units that are surveyed in a content analysis can range from newspapers, books, films, nomination speeches, or TV shows to persons referred to in other communications, themes expressed in documents, or propositions made in different statements. Often, a comprehensive archive can provide the primary data for the analysis (Neuendorf 2002:76–77). For a fee, the LEXIS/NEXIS service makes a large archive of newspapers available for analysis. Words or other features of these units are then coded in order to measure the variables involved in the research question. The content analysis proceeds through several stages (Weber 1985).

Identify a population of documents or other textual sources for study. This population should be selected so that it is appropriate to the research question of interest. Perhaps the population will be all newspapers published in the United States, college student newspapers, nomination speeches at political party conventions, or "state of the nation" speeches by national leaders.

Determine the units of analysis. These could be newspaper articles, whole newspapers, speeches, or political conventions.

Select a sample of units from the population. The simplest strategy might be a simple random sample of documents. However, a stratified sample might be needed to ensure adequate representation of community newspapers in large and small cities, or of weekday and Sunday papers, or of political speeches during election years and in off years (see Chapter 4).

Design coding procedures for the variables to be measured. This requires deciding what unit of text to code, such as words, sentences, themes, or paragraphs. Then the categories into which the units are to be coded must be defined. These categories may be broad, such as "supports democracy," or narrow, such as "supports universal suffrage."

Test and refine the coding procedures. Clear instructions and careful training of coders are essential.

Base statistical analyses on counting occurrences of particular items. These could be words, themes, or phrases. You will also need to test relations between different variables.

Developing reliable and valid coding procedures is not an easy task. The meaning of words and phrases is often ambiguous. Homographs (words such as "mine" that have different meanings in different contexts) create special problems, as do many phrases that have special meanings (such as "point of no return") (Weber 1985:30). As a result, coding procedures cannot simply categorize and count words; text segments in which the words are embedded must also be inspected before codes are finalized. Because different coders may perceive different meanings in the same text segments, explicit coding rules are required to ensure coding consistency. Special dictionaries can be developed to keep track of how the categories of interest are defined in the study (Weber 1985:24–34).

After coding procedures are developed, their reliability should be assessed by comparing different coders' codes for the same variables. Computer program content analysis can be used to enhance reliability (Weber 1985). Whatever rules the computer is programmed to use to code text will be applied consistently. The criteria for judging quantitative content analyses of text are the same standards of validity applied to data collected with other quantitative methods. We must review the sampling approach, the reliability and validity of the measures, and the controls used to strengthen any causal conclusions.

The various steps in a content analysis are represented in the flowchart in Exhibit 10.3. Note that the steps are comparable to the procedures in quantitative survey research. Use this flowchart as a checklist when you design or critique a content analysis project.

Case Study: Media Representations of Crime

Researchers interested in the media and crime have used content analysis in a number of ways. For example, scholars analyzing crime depictions presented in the media often conclude that newspaper and television coverage of crime is frequently inaccurate and misleading; stories disproportionately report violent crimes and reporters tend to focus attention on sensational matters such as the capture of a criminal or high status offenders.

One type of television program presents vignettes depicting actual crimes in which theories of crime are dramatized. These dramatizations feature actors, actual photographs or film footage, and interviews conducted with participants and the police. Viewers are urged to call the police or program representatives with information related to the crime, and police officers are on standby in the television studio to take these calls. Two such programs are *America's Most Wanted* (AMW) and *Unsolved Mysteries* (UM).

To analyze the images these programs were sending to viewers, Cavender and Bond-Maupin (2000) conducted a content analysis of AMW and UM programs that aired between January 25 and May 31, 1989. All programs that aired during this time period were video-taped and a subsample of 16 episodes were randomly selected that contained 77 crime vignettes. The crime vignettes served as the study's units of analysis. The coding protocol used for data collection was focused on three aspects of the programs:

1. *Demographics.* Types of crime and general information on the crime, criminals, and victims;

2. *Characterizations.* Specific depictions of crime, criminals, and victims, such as brutality, dangerousness, or a victim's vulnerability; and

3. *Worldview.* The relative safety of people and places, the terror and randomness of crime, and what the audience should do about crime.

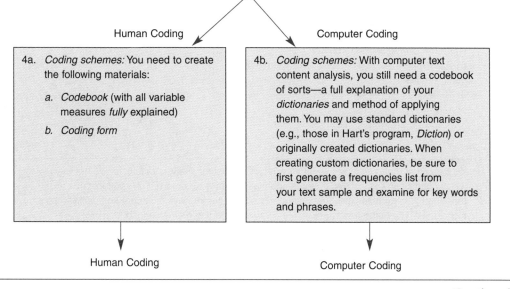

EXHIBIT 10.3   Flowchart for the Typical Process of Content Analysis Research

1. *Theory and rationale:* What *content* will be examined, and *why*? Are there certain *theories* or perspectives that indicate that this particular message content is important to study? Library work is needed here to conduct a good literature review. Will you be using an integrative model, linking content analysis with other data to show relationships with source or receiver characteristics? Do you have *research questions*? *Hypotheses?*

2. *Conceptualizations:* What *variables* will be used in the study, and how do you define them *conceptually* (i.e., with dictionary-type definitions)? Remember, you are the boss! There are many ways to define a given construct, and there is no one right way. You may want to screen some examples of the content you're going to analyze, to make sure you've covered everything you want.

3. *Operationalizations (measures):* Your measures should match your conceptualizations . . . What *unit of data collection* will you use? You may have more than one unit (e.g., a by-utterance coding scheme and a by-speaker coding scheme). Are the variables measured well (i.e., at a high *level of measurement,* with categories that are *exhaustive and mutually exclusive)?* An *a priori* coding scheme describing all measures must be created. Both face validity and content validity may also be assessed at this point.

Human Coding   Computer Coding

4a. *Coding schemes:* You need to create the following materials:

    a. *Codebook* (with all variable measures *fully* explained)

    b. *Coding form*

4b. *Coding schemes:* With computer text content analysis, you still need a codebook of sorts—a full explanation of your *dictionaries* and method of applying them. You may use standard dictionaries (e.g., those in Hart's program, *Diction*) or originally created dictionaries. When creating custom dictionaries, be sure to first generate a frequencies list from your text sample and examine for key words and phrases.

Human Coding   Computer Coding

(Continued)

EXHIBIT 10.3 (Continued)

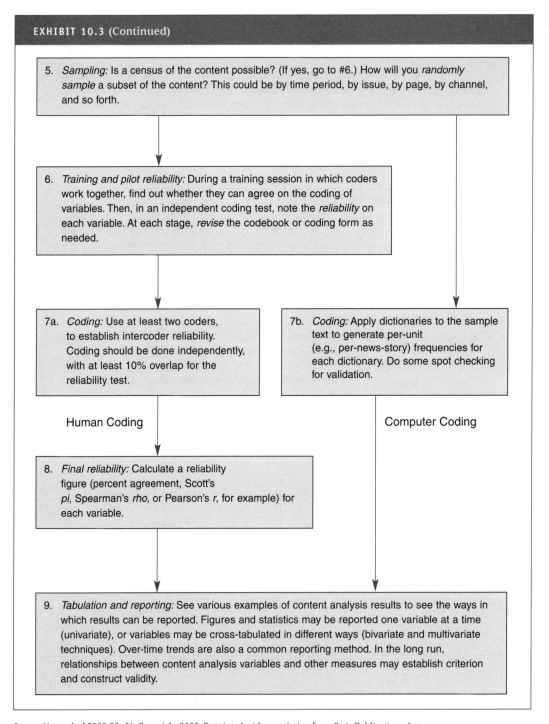

5. *Sampling:* Is a census of the content possible? (If yes, go to #6.) How will you *randomly sample* a subset of the content? This could be by time period, by issue, by page, by channel, and so forth.

6. *Training and pilot reliability:* During a training session in which coders work together, find out whether they can agree on the coding of variables. Then, in an independent coding test, note the *reliability* on each variable. At each stage, *revise* the codebook or coding form as needed.

7a. *Coding:* Use at least two coders, to establish intercoder reliability. Coding should be done independently, with at least 10% overlap for the reliability test.

7b. *Coding:* Apply dictionaries to the sample text to generate per-unit (e.g., per-news-story) frequencies for each dictionary. Do some spot checking for validation.

Human Coding

Computer Coding

8. *Final reliability:* Calculate a reliability figure (percent agreement, Scott's *pi*, Spearman's *rho*, or Pearson's *r*, for example) for each variable.

9. *Tabulation and reporting:* See various examples of content analysis results to see the ways in which results can be reported. Figures and statistics may be reported one variable at a time (univariate), or variables may be cross-tabulated in different ways (bivariate and multivariate techniques). Over-time trends are also a common reporting method. In the long run, relationships between content analysis variables and other measures may establish criterion and construct validity.

Source: Neuendorf 2002:50–51. Copyright 2002. Reprinted with permission from Sage Publications, Inc.

EXHIBIT 10.4   Number and Percentage of Vignettes Depicted by the Programs *Unsolved Mysteries* and *America's Most Wanted*

	%	n
Murder	52	40
Theft or fraud extortion	13	10
Escape	12	9
Bank robbery or armed robbery	10	8
Illegal arms or terrorism	3	2
Unexplained death or missing	7	5
Rape	5	4
Kidnapping	5	4
Child molestation	5	4
Attempted murder	3	2
Drug dealing	3	2
Other	7	5

Source: Adapted from Cavender & Bond-Maupin 2000.

Exhibit 10.4 presents the percentage of program vignettes that depicted offense types for both AMW and UM. Consistent with previous research, Cavender and Bond-Maupin (2000) found that these shows focused a majority of their attention on violent crime directed at persons. Corporate and political criminals were notably absent from the vignettes.

Cavender and Bond-Maupin (2000) concluded that the overwhelming majority of criminals on these shows were depicted as dangerous people who were beyond the help of social control or rehabilitation. In addition, both shows regularly featured criminals who were characterized with deviant psychological labels such as a "crazed killer, a psycho, a maniac," schizophrenic, emotionally disturbed, or showing no emotion. Other criminals were portrayed as Satanists, gang members, and drug dealers. On the other hand, victims were usually presented as upstanding citizens who you would like to have as a neighbor. The camera most often took the victims' perspective as well, achieving what Cavender and Bond-Maupin called the "good/evil dichotomy," with the victim and the audience aligned against the criminal.

Cavender and Bond-Maupin (2000) were particularly interested in how the shows' cinematographic techniques were used to make the dramas entertaining. They found that in a majority of the violent crimes, graphic details of the incident were depicted, including the number and location of a victim's wounds. For example, one victim was stabbed 30 times through the heart, and another victim was shot five times and ultimately died of massive bleeding. To sensationalize these acts, in some vignettes the film speed was reduced to slow motion to capture the muzzle flash as a criminal drew a gun and shot the victim. What Cavender and Bond-Maupin called "fear cues" were also frequently employed by the shows to capture the attention of viewers. For example, in one vignette, the host emerges from the night on a deserted road and makes reference to a common fear of driving on a lonely stretch of highway where one's cries for help would go unheeded. In sum, Cavender and Bond-Maupin concluded that the messages sent to viewers by both programs was that danger lurks everywhere, awaiting the victim, and by implication, the viewer.

CRIME MAPPING

The image of crime mapping most of you probably have in your head is one of a large map on a police precinct wall with push pins stuck all over it identifying the location of crime incidents. Although applied crime mapping like this has been used for over 100 years in policing, the type of crime mapping we will discuss here is related to mapping techniques used for traditional research purposes (e.g., testing theory about the causes of crime), not for investigative purposes.

Case Study: Social Disorganization and the Chicago School

Crime mapping for general research purposes has a long history in criminological research. **Crime mapping** for research purposes is generally used to identify the spatial distribution of crime along with the social indicators such as poverty and social disorganization that are similarly distributed across areas (e.g., neighborhood, census tracks). Although they were not the first researchers to use crime mapping. Shaw and McKay (1942) conducted a landmark analysis in criminology on juvenile delinquency in Chicago neighborhoods back in the 1930s. These researchers mapped thousands of incidents of juvenile delinquency and analyzed relationships between delinquency and various social conditions such as social disorganization. After analyzing rates of police arrests for delinquency, using police records to determine the names and addresses of those arrested in Chicago between 1927 and 1935, Shaw and McKay observed a striking pattern that persisted over the years.

Exhibit 10.5 displays one of the maps Shaw and McKay (1942) created that illuminates the spatial distribution of delinquency within concentric circles of Chicago that spread out to the suburbs from the city's center. As noted in Exhibit 10.5, there is a linear decrease in rates of delinquency as the distance from the Loop (city center) increases. When rates of other community characteristics were similarly mapped (e.g., infant mortality, tuberculosis cases, percentage of families who own their own homes, percentage of foreign born

EXHIBIT 10.5   Zone Rates of Police Arrests in Chicago, 1931

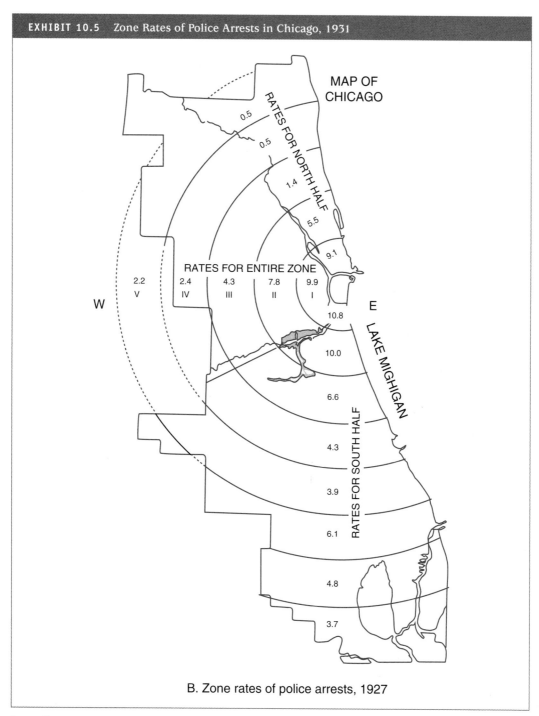

B. Zone rates of police arrests, 1927

Source: Shaw & McKay 1984.

residents, percentage of families on relief), the conclusions were obvious. Shaw and McKay concluded,

> It may be observed, in the first instance, that the variations in rates of officially recorded delinquents in communities of the city correspond very closely with variations in economic status. The communities with the highest rates of delinquents are occupied by these segments of the population whose position is most disadvantageous in relation to the distribution of economic, social, and cultural values. Of all the communities in the city, these have the fewest facilities for acquiring the economic goods indicative of status and success in our conventional culture. . . . In the low-income areas, where there is the greatest deprivation and frustration, where, in the history of the city, immigrant and migrant groups have brought together the widest variety of divergent cultural traditions and institutions, and where there exists the greatest disparity between the social values to which the people aspire and the availability of facilities for acquiring these values in conventional ways, the development of crime as an organized way of life is most marked. (Pp. 318–319)

Case Study: Gang Homicides in St. Louis

Contemporary researchers interested in issues related to crime and criminology have access to more sophisticated computer technology that allows the creation of more enhanced crime maps. The purpose of crime maps, however, remains the same: to illuminate the relationship between some category of crime and corresponding characteristics such as poverty and disorganization across given locations. Rosenfeld, Bray, and Egley (1999) recently examined the mechanisms through which gangs facilitate violent offending in St. Louis, Missouri. The primary purpose of this research was to examine whether gang membership impels members to engage in violence or merely exposes them to violent persons and situations. Rosenfeld, Bray, and Egley compared gang-affiliated, gang-motivated, and nongang youth homicides occurring in St. Louis between 1985 and 1995. In one part of their analysis, they examined the spatial distribution of gang and nongang youth homicide in relation to attributes of the neighborhood context (e.g., economic deprivation) in which the incidents occurred.

Rosenfeld, Bray, and Egley (1999) did not rely on the police classification of homicides as gang-related; they coded homicide reports themselves. Operationally, a homicide was coded gang-motivated if it resulted from gang behavior or relationships, such as an initiation ritual, the "throwing" of gang signs, or a gang fight. A case was coded as gang-affiliated if the homicide involved a suspect or victim identified in the police report as a gang member but did not arise from gang activity. For example, an incident in which a gang member was killed during a robbery would be gang-affiliated but not gang-motivated. A youth homicide was coded as nongang related if no indication of gang activity or affiliation was present in the case file, and the suspect was between 10 and 24 years of age. To examine the spatial relationship between these three types of homicides and the relationship they had to neighborhood disadvantage (such as economic deprivation or social disorganization) and instability (such as residents moving in and out of the neighborhood), the researchers used

EXHIBIT 10.6   Neighborhoods in St. Louis Where Gang-Affiliated and Gang-Motivated Homicides Occurred

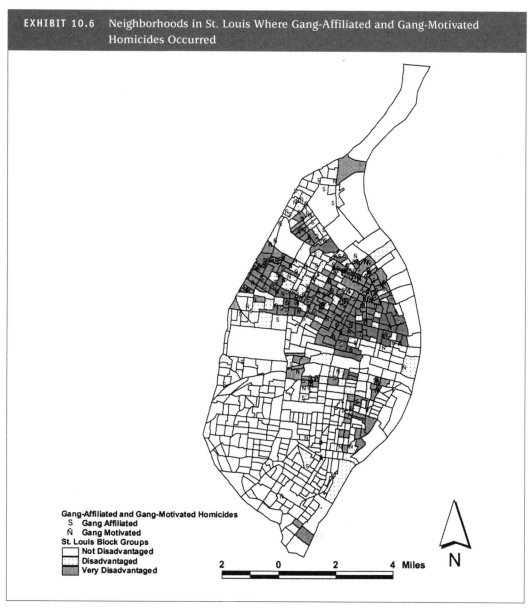

Source: Rosenfeld et al. 1999.

the census block group as their units of analysis. Exhibit 10.6 displays the map of block groups where gang-affiliated and gang-motivated homicides occurred, along with the extent of neighborhood disadvantage in each block for 1990–1995 (Rosenfeld, Bray & Egley 1999). Exhibit 10.7 displays this same information for nongang youth homicides.

EXHIBIT 10.7 Neighborhoods in St. Louis Where Nongang Homicides Occurred

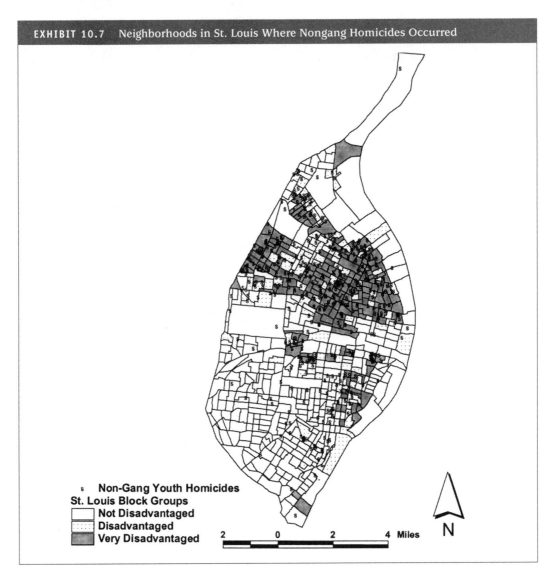

Source: Rosenfeld et al. 1999.

From these maps, you can see that both gang and nongang homicides were concentrated in disadvantaged areas. Other maps revealed a similar finding with regard to neighborhood instability; homicides were concentrated in neighborhoods with moderate levels of instability regardless of whether they were gang-affiliated, gang-motivated, or nongang related. Rosenfeld, Bray, and Egley (1999) conclude,

> Our results offer powerful evidence of the clustering of both gang and nongang youth homicides in areas characterized by high levels of social-economic disadvantage and

racial isolation. Although the accumulated evidence for gang facilitation of violence is quite compelling, our results serve as a reminder that concentrated disadvantage and racial isolation remain the fundamental sources of lethal violence in urban areas. (P. 514)

These conclusions echo those made over a half century earlier by Shaw and McKay (1942). As such, we are compelled to note Ernest W. Burgess' words in the introduction of Shaw and McKay's original work:

We must realize that the brightest hope in reformation is in changing the neighborhood and in control of the gang in which the boy moves, lives, and has his being and to which he returns after this institutional treatment. . . . We must reaffirm our faith in prevention, which is so much easier, cheaper, and more effective than cure and which begins with the home, the play group, the local school, the church, and the neighborhood. (P. xiii)

METHODOLOGICAL ISSUES WHEN USING SECONDARY DATA

Each of the methods we have discussed in this chapter presents unique methodological challenges. For example, in historical events research and comparative research, small numbers of cases, spotty historical records, variable cross-national record-keeping practices, and different cultural and linguistic contexts limit the confidence that can be placed in measures, samples, and causal conclusions. Just to identify many of the potential problems for a comparative historical research project requires detailed knowledge of the times and of the nations or other units investigated (Kohn 1987). This requirement often serves as a barrier to in-depth historical research and to comparisons between nations.

Analysis of secondary data presents several challenges, ranging from uncertainty about the methods of data collection to the lack of maximal fit between the concepts that the primary study measured and each of the concepts that are the focus of the current investigation. Responsible use of secondary data requires a good understanding of the primary data source. The researcher should be able to answer the following questions (most of which were adopted from Riedel 2000:55–69; Stewart 1984:23–30):

1. What were the agency's goals in collecting the data? If the primary data were obtained in a research project, what were the project's purposes?

2. Who was responsible for data collection, and what were their qualifications? Are they available to answer questions about the data? Each step in the data-collection process should be charted and the personnel involved identified.

3. What data were collected, and what were they intended to measure?

4. When was the information collected?

5. What methods were used for data collection? Copies of the forms used for data collection should be obtained and the way in which these data are processed by the agency or agencies should be reviewed.

6. How is the information organized (by date, event, etc.)? Are there identifiers that are used to identify the different types of data available on the same case? In what form are the data available (computer tapes, disks, paper files)? Answers to these questions can have a major bearing on the work that will be needed to carry out the study.

7. How consistent are the data with data available from other sources?

8. What is known about the success of the data-collection effort? How are missing data indicated? What kind of documentation is available?

Answering these questions helps to ensure that the researcher is familiar with the data he or she will analyze and can help to identify any problems with it.

Data quality is always a concern with secondary data, even when the data are collected by an official government agency. The need for concern is much greater in research across national boundaries, because different data-collection systems and definitions of key variables may have been used (Glover 1996). Census counts can be distorted by incorrect answers to census questions as well as by inadequate coverage of the entire population (see Chapter 4; Rives & Serow 1988:32–35). Social and political pressures may influence the success of a census in different ways in different countries. These influences on records are particularly acute for crime data. For example, Archer and Gartner (1984) note, "It is possible, of course, that many other nations also try to use crime rate fluctuations for domestic political purposes—to use 'good' trends to justify the current administration or 'bad' trends to provide a mandate for the next" (p. 16).

Researchers who rely on secondary data inevitably make trade-offs between their ability to use a particular data set and the specific hypotheses they can test. If a concept that is critical to a hypothesis was not measured adequately in a secondary data source, the study might have to be abandoned until a more adequate source of data can be found. Alternatively, hypotheses or even the research question itself may be modified in order to match the analytic possibilities presented by the available data (Riedel 2000:53).

Measuring Across Contexts

One problem that historical and comparative research projects often confront is the lack of data from some historical periods or geographical units (Rueschemeyer, Stephens, & Stephens 1992:4; Walters, James, & McCammon 1997). The widely used U.S. Uniform Crime Reporting System did not begin until 1930 (Rosen 1995). Missing data prevented Olzak, Shanahan, and McEneaney (1996:598) from including more than 55 out of a total of 212 standard metropolitan statistical areas (SMSAs) in their analysis of race riots. Sometimes alternative sources of documents or estimates for missing quantitative data can fill in gaps (Zaret 1996), but even when measures can be created for key concepts, multiple measures of the same concepts are likely to be out of the question; as a result, tests of reliability and validity may not be feasible. Whatever the situation, researchers must assess the problem honestly and openly (Bollen, Entwisle, & Alderson 1993).

Those measures that are available are not always adequate. What remains in the historical archives may be an unrepresentative selection of materials from the past. At various times, some documents could have been discarded, lost, or transferred elsewhere for a variety of reasons. "Original" documents may be transcriptions of spoken words or handwritten pages and could have been modified slightly in the process; they could also be outright distortions (Erikson 1966:172, 209–210; Zaret 1996). When relevant data are obtained from previous publications, it is easy to overlook problems of data quality, but this simply makes it all the more important to evaluate the primary sources. Developing a systematic plan for identifying relevant documents and evaluating them is very important.

A somewhat more subtle measurement problem is that of establishing measurement equivalence. The meaning of concepts and the operational definition of variables may change over time and between nations or regions (Erikson 1966:xi). The concept of being a "good son or daughter" refers to a much broader range of behaviors in China than in most Western countries (Ho 1996). Individuals in different cultures may respond differently to the same questions (Martin & Kinsella 1995:385). Alternatively, different measures may have been used for the same concepts in different nations and the equivalence of these measures may be unknown (van de Vijver & Leung 1997:9).

The value of statistics for particular geographic units like counties may vary over time simply due to change in the boundaries of these units (Walters, James, & McCammon 1997). As Archer and Gartner (1984) note,

> These comparative crime data were recorded across the moving history of changing societies. In some cases, this history spanned gradual changes in the political and social conditions of a nation. In other cases, it encompassed transformations so acute that it seems arguable whether the same nation existed before and after. (P. 15)

Such possibilities should be considered, and any available opportunity should be taken to test for their effects.

A different measurement concern can arise as a consequence of the simplifications made to facilitate comparative analysis. In many qualitative comparative analyses, the values of continuous variables are dichotomized. For example, nations may be coded as "democratic" or "authoritarian." This introduces an imprecise and arbitrary element into the analysis (Lieberson 1991). On the other hand, for some comparisons, qualitative distinctions such as "simple majority rule" or "unanimity required" may capture the important differences between cases better than quantitative distinctions. It is essential to inspect carefully the categorization rules for any such analysis and to consider what form of measurement is both feasible and appropriate for the research question being investigated (King, Keohane, & Verba 1994:158–163).

Sampling Across Time and Place

The simplest type of sample is the selection of one case for a historical study. Although a great deal can be learned from the intensive focus on one nation or other unit, the lack of a comparative element shapes the type of explanations that are developed. Explanations for change within one nation are likely to focus on factors that have changed within it, such as the decisions of a political leader, rather than factors like political systems that vary between nations (Rueschemeyer, Stephens, & Stephens 1992:31–36). However, comparisons between nations may reveal that differences in political systems are much more important than voluntary decisions by individual political leaders. It is only when the political system changes over time within one nation that its impact can be evaluated empirically (Rueschemeyer, Stephens, & Stephens 1992:29).

Qualitative comparative historical studies are likely to rely on availability samples or purposive samples of cases. In an availability sample, researchers study a case or multiple cases simply because they are familiar with or have access to them. When using a purposive sampling strategy, researchers select cases because they reflect theoretically important distinctions. Quantitative historical comparative researchers often select entire populations of cases for which the appropriate measures can be obtained.

When geographic units like nations are sampled for comparative purposes, it is assumed that the nations are independent of each other in terms of the variables examined. Each nation can then be treated as a separate case for identifying possible chains of causes and effects. However, in a very interdependent world, this assumption may be misplaced; nations may develop as they do because of how other nations are developing (and the same can be said of cities and other units). As a result, comparing the particular histories of different nations may overlook the influence of global culture, international organizations, or economic dependency. These common international influences may cause the same pattern of changes to emerge in different nations; looking within the history of these nations for the explanatory influences would lead to spurious conclusions. The possibility of such complex interrelations should always be considered when evaluating the plausibility of a causal argument based on a comparison between two apparently independent cases (Jervis 1996).

Identifying Causes

The restriction of historical research to one setting in the past limits possibilities for testing causal connections, because explanations applied to these events can seem arbitrary (Skocpol 1984:365). The researcher can reduce this problem by making some comparisons to other events in the past or present. However, the interdependence of sampled units (noted in the previous section) may mean that several cases available for comparison should be treated as just one larger case. There may also be too many possible combinations of causal factors to test and not enough cases to test them with.

The inductive approach taken by many qualitative comparative researchers can also make it seem that whatever the causal explanation developed, it has been tailored to fit the particulars of the historical record and so is unlikely to be confirmed by other cases (Rueschemeyer, Stephens, & Stephens 1992:30).

Some comparative researchers use a systematic method for identifying causes, developed by the English philosopher John Stuart Mill (1872), called the **method of agreement** (see Exhibit 10.8). The core of this approach is the comparison of nations (*cases*) in terms of similarities and differences on potential causal variables and the phenomenon to be explained. For example, suppose three nations that have all developed democratic political systems are compared in terms of four socioeconomic variables hypothesized by different theories to influence violent crime. If the nations differ in terms of three of the variables but are similar in terms of the fourth, this is evidence that the fourth variable influences violent crime.

The features of the cases selected for comparison have a large impact on the ability to identify influences using the method of agreement. Cases should be chosen for their difference in terms of key factors hypothesized to influence the outcome of interest and their similarity on other, possibly confounding, factors (Skocpol 1984:383). For example, in order to understand how unemployment influences violent crime, you would need to select cases

EXHIBIT 10.8	John Stuart Mill's Method of Agreement		
Variable	*Case 1*	*Case 2*	*Case 3*
A	Different	Different	Different
B	Different	Same	Same
C	Different	Different	Different
D[a]	Same	Same	Same
Outcome	Same	Same	Same

Source: Adapted from Skocpol 1984:379.

a. D is considered the cause of the outcome.

for comparison that differ in unemployment rates, so that you could then see if they differ in rates of violence (King, Keohane, & Verba 1994:148–152).

This **deterministic causal approach** (Ragin 1987:44–52) requires that there be no deviations from the combination of factors that are identified as determining the outcome for each nation. Yet there are likely to be exceptions to any explanatory rule that we establish (Lieberson 1991). A careful analyst will evaluate the extent to which exceptions should be allowed in particular analyses.

With these cautions in mind, the combination of historical and comparative methods allows for rich descriptions of social and political processes in different nations or regions as well as for causal inferences that reflect a systematic, defensible weighing of the evidence. Data of increasingly good quality are available on a rapidly expanding number of nations, creating many opportunities for comparative research. We cannot expect one study comparing the histories of a few nations to control adequately for every plausible alternative causal influence, but repeated investigations can refine our understanding and lead to increasingly accurate causal conclusions (King, Keohane, & Verba 1994:33).

COMBINING RESEARCH DESIGNS

Designing research means deciding how to measure empirical phenomena, how to identify causal connections, and how to generalize findings not as separate decisions, but in tandem, with each decision having implications for the others. A research design is an integrated whole involving both fruitful conjunctions and necessary compromises. The carefully controlled laboratory conditions that increased the causal validity of Bushman's (1995) experiments on media violence decreased the generalizability of his conclusions (see Chapter 6). The representative sampling plan that increased the generalizability of Tjaden and Thoennes (2000) national study of violent victimization limited their options for estimating causal effects (see Chapter 7). The observations that underlay Decker and Van Winkle's (1996) descriptions of gang members would not be feasible in a national sample of gang members (see Chapter 8).

Comparing Research Designs

It is not enough to ask of a study you critique, or one that you plan, such questions as, "Were the measures valid?" and "Were the causal conclusions justified?" You must also consider how the measurement approach might have affected the causal validity of the researcher's conclusions and how the sampling strategy might have altered the quality of measures. In fact, you must be concerned with how each component of the research design influenced the other components.

In the real world of social and criminological research, the boundaries separating different methods of data collection often overlap: experiments may be conducted in the field; surveys may involve some intensive open-ended questioning; field research may utilize quantitative counts of phenomena or random samples of events. The central features of experiments, surveys, and qualitative methods provide distinct perspectives even when used to study the same social processes. Comparing subjects randomly assigned to a treatment and a comparison group, asking standard questions of the members of a random sample, and observing while participating in a natural social setting involve markedly different decisions about measurements, causality, and generalizability. No method can reasonably be graded as superior to the others, and each varies in its suitability to different research questions and goals.

In general, experimental designs are strongest for testing causal hypotheses and most appropriate for studies of treatment (see Chapter 6). Because random assignment reduces the possibility of preexisting differences between treatment and comparison groups to small, specifiable, chance levels, many of the variables are controlled that might otherwise create a spurious association. But in spite of this clear advantage, experimental designs require a degree of control that cannot always be achieved outside of the laboratory. It can be difficult to ensure in real-world settings that a treatment was delivered as intended and that other influences did not intrude. As a result, what appears to be a treatment effect or noneffect may be something else altogether. Field experiments thus require careful monitoring of the treatment process.

Laboratory experiments permit much more control over conditions, but at the cost of less generalizable findings. People must volunteer for most laboratory experiments, and so there is a good possibility that experimental subjects differ from those who do not volunteer. The problem of generalizability in an experiment using volunteers lessens only when the object of investigation is an orientation, a behavior, or a social process that is relatively invariable among people.

Both surveys and experiments typically use standardized, quantitative measures of attitudes, behaviors, or social processes. Closed-ended questions are most common and are well suited for the reliable measurement of variables that have been studied in the past and whose meaning is well understood (see Chapter 3). Of course, surveys often include measures of many more variables than are included in an experiment, but this quality is not inherent in either design. Phone surveys may be quite short, and some experiments can involve very lengthy sets of measures (see Chapter 7).

Most social science surveys rely on random sampling for their selection of cases from a larger population, and it is this feature that makes them preferable for maximizing generalizability (see Chapter 4). When description of a particular large population is a key concern, survey research is likely to be the method of choice. Surveys are also often used to investigate hypothesized causal relationships, because many variables can be measured within a given survey, and repeated measurements are possible over time. If variables that might create spurious relations are included in the survey, they can be controlled statistically in the analysis, thus strengthening the evidence for or against a hypothesized causal relationship.

Qualitative methods presume an intensive measurement approach in which indicators of concepts are drawn from direct observation or in-depth commentary (see Chapter 8). This approach is most appropriate when it is not clear what meaning people attach to a concept or what sense they might make of particular questions about it. Qualitative methods are also admirably suited to the exploration of new or poorly understood social settings, when it is not even clear what concepts would help to understand the situation. For these reasons, qualitative methods tend to be preferred when exploratory research questions are posed. But, of course, intensive measurement necessarily hampers the study of large numbers of cases or situations, resulting in the limitation of many field research efforts to small numbers of people or unique social settings.

When qualitative methods are used to study several individuals or settings that provide marked contrasts in a presumed independent variable, it becomes possible to evaluate causal hypotheses with these methods. However, for hypothesis testing, the impossibility of taking into account many possible extraneous influences in such limited comparisons makes qualitative methods a weak approach. Qualitative methods are more suited to the elucidation of causal mechanisms, and can be used to identify the multiple successive events that might have led to some outcome.

Historical and comparative methods range from cross-national quantitative surveys to qualitative comparisons of social features and political events (this chapter). Their suitability for exploration, description, explanation, and evaluation varies in relation to the particular method used, but they are essential for research on historical processes and national differences. If the same methods are used to study multiple eras or nations rather than just one nation at one time, the results are likely to be enhanced generalizability and causal validity.

In reality, none of these methods of data collection provides a foolproof means for achieving measurement validity, causal validity, or generalizability. Each will have some liabilities in a specific research application, and all can benefit from combination with one or more other methods (Sechrest & Sidani 1995). To benefit from this multiple-method approach, researchers have been using a triangulated methodology, that is, combining different methods in the same project to reveal different dimensions of the same phenomenon. It is to this subject that we now turn.

Triangulating Research Designs

The use of multiple methods to study one research question is called **triangulation**. The term suggests that a researcher can get a clearer picture of the social reality being studied by viewing it from several different perspectives. The term is actually derived from land surveying, where knowing a single landmark allows you to locate yourself only somewhere along a line in a single direction from the landmark. However, with two landmarks you can take bearings on both and locate yourself at their intersection.

Triangulation The use of multiple methods to study one research question.

Triangulation in Action: American Indian Homicide

When studying the social causes of homicide and other violence within contemporary American Indian communities, Bachman (1992) used triangulation. To examine the social causes of homicide within this population, she conducted in-depth interviews with American Indian homicide offenders who were incarcerated in state prisons in two states. In addition, she conducted national-, state-, and reservation-level statistical analyses to determine the social factors, such as levels of poverty and social disorganization, that were related to American Indian homicide rates at the aggregate level. When she focused her attention on intimate partner violence within American Indian families, she performed intensive interviews with the staff and residents of three battered women's shelters located on reservations. She also performed a statistical analysis of data obtained from the National Family Violence Survey (Straus & Gelles 1990).

To illustrate triangulation, let us take a closer look at the process of Bachman's (1992) work on American Indian homicide offending. Bachman found that compared to all other racial or ethnic groups, American Indians had some of the highest rates of homicide, sometimes reaching over 100 per 100,000 population on some reservations (e.g., Douglas, Nevada, 103.5; Harney, Oregon, 127.5). Although other researchers had noted high rates of American Indian homicide, virtually no attempts were made to explore the factors that contributed to such high rates. Are the causal mechanisms that create these high rates the same as those that were documented in the population in general? Or are the social forces that were identified as increasing both African American and white homicide rates different from those forces that contribute to American Indian homicide? Because there was virtually no research examining the etiology of American Indian homicide, Bachman first wanted to gain some insight into and understanding of the structural and cultural conditions that may contribute to lethal violence at the individual level. Accordingly, she selected the qualitative method of intensive interviewing as her first method.

Bachman (1992) conducted face-to-face interviews with homicide offenders at three midwestern state prisons. Offenders responded to a lengthy set of open-ended questions and probes concerning the circumstances surrounding their crime, their life before the crime, and their attitudes about crime in general. Bachman characterized the interview process as

> a conversation between two people getting to know one another rather than a rigid structure in which questions were formally addressed. An interview was usually ended not because there was nothing left to talk about, but because there was another interview scheduled or because the prisoner had to return to his cell for count. (P. 32)

Thus, Bachman (1992) started with inductive research, beginning with qualitative interview data, which were then used to develop (induce) a general explanation (a theory) to

EXHIBIT 10.9   Theoretical Model for American Indian Homicide That Emerged From Bachman's Qualitative Interviews With American Indian Homicide Offenders

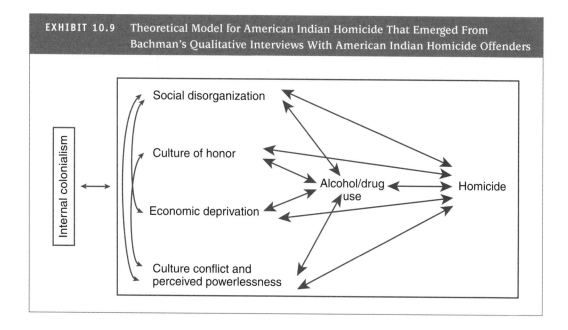

account for the data. The theoretical model that emerged from her interview data is displayed in Exhibit 10.9. The model combines the causal forces of social disorganization, economic deprivation, a culture of honor or violence, and the psychological mechanisms of culture conflict and perceived powerlessness with the intervening variable of alcohol or drug abuse. Also included in the theoretical guide is the antecedent variable to each of these concepts, internal colonialism. Although this concept was not explicitly derived from the qualitative analysis, Bachman explained, "No model explaining any phenomenon with regard to American Indians would be complete without acknowledgment of the colonization process to which our government has subjected this population" (p. 36).

This model assisted Bachman (1992) in subsequent quantitative analysis by providing a theoretical framework from which to start. Using this model as a guide along with the other relevant literature on the etiology of homicide, Bachman then performed deductive research in which she obtained American Indian homicide rates (dependent variable) and indicators of some of these theoretical constructs (independent variables) at the state and reservation levels to test three hypotheses:

1. The higher the level of social disorganization within a reservation community, the higher the rate of homicide.

2. The higher the level of economic deprivation within a reservation community, the higher the rate of homicide.

3. The more traditional and integrated a reservation community, the lower the rate of homicide.

Results indicated that American Indian communities with higher levels of both social disorganization and economic deprivation also had higher rates of homicide. She did not find support in her data for the third hypothesis. In addition to guiding her deductive research, the qualitative interview data also provided Bachman's (1992) research with a wealth of narratives that added meaning and depth to the statistical relationships that were found at the aggregate level. For example, many homicide offenders revealed what conditions of social disorganization and economic deprivation were like in their own lives. One offender talked about the reality of his disorganized childhood, being placed in multiple foster homes:

> I was pretty much living in foster homes since I was about two. I guess my parents weren't doing too good—you know, drinking and partying. My brother got put into a different foster home, but my sister and I got in the same one. After several moves, we all eventually got placed in the same home. We moved around quite a bit. (P. 38)

Another talked about his existence living on a reservation without a job:

> Without a job, you sit in your house day after day—nothing to do. You listen to that same car drive by your house again and again, and pretty soon you hate those people in that car. Kids go running through your yard—back and forth and back and forth—and pretty soon you hate those kids. Pretty soon you want to hurt somebody. (Bachman 1992:50)

As you can see, using triangulation in which both quantitative and qualitative methods are used to investigate the same research question has many advantages. In addition, finding support for relationships between independent and dependent variables at two different levels—the individual level and the community level—also provided validity to Bachman's (1992) conclusions that social disorganization and economic deprivation are both related to homicide in American Indian communities. Campbell and Fiske (1959) explain the validity-enhancing qualities of triangulation:

> If a hypothesis can survive the confrontation of a series of complementary methods of testing, it contains a degree of validity unattainable by one tested within the more constricted framework of a single method. . . . Findings from this latter approach must always be subject to the suspicion that they are method-bound: Will the comparison totter when exposed to an equally prudent but different testing method? (P. 82)

Choice of a data-collection method should be guided in part by the aspect of validity that is of most concern, but each aspect of validity must be considered in attempting to answer every research question. Experiments may be the preferred method when causal validity is a paramount concern, and surveys may be the natural choice if generalizability is critical. But generalizability must still be a concern when assessing the results of an experiment, and causal validity is a key concern in most social science surveys. Field research has unique value for measuring social processes as they happen, but the causal validity and generalizability of field research results are often open to question. A researcher should always consider whether data of another type should be collected in what is basically a single-method study and whether additional research using different methods is needed before the research question can be answered with sufficient confidence.

The ability to apply diverse techniques to address different aspects of a complex research question is one mark of a sophisticated social researcher. Awareness that one study's findings must be understood in the context of a larger body of research, and the ability to speculate on how the use of different methods might alter a study's findings are prerequisites for informed criticism of social research. As social research methods and substantive findings continue to grow in number, these insights should stimulate more ambitious efforts to combine research methods and integrate many studies' findings.

But the potential for integrating methods and combining findings does not decrease the importance of single studies using just one method of data collection. The findings from well-designed studies in carefully researched settings are the necessary foundation for broader, more integrative methods. There is little point in combining methods that are poorly implemented or in merging studies that produced invalid results. Whatever the research question, we should consider the full range of methodological possibilities, make an informed and feasible choice, and then carefully carry out our strategy.

Finally, realistic assessment of the weaknesses as well as the strengths of each method of data collection should help you to remember that humility is a virtue in science. Advancement of knowledge and clear answers to specific research questions are attainable with the tools you now have in your methodological toolbox. Perfection, however, is not a realistic goal. No matter what research method we use, our mental concepts cannot reflect exactly what we measured, our notions of causation cannot reveal a tangible causal force, and our generalizations cannot always extend beyond the cases that were actually studied. This is not cause for disillusionment, but it should keep us from being excessively confident in our own interpretations or unreasonably resistant to change. As you saw in Chapter 2, research questions are never completely answered. Research findings regarding a particular question are added to the existing literature to inform future research, and the research circle continues.

ETHICAL ISSUES WHEN ANALYZING AVAILABLE DATA AND CONTENT

Analysis of historical documents or quantitative data collected by others does not create the potential for harm to human subjects that can be a concern when collecting primary data. It is still important to be honest and responsible in working out arrangements for data access when data must be obtained from designated officials or data archivists, but of course many data are easily accessed in libraries or on the Web. Researchers who conclude that they are being denied access to public records of the federal government may be able to obtain the data by filing a Freedom of Information Act request. The FOIA stipulates that all persons have a right to access all federal agency records unless the records are specifically exempted (Riedel 2000:130–131). Researchers who review historical or government documents must also try to avoid embarrassing or otherwise harming named individuals or their descendants by disclosing sensitive information.

Subject confidentiality is a key concern when original records are analyzed. Whenever possible, all information that could identify individuals should be removed from the records to be analyzed so that no link is possible to the identities of living subjects or the living descendants of subjects (Huston & Naylor 1996:1698). When you used data that have already been archived, you need to find out what procedures were used to preserve subject confidentiality. The work required to ensure subject confidentiality probably will have been done for you by the data archivist. For example, the ICPSR examines carefully all data deposited in the archive for the possibility of disclosure risk. All data that might be used to identify respondents is altered to ensure confidentiality, including removal of information such as birth dates or service dates, specific incomes, or place of residence that could be used to identify subjects indirectly (see www.icpsr.edu/irb/statement/htm1). If all information that could be used in any way to identify respondents cannot be removed from a dataset without diminishing dataset quality (such as by preventing links to other essential data records), ICPSR restricts access to the data and requires that investigators agree to conditions of use that preserve subject confidentiality.

It is not up to you to decide whether there are any issues of concern regarding human subjects when you acquire a dataset for secondary analysis from a responsible source. The Institutional Review Board (IRB) for the Protection of Human Subjects at your college or university or other institution has the responsibility to decide whether they need to review and approve proposals for secondary data analysis. The federal regulations are not entirely clear on this point, so the acceptable procedures will vary between institutions based on what their IRBs have decided.

Ethical concerns are multiplied when surveys are conducted or other data are collected in other countries. If the outside researcher lacks much knowledge of local norms, values, and routine activities, the potential for inadvertently harming subjects is substantial. For this reason, cross-cultural researchers should spend time learning about each of the countries in which they plan to collect primary data and strike up collaborations with researchers in those countries (Hantrais & Mangen 1996). Local advisory groups may also be formed in each country so that a broader range of opinion is solicited when key decisions must be made. Such collaboration can also be invaluable when designing instruments, collecting data, and interpreting results.

CONCLUSION

Historical and comparative social science investigations use a variety of techniques that range from narrative histories having much in common with qualitative methods to analyses of secondary data that in many respects are like traditional survey research. Each of these methods can help researchers gain new insights into processes like factors related to homicide offending. They encourage intimate familiarity with the course of development of the nations studied and thereby stimulate inductive reasoning about the interrelations among different historical events. Systematic historical and comparative techniques can be used to test deductive hypotheses concerning international differences as well as historical events.

Most historical and comparative methods encourage causal reasoning. They require the researcher to consider systematically the causal mechanism, or historical sequences of events, by which earlier events influence later outcomes. They also encourage attention to causal context, with a particular focus on the ways in which different cultures and social structures may result in different effects on other variables. There is much to be gained by continuing to use and develop these methods.

KEY TERMS

Comparative research
Content analysis
Crime mapping
Deterministic causal approach

Historical events research
Method of agreement
Triangulation

HIGHLIGHTS

- Secondary data analysis is the act of collecting or analyzing data that was originally collected for another purpose.

- In general, there are four major types of secondary data: surveys, official statistics, official records, and other historical documents including written text or media representations (e.g., trial transcripts, newspaper articles, television shows).

- The central insight behind historical and comparative methods is that we can improve our understanding of social processes when we make comparisons to other times and places.

- When research on past events does not follow processes for some long period of time, when it is basically cross-sectional, then it is usually referred to as historical events research.

- Content analysis is a tool for systematic analysis of documents and other textual data. It requires careful testing and control of coding procedures to achieve reliable measures.

- Crime mapping for research purposes is generally used to identify the spatial distribution of crime along with the social indicators such as poverty and social disorganization that are similarly distributed across areas (e.g., neighborhoods, census tracts).

- Secondary data for historical and comparative research are available from many sources. The Inter-University Consortium for Political and Social Research (ICPSR) provides the most comprehensive data archive.

- Using multiple research designs to answer the same research question, called triangulation, enhances generalizability, measurement validity, and causal validity.

EXERCISES

1. Read the original article reporting one of the studies described in this chapter. Critique the article, using the article review questions presented in Appendix B as your guide. Focus particular attention on procedures for measurement, sampling, and establishing causal relations.

2. What historical events have had a major influence on social patterns in the nation? The possible answers are too numerous to list, ranging from any of the wars to major internal political conflicts, economic booms and busts, scientific discoveries, and legal

changes. Pick one such event in your own nation for this exercise. Find one historical book on this event and list the sources of evidence used. What additional evidence would you suggest for a social science investigation of the event?

3. Using your library's government documents collection or the U.S. Census site on the Web (www.census.gov), select one report by the U.S. Bureau of the Census about the population of the United States or some segment of it. Outline the report and list all the tables included in it. Summarize the report in two paragraphs. Suggest a historical or comparative study for which this report would be useful.

4. Review the survey data sets available through the Inter-University Consortium for Political and Social Research (ICPSR), using either its published directory or its Internet site (www.icpsr.umich.edu). Select two data sets that might be used to study a research question in which you are interested. Use the information ICPSR reports about them to answer Questions 1–5 in the section "Methodological Issues When Using Secondary Data" earlier in this chapter. Is the information adequate to answer these questions? What are the advantages and disadvantages of using one of these data sets to answer your research question compared to designing a new study?

5. Find a magazine or newspaper report on crime in the United States or in some city or state. Explain how one of the key demographic concepts could be used or was used to improve understanding of this issue.

6. Select a current crime or political topic that has been the focus of news articles. Propose a content analysis strategy for this topic, using newspaper articles or editorials as your units of analysis. Your strategy should include a definition of the population, selection of the units of analysis, a sampling plan, and coding procedures for key variables. Now find an article on this topic and use it to develop your coding procedures. Test and refine your coding procedures with another article on the same topic.

DEVELOPING A RESEARCH PROPOSAL

Add a historical or comparative dimension to your proposed study.

1. Consider which of the four types of comparative-historical methods would be most suited to an investigation of your research question. Think of possibilities for qualitative and quantitative research on your topic with the method you prefer. Will you conduct a variable-oriented or case-oriented study? Write a brief statement justifying the approach you choose.

2. Review the possible sources of data for your comparative-historical project. Search the Web and relevant government, historical, and international organization sites or publications. Search the social science literature for similar studies and read about the data sources that they used.

3. Specify the hypotheses you will test or the causal sequences you will investigate. Describe what your cases will be (nations, regions, years, etc.). Explain how you will select cases. List the sources of your measures and describe the specific type of data you expect to obtain for each measure.

4. Review the list of potential problems in comparative-historical research and discuss those that you believe will be most troublesome in your proposed investigation. Explain your reasoning.

> **Student Study Site**
>
> The companion Web site for *The Practice of Research in Criminology and Criminal Justice*, Third Edition
>
> **http://www.sagepub.com/prccj3**
>
> Visit the Web-based Student Study Site to enhance your understanding of the chapter content and to discover additional resources that will take your learning one step further. You can enhance your understanding of the chapters by using the comprehensive study material, which includes e-flashcards, Web exercises, practice self-tests, and more. You will also find special features, such as Learning from Journal Articles, which incorporates SAGE's online journal collection.

WEB EXERCISES

1. The Bureau of Justice Statistics (BJS) home page can be found at www.ojp.usdoj.gov/bjs. The site contains an extensive list of reports published by the BJS along with links to the data used in the reports. You can access tables of data that display phenomena over time and across various geographical locations.
   a. Using data and/or reports available from the National Crime Victimization Survey, conduct an analysis of the rates of violent victimization from 1993 until the most recent date the data report available. What do you conclude about the trends of violent victimization?
   b. Find a report available on the site that makes cross-national comparisons. Summarize the methods by which the data were collected. Now summarize the findings of the report.

2. The National Institute of Justice has a wealth of information on crime mapping located at www.nicrs.org. Search the site for information on "crime mapping" and you will find that the site contains a multitude of information, including the latest technological advances in crime mapping strategies for police departments as well as full-text articles discussing recent research that uses crime mapping techniques. Select a report available online and summarize its findings.

3. The World Bank offers numerous resources that are useful for comparative research. Visit the World Bank Web site at www.worldbank.org. You will find data listed by countries; select a random sample of countries along with social indicators of your choice. Now summarize the differences and similarities you have identified between the countries.

4. The U.S. Bureau of Labor Statistics (BLS) provides extensive economic indicator data on the Web for regions, states, and cities. Go to the BLS Web page, which offers statistics by location: http://stats.bls.gov/eag/. Click on a region and explore the types of data that are available. Write out a description of the steps you would have to take to conduct a comparative analysis using the data available from the BLS Web site.

5. The U.S. Census Bureau's home page can be found at www.census.gov. This site contains extensive reporting of census data including population data, economic indicators, and other information acquired through the U.S. Census. This Web site allows you to collect information on numerous subjects and topics, which can be used to make comparisons between different states or cities. Find the "State and County Quick Facts" option and choose your own state. Now pick the county in which you live and copy down several statistics of interest. Repeat this process for other counties in your state. Use the data you

have collected to compare your county with other counties in the state. Write a one-page report summarizing your findings.

6. Go to the International Association of Law Enforcement Intelligence Analysts Web site at www.ialeia.org. On the site you can download papers and information regarding numerous topics, including crime mapping. Provide a review of at least one of these applications. You can also link to the IALEIA's journal (Hank Brightman, Editor, Saint Peter's College, The Jesuit College of New Jersey) to read the latest publications in this area.

ETHICS EXERCISES

1. Bachman (1992) found very high rates of homicide in her study of American Indian reservations. She reported this and other important and troubling findings in her book, *Death and Violence on the Reservation*. In your opinion, does a researcher have an ethical obligation to urge government officials or others to take action in response to social problems that they have identified? Why or why not?

2. Should any requirements be imposed on researchers who seek to study other cultures, in order to ensure that procedures are appropriate and interpretations are culturally sensitive? What practices would you suggest for cross-cultural researchers in order to ensure that ethical guidelines are followed? (Consider the wording of consent forms and the procedures for gaining voluntary cooperation.)

SPSS EXERCISES

1. Using the data set STATE2000, examine the rates of violence and murder across the four regions of the country (REGION). The best procedure to use for this is called MEANS. You can access this under ANALYZE, then COMPARE MEANS, then click on MEANS. In the Mean dialog box, place the rate of violence you wish to compare first (VIOLENT or MURDER) in the dependent variables box and REGION in the independent variable box. Summarize your findings.

   Now, using this same procedure, examine whether there are similar differences across regions for rates of poverty, which are often found to be associated with rates of violence. That is, geographical locations that have higher rates of poverty and overdeprivation also tend to have higher rates of violence. What do you conclude?

2. Because the GSS file is cross-sectional, we cannot use it to conduct historical research. However, we can develop some interesting historical questions by examining differences in the attitudes of Americans in different birth cohorts.
   a. Inspect the distributions of the same set of variables. Would you expect any of these attitudes and behaviors to have changed over the twentieth century? State your expectations in the form of hypotheses.
   b. Request a crosstabulation of these variables by birth COHORT.
   c. What appear to be the differences among the cohorts? Which differences do you think are due to historical change, and which do you think are due to the aging process? Which attitudes and behaviors would you expect to still differentiate the baby-boom generation and the post-Vietnam generation in 20 years?

Evaluation Research and Policy

In the beginning, God created the heaven and the earth. And God saw everything that he made. "Behold," God said, "it is very good." . . . And on the seventh day God rested from all His work. His archangel came then unto Him asking, "God, how do you know that what you have created is 'very good'? What are your criteria? On what data do you base your judgment. . . . And aren't you a little close to the situation to make a fair and unbiased evaluation?" God thought about these questions all that day and His rest was greatly disturbed. On the eighth day God said, "Lucifer, go to hell." Thus was evaluation born in a blaze of glory.

—Halcolm, the evaluation sage,
aka Michael Quinn Patton (1997:1)

Every year, the U.S. Department of Justice (DOJ) spends an average of $3 billion in grants to help state and local law enforcement and communities prevent and ameliorate the consequences of crime. But similar to virtually every government agency, very little money has been allocated by the DOJ to determine whether these prevention programs actually work. In 1996, Congress required the attorney general to provide a "comprehensive evaluation of the effectiveness" of the programs that had been implemented with this money. Do not be alarmed if you think $3 billion is too much money to spend on prevention. By definition, virtually every aspect of the criminal justice system as well as other government and private institutions are considered preventive in nature. Not only does the money go for such stereotypical prevention programs as after-school recreation programs, but it also supports such efforts as community-oriented policing, drug raids, prisoner rehabilitation, boot camps, home confinement and electronic monitoring, job corps and vocational training for prison inmates, preschool education programs, and other programs too numerous to mention. Most of the Department of Justice money set aside for prevention, however, is allocated most heavily in police and prisons, with very little support for prevention programs in other institutions (Sherman et al. 1997).

As the mandate from Congress suggests, evaluation plays an increasingly important role in the fields of criminology and criminal justice. Despite the increased attention given to evaluation research in the field, the results of evaluation studies have always influenced decisions made by criminal justice policy makers. The Manhattan Bail Project, for example, is a frequently cited example of how evaluation research led to massive policy change in the criminal justice system (Botein 1965). The Manhattan Bail Project was an experiment designed by the Vera Institute of Justice in New York City in 1961 to determine whether certain kinds of defendants could be released without bail and still be counted on to appear for their trials. The project was primarily undertaken because of the inequity many believed existed in the pretrial release of defendants; defendants with financial means were able to post bail to secure pretrial release, whereas indigent defendants who could not afford bail had to remain in custody. This, of course, is not fair. It was hypothesized by researchers that those with strong ties to the community (e.g., those married with children and stable employment) could be released without bail and still be expected to show up for future adjudication. To investigate this hypothesis, the Vera Institute identified a sample of defendants who had strong community ties and then randomly divided this group into a group recommended for release without bail (experimental group) and a group in which no recommendation was made (control group). The results of the initial study found that over 99% of those released without bail appeared in court for trial at the appropriate time. This was an even higher compliance rate than for those who were released on bail. Thus, the results suggested that releasing a person on the basis of verified information such as marital and employment status more effectively guaranteed appearance in court than did money bail. As a result of this study and research like it, almost every major jurisdiction in the nation today now allows eligible defendants to be released on their own recognizance (ROR) without bail.

In this chapter, we will provide you with an overview of evaluation research in the field of criminology and criminal justice. Although all the research we have discussed in this text is ultimately useful, evaluation research is inherently so because the results generally have an impact on policy choices in the immediate future. As such, evaluation research is often referred to as *applied research* because the findings can immediately be utilized and applied. We will first provide you with a brief history of evaluation research. Then, after describing the different types of evaluation research, we will provide you with case studies that illustrate the various methodologies used to assess the impacts of programs and policies in criminology and criminal justice.

A BRIEF HISTORY OF EVALUATION RESEARCH

Evaluation research is not a method of data collection, like survey research or experiments, nor is it a unique component of research designs, like sampling or measurement. Instead, evaluation research is social research that is conducted for a distinctive purpose: to investigate social programs (such as substance abuse treatment programs, welfare programs, criminal justice programs, or employment and training programs). For each project, an evaluation researcher must select a research design and method of data collection that are useful for answering the particular research questions posed and appropriate for the particular program investigated.

You can see why we placed this chapter after most of the others in the text: When you review or plan evaluation research, you have to think about the research process as a whole and how different parts of that process can best be combined.

Although scientific research methods had been used prior to the 1950s (in fact as early as the 1700s) to evaluate outcomes of particular social experiments and programs, it was not until the end of the 1950s that social research became immersed in the workings of government with the common goal of improving society. As Weiss (1977) observed, social science was expected to bring rationality to the untidy world of government:

> It would provide hard data for planning and give cause-and-effect theories for policy making, so that statesmen would know which variables to alter in order to effect the desired outcomes. . . . And once policies were in operation, it would provide objective evaluation of their effectiveness so that necessary modifications could be made to improve performance. (P. 4)

During the 1960s, the practice of evaluation research increased dramatically not only in the United States but also around the world. One of the main initiatives that spawned this growth of evaluation research in the United States was the so-called War on Poverty that was part of the Great Society legislation of the 1960s. When the federal government began to take a major role in alleviating poverty and the social problems associated with it, such as delinquency and crime, the public wanted accountability for the tax dollars spent on such programs. Were these programs actually having their intended effects? Did the benefits outweigh the costs? During this time, the methods of social science were utilized like never before to evaluate this proliferation of new programs.

In 1967, Edward Suchman put forth his seminal definition of evaluation research as the application of social research techniques to the study of large-scale human service programs. In fact, a version of this definition is still utilized by some scholars today. For example, Rossi and Freeman (1989) encompass his definition in their more recent version: "Evaluation research is the systematic application of social research procedures for assessing the conceptualization, design, implementation, and utility of social intervention programs" (p. 18). What exactly does *systematic* mean? Well, regardless of what treatment or program is being examined, evaluations are systematic because they employ social research approaches to gathering valid and reliable data. Note the plural *approaches* instead of the singular *approach*. Evaluation research covers the spectrum of research methods that we have discussed in this text.

By the mid-1970s, evaluators were called on not only to assess the overall effectiveness of programs, but also to determine whether programs were being implemented as intended and to provide feedback to help solve programming problems as well. As an indication of the growth of evaluation research during this time, several professional organizations emerged to assist in the dissemination of ideas from the burgeoning number of social scientists engaged in this type of research in the United States, along with similar organizations in other countries (Patton 1997).

Of course, despite the programs implemented by Presidents Kennedy and Johnson in the 1960s, poverty and its consequent social problems did not go away. By the 1990s, the public wanted even more accountability. Unfortunately, clear answers were not readily available. Few social programs could provide hard data on results achieved and outcomes obtained. Of course, government bureaucrats had produced a wealth of data on other things, including exactly how funds in particular programs were spent, for whom this money was spent, and for how many. However, these data primarily measured whether government staff were following the rules and regulations, not whether the desired results were being achieved. Instead of being rewarded for making their programs produce the intended outcomes (e.g., more jobs, fewer delinquents), the bureaucracy of government had made it enough simply to do the required paperwork of program monitoring. The results of evaluations that were conducted rarely saw the light of day. This, in turn, was interpreted by many politicians and individual citizens alike as "nothing works."

Professional evaluation researchers soon realized that it was not enough simply to perform rigorous experiments to determine program efficacy; they must also be responsible for making sure their results could be understood and utilized by the practitioners (e.g., government officials, corporations, and nonprofit agencies) to make decisions about scrapping or modifying existing programs. In addition, there was increased concern in the field regarding fiscal accountability, documenting the worth of social program expenditures in relation to their costs.

It was in this context that professional evaluators began discussing standards for the field of evaluation research. In 1981, a Joint Committee on Standards published the list of features all evaluations should have. These standards still exist today (Joint Committee 1994, as cited in Patton 1997:17):

Utility: The utility standards are intended to ensure that an evaluation will serve the practical information needs of intended users.

Feasibility: The feasibility standards are intended to ensure that an evaluation will be realistic, prudent, diplomatic, and frugal.

Propriety: The propriety standards are intended to ensure that an evaluation will be conducted legally, ethically, and with due regard for the welfare of those involved in the evaluation, as well as those affected by its results.

Accuracy: The accuracy standards are intended to ensure that an evaluation will reveal and convey technically adequate information about the features that determine worth or merit of the program being evaluated.

To the surprise of many in the field, utility was at the top of the list. This conveyed the idea that an evaluation only met the standards if, in fact, it served the needs of intended users. Professionals in the field responded accordingly. For example, Patton (1997) developed the practice of and coined the term "utilization-focused evaluation" as a way to fulfill the utility standard. Utilization-focused evaluation offered both a philosophy of evaluation and a practical framework for designing and conducting evaluations. The primary premise of this method was that evaluations should be judged by their utility and actual use. Use concerned how real people in the real world applied evaluation findings and experienced the evaluation process. As such, the focus in utilization-focused evaluation was on intended use by intended users. Like the field as a whole, utilization-focused evaluation does not advocate any particular evaluation content, model, method, or theory. Rather, it stresses an interactive process between an evaluator and primary intended users to select the most appropriate content, model, and methods for their particular situation. To stress this process, Patton offers the following definition of program evaluation:

> Program evaluation is the systematic collection of information about the activities, characteristics, and outcomes of programs to make judgments about the program, improve program effectiveness, and/or inform decisions about future programming. Utilization-focused evaluation (as opposed to program evaluation in general) is evaluation done for and with specific, intended primary users for specific, intended uses. (P. 23)

Evaluation Basics

Exhibit 11.1 illustrates the process of evaluation research as a simple systems model. First, clients, customers, students, or some other persons or units—cases—enter the program as **inputs**. (You will notice that this model treats programs like machines, with people functioning as raw materials to be processed.) Students may begin a new D.A.R.E. program, sex offenders may enter a new intense probation program, or crime victims may be sent to a victim advocate. Resources and staff required by a program are also program inputs.

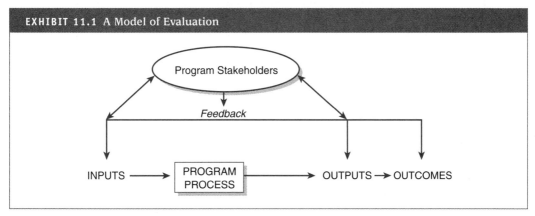

EXHIBIT 11.1 A Model of Evaluation

Source: Adapted from Martin, Lawrence L. & Peter M. Kettner, 1996. *Measuring the Performance of Human Service Programs.* Thousand Oaks, CA: Sage. Used with permission.

Inputs Resources, raw materials, clients, and staff that go into a program.

Next, some service or treatment is provided to the cases. This may be attendance in a class, assistance with a health problem, residence in new housing, or receipt of special cash benefits. The process of service delivery (program process) may be simple or complicated, short or long, but it is designed to have some impact on the cases, as inputs are consumed and outputs are produced.

Program process The complete treatment or service delivered by the program.

The direct product of the program's service delivery process is its output. Program outputs may include clients served, case managers trained, food parcels delivered, or arrests made. The program outputs may be desirable in themselves, but they primarily serve to indicate that the program is operating.

Outputs The services delivered or new products produced by the program process.

Program **outcomes** indicate the impact of the program on the cases that have been processed. Outcomes can range from improved test scores or higher rates of job retention to fewer criminal offenses and lower rates of poverty. There are likely to be multiple outcomes of any social program, some intended and some unintended, some positive and others that are viewed as negative.

Outcomes The impact of the program process on the cases processed.

Variation in both outputs and outcomes in turn influences the inputs to the program through a **feedback** process. If not enough clients are being served, recruitment of new clients may increase. If too many negative side effects result from a trial medication, the trials may be limited or terminated. If a program does not appear to lead to improved outcomes, clients may go elsewhere.

Feedback Information about service delivery system outputs, outcomes, or operations that is available to any program inputs.

Evaluation research enters into this picture as a more systematic approach to feedback that strengthens the feedback loop through credible analyses of program operations and outcomes. Evaluation research also broadens this loop to include connections to parties outside of the program itself. A funding agency or political authority may mandate the research, outside experts may be brought in to conduct the research, and the evaluation research findings may be released to the public, or at least funders, in a formal report.

The evaluation process as a whole, and the feedback process in particular, can only be understood in relation to the interests and perspectives of program stakeholders. **Stakeholders** are those individuals and groups who have some basis of concern with the program. They might be clients, staff, managers, funders, or the public. The board of a program or agency, the parents or spouses of clients, the foundations that award program grants, the auditors who monitor program spending, the members of Congress, each is a potential program stakeholder, and each has an interest in the outcome of any program evaluation. Some may fund the evaluation, some may provide research data, some may review, even approve, the research report (Martin & Kettner 1996:3). Who the program stakeholders are and what role they play in the program evaluation will have tremendous consequences for the research.

Stakeholders Individuals and groups who have some basis of concern with the program.

Can you see the difference between evaluation research and traditional social science research? Unlike explanatory social science research, evaluation research is not designed to test the implications of a social theory; the basic issue is often just, "What is the program's impact?" Process evaluation often uses qualitative methods, but unlike traditional exploratory research, the goal is not to create a broad theoretical explanation for what is discovered; instead, the question is, "How does the program do what it does?" Unlike social science research, the researchers cannot design evaluation studies simply in accord with the highest scientific standards and the most important research questions; instead, it is program stakeholders who set the agenda but there is no sharp boundary between the two: In their attempt to explain how and why the program has an impact, and whether the program is needed, evaluation researchers often bring social theories into their projects.

Evaluation Alternatives

Today, the field of evaluation research remains somewhat of a dichotomy. There are those who conduct evaluation research for the sake of knowledge alone (the more general evaluation research, such as the research performed by Horney and Spohn, 1991, discussed in Chapter 5), whereas others strive to make their products useful for some action (the more specific program evaluation research, such as that performed by Sherman and Berk, 1984, discussed in Chapter 2). Regardless of the emphasis, however, all evaluation is empirical and data-driven. Both program evaluation and evaluation research in general offer empirical answers to questions of policy and policy effectiveness. Objective and empirical assessments of policies and programs are the cornerstone of the evaluation field. Because there will never be an infinite amount of financial resources to fund all the programs we think will work, social scientists will always have work to do. As Patton (1997) so eloquently states, "As not everything can be done, there must be basis for deciding which things are worth doing. Enter evaluation" (p. 11).

Entire manuscripts have been devoted to delineating the various theories and methodologies utilized in evaluation research. Because this is a methods text, we are going to concentrate on the research methodologies most often utilized in criminology for evaluation purposes. Methods are always driven by research questions, however, so it is important to say something about the types of questions addressed by evaluation research. Patton (1997) distinguishes these three general question types:

Judgment-oriented evaluation. Evaluations aimed at determining the overall merit, worth, or value of something are **judgment-oriented evaluation**. Merit refers to the intrinsic value of a program. Questions addressed in this type of research include "How effective is the program in meeting the needs of those it is intended to help?" and "Does the program have the desired outcomes?"

Improvement-oriented evaluation. Using evaluation results to improve a program turns out, in practice, to be fundamentally different from rendering judgment about overall effectiveness, merit, or worth. **Improvement-oriented evaluation** focuses on making things better, rather than just rendering summative judgment. Although the questions addressed in this research are similar to those addressed in a judgment-oriented study, they are somewhat different: "What are the program's strengths and weaknesses?" and "To what extent are participants progressing toward the desired outcomes (e.g., are there any barriers to overcome in the delivery of services)?"

Knowledge-oriented evaluation. Contrary to the instrumental use of results of both judgment- and improvement-oriented evaluations, **knowledge-oriented evaluation** is used to influence thinking about issues in a general way. This knowledge can be as specific as clarifying a program's model, testing a theory, or figuring out how to measure outcomes. Academic researchers often initiate this type of an evaluation simply because the knowledge gained may fill a gap or add significantly to a particular research literature. For example, when Sherman and Berk (1984) began their experiment that examined the effect of arrest on recidivism rates of men who had assaulted their partners, they were not only concerned with the specific effect of arrest, but also with testing some basic tenets of deterrence theory such as "Are would-be offenders really deterred by the threat of formal sanctions like arrest and imprisonment?" and "Does the severity of the sanction effect the deterrent effect of formal sanctions?"

As these questions illustrate, evaluation research may be undertaken for a variety of reasons: for management and administrative purposes, to test hypotheses derived from theory, to identify ways to improve the delivery of services, or to decide whether to continue, cut, or modify a particular program. The goal of evaluation research, however, is primarily the same as the goal for all social science research, to design and implement a study that is objective and grounded in the rules of scientific methodology. These methods span the gamut of the methods we have discussed in this text. They can range from the strictly quantitative experimental and quasi-experimental designs to the qualitative methodologies of observation and intensive interviewing.

Evaluation projects can focus on several questions related to the operation of social programs and the impact they have:

- Is the program needed? (evaluation of need)
- Can the program be evaluated? (evaluability assessment)

- How does the program operate? (process evaluation)
- What is the program's impact? (impact evaluation)
- How efficient is the program? (efficiency analysis)

The specific methods used in an evaluation research project depend, in part, on which of these questions is being addressed.

The Evaluation of Need, or Needs Assessment

Is a new program needed or an old one still required? Is there a need at all? A needs assessment attempts to answer these questions with systematic, credible evidence. The initial impetus for implementing programs to alleviate social problems and other societal ailments typically comes from a variety of sources, including advocacy groups, moral leaders, community advocates, and political figures. Before a program is designed and implemented, however, it is essential to obtain reliable information on the nature and the scope of the problem, and the target population in need of the intervention. Evaluation researchers often contribute to these efforts by applying research tools to answer such questions as "What is the magnitude of this problem in this community?" "How many people in this community are in need of this program?" "What are the demographic characteristics of these people (e.g., age, gender, and race or ethnicity)?" and "Is the proposed program or intervention appropriate for this population?"

Rossi and Freeman (1989) provide several examples of the problems encountered in program implementation when these questions are not adequately answered. For example, during the early 1990s, when juvenile crime and delinquency were increasing, a suburban community in the Midwest implemented a program designed to prevent adolescent criminal behavior. However, after the program was started, it was discovered that there was virtually no juvenile crime in that particular community to begin with. The community planners had wrongly assumed that because juvenile delinquency was a social problem nationally, it was a problem in their community as well. Consequently, a lot of time, effort, and financial resources were wasted.

An empirical assessment by a social researcher of the community's needs in this area could have avoided this costly mistake. This research could have taken several forms, including doing an in-depth records search of local police agencies to determine the extent of crime and delinquency committed by juveniles in the area. A survey is another appropriate method used to determine the incidence of crime among juveniles in the community. For example, questionnaires could have been distributed in the schools to determine whether the rate of juvenile offending was serious enough in the community to warrant allocating the financial resources for the prevention programs.

Evaluability Assessment

Evaluation research will be pointless if the program itself cannot be evaluated. Yes, some type of study is always possible, but a study specifically to identify the effects of a particular program may not be possible within the available time and resources. So researchers may conduct an **evaluability assessment** to learn this in advance, rather than expend time and effort on a fruitless project.

Why might a social program not be evaluable?

To test your understanding of evaluation questions, go to the Evaluation Research Interactive Exercises on the Student Study Site.

- Management only wants to have its superior performance confirmed and does not really care whether the program is having its intended effects. This is a very common problem.
- Staff are so alienated from the agency that they do not trust any attempt sponsored by management to check on their performance.
- Program personnel are just "helping people" or "putting in time" without any clear sense of what the program is trying to achieve.
- The program is not clearly distinct from other services delivered from the agency and so cannot be evaluated by itself. (Patton 2002:164)

An evaluability assessment can help to solve the problems identified. Discussion with program managers and staff can result in changes in program operations. The evaluators may use the evaluability assessment to "sell" the evaluation to participants and sensitize them to the importance of clarifying their goals and objectives. Knowledge about the program gleaned through the evaluability assessment can be used to refine evaluation plans.

Because they are preliminary studies to "check things out," evaluability assessments often rely on qualitative methods. Program managers and key staff may be interviewed in depth, or program sponsors may be asked about the importance they attach to different goals. These assessments also may have an "action research" aspect, because the researcher presents the findings to program managers and encourages changes in program operations.

Process Evaluation (Program Monitoring)

What actually happens in a program? Once a program has been started, evaluators are often called on to document the extent to which implementation has taken place, whether the program is reaching the target individuals or groups, whether the program is actually operating as expected, and what resources are being expended (e.g., bucks spent!) in the conduct of the program. Rossi and Freeman (1989) define program monitoring as the

> systematic attempt by evaluation researchers to examine program coverage and delivery. Assessing program coverage consists of estimating the extent to which a program is reaching its intended target population; evaluating program delivery consists of measuring the degree of congruence between the plan for providing services and treatments and the ways they are actually provided. (P. 170)

Process evaluations are extremely important primarily because there is no way to reliably determine whether the intended outcomes have occurred without being certain the program is working according to plan. For example, imagine you are responsible for determining whether an anti-bullying curriculum given in a school has been successful in decreasing the amount of bullying behavior by the students. You conduct a survey of the students both before and after the curriculum was given and determine that rates of bullying have not significantly changed in the school since the curriculum was given. After you write your report, however, you find out that instead of being given in a 5-day series of

1-hour sessions as intended, the curriculum was actually crammed into a 2-hour format delivered on a Friday afternoon. A process evaluation would have revealed this implementation problem. If a program has not been implemented as intended, there is obviously no need to ask whether it had the intended outcomes.

A process evaluation can take many forms. Because most government and private organizations inherently monitor their activities through such things as application forms, receipts, and stock inventories, it should be relatively easy to obtain quantitative data for monitoring the delivery of services. This information can be summarized to describe things such as the clients served and the services provided. In addition to this quantitative information, a process evaluation will also likely benefit from qualitative methodologies such as unstructured interviews with people using the service or program. Interviews can also be conducted with staff to illuminate what they perceive to be obstacles to their delivery of services. As Posavac and Carey (1997) state, "Such qualitative information often provides points of view that neither evaluators nor service providers had considered" (p. 8).

Process evaluation is even more important when more complex programs are evaluated. Many social programs comprise multiple elements and are delivered over an extended period of time, often by different providers in different areas. Due to this complexity, it is quite possible that the program as delivered is not the same for all program recipients nor consistent with the formal program design.

The evaluation of D.A.R.E. by Research Triangle Institute researchers Ringwalt et al. (1994) included a process evaluation with three objectives:

1. Assess the organizational structure and operation of representative D.A.R.E. programs nationwide.

2. Review and assess factors that contribute to the effective implementation of D.A.R.E. programs nationwide.

3. Assess how D.A.R.E. and other school-based drug prevention programs are tailored to meet the needs of specific populations. (P. 7)

The process evaluation (they called it an "implementation assessment") was an ambitious research project in itself, with site visits, informal interviews, discussions, and surveys of D.A.R.E. program coordinators and advisors. These data indicated that D.A.R.E. was operating as designed and was running relatively smoothly. As shown in Exhibit 11.2, drug prevention coordinators in D.A.R.E. school districts rated the program components as much more satisfactory than did coordinators in school districts with other types of alcohol and drug prevention programs.

Process evaluation also can be used to identify the specific aspects of the service delivery process that have an impact. This, in turn, will help to explain why the program has an effect and which conditions are required for these effects. Implementation problems identified in site visits included insufficient numbers of officers to carry out the program as planned and a lack of Spanish-language D.A.R.E. books in a largely Hispanic school. Classroom observations indicated engaging presentations and active student participation (Ringwalt et al. 1994:58).

Process analysis of this sort can also help to show how apparently unambiguous findings may be incorrect. The apparently disappointing results of the Transitional Aid Research

EXHIBIT 11.2 Components of D.A.R.E. and Other Alcohol and Drug Prevention Programs Rated as Very Satisfactory (in Percentages)

Component	D.A.R.E. Program (N = 222)	Other AOD Programs (N = 406)
Curriculum	67.5	34.2
Teaching	69.7	29.8
Administrative requirements	55.7	23.1
Receptivity of students	76.5	34.6
Effects on students	63.2	22.8

Source: Ringwalt et al. 1994:58.

Project (TARP) provide an instructive lesson of this sort. TARP was a social experiment designed to determine whether financial aid during the transition from prison to the community would help released prisoners to find employment and avoid returning to crime. A total of 2,000 participants in Georgia and Texas were randomized to receive either a particular level of benefits over a particular period of time or no benefits at all (the control group). Initially, it seemed that the payments had no effect: The rate of subsequent arrests for both property and nonproperty crimes was not affected by TARP treatment condition.

But this wasn't all there was to it. Peter Rossi tested a more elaborate causal model of TARP effects that is summarized in Exhibit 11.3 (Chen 1990). Participants who received

EXHIBIT 11.3 Model of TARP Effects

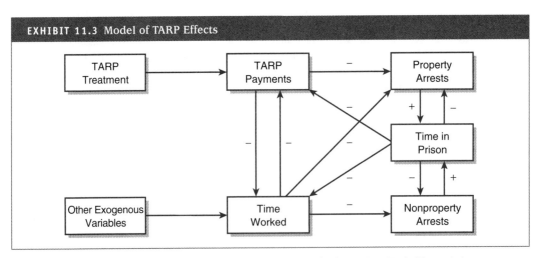

Source: From Chen, Huey-Tsyh 1990:210. *Theory-Driven Evaluations.* Thousand Oaks, CA: Sage. Used with permission.

TARP payments had more income to begin with and so had more to lose if they were arrested; therefore, they were less likely to commit crimes. However, TARP payments also created a disincentive to work and therefore increased the time available in which to commit crimes. Thus, the positive direct effect of TARP (more to lose) was cancelled out by its negative indirect effect (more free time).

The term **formative evaluation** may be used instead of process evaluation when the evaluation findings are used to help shape and refine the program (Rossi & Freeman 1989). Formative evaluation procedures that are incorporated into the initial development of the service program can specify the treatment process and lead to changes in recruitment procedures, program delivery, or measurement tools (Patton 2002:220).

Process evaluation can employ a wide range of indicators. Program coverage can be monitored through program records, participant surveys, community surveys, or utilizers versus dropouts and ineligibles. Service delivery can be monitored through service records completed by program staff, a management information system maintained by program administrators, or reports by program recipients (Rossi & Freeman 1989).

Qualitative methods are often a key component of process evaluation studies because they can be used to elucidate and understand internal program dynamics, even those that were not anticipated (Patton 2002:159; Posavac & Carey 1997). Qualitative researchers may develop detailed descriptions of how program participants engage with each other, how the program experience varies for different people, and how the program changes and evolves over time.

The Evaluation of Impact or Outcomes

If a process study shows that the implementation of the program has been delivered to the target population as planned, the next role for an evaluator is to assess the extent to which the program achieved its goals. "Did the program work?" "Did the program have the intended consequences?" This question should by now be familiar to you; stated more like a research question we are used to, "Did the treatment or program (independent variable) effect change in the dependent variable?" It all comes back to the issue of causality. This part of the research is variously called impact analysis or impact evaluation.

Impact evaluation (or analysis) Analysis of the extent to which a treatment or other service has an effect.

The bulk of the published evaluation studies in our field are devoted to some type of impact assessment. Have new seat-belt laws increased rates of seat-belt usage? Have rape reform statutes increased the willingness of rape victims to report their victimizations to police? Are boot camps more likely to reduce recidivism among juveniles compared to more traditional juvenile detention settings? Have mandatory minimum sentencing guidelines decreased the probability that extra legal factors such as sex and race will affect an individual's sentence? The list could go on and on.

The D.A.R.E. program (independent variable), for instance, tries to reduce drug use (dependent variable). When the program is present, we expect less drug use. In a more elaborate study, we might have multiple values of the independent variable; for instance, we might look at "no program," "D.A.R.E. program," and "other drug or alcohol education" conditions and compare the results of each.

To test your understanding of impact evaluation, go to the Evaluation Research Interactive Exercises on the Student Study Site.

D'Amico and Fromme's (2002) study of a new Risk Skills Training Program (RSTP) is a good example of a more elaborate study. They compared the impact of RSTP on children 14 to 19 years of age to that of an abbreviated version of D.A.R.E. and to a control group. The impacts they examined included positive and negative "alcohol expectancies" (the anticipated effects of drinking) as well as perception of peer risk-taking and actual alcohol consumption. D'Amico and Fromme (2002:568–570) found that negative alcohol expectancies increased for the RSTP group in the posttest but not for the D.A.R.E. group or the control group, while weekly drinking and "positive expectancies" for drinking outcomes actually *increased* for the D.A.R.E. group and/or the control group by the 6-month follow-up but not for the RSTP group (see Exhibit 11.4).

As in other areas of research, an experimental design is the preferred method for maximizing internal validity, that is, for making sure your causal claims about program impact are justified. Cases are assigned randomly to one or more experimental treatment groups and to a control group so that there is no systematic difference between the groups at the outset (see Chapter 6). The goal is to achieve a fair, unbiased test of the program itself, so that the judgment about the program's impact is not influenced by differences between the types of people who are in the different groups. It can be a difficult goal to achieve, because the usual practice in social programs is to let people decide for themselves whether they want to enter a program or not and also to establish eligibility criteria that ensure that people who enter the program are different from those who do not (Boruch 1997). In either case, a selection bias is introduced.

Of course, program impact may also be evaluated with quasi-experimental designs (see Chapter 7) or survey or field research methods, without a randomized experimental design. But if current participants who are already in a program are compared to nonparticipants, it is unlikely that the treatment group will be comparable to the control group. Participants will probably be a selected group, different at the outset from nonparticipants. As a result, causal conclusions about program impact will be on much shakier ground. For instance, when a study at New York's maximum-security prison for women found that "Income Education [i.e., classes] Is Found to Lower Risk of New Arrest," the conclusions were immediately suspect: The research design did not ensure that the women who enrolled in the prison classes were the same as those who had not enrolled in the classes, "leaving open the possibility that the results were due, at least in part, to self-selection, with the women most motivated to avoid reincarceration being the ones who took the college classes" (Lewin 2001).

Impact analysis is an important undertaking that fully deserves the attention it has been given in government program funding requirements. However, you should realize that more rigorous evaluation designs are less likely to conclude that a program has the desired effect; as the standard of proof goes up, success is harder to demonstrate. The prevalence of "null findings" (or "we can't be sure it works") has led to a bit of gallows humor among evaluation researchers. We will provide other case studies of impact evaluations at the end of the chapter.

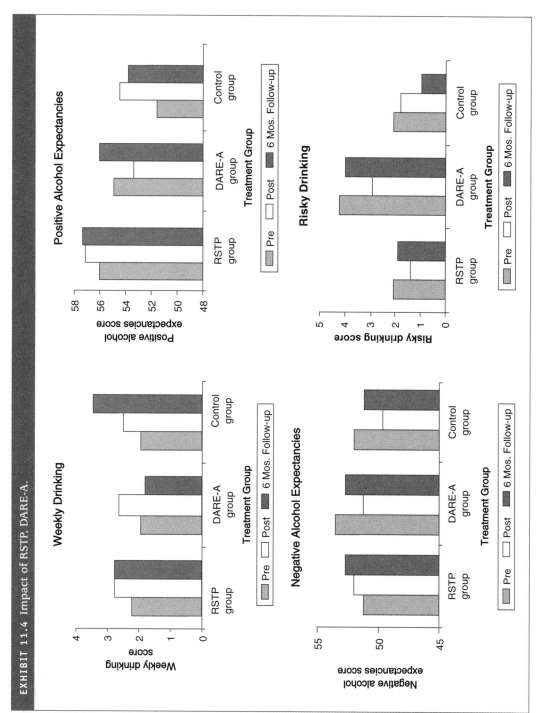

EXHIBIT 11.4 Impact of RSTP, DARE-A.

Source: Based on D'Amico & Fromme 2002:569.

375

The Evaluation of Efficiency

Whatever the program's benefits, are they sufficient to offset the program's costs? Are the taxpayers getting their money's worth? What resources are required by the program? These efficiency questions can be the primary reason that funders require evaluation of the programs they fund. As a result, **efficiency analysis**, which compares program effects to costs, is often a necessary component of an evaluation research project.

A **cost-benefit analysis** must identify the specific program costs and the procedures for estimating the economic value of specific program benefits. This type of analysis also requires that the analyst identify whose perspective will be used in order to determine what can be considered a benefit rather than a cost.

A **cost-effectiveness analysis** focuses attention directly on the program's outcomes rather than on the economic value of those outcomes. In a cost-effectiveness analysis, the specific costs of the program are compared to the program's outcomes, such as the number of jobs obtained, the extent of improvement in reading scores, or the degree of decline in crimes committed. For example, one result might be an estimate of how much it cost the program for each job obtained by a program participant.

Efficiency analysis A type of evaluation research that compares program costs to program effects. It can be either a cost-benefit analysis or a cost-effectiveness analysis.

Cost-benefit analysis A type of evaluation research that compares program costs to the economic value of program benefits.

Cost-effectiveness analysis A type of evaluation research that compares program costs to actual program outcomes.

Social science training often does not include much attention to cost-benefit analysis, so it can be helpful to review possible costs and benefits with an economist or business school professor or student. Once potential costs and benefits have been identified, they must be measured. It is a need highlighted in recent government programs:

> The Governmental Accounting Standards Board's (GASB) mission is to establish and improve standards of accounting and financial reporting for state and local governments in the United States. In June 1999, the GASB issued a major revision to current reporting requirements ("Statement 34"). The new reporting will provide information that citizens and other users can utilize to gain an understanding of the financial position and cost of programs for a government and a descriptive management's discussion and analysis to assist in understanding a government's financial results. (Campbell 2002:1)

In addition to measuring services and their associated costs, a cost-benefit analysis must be able to make some type of estimation of how clients benefited from the program. Normally, this will involve a comparison of some indicators of client status before and after clients received program services, or between clients who received program services and a comparable group who did not.

A recent study of therapeutic communities provides a clear illustration. A therapeutic community is an alternative to the traditional correctional response to drug addiction, which is typically incarceration for those convicted of either the possession or trafficking of illegal substances. In therapeutic communities, abusers participate in an intensive, structured living experience with other addicts who are attempting to stay sober. Because the treatment involves residential support as well as other types of services, it can be quite costly. Are those costs worth it?

Sacks et al. (2002) conducted a cost-benefit analysis of a modified therapeutic community (TC). A total of 342 homeless mentally ill chemical abusers were randomly assigned to either a TC or a "treatment-as-usual" comparison group. Employment status, criminal activity, and utilization of health care services were each measured for the 3 months prior to entering treatment and the 3 months after treatment. Earnings from employment in each period were adjusted for costs incurred by criminal activity and utilization of health care services.

Was it worth it? The average cost of TC treatment for a client was $20,361. In comparison, the economic benefit (based on earnings) to the average TC client was $305,273, which declined to $273,698 after comparing post- to pre-program earnings, but was still $253,337 even after adjustment for costs. The resulting benefit-cost ratio was 13:1, although this ratio declined to only 5.2:1 after further adjustments (for cases with extreme values). Nonetheless, the TC program studied seems to have had a substantial benefit relative to its costs.

The goal of most policy makers, of course, is to offer services that will justify the investment of funds. In a business, these cost-benefit analyses are relatively straightforward. When delivering human services, however, the evaluation of efficiency is a bit more tricky, because the benefits are usually hard to convert into a dollar figure. The procedures involved in these types of analyses are highly technical and, as such, are beyond the scope of this book. It is important, however, to understand the concept of efficiency analysis. Sherman (1997) succinctly describes the importance of this type of evaluation:

> Even though scientific evaluation results are a key part of rational policy analysis, those results cannot automatically select the best policy. This is due not just to the scientific limitations of generalizing results from one setting to the next. Another reason is that evaluations often omit key data on cost-benefit ratios; the fact that a program is "effective" may be irrelevant if the financial or social costs are too high. (P. 11)

DESIGN DECISIONS

Once we have decided on, or identified, the goal or focus for a program evaluation, there are still important decisions to be made about how to design the specific evaluation project. The most important decisions are the following:

- Black box or program theory: Do we care how the program gets results?
- Researcher or stakeholder orientation: Whose goals matter most?
- Quantitative or qualitative methods: Which methods provide the best answers?
- Simple or complex outcomes: How complicated should the findings be?

Black Box Evaluation or Program Theory

The "meat and potatoes" of most evaluation research involves determining whether a program has the intended effect. If the effect occurred, the program "worked"; if the effect didn't occur, then, some would say, the program should be abandoned or redesigned. In this approach, the process by which a program has an effect on outcomes is often treated as a **black box**; that is, the focus of the evaluation researcher is on whether cases seem to have changed as a result of their exposure to the program, between the time they entered the program as inputs and when they exited the program as outputs (Chen 1990). The assumption is that program evaluation requires only the test of a simple input/output model. There may be no attempt to open the black box of the program process.

If an investigation of program process is conducted, a **program theory** may be developed. A program theory describes what has been learned about how the program has its effect. When a researcher has sufficient knowledge before the investigation begins, outlining a program theory can help to guide the investigation of program process in the most productive directions. This is termed a **theory-driven evaluation**.

A program theory specifies how the program is expected to operate and identifies which program elements are operational (Chen 1990:32). In addition, a program theory specifies how a program is to produce its effects and so improves understanding of the relationship between the independent variable (the program) and the dependent variable (the outcome or outcomes). For example, Exhibit 11.5 illustrates the theory for an alcoholism treatment program. It shows that persons entering the program are expected to respond to the combination of motivational interviewing and peer support. A program theory can also decrease the risk of failure when the program is transported to other settings, because it will help to identify the conditions required for the program to have its intended effect.

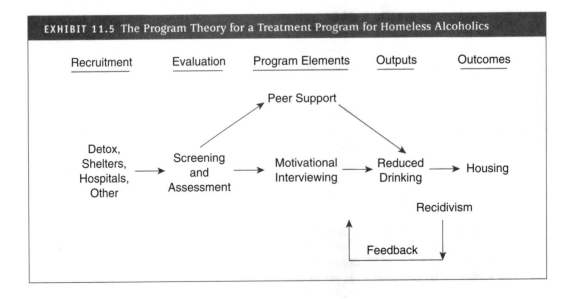

EXHIBIT 11.5 The Program Theory for a Treatment Program for Homeless Alcoholics

> ***Program theory*** A descriptive or prescriptive model of how a program operates and produces effects.

Program theory can be either descriptive or prescriptive (Chen 1990). Descriptive theory specifies what impacts are generated and how they occur. It suggests a causal mechanism, including intervening factors, and the necessary context for the effects. Descriptive theories are generally empirically based. On the other hand, prescriptive theory specifies what ought to be done by the program and is not actually tested. Prescriptive theory specifies how to design or implement the treatment, what outcomes should be expected, and how performance should be judged. Comparison of the descriptive and prescriptive theories of the program can help to identify implementation difficulties and incorrect understandings that can be corrected (Patton 2002:162–164).

Researcher or Stakeholder Orientation

Whose prescriptions specify how the program should operate, what outcomes it should try to achieve, or who it should serve? Most social science research assumes that the researcher specifies the research questions, the applicable theory or theories, and the outcomes to be investigated. Social science research results are most often reported in a professional journal or at professional conferences where scientific standards determine how the research is received. In program evaluation, however, the research question is often set by the program sponsors or the government agency that is responsible for reviewing the program. In consulting projects for businesses, the client—a manager, perhaps, or a division president—decides what question researchers will study. It is to these authorities that research findings are reported. Most often this authority also specifies the outcomes to be investigated. The first evaluator of the evaluation research is the funding agency, then, not the professional social science community. Evaluation research is research for a client, and its results may directly affect the services, treatments, or even punishments (e.g., in the case of prison studies) that program users receive. In this case, the person who pays the piper gets to call the tune.

Should the evaluation researcher insist on designing the evaluation project and specifying its goals or should she accept the suggestions and adopt the goals of the funding agency? What role should the preferences of program staff or clients play? What responsibility does the evaluation researcher have to politicians and taxpayers when evaluating government-funded programs? The different answers that various evaluation researchers have given to these questions are reflected in different approaches to evaluation (Chen 1990:66–68).

Stakeholder approaches encourage researchers to be responsive to program stakeholders (so this approach is also termed **responsive evaluation**). Issues for study are to be based on the views of people involved with the program, and reports are to be made to program participants (Shadish, Cook, & Leviton 1991:275–276). The program theory is developed by the researcher to clarify and develop the key stakeholders' theory of the program (Shadish, Cook, & Leviton 1991:254–255). In one stakeholder approach, termed *utilization-focused evaluation*, the evaluator forms a task force of program stakeholders who help to shape the evaluation project so that they are most likely to use its results (Patton 2002:171–175). In evaluation research termed *action research* or *participatory research*, program participants are engaged with the researchers as coresearchers and help to design, conduct, and report the research. One research approach that has been termed *appreciative inquiry* eliminates the professional researcher altogether in favor of a structured dialogue about needed changes among program participants themselves (Patton 2002:177–185).

In their book, *Fourth Generation Evaluation*, Guba and Lincoln (1989) argue for evaluations oriented toward stakeholders:

> The stakeholders and others who may be drawn into the evaluation are welcomed as equal partners in every aspect of design, implementation, interpretation, and resulting action of an evaluation—that is, they are accorded a full measure of political parity and control . . . determining what questions are to be asked and what information is to be collected on the basis of stakeholder inputs. (P. 11)

Because different stakeholders may differ on their reports about or assessment of the program, there is not likely to be one conclusion about program impact. The evaluators are primarily concerned with helping participants to understand the views of other stakeholders and to generate productive dialogue. Abma (2005) took this approach in a study of an injury prevention program at a dance school in the Netherlands:

> The evaluators acted as facilitators, paying deliberate attention to the development of trust and a respectful, open and comfortable climate. . . . Furthermore, the evaluation stimulated a public discourse about issues that were taboo, created a space for reflection fostered dynamics and motivated participants to think about ways to improve the quality of their teaching practice. (Pp. 284–285)

Social science approaches emphasize the importance of researcher expertise and maintenance of some autonomy in order to develop the most trustworthy, unbiased program evaluation. It is assumed that "evaluators cannot passively accept the values and views of the other stakeholders" (Chen 1990:78). Evaluators who adopt this approach derive a program theory from information they obtain on how the program operates and extant social science theory and knowledge, not from the views of stakeholders. In one somewhat extreme form of this approach, *goal-free evaluation*, researchers do not even permit themselves to learn what goals the program stakeholders have for the program. Instead, the researcher assesses and then compares the needs of participants to a wide array of program outcomes (Scriven 1972b). The goal-free evaluator wants to see the unanticipated outcomes and to remove any biases caused by knowing the program goals in advance.

Of course, there are disadvantages to both stakeholder and social science approaches to program evaluation. If stakeholders are ignored, researchers may find that participants are uncooperative, that their reports are unused, and that the next project remains unfunded. On the other hand, if social science procedures are neglected, standards of evidence will be compromised, conclusions about program effects will likely be invalid, and results are unlikely to be generalizable to other settings. These equally undesirable possibilities have led to several attempts to develop more integrated approaches to evaluation research.

Integrative approaches attempt to cover issues of concern to both stakeholders and evaluators, and to include stakeholders in the group from which guidance is routinely sought

(Chen & Rossi 1987:101–102). The emphasis given to either stakeholder or social concern is expected to vary with the specific project circumstances. Integrated approaches seek to balance the goal of carrying out a project that is responsive to stakeholder concerns with the goal of objective, scientifically trustworthy and generalizable results. When the research is planned, evaluators are expected to communicate and negotiate regularly with key stakeholders and to take stakeholder concerns into account. Findings from preliminary inquiries are reported back to program decision makers so that they can make improvements in the program before it is formally evaluated. When the actual evaluation is conducted, the evaluation research team is expected to operate more autonomously, minimizing intrusions from program stakeholders.

Many evaluation researchers now recognize that they must take account of multiple values in their research and be sensitive to the perspectives of different stakeholders, in addition to maintaining a commitment to the goals of measurement validity, internal validity, and generalizability (Chen 1990). Ultimately, evaluation research takes place in a political context, in which program stakeholders may be competing or collaborating to increase program funding or to emphasize particular program goals. It is a political process that creates social programs, and it is a political process that determines whether these programs are evaluated and what is done with evaluation findings (Weiss 1993:94). Developing supportive relations with stakeholder groups will increase the odds that political processes will not undermine evaluation practice. You do not want to find out after you are all done that "people operating ineffective programs who depend on them for their jobs" are able to prevent an evaluation report from having any impact ("'Get Tough' Youth Programs Are Ineffective, Panel Says" 2004:25).

Simple or Complex Outcomes

Does the program have only one outcome? Unlikely. How many outcomes are anticipated? How many might be unintended? Which are direct consequences of program action and which are indirect effects that occur as a result of the direct effects (Mohr 1992)? Do the longer-term outcomes follow directly from the immediate program outputs? Does the output (the increase in test scores at the end of the preparation course) result surely in the desired outcomes (increased rates of college admission)? Due to these and other possibilities, the selection of outcome measures is a critical step in evaluation research.

The decision to focus on one outcome rather than another, on a single outcome or on several, can have enormous implications. When Sherman and Berk (1984) evaluated the impact of an immediate arrest policy in cases of domestic violence in Minneapolis, they focused on recidivism as the key outcome. Similarly, the reduction of recidivism was the single desired outcome of prison "boot camps" opened in the 1990s. Boot camps are military style programs for prison inmates that provide tough, highly regimented activities and harsh punishment for disciplinary infractions, with the goal of scaring inmates "straight." They were quite the rage in the 1990s, and the researchers who evaluated their impact understandably focused on criminal recidivism.

But these single-purpose programs turned out not to be quite so simple to evaluate. The Minneapolis researchers found that there was no adequate single source for records of recidivism in domestic violence cases, so they had to hunt for evidence from court and police records, follow-up interviews with victims, and family member reports. More easily measured variables, such as partners' ratings of the accuseds' subsequent behavior, eventually received more attention. Boot camp researchers soon concluded that the experience did not reduce recidivism: "Many communities are wasting a great deal of money on those types of programs" (Robert L. Johnson quoted in "'Get Tough' Youth Programs Are Ineffective, Panel Says" 2004:25). However, some participants felt the study had missed something:

> [A staff member] saw things unfold that he had never witnessed among inmates and their caretakers. Those experiences profoundly affected the drill instructors and their charges, who still call to talk to the guards they once saw as torturers. Graduation ceremonies routinely reduced inmates, relatives, and sometimes even supervisors to tears. (Latour 2002:B7)

Former Boot Camp Superintendent Michael Corsini compared the Massachusetts boot camp to other correctional facilities and concluded, "Here, it was a totally different experience" (Latour 2002:B7).

Some now argue that the failure of boot camps to reduce recidivism was due to the lack of post-prison support rather than a failure of the camps to promote positive change in inmates. Looking only at recidivism rates would ignore some important positive results.

So in spite of the additional difficulties introduced by measuring multiple outcomes, most evaluation researchers attempt to do so (Mohr 1992). The result usually is a much more realistic, and richer, understanding of program impact.

Of course, there is a potential downside to the collection of multiple outcomes. Policy makers may choose to publicize those outcomes that support their own policy preferences and ignore the rest. Often, evaluation researchers themselves have little ability to publicize a more complete story.

In a sense, all these choices (black box or program theory, researcher or stakeholder interests, and so on) hinge on (1) what your real goals are in doing the project and (2) how able you will be in a "research for hire" setting to achieve those goals. Not every agency really wants to know if its programs work, especially if the answer is "no." Dealing with such issues, and the choices they require, is part of what makes evaluation research both scientifically and politically fascinating.

EVALUATION IN ACTION

Case Study: Problem-Oriented Policing in Violent Crime Areas—A Randomized Controlled Experiment

Several studies have found that over half of all crimes in a city are committed at a few criminogenic places within communities. These places have been called "hot spots" by some criminologists (Sherman, Gartin, & Buerger 1989; Weisburd, Maher, & Sherman 1992). Even within the most crime-ridden neighborhoods, it has been found that crime clusters at a few discrete locations while other areas remain relatively crime-free. The clustering of violent crime at particular locations suggests that there are important features or dynamics at these locations that give rise to violent situations. As such, focused crime prevention efforts should be able to modify these criminogenic conditions and reduce violence.

Problem-oriented policing strategies are increasingly utilized by urban jurisdictions to reduce crime in these high-activity crime places. Problem-oriented policing challenges officers to identify and analyze the causes of problems behind a string of criminal incidents. Once the underlying conditions that give rise to crime problems are known, police officers can then develop and implement appropriate responses. Despite the increasing use of problem-oriented policing strategies, however, there has been very little evaluation research conducted to determine whether such strategies are actually effective in decreasing crime rates.

Braga et al. (1999) created a novel experiment designed to determine the effectiveness of problem-oriented policing to decrease the incidence of violent street crime in Jersey City, New Jersey, The methodology they employed for their study was a true randomized experimental design. Recall from Chapter 5 that this design allows researchers to assume that the only systematic difference between a control and an experimental group is the presence of the intervention, in this case, the presence or absence of problem-oriented policing strategies.

To determine which places would receive the problem-oriented strategies and which places would not, Braga et al. (1999) used computerized mapping technologies to analyze all 1993 robbery and assault incidents and emergency citizen calls for services in Jersey City, New Jersey. These incidents were than matched to intersection areas (the intersection and its four adjoining street segments) and counted. Braga et al. then identified 56 discrete high-activity violent crime places. These 56 places were then matched into 28 pairs. Matching was done on a number of variables, including the primary offenses in each place (e.g., places with lower numbers of robberies and higher numbers of assaults were grouped together), types of problems at a place (e.g., robberies of commuters versus robberies of convenience stores), known dynamics of the place (e.g., the presence of active drug markets), and physical characteristics (e.g., presence of a park or school). A coin was flipped by the researchers to randomly determine which of the places within the pair would receive the problem-oriented policing treatment (experimental places). Remember that a key feature of true experimental designs is this **random assignment**. The places that were not selected from the flip in each pair did not receive the new policing strategies (control places). The design of this experimental evaluation is illustrated in Exhibit 11.6

EXHIBIT 11.6 Randomized Experimental Design Used to Evaluate Problem-Oriented Policing Strategies

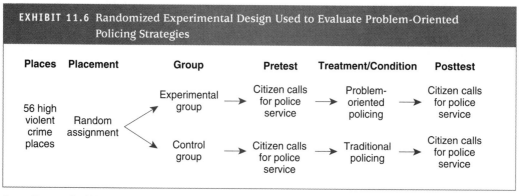

Source: Braga et al. 1999.

In each of the experimental places, police officers from the Violent Crime Unit (VCU) of the Jersey City police department established networks consistent with problem-oriented policing. For example, community members were used as information sources to discuss the nature of the problems the community faced, the possible effectiveness of proposed responses, and the assessment of implemented responses. In most places, the VCU officers believed that the violence that distinguished these places from other areas of the city was closely related to the disorder of the place. Although specific tactics varied from place to place, most attempts to control violence in these places were actually targeted at the social disorder problems. For example, some tactics included cleaning up the environment of the place through aggressive order maintenance and making physical improvements such as securing vacant lots or removing trash from the street. Other tactics included aggressive interventions to control the social disorder of a place, such as repeat foot and radio car patrols, dispersing groups of loiterers, and stop-and-frisks of suspicious persons. In addition, in places where drug markets were located, VCU officers increased their investigations to disrupt both the selling and the consumption of illicit drugs. The independent variable or treatment, then, was the use of problem-oriented policing, which comprised a number of specific tactics implemented by police officers to control the physical and social disorder at experimental violent places. In contrast, control places did not receive these problem-solving efforts; they received traditional policing strategies such as arbitrary patrol interventions and routine follow-up investigations by detectives. No problem-oriented strategies were employed.

Braga et al. (1999) examined the efficacy of these problem-oriented policing strategies by using three separate dependent variables. The first two were traditional indicators of crime, incident report data and citizen emergency calls for service within each place. The third variable was a physical observation of each place during the pretest and posttest periods. This variable was used to indicate changes in both physical incivilities at places such as vacant lots, trash, graffiti, or broken windows and social incivilities such as drinking in public and loitering. These variables were measured for 6-month preintervention and postintervention periods. If the problem-oriented policing approach was effective, then Braga et al. should have seen a decrease in incidents and emergency calls for service in the experimental areas in the posttest compared to the control areas. They should also have seen decreased signs of physical and social incivilities in the experimental areas compared to the control areas.

Statistical results indicated that the total number of criminal incidents and the total numbers of citizen calls for service were significantly reduced at the experimental places relative to the control places. This effect was true for most specific crime types examined as well. Examination of the observational data indicated that both physical and social disorder was reduced in 91% of the experimental places compared to the control places. Thus, the problem-oriented policing strategies examined in this evaluation research appear to have had a great deal of success in controlling and preventing crime.

Recall from Chapter 6 that although randomized experiments do allow us to determine whether the independent variable had an effect on the dependent variable, they do not tell us why. That is, they do not provide us with information about the causal **mechanisms** that may be producing this effect. Braga et al. (1999), however, utilized several theories to speculate on how these problem-oriented strategies may have changed the dynamics of a place in ways that resulted in a decrease in crime. For example, based on *rational choice theory*, which asserts that offenders consider risks, effort, and reward when contemplating criminal acts, the researchers speculated that the increased presence and order maintenance activities at the experimental places may have served as a powerful deterrent by communicating to offenders that disorderly behavior would no longer be tolerated. Braga et al. correctly pointed out, however, that the study did not gather data to determine the specific causal mechanisms responsible for the reductions found in crime. To do this, other forms of research would be necessary, such as interviewing offenders to determine their perceptions regarding their decisions to offend or to desist from offending in particular places.

What is certain from this research is that problem-oriented policing strategies are more effective in reducing both incidents of violent crime and citizen calls for emergency service than traditional policing strategies. Braga et al. (1999) concluded,

> Law enforcement agencies interested in controlling violence should consider implementing problem-oriented policing programs that focus on the places where violence clusters by developing tailored interventions addressing the underlying conditions and dynamics that give rise to violent situations. (P. 571)

Strengths of Randomized Experimental Design in Impact Evaluations

The research design used by Braga et al. (1999) meets all three criteria for a true experimental design. First, they used at least two comparison groups. In the Braga et al. research, some communities received the problem-oriented patrol strategies (experimental groups) while the other comparison communities received traditional police patrol (control groups). And finally, the assessment of change in the dependent variables used in their study was performed after the experimental condition (problem-oriented strategies) had been delivered.

Recall that the three criteria necessary for establishing a causal relationship between independent and dependent variables are (see Chapter 5):

1. Association between the independent and dependent variables
2. Correct time order (independent variable precedes the dependent)
3. Nonspuriousness (rule out influence of other variables)

Data obtained from a true experimental design is the best way to determine that all three of these criteria have been met.

Because the communities in Braga et al.'s (1999) study were randomly assigned to receive either problem-oriented or traditional patrol strategies, the relationship found between these strategies (independent variable) and the dependent variables (incidents of crime, etc.) are unambiguous. Braga et al. monitored their dependent variables both before and after the different policing strategies were implemented, so there is also no question that the strategies came before the change in the dependent variable. Finally, the random assignment of the communities to either the problem-oriented or traditional police conditions controlled for a host of possible extraneous influences that may have created spurious relationships.

The extent to which the findings of these studies can be generalized to the larger population, however, is another issue. Can Braga et al.'s (1999) findings be generalized to the larger population in New Jersey (sample generalizability) or to other states and communities (external validity)? Issues of sample generalizability, you will recall, are related to selecting a random sample from the population in the first place (random selection), not random assignment. However, because Braga et al.'s study utilized several experimental and control communities, this increases the likelihood that their findings are generalizable to their respective populations. In addition, because Braga et al.'s study was performed in the field (i.e., real world) and not in a laboratory, their findings are more likely to be generalizable to the larger population.

QUASI-EXPERIMENTAL DESIGNS IN EVALUATION RESEARCH

We have already learned that many research questions or situations are not amenable to a true experimental design. The same is true in evaluation research. There are many reasons why it may be impossible to use randomized experiments to evaluate the impacts of programs and policies. The primary reason why randomization is not possible in evaluation research is that the program is usually outside the control of the evaluator. As Rossi and Freeman (1989) state, "for political, human subject or other considerations, program staff, sponsors, or other powerful stakeholders resist randomization" (p. 313). Obtaining the cooperation of the program staff and sponsors is often an obstacle in evaluation research. In fact, when there are several locations implementing similar programs, the sites included in an evaluation study may be selected based primarily on the cooperation of the staff and sponsors. For example, in their evaluation study of Treatment Alternatives to Street Crime (TASC), Anglin, Longshore, and Turner (1999) state, "The program's ability to negotiate successfully with local officials to ensure their cooperation with evaluation activities was also a condition of study inclusion" (p. 172).

In general, quasi-experimental designs are the most powerful alternatives to true randomized experimental designs. The more alike the experimental and control groups are to each other, particularly on characteristics thought to be related to the intervention or treatment, the more confident we can be in a study's findings. In this section, we are going to discuss two types of quasi-experimental designs that are often utilized in criminology and criminal justice to determine the impacts of programs. To illustrate the first type, a **nonequivalent control group design**, we will use research that has been done to evaluate whether boot camps are more likely to reduce recidivism after release compared to traditional incarceration. The second type of quasi-experimental design we will highlight is the **time series design**. To illustrate this design, we will use a study that evaluated the impact of mandatory sentences for gun crimes.

Case Study: Boot Camps

Since their beginning in 1983 in Georgia, the use of correctional boot camps, sometimes called shock incarceration programs, has skyrocketed in both state and federal prison systems. For politicians and citizens alike, the notion of a strict, military-style punishment as an alternative to extended incarceration is attractive. Many questions remain, however, about the value of boot camps. Do they work? Because there are so many different types of boot camps (e.g., some have a military-style atmosphere whereas others do not) there is no easy answer to this question. Moreover, despite their increased popularity, there have been only a few systematic attempts to evaluate their efficacy in reducing recidivism.

In 1990, Doris MacKenzie was the principal investigator of a National Institute of Justice (NIJ) grant, which funded evaluation of boot camp program efficacy in eight states: Florida, Georgia, Illinois, Louisiana, New York, Oklahoma, South Carolina, and Texas. These sites were selected because they incorporated the core elements of a boot camp program, including (1) strict rules, discipline, and a military boot camp–like atmosphere; (2) mandatory participation in military drills and physical training; and (3) separation of program participants from other prison inmates.

The evaluation was extensive and resulted in many publications (Brame & MacKenzie 1996; MacKenzie 1994; MacKenzie et al. 1995; MacKenzie & Piquero 1994; MacKenzie & Souryal 1995). Rather than focusing on one measure of program success, such as recidivism reduction, MacKenzie et al. developed five different components to their evaluation: a process evaluation of the programs based on interviews with staff and inmates, official program materials, and observation; a study of inmate attitudinal change during incarceration; a study of offender recidivism; a study of positive adjustment during community supervision; and a study of prison bed-space savings.

We will highlight the results of the impact assessments for recidivism here. To determine whether boot camps were more likely than traditional prison time to reduce recidivism after release, MacKenzie et al. utilized a quasi-experimental design. In each state, a sample of male boot camp program graduates was used as the experimental condition. Two other offender samples were used as comparison groups and usually included regular prison parolees, probationers, and boot camp dropouts. The individuals from all samples, however, met the eligibility requirements for the boot camp programs. This nonequivalent control group design is illustrated for you in Exhibit 11.7.

After being sentenced to probation or released from boot camps and prison, subjects were tracked for a follow-up period of up to 2 years of community supervision. Recidivism was measured by arrests and revocations for new crimes and/or for technical violations.

Were boot camps successful in meeting their goals? Well, in five states (Florida, Georgia, Oklahoma, South Carolina, and Texas), MacKenzie's findings suggest that boot camp

EXHIBIT 11.7 Nonequivalent Control Group Design Used to Evaluate Boot Camps

Subjects	Group	Treatment/ Condition	Posttest
Boot camp-eligible offenders	Experimental group →	Boot camp →	Recidivism
	Comparison →	Prison →	Recidivism
	Comparison →	Probation →	Recidivism

Source: Brame & MacKenzie 1996.

graduates appeared to recidivate at about the same rate as those from the comparison groups. In Illinois, Louisiana, and New York, however, boot camp graduates had lower recidivism rates on particular recidivism measures. Graduates in Illinois and Louisiana, for example, were less likely to have their supervision status revoked as a result of a new crime revocation than were comparison samples. And although they did not differ from the comparison sample on other measures of recidivism, New York boot camp graduates were less likely to be returned to prison as a result of a technical violation.

The weakness of this quasi-experimental design lay in the fact that MacKenzie was not able to randomly assign offenders to the boot camps or the comparison treatments. Therefore, there is less confidence in the findings, primarily because there is no assurance that the offenders were equivalent before they received the treatments. To control for these differences, MacKenzie and colleagues utilized strong statistical evidence that served to increase the internal validity of their findings.

In addition, even for the modest effects found in the three states, there is still some ambiguity about how the independent variable, in this case the type of incarceration, influenced the dependent variable (recidivism). Because this evaluation was not performed using a true experimental design with random assignment to the boot camps, prison, or probation, there is even more ambiguity regarding the extent to which boot camps actually affected recidivism rates. MacKenzie does, however, speculate about the reasons why the Illinois, Louisiana, and New York programs showed reduced recidivism rates compared to their respective comparison programs. In fact, MacKenzie contends that one of the reasons may have had little to do with the boot camps themselves:

> These programs stand out as the only boot camps in the study that have developed an intensive supervision phase for boot camp graduates. It is therefore very likely that differences in recidivism rates in these states were due to the type of community supervision provided to graduates, not in-prison programming. (MacKenzie & Souryal 1996:293)

In addition, these boot camps also devoted a considerable amount of time per day to rehabilitative activities such as drug treatment services.

In summary, MacKenzie's (1994) results were not very favorable for boot camps. She states,

> Results clearly show that the core elements of boot camp program—military-style discipline, hard labor, and physical training—by themselves did not reduce offender recidivism—but it is likely that some mixture of rehabilitation and intensive follow-up supervision plays an important role. (P. 65)

Case Study: Mandatory Sentencing Laws

In this era of "get tough on crime" policies, the use of mandatory and enhanced sentences at both the federal and state levels has dramatically increased. The effects of most of these policies, however, remain unevaluated (this is becoming a theme, huh?). The most obvious impact they have had, particularly those policies related to the use and sale of drugs, is to increase the prison population in the United States to the highest levels in history.

One category of mandatory sentences that has received research attention is the mandatory sentence enhancements for gun crimes. By providing stiff and certain penalties when a gun is involved in an offense, sentence enhancement laws are meant to reduce the use of firearms by criminals. The most comprehensive evaluation of these laws to date was performed by McDowall, Loftin, and Wiersema (1992). These researchers conducted a time series analysis based on six cities that had enacted similar mandatory sentence enhancements for gun crimes: Detroit, Jacksonville, Tampa, Miami, Philadelphia, and Pittsburgh. The key features of the law were the same in each city: (1) judges were required to impose a specified sentence on defendants convicted of an offense that involved a gun; (2) probation, suspended sentences, and parole under these sentences were prohibited; and (3) all areas used advertising campaigns involving radio and television commercials, posters, bumper stickers, and billboards to communicate the message that offenders would receive additional punishment if they used a gun to commit a crime.

McDowall, Loftin, and Wiersema (1992) examined the law's effects in each city separately and in a pooled analysis to obtain an overall estimate of the effect of the statutes on three types of violent crimes: homicide, robbery, and assault. We will summarize the findings from the combined study here. The statistical methods used in this analysis were quite complex, so we will not burden you with them; those of you who are interested can go right to the source! In general, what McDowall did was examine rates of gun-related violent crime both before (pretest) and after (posttest) the mandatory sentencing laws were implemented. If the laws were effective in reducing firearm crimes, the number of gun offenses should decrease during the posttest period. The dependent variables monitored were monthly rates of various violent crimes.

To enhance the research design, the researchers did not just examine a few months before and after statute implementation. McDowall, Loftin, and Wiersema (1992) examined many months of data (up to 150 depending on data availability in each city) both before and after implementation of the law. Recall that this is similar to the research design used by Horney and Spohn (1991) that evaluated the impacts of rape reform legislation (see Chapter 6).

Were the sentencing enhancements effective in reducing violent crime? Well, the results were somewhat mixed. The only crimes for which the laws had a consistent impact across all cities were homicides committed with guns. For example, McDowall, Loftin, and Wiersema (1992) offer a statistical illustration of the law's efficacy in Detroit:

> Consider Detroit, a city with a pre-intervention mean of forty gun homicides per month and a standard deviation of eight. Here, a decrease of .69 standard deviation units represents an average of 5.5 lives saved each month, a fourteen percent reduction. (P. 386)

The analyses of gun-related assaults and armed robberies, however, were not so positive. The only city in which gun assaults significantly decreased was Jacksonville. Although gun assaults also decreased in Detroit, Miami, Philadelphia, and Pittsburgh, the decrease was not significantly different from zero (no decrease at all). Furthermore, gun assaults actually increased in Tampa. Armed robberies (armed with any weapon) did not significantly decrease in any city and only showed small decreases in two cities. Tampa and Miami, however, showed large and significant increases in unarmed robbery following the law's adoption.

The researchers argue that this may illustrate that enhanced sentencing laws may have prevented armed robberies from increasing in the same way as unarmed robberies.

So, you may be asking yourself, what did they conclude about the effects of the laws? McDowall, Loftin, and Wiersema (1992) were confident in their results regarding the benefits of the sentence enhancements in reducing gun homicides. In fact, they state,

> The consistency of the effects on gun homicide virtually rules out the possibility that factors confounded with mandatory sentencing could account for the reductions in this crime. . . . There is clear and convincing evidence of preventive effects for homicide. (P. 389)

However, the researchers remained appropriately hesitant about conclusions for the other types of violent crime.

NONEXPERIMENTAL DESIGNS

Much research that is conducted to evaluate the effectiveness of programs or policies uses a **nonexperimental design** in which a single group is studied only once, after the treatment or program has been delivered. Unfortunately, such studies have absolutely no form of control or comparison. As such, they are "of almost no scientific value" (Campbell & Stanley 1966:6). Nevertheless, because they are so common in evaluation research, we provide an illustration of their use.

Case Study: Vocational Education for Serious Juvenile Offenders—A One-Shot Design

Some of the earliest programs for crime reduction have been those that provide legitimate employment or employment skills to offenders or at-risk individuals. Most of these programs target adolescents. One of the myriad programs to provide vocational education for juvenile offenders is New Pride, Inc. This program has been providing vocational services to serious young offenders (14 through 18 years of age) for almost two decades. Designed as an alternative to institutionalization in state correctional facilities, New Pride uses a holistic approach and provides a wide array of services, including diagnostic and needs assessment, intensive supervision in the community, individual and family counseling, prevocational training (e.g., basic academic and social interactional skills), vocational training and job placement, structured recreation, and cultural education. Based on results of the diagnostic and needs assessment, services are individualized to meet the specific needs of each youth and are integrated into a single treatment plan.

The planners of New Pride, understandably, have a lot of faith in their program. Unfortunately, there have been few systematic attempts to evaluate the effectiveness of this program and others like it in either increasing employment or reducing recidivism (Bushway & Reuter 1997). For example, the results of the one-shot design methodology evaluation performed by James and Granville (1984) of the New Pride program are too scientifically deficient to be of value. This methodology is displayed in Exhibit 11.8.

EXHIBIT 11.8 One-Shot Design Used to Evaluate the New Pride Program

Subjects	Treatment/ Condition	Posttest
Serious young offenders (14–18 years old)	Received New Pride program	→ Recidivism and school placement

Source: James & Granville 1984.

Basically, a one-shot design measures the dependent variable after the treatment has been delivered for only those who received the treatment. James and Granville (1984) found that over 70% of those who completed the New Pride program were successfully reintegrated into the public school system, continued part-time employment, and were not rearrested within 6 months of program completion. Unfortunately, because there was no random assignment to another comparison group—in fact, no comparison group whatsoever—these results tell us very little. This type of design is vulnerable to all the confounding effects described in Chapter 6 including selection bias, endogenous change, and history effects. Perhaps the results would have been the same for youths who had received no vocational training. The problem is, we can never be sure when a one-shot design such as this is used.

QUALITATIVE AND QUANTITATIVE METHODS

Evaluation research that attempts to identify the effects of a treatment, law, or program typically is quantitative. It is fair to say that when there is an interest in comparing outcomes between an experimental and a control group, or tracking change over time in a systematic manner, quantitative methods are favored.

But qualitative methods can add much to quantitative evaluation research studies, including more depth, detail, nuance, and exemplary case studies (Patton 1997). Perhaps the greatest contribution qualitative methods can make in many evaluation studies is investigating program process, finding out what is "inside the black box." Although it is possible, even recommended, to track the process of service delivery with quantitative measures like staff contact hours, frequency of complaints, and the like, the goal of finding out what is happening to program clients and how clients experience the program can often best be achieved by observing program activities and interviewing staff and clients intensively.

Another good reason for using qualitative methods in evaluation research is the importance of learning how different individuals react to the treatment. Qualitative methods can also help in understanding how social programs actually operate. Complex social programs have many different features, and it is not always clear whether it is the combination of those features or some particular features that are responsible for the program's effect, or for the absence of an effect. Lisbeth B. Schorr, Director of the Harvard Project on Effective Interventions, and Daniel Yankelovich, President of Public Agenda, put it his way: "Social programs are sprawling efforts with multiple components requiring constant mid-course corrections, the involvement of committed human beings, and flexible adaptation to local circumstances" (Schorr & Yankelovich 2000).

The more complex the social program, the more value that qualitative methods can add to the evaluation process. Schorr and Yankelovich (2000) point to the Ten Point Coalition, an alliance of African American ministers who helped to reduce gang warfare in Boston through multiple initiatives, "ranging from neighborhood probation patrols to safe havens for recreation." Qualitative methods would help to describe a complex, multifaced program like this.

For the most part, the strengths and weaknesses of methodologies used in evaluation research are those of the methods we have already discussed throughout this text. As Patton (1997) contends, "There are no perfect [evaluation] studies. And there cannot be, for there is no agreement on what constitutes perfection" (p. 23). There are, however, methods that are better able to infer cause and effect. Because the basic question of most impact evaluations is "Does program (cause) have its intended consequences (effect)?" there are basic methodological criteria we can use to judge the methods discussed in this chapter.

In their review of the crime prevention evaluation literature, Sherman et al. (1997) adopted a way to judge the "methodological rigor" of evaluation studies using a 5-point scale. The researchers gave each study examined in their review a "scientific methods score" of 1 to 5, with 5 being the strongest scientific evidence. Studies were assigned a score based on the following criteria:

1. Correlation between a crime prevention program and its intended outcome (e.g., a measure of crime or crime risk factors)

2. Temporal sequence between the program and the crime or risk outcome clearly observed (e.g., program before outcome), or a comparison group present without demonstrated comparability to the treatment group

3. A comparison between two or more groups or units, one with and one without the program and/or a long time series analysis

4. A comparison between multiple groups or units with and without the program, controlling for other factors, or a nonequivalent comparison group that has only minor differences evident between the groups

5. Random assignment to program (experimental group) and comparison (control groups)

As you can see, a study with a score of 1 simply demonstrated a correlation between a crime prevention program and its intended outcome. But as you learned in Chapter 5, correlation does not mean causation. Hence the low rating of 1 for studies only demonstrating correlation. To reach level 3, a study had to employ some kind of control or comparison group to determine what would have happened without the prevention program. If that comparison group consisted of a large number of matched or almost randomized cases, the study could achieve a score of 4. If the comparison was to a large number of comparable units selected at random to receive the program or not, the study received the highest score of 5 because, as Sherman et al. state (1997), "Random assignment offers the most effective means available of eliminating competing explanations for whatever outcome is observed" (p. 18).

Despite the fact that a randomized experiment is the best way to determine the impact of a program, it is also important to remember that the methodology selected for an evaluation project should be relevant to the policy makers and planners who intend to use the results. Although some researchers vehemently contend that only a randomized experimental design can provide reliable information on the impacts of a program, others just as vehemently disagree. For example, in his utilization-focused method of evaluation, Patton (1997) believes that when making evaluation methods decisions, the primary focus should be on getting the best possible data to answer primary users' questions. This cannot be done without involving them in the decision-making process to begin with. For methods decisions, the emphasis is on the appropriateness of the method to answer the practitioner's questions. Patton contends that only when these decisions are made in collaboration with the primary users will they be perceived as credible and, thus, more likely to be used.

It should also be noted that evaluation research faces obstacles not faced by general social research. We have already highlighted a few of these including the obstacle of obtaining support and cooperation from program staff and sponsors. In their review of evaluation research on crime prevention programs, Sherman et al. (1997) delineate several other obstacles. For example, one obstacle noted is the structural separation of research and program funding. In everyday life, this often means that an evaluator is brought in after a program has been implemented and asked to assess its consequences. This, of course, leads to several problems. How can the consequences be accurately assessed if the evaluator did not have the opportunity to monitor the dependent variable (outcomes) before the program was started? How can the consequences be accurately assessed if the evaluator had no chance to obtain a comparison group by randomly assigning some units (e.g., individuals or communities) to the program or treatment condition and others elsewhere? To ameliorate this problem, Sherman recommends that Congress produce financial incentives for partnerships to be created between local agencies implementing programs and researchers evaluating them.

ETHICS IN EVALUATION

Evaluation research can make a difference in people's lives while it is in progress, as well as after the results are reported. Educational and vocational training opportunities in prison, the availability of legal counsel, and treatment for substance abuse, each is a potentially important benefit, and an evaluation research project can change both their type and their availability. This direct impact on research participants and, potentially, their families, heightens the attention that evaluation researchers have to give to human subjects concerns. Although the particular criteria that are at issue and the decisions that are most ethical vary with the type of evaluation research conduced and the specifics of a particular project, there are always serious ethical as well as political concerns for the evaluation researcher (Boruch 1997:13; Dentler 2002:166).

Assessing needs, determining evaluability, and examining the process of treatment delivery have few special ethical dimensions. Cost-benefit analyses in themselves also raise few ethical concerns. It is when program impact is the focus that human subjects considerations multiply. What about assigning persons randomly to receive some social program or benefit? One justification given by evaluation researchers has to do with the scarcity of these resources. If not everyone in the population who is eligible for a program can receive it, due to resource limitations, what could be a fairer way to distribute the program benefits than through a lottery? Random assignment also seems like a reasonable way to allocate potential program benefits when a new program is being tested with only some members of the target recipient population. However, when an ongoing entitlement program is being evaluated and experimental subjects would normally be eligible for program participation, it may not be ethical simply to bar some potential participants from the programs. Instead, evaluation researchers may test alternative treatments or provide some alternative benefit while the treatment is being denied.

There are many other ethical challenges in evaluation research:

- How can confidentiality be preserved when the data are owned by a government agency or are subject to discovery in a legal proceeding?
- Who decides what level of burden an evaluation project may tolerably impose upon participants?
- Is it legitimate for research decisions to be shaped by political considerations?
- Must evaluation findings be shared with stakeholders rather than only with policy makers?
- Is the effectiveness of the proposed program improvements really uncertain?
- Will a randomized experiment yield more defensible evidence than the alternatives?
- Will the results actually be used?

The Health Research Extension Act of 1985 (Public Law 99–158) mandated that the Department of Health and Human Services require all research organizations receiving federal funds to have an Institutional Review Board (IRB) to assess all research for adherence to ethical practice guidelines. We have already reviewed the federally mandated criteria (Boruch 1997:29–33):

- Are risks minimized?
- Are risks reasonable in relation to benefits?
- Is the selection of individuals equitable? (randomization implies this)
- Is informed consent given?
- Are the data monitored?
- Are privacy and confidentiality assured?

Evaluation researchers must consider whether it will be possible to meet each of these criteria long before they even design a study.

The problem of maintaining subject confidentiality is particularly thorny, because researchers, in general, are not legally protected from the requirements that they provide evidence requested in legal proceedings, particularly through the process known as "discovery." However, it is important to be aware that several federal statutes have been passed specifically to protect research data about vulnerable populations from legal disclosure requirements. For example, the Crime Control and Safe Streets Act (28 CFR Part 11) includes the following stipulation:

Copies of [research] information [about persons receiving services under the act or the subjects of inquiries into criminal behavior] shall be immune from legal process and shall not, without the consent of the persons furnishing such information, be admitted as evidence or used for any purpose in any action, suit, or other judicial or administrative proceedings. (Boruch 1997:60)

When it appears that it will be difficult to meet the ethical standards in an evaluation project, at least from the perspective of some of the relevant stakeholders, modifications should be considered in the study design. Several steps can be taken to lessen any possibly detrimental program impact (Boruch 1997:67–68):

- Alter the group allocation ratios to minimize the number in the untreated control group.
- Use the minimum sample size required to be able to adequately test the results.
- Test just parts of new programs rather than the entire program.
- Compare treatments that vary in intensity (rather than presence or absences).
- Vary treatments between settings rather than among individuals within a setting.

Essentially, each of these approaches limits the program's impact during the experiment and so lessens any potential adverse effects on human subjects. It is also important to realize that it is costly to society and potentially harmful to participants to maintain ineffective programs. In the long run, at least, it may be more ethical to conduct an evaluation study than to let the status quo remain in place.

CONCLUSION

In recent years, the field of evaluation research has become an increasingly popular and active research specialty within the fields of criminology and criminal justice. Many social scientists find special appeal in evaluation research because of its utility.

The research methods applied to evaluation research are not different from those covered elsewhere in this text; they can range from qualitative intensive interviews to rigorous randomized experimental designs. In process evaluations, qualitative methodologies can be particularly advantageous. However, the best method for determining cause and effect, or for determining whether a program had its intended consequences (impacts), is the randomized experimental design. Although this may not always be possible in the field, it is the "gold standard" with which to compare methodologies used to assess the impacts of all programs and/or policies.

KEY TERMS

Accuracy	Efficiency analysis
Black box evaluation	Evaluation of efficiency
Comparison group	Evaluation of impact
Control group	Evaluation of need
Cost-benefit analysis	Evaluation of process
Cost-effectiveness analysis	Experimental designs

Feasibility

Feedback

Formative evaluation

Impact evaluation

Improvement-oriented evaluation

Inputs

Judgment-oriented evaluation

Knowledge-oriented evaluation

Mechanisms

Needs assessment

Nonequivalent control group design

One-shot design

Outcomes

Outputs

Program process

Program theory

Propriety

Quasi-experimental design

Random assignment

Stakeholders

Time series design

Utility

HIGHLIGHTS

- Whenever people implement a program for a specific purpose, they pay attention to the consequences of it, even if this attention is not in the form of rigorous scientific methods. Evaluation research, however, has become increasingly used to assess the impacts of programs and policies since the 1950s.

- The evaluation process can be modeled as a feedback system, with inputs entering the program, which generates outputs and then outcomes, which feed back to program stakeholders and affect program inputs.

- The evaluation process as a whole, and the feedback process in particular, can be understood only in relation to the interests and perspectives of program stakeholders.

- The process by which a program has an effect on outcomes is often treated as a "black box," but there is good reason to open the black box and investigate the process by which the program operates and produces, or fails to produce, an effect.

- A program theory may be developed before or after an investigation of program process is completed. It may be either descriptive or prescriptive.

- Evaluation research is done for a client, and its results may directly affect the services, treatments, or punishments that program users receive. Evaluation researchers differ in the extent to which they attempt to orient their evaluations to program stakeholders.

- There are five primary types of program evaluation: needs assessment, evaluability assessment, process evaluation (including formative evaluation), impact evaluation (also termed summative evaluation), and efficiency (cost-benefit) analysis.

- The evaluation of need generally aims to describe the nature and scope of the problem along with the target population in need of services.

- The evaluation of process determines the extent to which implementation has taken place, whether the program is reaching the target population, whether the program is actually operating as expected, and what resources are being expended. Qualitative methods are typically utilized in process evaluations.

- Impact evaluations determine whether the program or policy produced the desired outcomes. Quantitative methods are typically utilized in impact evaluations.

- The evaluation of efficiency relates program outcomes to program costs.

- True randomized experiments are the most appropriate method for determining cause and effect, and as such, for determining the impact of programs and policies.

- Evaluation research raises complex ethical issues because it may involve withholding desired social benefits.

EXERCISES

1. Read one of the articles reviewed in this chapter. Fill in the answers to the article review questions (Appendix B) not covered in the chapter. Do you agree with the answers to the other questions discussed in the chapter? Could you add some points to the critique provided by the author of the text or to the lessons on research design drawn from these critiques?

2. Evaluate the ethics of one of the studies reviewed in which human subjects were used. Sherman and Berk's (1984) study of domestic violence raises some interesting ethical issues, but there are also points to consider in most of the other studies. Which ethical guidelines (see Chapter 2) seem most difficult to adhere to? Where do you think the line should be drawn between not taking any risks at all with research participants and developing valid scientific knowledge? Be sure to consider various costs and benefits of the research.

3. Propose a randomized experimental evaluation of a social program with which you are familiar. Include in your proposal a description of the program and its intended outcomes. Discuss the strengths and weaknesses of your proposed design.

4. Identify the key stakeholders in a local social or educational program. Interview several stakeholders to determine their goals for the program and what tools they use to assess goal achievement. Compare and contrast the views of each stakeholder and try to account for any differences you find.

DEVELOPING A RESEARCH PROPOSAL

Imagine that you are submitting a proposal to the U.S. Justice Department to evaluate the efficacy of a new treatment program for substance abusers within federal correctional institutions.

1. What would your research question be if you proposed a process evaluation component to your research?

2. For the outcome evaluation, what is your independent variable and what would your dependent variable be? How would you operationalize both?

3. What type of research design would you propose to answer both the process evaluation and outcome evaluation components in your proposal?

Student Study Site

The companion Web site for *The Practice of Research in Criminology and Criminal Justice*, Third Edition

> **http://www.sagepub.com/prccj3**

Visit the Web-based Student Study Site to enhance your understanding of the chapter content and to discover additional resources that will take your learning one step further. You can enhance your understanding of the chapters by using the comprehensive study material, which includes e-flashcards, Web exercises, practice self-tests, and more. You will also find special features, such as Learning from Journal Articles, which incorporates SAGE's online journal collection.

WEB EXERCISES

1. Go to the American Evaluation Association Web site at www.eval.org. Choose "Publications" and then "Guiding Principles for Evaluator." What are the five guiding principles discussed in this document? Provide a summary of each principle.

2. How adequate are juvenile court records? Go to the Bureau of Justice Statistics Web site at http://blackstone.ojp.usdoj.gov/bjs/pub/ascii/pjjr.txt.

   a. Read the report and write a brief summary.

   b. The Bureau of Justice Statistics would like to find out if any improvements have been made in the collection and maintenance of juvenile records. Propose a research project to evaluate the adequacy of the juvenile justice system in the collection and maintenance of juvenile law enforcement records and/or juvenile court records.

3. Describe the resources available for evaluation researchers at one of the following three Web sites:

   www.wmich.edu/evalctr/
   www.stanford.edu/ ~ davidf/empowermentevaluation.html
   www.worldbank.org/oed/

4. You can check out the latest information regarding the D.A.R.E. program at www.dare.com. What is the current approach? Can you find information on the Web about current research on D.A.R.E.?

ETHICS EXERCISES

1. The Manhattan Bail Project randomly assigned some defendants with strong community ties to be recommended for release without bail and some not to be recommended. You learned in this chapter that they found a very high (99%) rate of appearance at trial for those released without bail, but of course the researchers did not know that this would be the case before they conducted the study. Would you consider this study design ethical? What if one of the persons in the 1% who did not come to trial murdered someone when they should have been at court? Are there any conditions when randomization should not be permitted when evaluating criminal justice programs?

2. A large body of evaluation research suggested that the DARE program was not effective in reducing drug use in schools, but many school and police officials and parents and students insisted on maintaining the program without change for many years. Do you think that government agencies should be allowed to de-fund programs that numerous evaluation research studies have shown to be ineffective? Should government agencies be *required* to justify funding for social programs on the basis of evaluation research results?

SPSS EXERCISES

Do hospitals provide equal treatment across race and social class? Although the data available do not provide enough information for an in-depth evaluation of hospital procedures or quality of care, NCS93.POR enables you to explore the relationship between the total amount of medical expenses incurred as a result of injuries sustained from rape, robbery, or assault (MEDEXP) and the victim's family income (INCOME), race or ethnicity (PPRACE), and coverage by any medical insurance or other type of health benefits program (INSURANC).

1. Write at least three hypotheses about the relationship between medical expenses incurred from rape, robbery, or assault across race and social class based on the variables discussed above.

2. Obtain a frequency distribution for INCOME. Recode INCOME to measure victim's family income into three or four categories.

3. Select all cases in which the victim sustained injuries (YES = 1). Compare the mean amount of expenses incurred as a result of the victim's injuries across categories of each independent variable. Describe the results; be sure to note the number of victims for each category when interpreting your results.

4. Were your hypotheses supported? What conclusions do you make about these relationships? Can you determine whether hospitals provide equal treatment across race and social class based on these data? Explain.

C H A P T E R   1 2

Quantitative Data Analysis

"Oh no, not data analysis and statistics!" Perhaps most of you have been sailing along in this textbook understanding almost everything you have been reading and studying, if not immediately, then at least after a second reading or a classroom lecture. You probably feel somewhat confident that you can design your own criminological research project, develop a measuring instrument, and collect your own data. Now, however, you hit the chapter that you may have been fearing all along, the chapter on data analysis and the use of statistics. This chapter describes what you need to do after your data have been collected. You now need to analyze what you have found, interpret it, and decide how to present your data so that you can most clearly make the points you wish to make.

What you probably dread about this chapter is something that you either sense or know from a previous course: The studying of data analysis and statistics will lead you into that feared world of mathematics. You may not be too crazy about this if you lack confidence in your mathematical abilities. You may, therefore, be going into this chapter with a great deal of apprehension. We would like to state at the beginning, however, that you have relatively little to fear. The kind of mathematics required to perform the data analysis tasks in this chapter are minimal. If you can add, subtract, multiply, and divide, and are willing to put some effort into carefully reading the chapter, you will do well in the statistical analysis of your data. In fact, it is our position that the analysis of your data will require more in the way of careful and logical thought than in mathematical skill. One helpful way to think of statistics is that it consists of a set of tools that you will use to examine your data to help you answer the questions that motivated your research in the first place. Right now, the toolbox that holds your statistical tools is fairly empty (or completely empty). In the course of this chapter, we will add some fundamental tools to that toolbox. We would also like to note at the beginning that the kinds of statistics you will use on criminological data are very much the same as those used by economists, psychologists, political scientists, sociologists, and other social scientists. In other words, statistical tools are statistical tools, and all that changes is the nature of the problem to which those tools are applied.

This chapter will introduce several common statistics in social research and highlight the factors that must be considered in using and interpreting statistics. Think of it as a review of fundamental social statistics, if you have already studied them, or as an introductory overview, if you have not.

Two preliminary sections lay the foundation for studying statistics. In the first we will discuss the role of statistics in the research process, returning to themes and techniques you already know. In the second preliminary section, we will outline the process of acquiring data for statistical analysis. In the rest of the chapter, we will explain how to describe the distribution of single variables and the relationship between variables. Along the way, we will address ethical issues related to data analysis. This chapter will be successful if it encourages you to see statistics responsibly and evaluate them critically, and gives you the confidence necessary to seek opportunities for extending your statistical knowledge.

It should be noted that in this chapter we focus primarily on the use of statistics for descriptive purposes. Those of you seeking a more advanced discussion of statistical methods used in criminal justice and criminology should seek other textbooks (see, e.g., Bachman & Paternoster 2003).

INTRODUCING STATISTICS

Statistics play a key role in achieving valid research results, in terms of measurement, causal validity, and generalizability. Some statistics are useful primarily to describe the results of measuring single variables and to construct and evaluate multi-item scales. These statistics include frequency distributions, graphs, measures of central tendency and variation, and reliability tests. Other statistics are useful primarily in achieving causal validity, by helping us describe the association among variables and to control for, or otherwise take into account, other variables.

Crosstabulation is one technique for measuring association and controlling other variables and is introduced in this chapter. All these statistics are called **descriptive statistics** because they are used to describe the distribution of and relationship among variables.

You learned in Chapter 4 that is possible to estimate the degree of confidence that can be placed in generalizations for a sample and for the population from which the sample was selected. The statistics used in making these estimates are called inferential statistics, and they include confidence intervals, to which you were exposed in Chapter 4. In this chapter we will refer only briefly to inferential statistics, but we will emphasize later in the chapter their importance for testing hypotheses involving sample data.

Criminological theory and the results of prior research should guide our statistical plan or analytical strategy, as they guide the choice of other research methods. In other words, we want to use the statistical strategy that will best answer our research question. There are so many particular statistics and so many ways for them to be used in data analysis that even the best statistician can become lost in a sea of numbers if she is not using prior research and theorizing to develop a coherent analysis plan. It is also important for an analyst to choose statistics that are appropriate to the level of measurement of the variables to be analyzed. As you learned in Chapter 3, numbers used to represent the values of variables may not actually signify different quantities, meaning that many statistical techniques will be inapplicable. Some statistics, for example, will only be appropriate when the variable you are examining is measured at the nominal level. Other kinds of statistics will require interval-level measurement. To use the right statistic, then, you must be very familiar with the measurement properties of your variables (and you thought that stuff would go away!).

Case Study: The Causes of Delinquency

In this chapter we will use research on the causes of delinquency for our examples. More specifically, our data will be a subset of a much larger study of a sample of approximately 1,200 high school students selected from the metropolitan and suburban high schools of a city in South Carolina. These students, all of whom were in the tenth grade, completed a questionnaire that asked about such things as how they spent their spare time; how they got along with their parents, teachers, and friends; their attitudes about delinquency; whether their friends committed delinquent acts; and their own involvement in delinquency. The original research study was designed to test specific hypotheses about the factors that influence delinquency. It was predicted that delinquent behavior would be affected by such things as the level of supervision provided by parents, the students' own moral beliefs about delinquency, their involvement in conventional activities such as studying and watching TV, their fear of getting caught, their friends' involvement in crime, and whether these friends provided verbal support for delinquent acts. All these hypotheses were derived from extant criminological theory, theories we have referred to throughout this book. One specific hypothesis, derived from deterrence theory, predicts that youths who believe they are likely to get caught by the police for committing delinquent acts are less likely to commit delinquency than others. This hypothesis is shown in Exhibit 12.1. The variables from this study that we will use in our chapter examples are displayed in Exhibit 12.2.

| EXHIBIT 12.1 | Perceived Fear of Punishment and Delinquency |

Fear of getting caught by police → Delinquency

PREPARING DATA FOR ANALYSIS

Our analysis of the causes of delinquency in this chapter is an example of **secondary data analysis**. As we described in Chapter 10, it is secondary because we received the data secondhand; we did not design the data-collection instrument ourselves. Using secondary data in this way has a major disadvantage; if you did not design the study yourself, it is unlikely that all the variables you think should have been included actually were included, and that they were measured in the way that you prefer. In addition, the sample may not represent the population in which you are interested, and the study design may be only partially appropriate to your research question. For example, this particular study was completed before recent theoretical work in criminology on self-control. It also contains only self-reported delinquency and no measures as to whether these delinquent acts resulted in an arrest or juvenile court appearance. Furthermore, because it is a survey of individuals, the study has no measures of the kinds of communities within which the students lived; for example, the level of crime in the community or measures of informal community control. Because the survey sample is selected only from one city in South Carolina, we cannot make generalizations to youths in other cities or in other states.

Although it comes with many disadvantages, there is one important advantage of secondary data: they are readily available. It is the availability of secondary data that makes their use preferable for many purposes. A great many high-quality data sets (such as the data set on delinquency we are using) are available for reanalysis from the Inter-University Consortium for Political and Social Research (ICPSR) at the University of Michigan. Other data sets for secondary data analysis can be obtained from the government, individual researchers, and other research organizations. Most of these data sets are stored on computer disks or CD-ROMs, ready for use without any further effort. Therefore, for a great many research problems, a researcher should first check what data sets on the question already are available. An enormous savings in time and resources may be the result.

If you have conducted your own survey or experiment, your quantitative data must be prepared in a format suitable for computer entry. You learned in Chapter 7 that questionnaires and interview schedules can be precoded to facilitate data entry by representing each response with a unique number (see Exhibit 12.3). This method allows direct entry of the precoded responses into a computer file, after responses are checked to ensure that only one valid answer code has been circled (extra written answers can be assigned their own numerical codes). Most survey research organizations now use a database management program to control data entry. The program prompts the data entry clerk for each response, checks the response to ensure that it is a valid response for that variable, and then saves

EXHIBIT 12.2 List of Variables for Class Examples of Causes of Delinquency

Variable	SPSS Variable Name	Description
Gender	V1	Sex of respondent.
Age	V2	Age of respondent.
TV	V21	Number of hours per week the respondent watches TV.
Study	V22	Number of hours per week the respondent spends studying.
Supervision	V63	Do parents know where respondent is when he or she is away from home?
Friends think theft wrong	V77	How wrong respondents' best friends think it is to commit petty theft.
Friends think drinking wrong	V79	How wrong respondents' best friends think it is to drink liquor under age.
Punishment for drinking	V109	If respondent was caught drinking liquor under age and taken to court, how much of a problem it would be.
Cost of vandalism	V119	How much would respondent's chances of having good friends be hurt if he or she was arrested for petty theft.
Parental supervision	PARSUPER	Added scale from items that ask respondent if parents know where he or she is and who he or she is with when away from home. A high score means more supervision.
Friend's opinion	FROPINON	Added scale that asks respondent if his or her best friend thought that committing various delinquent acts was all right. A high score means more support for delinquency from friends.
Friend's behavior	FRBEHAVE	Added scale that asks respondent how many of his or her best friends commit delinquent acts.
Certainty of punishment	CERTAIN	Added scale that measures how likely respondent thinks it is that he or she will be caught by police if he or she were to commit delinquent acts. A high score means more fear of getting caught.
Morality	MORAL	Added scale that measures how morally wrong respondent thinks it is to commit diverse delinquent acts. A high score means strong moral inhibitions against committing delinquency.
Delinquency	DELINQ1	An additive scale that counts the number of times respondent admits to committing a number of different delinquent acts in the past year. The higher the score, the more delinquent acts committed.

EXHIBIT 12.3   Section of the National Violence Against Men and Women Survey on Stalking Victimizations Given to Female Respondents

SECTION H: STALKING VICTIMIZATION

H1. Now I'd like to ask you some questions about following or harassment you may have experienced on more than one occasion by strangers, friends, relatives, or even husbands and partners. Not including bill collectors, telephone solicitors or other sales people, has anyone, male or female, ever. . . . MARK ALL THAT APPLY

   01 Followed you or spied on you?
   02 Sent you unsolicited letters or written correspondence?
   03 Made unsolicited phone calls to you?
   04 Stood outside your home, school, or workplace?
   05 Showed up at places you were even though he or she had no business being there?
   06 Left unwanted items for you to find?
   07 Tried to communicate with you in other ways against your will?
   08 Vandalized your property or destroyed something you loved?
   09 (Volunteered) Don't Know   GO TO SECTION I
   10 (Volunteered) Refused   GO TO SECTION I
   11 (Volunteered) None   GO TO SECTION I

H2. If H1 = ANY OF 1–8 (RESPONDENT HAS BEEN STALKED) GO TO H3, ELSE GO TO SECTION I.

H3. Has anyone ever done any of these things to you on more than one occasion?

   1 Yes
   2 No   GO TO SECTION I
   3 (Volunteered) Don't Know   GO TO SECTION I
   4 (Volunteered) Refused   GO TO SECTION I

H4. How many Different people have ever done this to you on more than one occasion?

   ____ Number of people [RANGE IS 1–97]
   98 (Volunteered) Don't Know   GO TO SECTION I
   99 (Volunteered) Refused   GO TO SECTION I

H5. Was this person/these persons. . . . MARK ALL THAT APPLY

   01 Your current spouse?
   02 An ex-spouse?
   03 A male live-in partner?
   04 A female live-in partner?
   05 A relative?
   06 Someone else you know?
   07 A stranger?
   08 (Volunteered) Don't Know
   09 (Volunteered) Refused

Source: Tjaden & Thoennes 2000.

EXHIBIT 12.4   An Example of a Data Entry Screen from SPSS

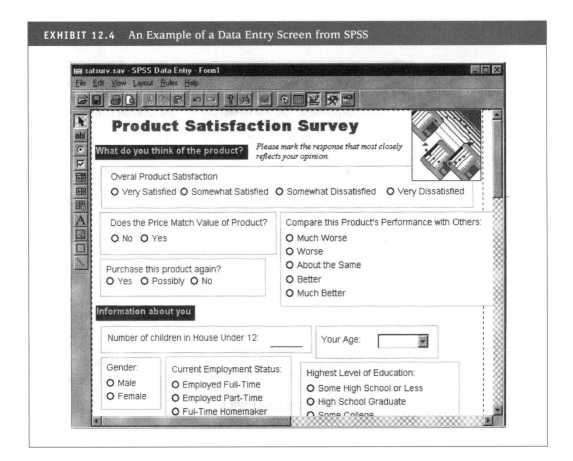

---

the response in the data file. Exhibit 12.4 is an example of a computer data entry screen. Not all studies have used precoded data entry, however, and individual researchers must enter the data themselves. This is an arduous and time-consuming task, but not for us if we use secondary data. After all, we get the data only after they have been coded and computerized.

Of course, numbers stored in a computer file are not yet numbers that can be analyzed with statistics. After the data are entered, they must be checked carefully for errors, a process called **data cleaning**. If a data entry program has been used and programmed to flag invalid values, the cleaning process is much easier. If data are read in from a text file, a computer program must be written that defines which variables are coded in which columns, attaches meaningful labels to the codes, and distinguishes values representing missing data. The procedures for doing so vary with each specific statistical package. We used the Windows version of the Statistical Package for the Social Sciences (SPSS) for the analysis in this chapter; you will find examples of SPSS commands required to define and analyze data on the Student Study Site for this text, http://www.sagepub.com/prccj3. More detailed information on using SPSS is contained in SPSS manuals and in Babbie, Halley, and Zaino (2000).

DISPLAYING UNIVARIATE DISTRIBUTIONS

The first step in data analysis is usually to display the variation in each variable of interest in what are called *univariate frequency distributions*. For many descriptive purposes, the analysis may go no further. Frequency distributions and graphs of frequency distributions are the two most popular approaches for displaying variation; both allow the analyst to display the distribution of cases across the value categories of a variable. Graphs have the advantage over numerically displayed frequency distributions because they provide a picture that is easier to comprehend. Frequency distributions are preferable when exact numbers of cases with particular values must be reported, and when many distributions must be displayed in a compact form.

No matter which type of display is used, the primary concern of the data analyst is to accurately display the distribution's shape; that is, to show how cases are distributed across the values of the variable. Three features of the shape of a distribution are important: **central tendency**, **variability**, and **skewness** (lack of symmetry). All three of these features can be represented in a graph or in frequency distribution.

Central tendency The most common value (for variables measured at the nominal level) or the value around which cases tend to center (for a quantitative variable).

Variability The extent to which cases are spread out through the distribution or clustered in just one location.

Skewness The extent to which cases are clustered more at one or the other end of the distribution of a quantitative variable rather than in a symmetric pattern around the center. Skew can be positive (a "right skew"), with the number of cases tapering off in the positive direction, or negative (a "left skew"), with the number of cases tapering off in the negative direction.

These features of a distribution's shape can be interpreted in several different ways, and they are not all appropriate for describing every variable. In fact, all three features of a distribution can be distorted if graphs, frequency distributions, or summary statistics are used inappropriately.

A variable's level of measurement is the most important determinant of the appropriateness of particular statistics. For example, we cannot talk about the skewness (lack of symmetry) of a qualitative variable (measured at the nominal level). If the values of a variable cannot be ordered from lowest to highest, if the ordering of the values is arbitrary, we cannot say whether the distribution is symmetric because we could just reorder the values to make the distribution more (or less) symmetric. Some measures of central tendency and variability are also inappropriate for qualitative variables.

The distinction between variables measured at the ordinal level and those measured at the interval or ratio level should also be considered when selecting statistics to use, but social researchers differ on just how much importance they attach to this distinction. Many social researchers think of ordinal variables as imperfectly measured interval-level variables, and believe that in most circumstances statistics developed for interval-level variables also provide useful summaries for ordinal variables. Other social researchers believe that variation in ordinal variables will often be distorted by statistics that assume an interval level of measurement. We will touch on some of the details of these issues in the following sections on particular statistical techniques.

We will now examine graphs and frequency distributions that illustrate these three features of shape. Summary statistics used to measure specific aspects of central tendency and variability will be presented in a separate section. There is a summary statistic for the measurement of skewness, but it is used only rarely in published research reports and will not be presented here.

Graphs

A picture often is worth some unmeasurable quantity of words. Even for the statistically uninitiated, graphs can be easy to read, and they very nicely highlight a distribution's shape. They are particularly useful for exploring data because they show the full range of variation and identify data anomalies that might be in need of further study. And good, professional-looking graphs can now be produced relatively easily with software available for personal computers. There are many types of graphs, but the most common and most useful are bar charts, histograms, and frequency polygons. Each has two axes, the vertical axis (y-axis) and the horizontal axis (x-axis), and labels to identify the variables and the values with tick marks showing where each indicated value falls along the axis. The vertical y-axis of a graph is usually in frequency or percent units, whereas the horizontal x-axis displays the values of the variable being graphed. There are different kinds of graphs you can use to descriptively display your data, depending upon the level of measurement of the variable.

A **bar chart** contains solid bars separated by spaces. It is a good tool for displaying the distribution of variables measured at the nominal level and other discrete categorical variables because there is, in effect, a gap between each of the categories. In our delinquency data, one of the questions asked of respondents was if their parents knew where they were when they were away from home. We graphed the responses to this question in a bar chart, which is shown in Exhibit 12.5. In this bar chart we report both the frequency count for each value and the percentage of the total that each value represents. The bar chart in Exhibit 12.5 indicates that very few of the respondents (only 16, or 1.3%) reported that their parents "never" knew where they were when they were not at home. Almost one-half (562, or 44.3%) of the youths reported that their parents "usually" knew where they were. What you can also see by noticing the height of the bars above "usually" and "always" is that most youths report that their parents provide very adequate supervision. You can also see that the most frequent response was "usually" and the least frequent was "never." Because the response "usually" is the most frequent value, it is called the mode or modal response. With ordinal data like these, the mode is the most appropriate measure of central tendency. (More about this later.)

Notice that the cases tend to cluster in the two values of "usually" and "always"; in fact, about 80% of all cases are found in those two categories. There is not much variability in this distribution, then. In Exhibit 12.6, we show another bar chart from the delinquency data. In this example, the variable is the attitude of one's friends toward drinking under age. Respondents were asked how wrong their best friends thought it was to drink liquor under

EXHIBIT 12.5   Bar Chart Showing Youths' Responses on Parents Knowing Where They Are

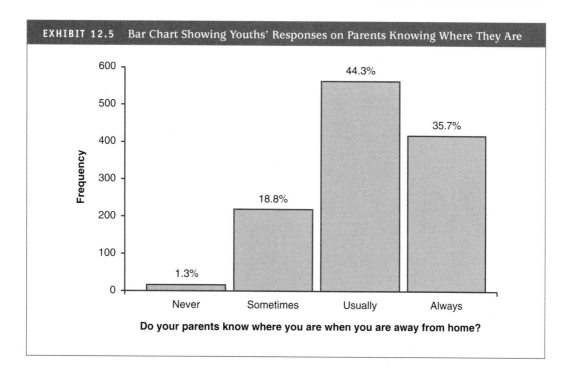

age. As you can see, unlike the case in Exhibit 12.5, the frequencies are more evenly spread out across the various values (the heights of the bars don't differ as much in this bar chart), so there is more variability in this variable than for the one shown in Exhibit 12.5.

A **histogram** is like a bar chart, but it has bars that are adjacent, or right next to each other, with no gaps. This is done to indicate that data displayed in a histogram, unlike the data in a bar chart, are quantitative variables that vary along a continuum (see the discussion of levels of measurement for variables in Chapter 3). Exhibit 12.7 shows a histogram from the delinquency data set we are using. The variable being graphed is the number of hours per week the respondent reported to be studying. Notice that the cases cluster at the low end of the values. In other words, there are a lot of youths who spend between 0 and 15 hours per week studying. After that, there are only a few cases at each different value with a "spike" occurring at 25, 30, 38, and 40 hours studied. This distribution is clearly not symmetric. In a symmetric distribution there is a lump of cases or a "spike" with an equal number of cases to the left and right of that spike. In the distribution shown in Exhibit 12.7, most of the cases are at the left end of the distribution (i.e., at low values) and the distribution trails off on the right side. The ends of a histogram like this are often called the tail of a distribution. In a symmetric distribution, the left and right tails are approximately the same length. As you can clearly see in Exhibit 12.7, however, the right tail is much longer than the left tail. When the tails of the distribution are uneven the distribution is said to be asymmetrical or skewed. A skew is either positive or negative. When the cases cluster to the left and the right tail of the distribution is longer than the left, as in Exhibit 12.7, our

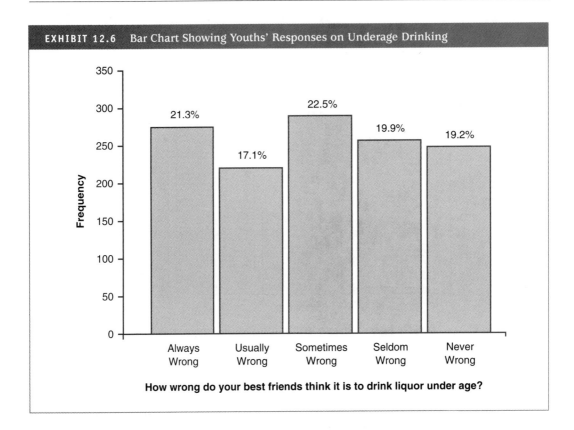

EXHIBIT 12.6 Bar Chart Showing Youths' Responses on Underage Drinking

variable distribution is **positively skewed**. When the cases cluster to the right side and the left tail of the distribution is long, our variable distribution is **negatively skewed**.

In a **frequency polygon**, a continuous line connects the points representing the number or percentage of cases with each value. The frequency polygon is an alternative to the histogram when the distribution of a quantitative, continuous variable must be displayed; this alternative is particularly useful when the variable has a wide range of values. We used the same number of hours studying data shown in Exhibit 12.7 to make the frequency polygon in Exhibit 12.8. Notice that the frequency polygon is constructed by making a point or dot to represent the frequency of each value and then drawing a line by connecting the dots. The frequency polygon gives us the same impression of the data as the histogram. These data cluster near the low values (representing fewer hours studied), and there are only a very few respondents who report studying more than 25 hours per week. Because the right tail is long, relative to the left tail of the distribution, we have a positive skew. We could have easily constructed both the histogram and the frequency polygon with percentages rather than raw frequencies.

If graphs are misused, they can distort, rather than display, the shape of a distribution. Compare, for example, the two graphs in Exhibit 12.9. The first graph shows that high school seniors reported relatively stable rates of lifetime use of cocaine between 1980 and 1985. The second graph, using exactly the same numbers, appeared in a 1986 *Newsweek*

EXHIBIT 12.7   Histogram

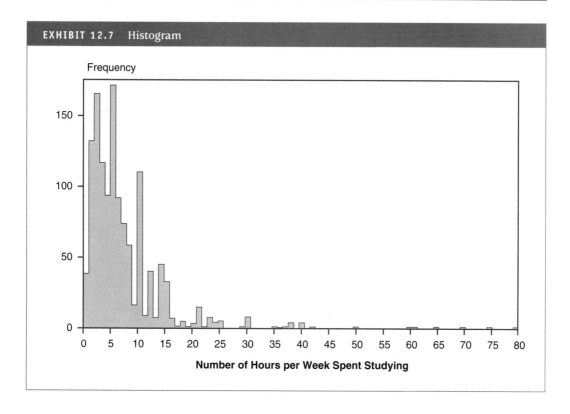

article on the coke plague (Orcutt & Turner 1993). To look at this graph, you would think that the rate of cocaine usage among high school seniors increased dramatically during this period. But, in fact, the difference between the two graphs is due simply to changes in how the graphs are drawn. In the "plague" graph (B), the percentage scale on the vertical axis begins at 15 rather than 0, making what was about a one-percentage point increase look very big indeed. In addition, omission from the plague graph of the more rapid increase in reported usage between 1975 and 1980 makes it look as if the tiny increase in 1985 were a new, and thus more newsworthy, crisis.

Adherence to several guidelines (Tufte 1983) will help you to spot these problems and to avoid them in your own work:

- The difference between bars can be exaggerated by cutting off the bottom of the vertical axis and displaying less than the full height of the bars. Instead, begin the graph of a quantitative variable at 0 on both axes. It may at times be reasonable to violate this guideline, as when an age distribution is presented for a sample of adults, but in this case be sure to mark the break clearly on the axis.
- Bars of unequal width, including pictures instead of bars, can make particular values look as if they carry more weight than their frequency warrants. Always use bars of equal width.

EXHIBIT 12.8 Frequency Polygon

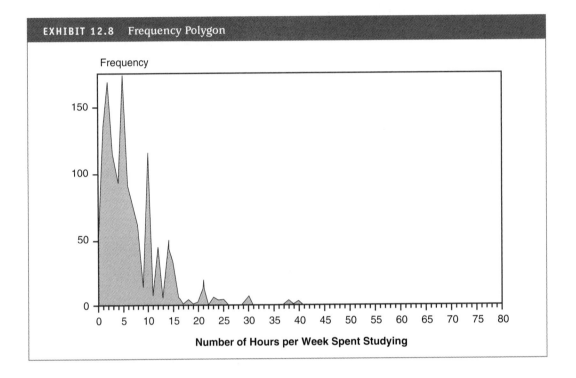

- Either shortening or lengthening the vertical axis will obscure or accentuate the differences in the number of cases between values. The two axes usually should be of approximately equal length.
- Avoid chart junk that can confuse the reader and obscure the distribution's shape (a lot of verbiage, numerous marks, lines, lots of crosshatching, etc.).

Frequency Distributions

A **frequency distribution** displays the number, the **percentage** (the relative frequencies), or both for cases corresponding to each of a variable's values or group of values. The components of the frequency distribution should be clearly labeled, with a title, a stub (labels for the values of the variable), a caption (identifying whether the distribution includes frequencies, percentages, or both), and perhaps the number of missing cases. If percentages are presented rather than frequencies (sometimes both are included), the total number of cases in the distribution (*the base N*) should be indicated (see Exhibit 12.10). Remember that a percentage is simply a relative frequency. A percentage shows the frequency of a given value relative to the total number of cases times 100.

Ungrouped Data

Constructing and reading frequency distributions for variables with few values is not difficult. In Exhibit 12.10, we created the frequency distribution from the variable "Punishment

EXHIBIT 12.9   Two Graphs of Cocaine Usage

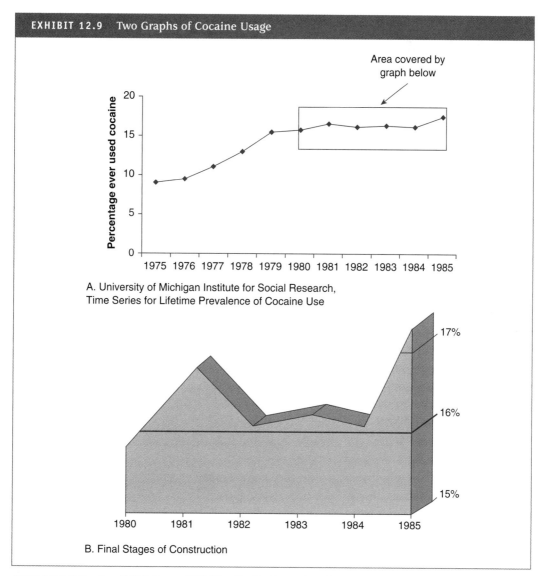

A. University of Michigan Institute for Social Research,
Time Series for Lifetime Prevalence of Cocaine Use

B. Final Stages of Construction

Source: James D. Orcutt and J. Blake Turner, 1993. "Shocking Numbers and Graphic Accounts." *Social Problems*, Vol. 49, No. 2, pp. 190–206. © 1993 by the Society for the Study of Social Problems. Reprinted by permission.

for Drinking" found in the delinquency data set (see Exhibit 12.2). This variable asked the youths to respond to the following question, "How much of a problem would it be if you went to court for drinking liquor under age?" The frequency distribution in Exhibit 12.10 shows the frequency for each value and its corresponding percentage.

As another example of calculating the frequencies and percentages, suppose we had a sample of 25 youths and asked them their gender. From this group of 25 youths, 13 were male and 12 were female. The frequency of males (symbolized here by f) would be 13 and the frequency of females would be 12. The percentage of males would be 52%, calculated by f / the total number of cases $\times$ 100 (13 / 25 $\times$ 100 = 52%). The percentage of females would be 12 / 25 $\times$ 100 = 48%.

In the frequency distribution shown in Exhibit 12.10, you can see that only a very small number (14 out of 1,272) of youths thought that they would experience "no problem" if they were caught and taken to court for drinking liquor under age. You can see that most, in fact 1,009 of these youths or 79.3% of them, thought that they would have either "a big problem" or "a very big problem" with this. If you compared Exhibit 12.10 to Exhibit 12.5, you can see that a frequency distribution (see Exhibit 12.10) can provide much of the same information as a graph about the number and percentage of cases in a variable's categories. Often, however, it is easier to see the shape of a distribution when it is graphed. When the goal of a presentation is to convey a general sense of a variable's distribution, particularly when the presentation is to an audience not trained in statistics, the advantages of a graph outweigh those of a frequency distribution.

Exhibit 12.10 is a frequency distribution of an ordinal-level variable; it has a very small number of discrete categories. In Exhibit 12.11, we provide an illustration of a frequency distribution with a continuous quantitative variable. This variable is one we have already looked at and graphed from the delinquency data, the number of hours per week the respondent spent studying. Notice that this variable, like many continuous variables in criminological research, has a large number of values. Although this is a reasonable frequency distribution to construct— you can, for example, still see that the cases tend to cluster in the low end of the distribution and are strung way out at the upper end—it is a little difficult to get a good sense of the distribution of the cases. The problem is that there are too many values to easily comprehend. It would be nice if we could simplify distributions like these that have a large number of different values. Well, we can. We can construct what is called a **grouped frequency distribution**.

EXHIBIT 12.10   Frequency Distribution

How much of a problem would it be if you went to court for drinking liquor under age?		
Value	*Frequency (f)*	*Percentage (%)*
No problem at all	14	1.1
Hardly any problem	53	4.2
A little problem	196	15.4
A big problem	421	33.1
A very big problem	588	46.2
Total	1,272	100.0

EXHIBIT 12.11   Frequency Distribution With Continuous Quantitative Data: Hours Studied per Week

Value	Frequency (f)	Percentage (%)
0	38	3.0
1	132	10.4
2	165	13.0
3	116	9.1
4	94	7.4
5	171	13.4
6	92	7.2
7	73	5.7
8	58	4.6
9	16	1.3
10	110	8.6
11	9	0.7
12	40	3.1
13	7	0.6
14	45	3.5
15	32	2.5
16	7	0.6
17	5	0.4
18	4	0.3
19	1	0.1
20	15	1.2
21	8	0.6
22	1	0.1
23	1	0.1
24	4	0.3
25	5	0.4
29	1	0.1
30	8	0.6
35	1	0.1
37	1	0.1
40	4	0.3
42	1	0.1
50	1	0.1
60	1	0.1
61	1	0.1
65	1	0.1
70	1	0.1
75	1	0.1
80	1	0.1
Total	1,272	100.0

Grouped Data

Many frequency distributions, such as those in Exhibit 12.11, and many graphs require grouping of some values after the data are collected. There are two reasons for grouping:

1. There are more than 15–20 values to begin with, a number too large to be displayed in an easily readable table.
2. The distribution of the variable will be clearer or more meaningful if some of the values are combined.

Inspection of Exhibit 12.11 should clarify these reasons. In this distribution it is very difficult to discern any shape, much less the central tendency. What we would like to now do to make the features of the data more visible is change the values into intervals of values, or a range of values. For example, rather than having five separate values of 0, 1, 2, 3, 4 hours studied per week, we can have a range of values or an interval for the first value, such as 0–4 hours studied. Then we can get a count or frequency of the number of cases (and percentage of the total) that fall within that interval.

Once we decide to group values, or categories, we have to be sure that in doing so we do not distort the distribution. Adhering to the following guidelines for combining values in a frequency distribution will prevent many problems:

- Categories should be logically defensible and preserve the distribution's shape.
- Categories should be mutually exclusive and exhaustive, so every case is classifiable in one and only one category.
- The first interval must contain the lowest value and the last interval must contain the highest value in the distribution.
- Each interval width, the number of values that fall within each interval, should be the same size.
- There should be between 7 and 13 intervals. This is a tough rule to follow. The key is not to have so few intervals that your data are clumped or clustered into only a few intervals (you will lose too much information about your distribution) and not to have so many intervals that the data are not much clearer than an ungrouped frequency distribution.

Let us use the data in Exhibit 12.11 on the number of hours studied by these youths to create a grouped frequency distribution. We will follow a number of explicit steps:

Step 1. Determine the number of intervals you think you want. This decision is arbitrary, but try to keep the number of intervals you have in the 7–13 range. For our example, let us say we initially decided we wanted to have 10 intervals. (Note, if you do your frequency distribution and it looks too clustered or there are too many intervals, redo your distribution with a different number of intervals.) Don't worry, there are no hard and fast rules for the correct number of intervals, and constructing a grouped frequency distribution is as much art as science. Just remember that the frequency distribution you make is supposed to convey information about the shape and central tendency of your data.

Step 2. Decide on the width of the interval (symbolized by w_i). The interval width is the number of different values that falls into your interval. For example, an interval width of 5 has five different values that fall into it, say, the values 0, 1, 2, 3, or 4 hours studied. There is a simple formula to approximate what your interval width should be given the number of intervals you decided on in the first step: determine the range of the data, where the range is simply the highest score in the distribution minus the lowest score. In our data, with the number of hours studied, the range is 80 because the high score is 80 and the low score is 0, so range = 80 – 0 = 80. Then determine the width of the interval by dividing the range by the number of intervals you want from Step 1. We wanted 10 intervals, so our interval width would be w_i = 80 / 10 = 8. We should therefore have an interval width of 8. If you use this simple formula for determining your interval width and you end up with a decimal, say 8.2 or 8.6, then simply round up or down to an integer.

Step 3. Make your first interval so that the lowest value falls into it. Our lowest value is 0 (for studied 0 hours per week), so our first interval begins with the value 0. Now, if the beginning of our first interval is 0 and we want an interval width of 8, is the last value of our interval 7 (with a first interval of 0–7 hours), or is the last value of our interval 8 (with a first interval of 0–8 hours)? One easy way to make a grouped frequency distribution is to do the following: Take the beginning value of your first interval (in our case, it is 0), and add the interval width to that value (8). This new value is the first value of your next interval. What we know, then, is that the first value of our first interval is 0, and the first value of our second interval is 8 (0–?, 8–?). This must mean that the last value to be included in our first interval is one less than 8, or 7. Our first interval, therefore, includes the ranges of values 0–7. If you count the number of different values in this interval you will find that it includes eight different values (0, 1, 2, 3, 4, 5, 6, 7). This is our interval width of 8.

Step 4. After your first interval is determined, the next intervals are easy. They must be the same width and not overlap (mutually exclusive). You must make enough intervals to include the last value in your variable distribution. The highest value in our data is 80 hours per week, so we construct the grouped frequency distribution as follows:

0–7
8–15
16–23
24–31
32–39
40–47
48–55
56–63
64–71
72–79
80–87

Notice that in order to include the highest value in our data (80 hours) we had to make 11 intervals instead of the 10 we originally decided upon in Step 1. No problem. Remember, the number of intervals is arbitrary and this is as much art as science.

Step 5. Count the number or frequency of cases that appear in each interval and their percentage of the total.

The completed grouped frequency distribution is shown in Exhibit 12.12. Notice that this grouped frequency distribution conveys the important features of the distribution of this data. Most of the data cluster at the low end of the number of hours studied. In fact, more than two-thirds of these youths studied fewer than 8 hours per week. Notice also that the frequency of cases thins out at each successive interval. In other words, there is a long right tail to this distribution indicating a positive skew because fewer youths studied a high number of hours. Notice also that the distribution was created in such a way that the interval widths are all the same and each case falls into one and only one interval (i.e., the intervals are exhaustive and mutually exclusive). We would have run into trouble if we had two intervals like 0–7 and 7–14 because we would not know where to place those youths who spent 7 hours a week studying. Should we put them in the first or second interval? If the intervals are mutually exclusive, as they are here, you will not run into these problems.

EXHIBIT 12.12   Example of a Grouped Frequency Distribution From Hours Studied

Value	Frequency (f)	Percentage (%)
0–7	881	69.26
8–15	317	24.92
16–23	42	3.30
24–31	18	1.42
32–39	2	0.16
40–47	5	0.39
48–55	1	0.08
56–63	2	0.16
64–71	2	0.16
72–79	1	0.08
80–87	1	0.08
Total	1,272	100.00

Note: Total may not equal 100.0% due to rounding error.

SUMMARIZING UNIVARIATE DISTRIBUTIONS

Summary statistics focus attention on particular aspects of a distribution and facilitate comparison among distributions. For example, suppose you wanted to report the rate of violent crimes for each city in the United States with over 100,000 in population. You could report each city's violent crime rate, but it is unlikely that two cities would have the same rate and you would have to report approximately 200 rates, one for each city. This would be a frequency distribution that many, if not most, people would find difficult to comprehend. One way to interpret your data for your audience would be to provide a summary measure that indicates what the average violent crime rate is in large U.S. cities. That is the purpose of the set of summary statistics called measures of central tendency. You would also want to provide another summary measure that shows the variability or heterogeneity in your data. In other words, a measure that shows how different the scores are from each other or from the central tendency. That is the purpose of the set of summary statistics called measures of *variation* or *dispersion*. We will discuss each type of measurement in turn.

Measures of Central Tendency

Central tendency is usually summarized with one of three statistics: the mode, the median, or the mean. For any particular application, one of these statistics may be preferable, but each has a role to play in data analysis. To choose an appropriate measure of central tendency, the analyst must consider a variable's level of measurement, the skewness of a quantitative variable's distribution, and the purpose for which the statistic is used. In addition, the analyst's personal experiences and preferences inevitably will play a role.

Mode

The mode is the most frequent value in a distribution. For example, refer to the data in Exhibit 12.12, which shows the grouped frequency distribution for the number of hours studied. The value with the greatest frequency in that data is the interval 0–7 hours; this is the mode of that distribution. Notice that the mode is the most frequently occurring value; it is not the frequency of that value. In other words, the mode in Exhibit 12.12 is 0–7 hours; the mode is not 881, which is the frequency of the modal category. To show how the mode can also be thought of as the value with the highest probability, refer again to Exhibit 12.12. Suppose you had this grouped frequency distribution, but knew nothing else about each of the 1,272 youths in the study. If you were to pick a case at random from the distribution of 1,272 youths, and were asked how many hours they studied per week, what would your best guess be? Well, since 881 of the 1,272 youths fall into the first interval of 0–7 hours studied, the probability that a randomly selected youth studied from 0 to 7 hours would be .696 (881 / 1,272). This is higher than the probability of any other interval. It is the interval with the highest probability because it is the interval with the greatest frequency or mode of the distribution. When a variable distribution has one case or interval that occurs more often than the others, it is called a **unimodal** distribution.

Sometimes a distribution has more than one mode because there are two values that have the highest frequency. This distribution would be called **bimodal**. Some distributions are trimodal in that there are three distinctively high frequency values. And some distributions

EXHIBIT 12.13   Frequency Distribution of Offense Convicted for 1,000 Offenders

Offense	Frequency (f)
Violent	125
Drug	210
Property	480
Public order	100
Other	85
Total	1,000

have no mode. Look back at the graph of the frequency distribution in Exhibit 12.6. There is no one value with a frequency much higher than another. In fact, if you calculate the probability of a randomly selected person falling into each of the different values, they would not be all that different. In other words, the outcomes are equiprobable, or there is no mode. In saying that there is no mode, though, you are communicating something very important about the data: that no case is more common than the others. Another potential problem with the mode is that it might happen to fall far from the main clustering of cases in a distribution. It would be misleading in this case, then, to say simply that the variable's central tendency was the same as the modal value.

Nevertheless, there are occasions when the mode is very appropriate. Most important, the mode is the only measure of central tendency that can be used to characterize the central tendency of variables measured at the nominal level. In Exhibit 12.13 we have the frequency distribution of the conviction offense for 1,000 offenders convicted in a criminal court. The central tendency of the distribution is property offense, because more of the 1,000 offenders were convicted of a property crime than any other crime. For the variable "type of offense convicted of," the most common value is property crime. The mode also is often referred to in descriptions of the shape of a distribution. The terms *unimodal* and *bimodal* appear frequently, as do descriptive statements such as "The typical (most probable) respondent was in her 30s." Of course, when the issue is determining the most probable value, the mode is the appropriate statistic.

Median

The **median** is the score in the middle of a rank-ordered distribution. It is, then, the score or point that divides the distribution in half (the 50th percentile). The median is inappropriate for variables measured at the nominal level because their values cannot be put in ranked order (remember, there is no "order" to nominal-level data), and so there is no meaningful middle position. To determine the median we simply need to do the following. First, rank-order the values from lowest to highest. If there is an odd number of scores then you can find the position of the median in the rank order of scores by using the following simple formula:

$$\frac{N+1}{2}$$

where N is equal to the total number of cases. If there is an even number of scores, then you can find the position of the median in the rank order of scores by finding the midpoint of the scores in the following two positions:

$$\text{Position 1} = \frac{N}{2}$$

$$\text{Position 2} = \frac{N+2}{2}$$

The midpoint of the two scores can be found by adding the scores in the two positions and dividing by 2. Please note that in both formulas you use the formula to find the position of the median; the formula does not give you the value of the median. Let us do an example.

In Exhibit 12.14, we first list a sample of nine U.S. cities and their rate of violent crime in 1994. In the bottom panel of Exhibit 12.14, we drop one of the cities (San Francisco) and have eight cities.

EXHIBIT 12.14   Rate of Violent Crime for Selected U.S. Cities

City	Number of Violent Crimes per 100,000
Atlanta	3,571
Boston	1,916
Cleveland	1,530
Dallas	1,589
Los Angeles	2,059
New Orleans	1,887
New York	1,861
Philadelphia	1,322
San Francisco	1,461
Atlanta	3,571
Boston	1,916
Cleveland	1,530
Dallas	1,589
Los Angeles	2,059
New Orleans	1,887
New York	1,861
Philadelphia	1,322

Let us determine the median for both lists. First, we rank-order the violent crime rates for the first list:

Rank	Crime Rate
1	1,322
2	1,461
3	1,530
4	1,589
5	1,861
6	1,887
7	1,916
8	2,059
9	3,571

To test your understanding of measures of central tendency, go to the Quantitative Data Analysis Interactive Exercises on the Student Study Site.

Because we have an odd number of scores ($N = 9$), the position of the median is determined by $(9 + 1)/2 = 10/2 = 5$. The median violent crime rate, then, is in the fifth position in this rank order. Starting either at the top of the scores and counting down to the fifth position or the bottom and counting up, we find that in the fifth position is the score 1,861 violent crimes per 100,000, which is the median violent crime rate for these nine U.S. cities.

Now, let us find the median in the second list. We first rank-order these eight scores:

Rank	Crime Rate
1	1,322
2	1,530
3	1,589
4	1,861
5	1,887
6	1,916
7	2,059
8	3,571

Because we now have an even number of scores (notice that the violent crime rate for San Francisco is missing from this list), we know that the median position is the midpoint in between the fourth ($8/2 = 4$) and fifth ($[8 + 2]/2$) positions. The score at the fourth position is 1,861 and at the fifth is 1,887. The value of the median can now be found by adding these two scores and dividing by two. The median rate of violent crime is equal to $(1,861 + 1,887)/2 = 1,874$ violent crimes per 100,000 population.

The median is the score at the 50th percentile, so in a frequency distribution it is determined by identifying the value corresponding to a cumulative percentage of 50. We show you how do to this in Exhibit 12.15. These data are a repeat of the data in Exhibit 12.11 and show the number of hours studied for the youths in the delinquency data set. To find the 50th percentile, we simply added a new column to these data, labeled cumulative percentage.

EXHIBIT 12.15 Frequency Distribution With Continuous Quantitative Data: Hours Studied per Week

Value	Frequency (f)	Percentage (%)	Cumulative Percentage
0	38	3.0	3.0
1	132	10.4	13.4
2	165	13.0	26.4
3	116	9.1	35.5
4	94	7.4	42.9
5	141	13.4	56.3 Ø 50th percentile
6	92	7.2	
7	73	5.7	
8	58	4.6	
9	16	1.3	
10	110	8.6	
11	9	0.7	
12	40	3.1	
13	7	0.6	
14	45	3.5	
15	32	2.5	
16	7	0.6	
17	5	0.4	
18	4	0.3	
19	1	0.1	
20	15	1.2	
21	8	0.6	
22	1	0.1	
23	1	0.1	
24	4	0.3	
25	5	0.4	
29	1	0.1	
30	8	0.6	
35	1	0.1	
37	1	0.1	
40	4	0.3	
42	1	0.1	
50	1	0.1	
60	1	0.1	
61	1	0.1	
65	1	0.1	
70	1	0.1	
75	1	0.1	
80	1	0.1	
Total	1,272	100.0	

Cumulative percentages are found by taking the percentage of the interval percentage plus all others below it. So the first value (3.0%) would be entered as the first cumulative percentage because there are no other intervals below the first. This cumulative percentage simply means that 3% of the youths studied for 0 hours per week. Then we add the percentage in the next value (10.4%) to this to arrive at a cumulative percentage of 13.4%. This means that 13.4% of the youths studied for 1 hour per week or less. This becomes the second entry in the cumulative percentage column. We continue adding each adjacent percentage value until we reach 50%. There is a cumulative percentage of 56.3% at the value of 5 hours per week. The median number of hours studied per week, then, is 5 hours. Of the respondents, 50% studied less than 5 hours per week, and 50% studied more than 5 hours per week.

Mean

The **mean** is simply the arithmetic average of all scores in a distribution. It is computed by adding up the value of all the cases and dividing by the total number of cases, thereby taking into account the value of each case in the distribution:

Mean = Sum of value of all cases / number of cases

The symbol for the mean is $\bar{X}$ (pronounced "x-bar"). In algebraic notation, the equation is,

$$\bar{X} = \frac{\sum_1^N x_i}{N}$$

where x_i is a symbol for each ith score and i's go from 1 to N; N is the total number of cases. What the algebraic equation says to do is to sum all scores, starting at the first score and continuing until the last, or Nth, score; then divide this sum by the total number of cases (N). We will calculate the mean rate of violent crime for the nine U.S. cities listed in Exhibit 12.14:

$$\bar{X} = \frac{(3,571 + 1,916 + 1,530 + 1,589 + 2,059 + 1,887 + 1,861 + 1,332 + 1,461)}{9}$$

$$= 1,910.7$$

The mean rate of violent crime for these nine U.S. cities, then, is 1,910.7 violent crimes per 100,000 population. Notice that in calculating the mean we do not have to first rank order the scores. The mean takes every score into account, so it does not matter if we add 3,571 first, in the middle, or last.

Computing the mean requires adding up the values of the cases, so it makes sense to compute a mean only if the values of the cases can be treated as actual quantities, that is, if they reflect an interval or ratio level of measurement, or if they are ordinal and we assume that ordinal measures can be treated as interval. It would make no sense, however, to calculate the mean for the variable racial or ethnic status. Imagine a group of four people in which there were two Caucasians, one African American, and one Hispanic. To calculate the mean you would need to solve the equation (Caucasian + Caucasian + African American + Hispanic) / 4 = ? Even if you decide that Caucasian = 1, African American = 2, and Hispanic = 3 for data entry purposes, it still does not make sense to add these numbers because they do not represent real numerical quantities. In other words, just because you code Caucasian as "1" and African American as "2," that does not mean that African Americans possess twice the race or ethnicity as Caucasians. To see how numerically silly this is, note that we would just as easily have coded African Americans as "1" and Caucasians as "2." Now, with one arbitrary flip of our coding scheme Caucasians have twice the race or ethnicity as African Americans. Thus, both the median and the mean are inappropriate measures of central tendency for variables measured at the nominal level.

Median or Mean?

Both the median and the mean are used to summarize the central tendency of quantitative variables, but their suitability for a particular application must be carefully assessed.

The key issues to be considered in this assessment are the variable's level of measurement, the shape of its distribution, and the purpose of the statistical summary. Consideration of these issues will sometimes result in a decision to use both the median and the mean, and will sometimes result in neither measure being seen as preferable. But in many other situations, the choice between the mean and median will be clear-cut as soon as the researcher takes the time to consider these three issues.

Level of measurement is a key concern because to calculate the mean, we must add up the values of all the cases, a procedure that assumes the variable is measured at the interval or ratio level. So even though we know that coding Agree as 2 and Disagree as 3 does not really mean that Disagree is one unit more of disagreement than Agree, the mean assumes this evaluation to be true. Calculation of the median requires only that we order the values of cases, so we do not have to make this assumption. Technically speaking, then, the mean is simply an inappropriate statistic for variables measured at the ordinal level (and you already know that it is completely meaningless for nominal variables). In practice, however, many social researchers use the mean to describe the central tendency of variables measured at the ordinal level, for the reasons outlined earlier.

The shape of a variable's distribution should also be taken into account when deciding whether to use the median or the mean. When a distribution is perfectly symmetric (i.e., when the distribution is bell shaped), the distribution of values below the median is a mirror image of the distribution of values above the median, and the mean and median will be the same. But the values of the mean and median are affected differently by skewness, or the presence of cases with extreme values on one side of the distribution but not the other side. The median takes into account only the number of cases above and below the median point, not the value of these cases, so it is not affected in any way by extreme values. The mean is based on adding the value of all the cases, so it will be pulled in the direction of exceptionally high (or low) values. When the value of the mean is larger than the median, we know that the distribution is skewed in a positive direction, with proportionately more cases with lower than higher values. When the mean is smaller than the median, the distribution is skewed in a negative direction.

This differential impact of skewness on the median and mean is illustrated in Exhibit 12.16. We show in the first list the crime rate per 1,000 residents for seven neighborhoods in a hypothetical city. The crime rates range from a low of 51.7 per 1,000 to 135.3

EXHIBIT 12.16 Rates of Crime in Selected Neighborhoods

Community	Crime Rate per 1,000
Valley View	51.7
Glen Commons	63.1
Springdale	75.3
Green Acres	98.7
Oceanside	113.4
Meadowbrook	125.9
King's Crossing	135.3
Valley View	51.7
Glen Commons	63.1
Springdale	75.3
Green Acres	98.7
Oceanside	113.4
Meadowbrook	125.9
West Chester	697.5
Pleasantville	5.3
Glen Commons	63.1
Springdale	75.3
Green Acres	98.7
Oceanside	113.4
Meadowbrook	125.9
King's Crossing	135.3

per 1,000. When we calculate the mean crime rate for these seven neighborhoods we find that it is 94.8 crimes per 1,000; the median is 98.7 crimes per 1,000. Notice that this is a relatively symmetrical distribution, with the values of the mean and the median very comparable to each other. With these data both the mean and the median provide an accurate description of the central tendency

In the second list of neighborhoods, we replace the neighborhood that had the highest crime rate (i.e., King's Crossing) with West Chester, which has a substantially higher crime rate: 697.5 per 1,000. It is, then, an unusually high score. When we calculate the mean for these data, we find that it is 175.1 per 1,000; it has increased from its previous value of 94.8. The median, however, remains the same; it is still 98.7 crimes per 1,000. When an unusually high score is added to a distribution, then, the mean gets inflated relative to the median.

test your
derstanding
measures
central
dency,
to the
antitative
ta Analysis
eractive
rcises on
 Student
dy Site.

In distributions that are positively skewed, the mean is of greater magnitude than the median. Notice also that with this unusually high crime rate added, the median is still an accurate picture of the central tendency, whereas the mean of 175.1 is no longer accurate.

In the third list of neighborhoods, we replace the neighborhood that has the lowest crime rate (i.e., Valley View) with Pleasantville, which has a substantially lower crime rate, 5.3 per 1,000. It is, then, an unusually low score. When we calculate the mean for these data, we find that it is 88.1 per 1,000; it has decreased from the value in the first list, which was 94.8. The median, however, remains the same; it is still 98.7 crimes per 1,000. When an unusually low score is added to a distribution, then, the mean gets deflated relative to the median. In distributions that are negatively skewed, the mean is less than the median. Notice also that even with this unusually low crime rate added, the median is still an accurate picture of the central tendency, whereas the mean of 88.1 is no longer as accurate.

This example illustrates one of the problems with using the mean as a measure of central tendency. If the data are skewed (either positively or negatively), then the mean will be affected by the extreme scores in the distribution and will not give as accurate an estimate of the central tendency of the data as the median. The lesson is that one should probably report both the mean and the median. In general, the mean is the most commonly used measure of central tendency for quantitative variables, because it takes into account the value of all cases in the distribution and it is the foundation for many other more advanced statistics. However, the mean's very popularity results in its use in situations for which it is inappropriate. Keep an eye out for this problem.

Measures of Variation

You have learned that central tendency is only one aspect of the shape of a distribution. Although the measure of center is the most important aspect for many purposes, it is still just a piece of the total picture. A summary of distributions based only on their central tendency can be very incomplete, even misleading. For example, three towns might have the same mean and median crime rate but still be very different in their social character due to the shape of the crime distributions. We show three distributions of community crime rates for three different towns in Exhibit 12.17. If you calculate the mean and median crime rate for each town, you will find that the mean and median crime rate is the same for all three. In terms of its crime rate, then, each community has the same central tendency. As you can see, however, there is something very different about these towns. Town A is a very heterogeneous town; crime rates in its neighborhoods are neither very homogeneous nor clustered at either the low or high end. Rather, the crime rates in its communities are spread out from one another. Crime rates in these neighborhoods are, then, very diverse. Town B is characterized by neighborhoods with very homogeneous crime rates; there are no real high or low crime areas because each neighborhood is not far from the overall mean of 62.4 crimes per 1,000. Town C is characterized by neighborhoods with either very low crime rates or very high crime rates. Crime rates in the first four neighborhoods are much lower than the mean (62.4 crimes per 1,000), whereas the last four neighborhoods are much higher than the mean. Although they share identical measures of central tendency, these three towns have neighborhood crime rates that are very different.

EXHIBIT 12.17   Neighborhood Crime Rates in Three Different Towns

Town A	Town B	Town C
19.5	58.1	8.9
28.2	59.7	15.4
35.7	60.1	18.3
41.9	62.7	21.9
63.2	63.2	63.2
75.8	63.9	103.5
92.0	64.2	104.2
95.7	64.5	110.7
109.4	65.0	105.3

The way to capture these differences is with statistical measures of variation. Four popular measures of variation are the range, the interquartile range, the variance, and the standard deviation (which is the most popular measure of variability). To calculate each of these measures, the variable must be at the interval or ratio level. Statistical measures of variation are used infrequently with qualitative variables, so these measures will not be presented here.

Range

The **range** is a simple measure of variation, calculated as the highest value in a distribution minus the lowest value:

$$Range = Highest\ value - Lowest\ value$$

It often is important to report the range of a distribution, to identify the whole range of possible values that might be encountered. However, because the range can be drastically altered by just one exceptionally high or low value (called an **outlier**), it does not do an adequate job of summarizing the extent of variability in a distribution. For our three towns in Exhibit 12.17, the range in crimes rates for Town A is 89.9 (109.4 – 19.5), for Town B it is 6.9 (65.0 – 58.7), and for Town C it is 106.4 (115.3 – 8.9).

Interquartile Range

A version of the range statistics, the **interquartile range**, avoids the problem created by unusually high or low scores in a distribution. It is the difference between the scores at the first quartile and the third quartile. **Quartiles** are the points in a distribution corresponding to the first 25% of the cases (the first quartile), the first 50% of the cases (the second quartile), and the first 75% of the cases (the third quartile). You already know how to determine the second quartile, corresponding to the point in the distribution covering half of the cases; it is another name for the median. The first and third quartiles are determined in the same way, but by finding the points corresponding to 25% and 75% of the cases, respectively.

Variance

If the mean is a good measure of central tendency, then it would seem that a good measure of variability would be the distance each score is away from the mean. Unfortunately, we cannot simply take the average distance of each score from the mean. One property of the mean is that it exactly balances negative and positive distances from it, so if we were to sum the difference between each score in a distribution and the mean of that distribution, it would always sum to zero. What we can do, though, is to square the difference of each score from the mean so the distance retains its value. This is the notion behind the variance as a measure of variability.

The **variance** is the average square deviation of each case from the mean, so it takes into account the amount by which each case differs from the mean. The equation to calculate the variance is:

$$\sigma^2 = \frac{\Sigma_1^N (x - \bar{X})^2}{N}$$

In words, this formula says to take each score and subtract the mean, then square this difference, then sum all these differences, then divide this sum by N or the total number of scores. We will calculate the variance for the crime rate data from Town A in Exhibit 12.17.

X	$(x - \bar{X})$	$(x - \bar{X})^2$
19.5	$(19.5 - 62.4 = -42.9$	1,840.41
28.2	$(28.2 - 62.4) = -34.2$	1,169.64
35.7	$(35.7 - 62.4) = -26.7$	712.89
41.9	$(41.9 - 62.4) = -20.5$	420.25
63.2	$(63.2 - 62.4) = -0.8$	0.64
75.8	$(75.8 - 62.4) = 13.4$	179.56
92.0	$(92.0 - 62.4) = 29.6$	876.16
95.7	$(95.7 - 62.4) = 33.3$	1,108.89
109.4	$(109.4 - 62.4) = 47.0$	2,209.00
	$\Sigma(x - \bar{X})$	8.517.44

We can now determine that the variance is:

$$\sigma^2 = \frac{8,517.44}{9} = 946.38$$

The variance of these data, then, is 946.38. The variance is used in many other statistics, although it is more conventional to measure variability with the closely related standard deviation than with the variance.

Standard Deviation

The **standard deviation** is simply the square root of the variance. It is the square root of the average squared deviation of each case from the mean:

$$\sigma = \sqrt{\frac{\Sigma_1^N (x - \bar{X})^2}{N}}$$

To find the standard deviation, then, simply calculate the variance and take the square root. For our example, the standard deviation is:

$$\sigma = \sqrt{946.38} = 30.76$$

When the standard deviation is calculated from sample data, the denominator is supposed to be $N - 1$, rather than N, an adjustment that has no discernible effect when the number of cases is reasonably large.

The standard deviation has mathematical properties that make it the preferred measure of variability in many cases. In particular, the calculation of confidence intervals around sample statistics, which you learned about in Chapter 4, relies on an interesting property of normal curves. Areas under the normal curve correspond to particular distances from the mean, expressed in standard deviation units. If a variable is normally distributed, 68% of the cases will lie between plus and minus 1 standard deviation from the distribution's mean, and 95% of the cases will lie between 1.96 standard deviations above and below the mean. Cases that fall beyond plus or minus 1.96 standard deviations from the mean are termed outliers. Because of this property, the standard deviation tells us quite a bit about a distribution, if the distribution is normal. This same property of the standard deviation enables us to infer how confident we can be that the mean (or some other statistic) of a population sampled randomly is within a certain range of the sample mean (see Chapter 4).

CROSSTABULATING VARIABLES

Most data analyses focus on relationships among variables to test hypotheses or just to describe or explore relationships. For each of these purposes, we must examine the association among two or more variables. **Crosstabulation (crosstab)** is one of the simplest methods for doing so. A crosstabulation displays the distribution of one variable for each category of another variable; it can also be called a *bivariate distribution*. Crosstabs also provide a simple tool for statistically controlling one or more variables while examining the associations among others. In this section you will learn how crosstabs used in this way can help to test for spurious relationships and evaluate causal models. Crosstabulations are usually used when both variables are measured at either the nominal or ordinal level. That is, when the values of both variables are categories.

We are going to provide a series of examples of crosstabulations from our delinquency data. In our first example, the independent variable we are interested in is the youth's gender (VI, see Exhibit 12.2), and the dependent variable is the youth's self-reported involvement in delinquent behavior (DELINQ1). To use the delinquency variable in a crosstabulation, however, we first need to recode it into a categorical variable. We will make three approximately equal categories of self-reported delinquency: low, medium, and high. Using the SPSS recode command, we will create another variable called DELINQ2, which was made from the following recode commands:

$$(0 - 2 = 1)$$
$$(3 - 13 = 2)$$
$$(14 - 118 = 3)$$

Anyone who reported from 0 to 2 delinquent acts is now coded as 1, or low delinquency; anyone reporting from 3 to 13 delinquent acts is now coded as 2, or medium delinquency; and anyone reporting more than 14 delinquent acts is now coded as 3, or high delinquency. If you were to do a frequency distribution of this new variable, DELINQ2, you would see that there are three approximately equal groups.

We are interested in the relationship between gender and delinquency because a great deal of delinquency theory would predict that males are more likely to be delinquent than females. The gender of the youth is the independent variable, and the level of self-reported delinquency is the dependent variable.

Exhibit 12.18 shows the crosstabulation of gender with DELINQ2. Some explanation of this table is in order. Notice that there are two values of gender (male and female) that comprise the two rows of the table, and three values of delinquency (low, medium, and high) that comprise the three columns of the table. Crosstabulations are usually referred to by the number of rows and columns the table has. Our crosstabulation in Exhibit 12.18 is a 2 × 3 (pronounced "two-by-three") table because there are two rows and three columns. Notice also that there are values at the end of each row and at the end of each column. These totals are referred to as the *marginals* of the table. These **marginal distributions** provide the sum of the frequencies for each column and each row of the table. For example, there are 680 females in the data and 592 males. These row marginals should sum to the total number of youths in the data set: 1,272. There are 450 youths who are low in delinquency, 348 youths who are

EXHIBIT 12.18   Crosstabulation of Respondents' Gender by Delinquency

		SELF-REPORTED DELINQUENCY			
		Low	*Medium*	*High*	*Total*
GENDER	*Female*	275 40.4%	182 26.8%	223 32.8%	680 100%
	Male	175 29.6%	166 28.0%	251 42.4%	592 100%
	Total	450	348	474	1,272

medium in delinquency, and 251 youths who are high on the delinquency variable. These column marginals should also sum to the total number of youths in the data set: 1,272.

Now notice that there are 2 × 3 or 6 data entries in the table (let us ignore the percentages for now). These data entries are called the *cells* of the crosstabulation and represent the joint distribution of the two variables: gender and delinquency. The table in Exhibit 12.18 has six cells for the joint distribution of two levels of gender with three levels of delinquency. In other words, notice where the value for female converges with the value of low for delinquency. You see a frequency number of 275 in this cell. This frequency is how many times there is the joint occurrence of a female and low delinquency; it shows that 275 females were also low in delinquency. Moving to the cell to the right of this, we see that there are 182 females who were medium in delinquency, and moving to the right again we see that there are 223 females who are high in delinquency. The sum of these three numbers is equal to the total number of females, 680. The row for the males shows the joint distribution of males with each level of delinquency.

What we would like to know is whether there is a relationship or association between gender and delinquency. In other words are males more likely to be delinquent than females? Because raw frequencies can provide a deceptive picture, we determine whether there is any relationship between our independent and dependent variables by looking at the percentages. Keep in mind that the idea in looking at relationships is that we want to know if variation on the independent variable has any effect on the dependent variable. To determine this, what we always do in crosstabulation tables is to calculate our percentages on each value of the independent variable. For example, notice that in Exhibit 12.18, gender is our independent variable. We calculated our percentages so that for each value of gender the percentages sum to 100% at the end of each row. The percentages for both females and males, therefore, sum to 100% at the end of the row. Now we take a given category of the dependent variable and ask what percentage of each independent variable value falls into that category of the dependent variable. Another way to say this is that we calculate our percentages on the independent variable and compare them on the dependent variable. We compare the percentages for different levels of the independent variable on the same category or level of the dependent variable.

In Exhibit 12.18, for example, notice that 40.4% of the female youths were low in delinquency, but only 29.6% of the males were low. This tells us that females are more likely to be low in delinquency than males. Now let us look at the high category. We can see that 32.8% of the females were high in delinquency and 42.4% of the males were high. Together, this tells us that females are more likely to be low in delinquency and males are more likely to be high in delinquency. There is, then, a relationship between gender and delinquency. Also notice that the independent variable was the row variable, and the dependent variable was the column variable. It does not always have to be this way; the independent variable could just as easily have been the column variable. The important general rule to remember is to always calculate your percentages on the levels of the independent variable (e.g., use marginal totals for the independent variable as denominators) and compare percentages on a level of the dependent variable.

In Exhibit 12.19 we report the same data as in Exhibit 12.18, this time switching the rows and the columns. Now, the independent variable (gender) is the column variable, so we calculate our percentage going down each of the two columns. We then compare percentages

EXHIBIT 12.19   Crosstabulation of Respondents' Delinquency by Gender

		GENDER		
		Female	Male	Total
	Low	275 40.4%	175 29.6%	450
SELF-REPORTED DELINQUENCY	Medium	182 26.8%	166 28.0%	348
	High	223 32.8%	251 42.4%	474
	Total	680 100%	592 100%	1,272

across rows. For example, we still see that 40.4% of the females were low in delinquency, whereas only 29.6% of the males were. And 42.4% of the males were high in delinquency, but only 32.8% of the females were high in delinquency.

Describing Association

A crosstabulation table reveals four aspects of the association between two variables:

- *Existence.* Do the percentage distributions vary at all between categories of the independent variable?
- *Strength.* How much do the percentage distributions vary between categories of the independent variable?
- *Direction.* For quantitative variables, do values on the dependent variable tend to increase or decrease with an increase in value of the independent variable?
- *Pattern.* For quantitative variables, are changes in the percentage distribution of the dependent variable fairly regular (simply increasing or decreasing), or do they vary (perhaps increasing, then decreasing, or perhaps gradually increasing, then rapidly increasing)?

In Exhibit 12.18, an association exists, although we can only say that it is a modest association. The percentage difference at the low and high ends of the delinquency variables is approximately 10 percentage points.

We provide another example of a crosstabulation in Exhibit 12.20. This is a 3 × 3 table that shows the relationship between how morally wrong a youth thinks delinquency is (the independent variable) and his or her self-reported involvement in delinquency (the dependent variable). This table reveals a very strong relationship between moral beliefs and delinquency. We can see that 5.6% of youths with weak moral beliefs are low on

EXHIBIT 12.20 Crosstabulation of Respondents' Morals by Delinquency

		SELF-REPORTED DELINQUENCY			
		Low	Medium	High	Total
MORALS	Weak	20 5.6%	79 22.3%	256 72.1%	355 100%
	Medium	170 33.8%	185 36.8%	148 29.4%	503 100%
	Strong	260 62.8%	84 20.3%	70 16.9%	414 100%
	Total	450	348	474	1,272

delinquency; this increases to 33.8% for those with medium beliefs and to 62.8% for those with strong moral beliefs. At the high end, over two-thirds (72.1%) of those youths with weak moral beliefs are high in delinquency, 29.4% of those with medium moral beliefs are high in delinquency, and only 16.9% of those youths with strong moral beliefs are high in delinquency. Clearly, then, having strong moral beliefs serves to effectively inhibit involvement in delinquent behavior. This is exactly what control theory would have us believe.

Exhibit 12.20 is an example of a negative relationship between an independent and dependent variable. As the independent variable increases (i.e., as one goes from weak to strong moral beliefs), the likelihood of delinquency decreases (one becomes less likely to commit delinquency). The independent and dependent variables move in opposite directions, so this is a negative relationship. The pattern in this table is close to what is called monotonic. In a **monotonic** relationship, the value of cases consistently increases (or decreases) on one variable as the value of cases increases (or decreases) on the other variable. Monotonic is often defined a bit less strictly, with the idea that as the value of cases on one variable increases (or decreases), the value of cases on the other variable tends to increase (or decrease), and at least does not change direction. This describes the relationship between moral beliefs and delinquency. Delinquency is most likely when moral beliefs are low, less likely when moral beliefs are medium, and least likely when moral beliefs are strong.

We present another crosstabulation table for you in Exhibit 12.21. This table shows the relationship between the variable "number of hours studied" and the variable "certainty of punishment" (see Exhibit 12.2). Both variables were originally continuous variables that we recoded into three approximately equal groups for this example. We hypothesize that those youths who study more will have a greater perceived risk of punishment than those who study less, so hours studied is our independent variable and certainty is the dependent variable. Comparing levels of hours studied for those with high certainty, we see that there is not much variation. Of those who did not study very much (0–3 hours), 39.2% were high in perceived certainty. Of those who studied from 4–6 hours, 35.6% were high in perceived certainty and

EXHIBIT 12.21 Crosstabulation of Respondents' Morals by Delinquency

		CERTAINTY OF PUNISHMENT			
		Low	Medium	High	Total
NUMBER OF HOURS STUDIED	0–3 Hours	126 27.9%	148 32.8%	177 39.2%	451 100%
	4–6 Hours	117 32.8%	113 31.7%	127 35.6%	357 100%
	7+ Hours	129 27.8%	148 31.9%	187 40.39%	464 100%
	Total	372	409	491	1,272

EXHIBIT 12.22 Example of an Intervening Variable

Gender	→	Parental supervision	→	Delinquency

40.3% of those who studied more than 7 hours per week were high in perceived certainty. Much the same levels prevail at low levels of perceived certainty. Those who do not study very much are no more or less likely to perceive a low certainty of punishment than those who study a lot. Variation in the independent variable, then, is not related to variation in the dependent variable. It looks like there is no association between the number of hours a youth studies and the extent to which he or she thinks punishment for delinquent acts is certain.

You will find when you read research reports and journal articles that social scientists usually make decisions about the existence and strength of association on the basis of more statistics than just percentage differences in a crosstabulation table. A **measure of association** is a type of descriptive statistic used to summarize the strength of an association. There are many measures of association, some of which are appropriate for variables measured at particular levels. One popular measure of association in crosstabular analyses with variables measured at the ordinal level is **gamma**. As with many measures of association, the possible values of gamma vary from −1, meaning the variables are perfectly associated in a negative direction; to 0, meaning there is no association of the type that gamma measures; to +1, meaning there is a perfect positive association of the type that gamma measures.

EXHIBIT 12.23   Crosstabulation of Respondents' Gender by Delinquency Within Levels of Parental Supervision

Weak Parental Supervision

GENDER		SELF-REPORTED DELINQUENCY			
		Low	Medium	High	Total
	Female	26.1%	27.9%	46.0%	337
	Male	23.2%	27.2%	49.6%	427
	Total				764

$\chi^2 = 1.220$ ($p > .05$) Gamma = .067

Strong Parental Supervision

GENDER		SELF-REPORTED DELINQUENCY			
		Low	Medium	High	Total
	Female	54.4%	25.7%	19.8%	343
	Male	46.1%	30.3%	23.6%	165
	Total				508

$\chi^2 = 3.193$ ($p > .05$) Gamma = .136

us to the conclusion that parental supervision intervenes in the relationship between gender and delinquency. A very important reason females are less delinquent than males, therefore, is that females are under stricter supervision from their parents than males, and strong parental supervision leads to a reduced risk of delinquency.

Extraneous Variables

Another reason for introducing a third variable into a bivariate relationship is to see whether the original relationship is spurious due to the influence of an **extraneous variable**, which is a variable that causes both the independent and dependent variables. The only reason why the independent and dependent variables are related, therefore, is that they both are the effects of a common cause (another independent variable).

Exhibit 12.24 shows what a spurious relationship would look like. In this case, the relationship between x and y exists only because both are the effects of the common cause z. Controlling for z, therefore, will eliminate the x-y relationship. Ruling out possible extraneous variables will help to considerably strengthen the conclusion that the relationship between the independent and dependent variables is causal, particularly if all the variables that seem to have the potential for creating a spurious relationship can be controlled.

EXHIBIT 12.24   A Spurious Relationship Between x and y

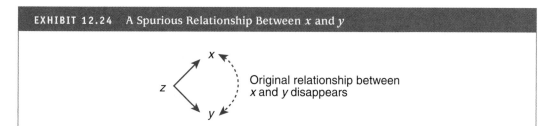

Original relationship between
x and y disappears

Notice that if a variable is acting as an extraneous variable, then controlling for it will cause the original relationship between the independent and dependent variables to disappear or substantially diminish. This was also the empirical test for an intervening variable. Therefore, the difference between intervening and extraneous variables is a logical one and not an empirical one. In both instances, controlling for the third variable will cause the original relationship to diminish or disappear. There should, therefore, be sound theoretical grounds for suspecting that a variable is acting as an intervening variable, explaining the relationship between the independent and dependent variables.

As an example of a possible extraneous relationship, we will look at the association between a youth's perception of the certainty of punishment and self-reported involvement in delinquency. Deterrence theory should lead us to predict a negative relationship between perceived certainty and delinquency. Indeed, this is exactly what we observe in our delinquency data. We will not show you the crosstabulation table, but when we looked at the relationship between perceived certainty and delinquency we found that 53.2% of youth who were low in certainty were high in delinquency; 39.1% of those who perceived medium certainty were high in delinquency; and only 23.6% of those who perceived a high certainty of punishment were high in delinquency. Youth who believed they would get caught if they engaged in delinquency, then, were less likely to be delinquent. The gamma value for this table was −.382, indicating a moderate negative relationship between perceived certainty and delinquency, exactly what deterrence theory would lead us to expect.

Someone may reasonably argue, however, that this discovered negative relationship may not be causal, but instead may be spurious. It could be suggested that what is actually behind this relationship is the extraneous variable, moral beliefs. The argument is that those with strong moral inhibitions against committing delinquent acts think that punishment for morally wrongful actions is certain, *and* refrain from delinquent acts. Thus, the observed negative relationship between perceived certainty and delinquency is really due to the positive effect of moral beliefs on perceived certainty, and the negative effect of moral beliefs on delinquency (see Exhibit 12.25). If moral beliefs are actually the causal factor at work, then controlling for them will eliminate or substantially reduce the original relationship between perceived certainty and delinquency.

To look at this possibility, we examined the relationship between perceived certainty and delinquency under three levels of moral beliefs (weak, medium, and strong). The crosstabulations are shown in Exhibit 12.26. What we can see is that in each of the subtables there is a negative and significant association between the perceived certainty of punishment and delinquency. In two of the three subtables, however, the relationship is weaker than what

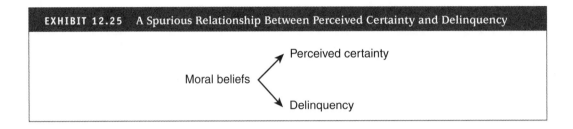

EXHIBIT 12.25   A Spurious Relationship Between Perceived Certainty and Delinquency

Moral beliefs

Perceived certainty

Delinquency

was in the original table (there the gamma was −.382); we obtained gammas of −.271 and −.197. Under the condition of strong moral beliefs, however, the original relationship is unchanged. What we would conclude from this elaboration analysis is that the variable "moral beliefs" is not acting as a very strong extraneous variable. Although some of the relationship between perceived risk and delinquency is due to their joint relationship with moral beliefs, we cannot dismiss the possibility that the perceived certainty of punishment has a causal influence on delinquent behavior.

Specification

By adding a third variable to an evaluation of a bivariate relationship, the data analyst can also specify the conditions under which the bivariate relationship occurs. A **specification** occurs when the association between the independent and dependent variables varies across the categories of one or more other control variables. That is, when the original relationship is stronger under some condition or conditions of a third variable and weaker under others.

In criminology, social learning theory would predict that youths who are exposed to peers who provide verbal support for delinquency are at greater risk for their own delinquent conduct. We found support for this hypothesis in our delinquency data set. We examined this relationship by recoding into two approximately equal groups the variable FROPINON (see Exhibit 12.2). The first group had weak verbal support from peers, whereas the second group had strong verbal support. Among those youths who reported that their peers provided only weak verbal support for delinquency, 15% were highly delinquent. Among those with strong verbal support from peers, nearly 58% were highly delinquent. The gamma value for this relationship was .711, a very strong positive relationship. Clearly, then, having friends give you verbal support for delinquent acts (for example, "it's OK to steal"), puts you at risk for delinquency.

It is entirely possible, however, that this relationship exists only when friends' verbal support is backed up by their own behavior. That is, verbal support from our peers might not affect our delinquency when they do not themselves commit delinquent acts, or commit only a very few. In this case, their actions (inaction in this case) speak louder than their words, and their verbal support does not influence us. When they also commit delinquent acts, however, the verbal support of peers carries great weight.

We looked at this possibility to examining the relationship between friends' verbal support for delinquency and a youth's own delinquency within two levels of friends' behavior (FRBEHAVE, see Exhibit 12.2). We recoded FRBEHAVE into two approximately equal groups.

EXHIBIT 12.26   Crosstabulation of Perceived Risk by Delinquency Within Levels of Moral Beliefs

Weak Moral Beliefs

		SELF-REPORTED DELINQUENCY			
		Low	Medium	High	Total
PERCEIVED CERTAINTY	Low	3.8%	14.7%	81.4%	156
	Medium	8.6%	27.3%	64.1%	128
	High	4.2%	29.6%	66.8%	71
	Total				355

$\chi^2 = 13.646$ ($p <.001$) Gamma = $-.271$

Medium Moral Beliefs

		SELF-REPORTED DELINQUENCY			
		Low	Medium	High	Total
PERCEIVED CERTAINTY	Low	22.0%	42.5%	35.4%	127
	Medium	33.9%	35.6%	30.5%	174
	High	41.1%	34.2%	24.8%	202
	Total				503

$\chi^2 = 13.646$ ($p <.001$) Gamma = $-.197$

Strong Moral Beliefs

		SELF-REPORTED DELINQUENCY			
		Low	Medium	High	Total
PERCEIVED CERTAINTY	Low	42.7%	28.1%	29.2%	89
	Medium	58.9%	17.7%	23.4%	107
	High	72.9%	18.3%	8.7%	218
	Total				414

$\chi^2 = 13.646$ ($p \leq .001$) Gamma = $-.393$

In the first group, fewer of one's friends are delinquent (few delinquent friends) than the other (many delinquent friends). This attempt to specify the relationship between friends' opinions and our own delinquency is shown in Exhibit 12.27. What we see is a little complex. When only a few of a youth's friends are also committing delinquent acts, their verbal support still has a significant and positive effect on self-reported delinquency. The gamma value in this subtable is .416, which is moderately strong, but less than the original gamma of .771. When many of a youth's friends are delinquent, however, the positive relationship between peers' verbal support and self-reported delinquency is much stronger, with a gamma of .608. The behavior of our peers, then, only weakly specifies the relationship between peer opinion and delinquency. Clearly, then, what our peers say about delinquency matters even if they are not committing delinquent acts all the time themselves.

Our goal in introducing you to crosstabulation has been to help you think about the association among variables and to give you a relatively easy tool for describing association. To read most statistical knowledge, in addition to the statistical techniques we have considered here and the inferential statistics we discussed previously, you should learn about the summary descriptive statistics used to indicate the strength and direction of association.

EXHIBIT 12.27 Crosstabulation of Friends' Verbal Support by Delinquency Within Levels of Friends' Delinquent Behavior

Few Delinquent Friends

FRIENDS' VERBAL SUPPORT		SELF-REPORTED DELINQUENCY			
		Low	*Medium*	*High*	*Total*
	Weak	67.3%	23.8%	8.9%	437
	Strong	44.3%	34.3%	21.4%	140
	Total				577

$\chi^2 = 27.374$ ($p > .001$) Gamma = .416

Many Delinquent Friends

FRIENDS' VERBAL SUPPORT		SELF-REPORTED DELINQUENCY			
		Low	*Medium*	*High*	*Total*
	Weak	30.5%	38.5%	31.0%	174
	Strong	7.5%	24.8%	67.4%	521
	Total				695

$\chi^2 = 87.508$ ($p > .001$) Gamma = .608

Until you do so, you will find it difficult to state with precision just how strong an association is. You also should learn about a different statistical approach to characterizing the association between two quantitative variables, called regression or correlation analysis. Statistics based on regression and correlation are used very often in social science and have many advantages over crosstabulation, as well as some disadvantages. You will need to take a course in social statistics, however, to become proficient in the use of statistics based on regression and correlation.

ANALYZING DATA ETHICALLY:
HOW NOT TO LIE ABOUT RELATIONSHIPS

When the data analyst begins to examine relationships among variables in some real data, social science research becomes most exciting. The moment of truth, it would seem, has arrived. Either the hypotheses are supported or not. But, in fact, this is also a time to proceed with caution, and to evaluate the analyses of others with even more caution. Once large data sets are entered into a computer, it becomes very easy to check out a great many relationships; when relationships are examined among three or more variables at a time, the possibilities become almost endless.

This range of possibilities presents a great hazard for data analysis. It becomes very temping to search around in the data until something interesting emerges. Rejected hypotheses are forgotten in favor of highlighting what's going on in the data. It is not wrong to examine data for unanticipated relationships; the problem is that inevitably some relationships between variables will appear just on the basis of chance association alone. If you search hard and long enough, it will be possible to come up with something that really means nothing.

A reasonable balance must be struck between deductive data analysis to test hypotheses and inductive analysis to explore patterns in a data set. Hypotheses formulated in advance of data collection must be tested as they were originally stated; any further analyses of these hypotheses that involve a more exploratory strategy must be labeled as such in research reports. Serendipitous findings do not need to be ignored, but they must be reported. Subsequent researchers can try to deductively test the ideas generated by our explorations.

We also have to be honest about the limitations of using survey data to test causal hypotheses. The usual practice for those who seek to test a causal hypothesis with nonexperimental survey data is to test for the relationship between the independent and dependent variables, controlling for other variables that might possibly create spurious relationships. This is what we did by examining the relationship between the perceived certainty of punishment and delinquency while controlling for moral beliefs (see Exhibit 12.26). But finding that a hypothesized relationship is not altered by controlling for just one variable does not establish that the relationship is causal; nor does controlling for two, three, or many more variables. There always is a possibility that some other variable that we did not think to control, or that was not even measured in the survey, has produced a spurious relationship between the independent and dependent variables in our hypothesis (Lieberson 1985). We must always think about the possibilities and be cautious in our causal conclusions.

CONCLUSION

This chapter has demonstrated how a researcher can describe phenomena in criminal justice and criminology, identify relationships among them, explore the reasons for these relationships, and test hypotheses about them. Statistics provide a remarkably useful tool for developing our understanding of the social world, a tool that we can use to test our ideas and generate new ones.

Unfortunately, to the uninitiated, the use of statistics can seem to end debate right there; you cannot argue with the numbers. But you now know better than that. The numbers will be worthless if the methods used to generate the data are not valid; and the numbers will be misleading if they are not used appropriately, taking into account the type of data to which they are applied. And even assuming valid methods and proper use of statistics, there is one more critical step, for the numbers do not speak for themselves. Ultimately, it is how we interpret and report the numbers that determines their usefulness. It is this topic we turn to in the next chapter.

KEY TERMS

Bar chart	Median
Base N	Mode
Bimodal distribution	Monotonic
Central tendency	Negatively skewed
Chi-square	Outlier
Crosstabulation (crosstab)	Percentage
Data cleaning	Positively skewed
Descriptive statistics	Probability average
Elaboration analysis	Quartile
Extraneous variable	Range
Frequency distribution	Secondary data analysis
Frequency polygon	Skewness
Gamma	Specification
Grouped frequency distribution	Standard deviation
Histogram	Statistical significance
Inferential statistics	Subtables
Interquartile range	Unimodal distribution
Marginal distributions	Variability
Mean	Variance
Measure of association	

HIGHLIGHTS

- Data-collection instruments should be precoded for direct entry, after verification, into a computer. Use of secondary data can save considerable time and resources, but may limit data analysis possibilities.
- Bar charts, histograms, and frequency polygons are useful for describing the shape of distributions. Care must be taken with graphic displays to avoid distorting a distribution's apparent shape.

- Frequency distributions display variation in a form that can be easily inspected and described. Values should be grouped in frequency distributions in a way that does not alter the shape of the distribution. Following several guidelines can reduce the risk of problems.

- Summary statistics are often used to describe the central tendency and variability of distributions. The appropriateness of the mode, mean, and median vary with a variable's level of measurement, the distribution's shape, and the purpose of the summary.

- The variance and standard deviation summarize variability around the mean. The interquartile range is usually preferable to the range to indicate the interval spanned by cases, due to the effect of outliers on the range. The degree of skewness of a distribution is usually described in words rather than with a summary statistic.

- Crosstabulations should normally be percentaged within the categories of the independent variable. A crosstabulation can be used to determine the existence, strength, direction, and pattern of an association.

- Elaboration analysis can be used in crosstabular analysis to test for spurious and intervening relationships and to identify the conditions under which relationships occur.

- Inferential statistics are used with sample-based data to estimate the confidence that can be placed in a statistical estimate of a population parameter. Estimates of the probability that an association between variables may have occurred on the basis of chance are also based on inferential statistics.

EXERCISES

1. Create frequency distributions from lists in the FBI Uniform Crime Reports on characteristics of arrestees in at least 100 cases (cites). You will have to decide on a grouping scheme for the distribution scheme of variables such as race, age, and crime committed, and how to deal with outliers in the frequency distribution.

    a. Decide what summary statistics to use for each variable of interest. How well were the features of each distribution represented by the summary statistics? Describe the shape of each distribution.

    b. Propose a hypothesis involving two of these variables, and develop a crosstabulation to evaluate the support for this hypothesis.

    c. Describe each relationship in terms of the four aspects of an association, after making percentages within each table within the categories of the independent variable. Which hypotheses appear to have been supported?

2. Become a media critic. For the next week, scan a newspaper or some magazines for statistics related to crime or criminal victimization. How many can you find using frequency distributions, graphs, and the summary statistics introduced in this chapter? Are these statistics used appropriately and interpreted correctly? Would any other statistics have been preferable or useful in addition to those presented?

DEVELOPING A RESEARCH PROPOSAL

Use the General Social Survey data to add a pilot study to your proposal. A pilot study is a preliminary effort to test out the procedures and concepts that you have proposed to research.

1. Review the GSSCRJ2K variable list and identify some variables that have at least some connection to your research problem. If possible, identify one variable that might be treated as independent in your proposed research and one that might be treated as dependent.

2. Request frequencies for these variables.

3. Request a crosstabulation of the dependent variable by the independent variable (if you were able to identify any). If necessary, recode the independent variable to three or fewer categories.

4. Write a brief description of your findings and comment on their implications for your proposed research. Did you learn any lessons from this exercise for your proposal?

Student Study Site

The companion Web site for *The Practice of Research in Criminology and Criminal Justice*, Third Edition
http://www.sagepub.com/prccj3
Visit the Web-based Student Study Site to enhance your understanding of the chapter content and to discover additional resources that will take your learning one step further. You can enhance your understanding of the chapters by using the comprehensive study material, which includes e-flashcards, Web exercises, practice self-tests, and more. You will also find special features, such as Learning from Journal Articles, which incorporates SAGE's online journal collection.

WEB EXERCISES

1. Search the Web for a crime-related example of statistics. Using the key terms from this chapter, describe the set of statistics you have identified. What phenomena does this set of statistics describe? What relationships if any, do the statistics identify?

2. Do a Web search for information on a criminological subject that interests you. How much of the information that you find relies on statistics as a tool for understanding the subject? How do statistics allow researchers to test their ideas about the subject and generate new ideas? Write your findings in a brief report, referring to the Web sites that you found.

ETHICS EXERCISES

1. Review the frequency distributions and graphs in this chapter. Change one of these data displays so that you are "lying with statistics." (You might consider using the graphic technique discussed by Orcutt & Turner 1993.)

2. Consider the relationship between gender and delinquency that is presented in Exhibit 12.19. What third variables do you think should be controlled in the analysis to better understand the basis for this relationship? How might criminal justice policies be affected by finding out that this relationship was due to differences in teacher expectations rather than to genetic differences in violence propensity?

SPSS EXERCISES

1. Develop a description of homicide defendants and their sentence length from HOMICIDE.POR. Examine each characteristic with three statistical techniques: a graph, a frequency distribution, and a measure of central tendency.

2. Describe the distribution of each variable, noting any skewness and/or outliers.

3. Collapse the categories for defendant's age (DAGE) and sentence length (PRITIME). Examine the data distributions using a graph and frequency distribution. Does the general shape of the distributions change as a result of changing how the data are measured?

4. Look at the relationship between these two variables. Indicate which variable is the dependent variable and which variable is the independent variable.

5. Run a crosstabulation on these data (use your newly created categorical variables). Be sure to obtain percentiles for categories of the independent variable (either row or column).

6. Propose two variables that might create a spurious relationship between defendant's age and sentence length. Explain your thinking. Propose a variable that might result in a conditional effect of age on sentence length, so that the relationship between age and sentence length would vary across the categories of the other variable. Test these propositions with three variable crosstabulations. Were any supported? How would you explain your findings?

CHAPTER 13

Reporting Research Results

The goal of research is not just to discover something but to communicate that discovery to a larger audience—other social scientists, government officials, your teachers, the general public—perhaps several of these audiences. Whatever the study's particular outcome, if the research report enables the intended audience to comprehend the results and learn from them, the research can be judged a success. If the intended audience is not able to learn about the study's results, the research should be judged a failure no matter how expensive the research, how sophisticated its design, or how much of yourself you invested in it.

This conclusion may seem obvious, and perhaps a bit unnecessary. After all, you may think that all researchers write up their results for other people to read. But the fact is that many research projects fail to produce a research report. Sometimes the problem is that the research is poorly designed to begin with and cannot be carried out in a satisfactory manner; sometimes unanticipated difficulties derail a viable project. But too often the researcher just never gets around to writing a report. And then there are many research reports that are very incomplete or poorly written, or that speak to only one of several interested audiences. The failure may not be complete, but the project's full potential is not achieved.

Remember that the time for congratulations is when credible results are released and they serve some useful function, not when the research project is first approved. Give careful attention to the research report, whatever type it is, before breaking out the bottle of champagne. And if you have conducted a worthwhile research project, consider whether some of the results might be appropriate for more than one report to different audiences.

The primary goals of this chapter are to help you develop worthwhile reports for any research you conduct and guide you in evaluating reports produced by others. We begin by teaching how to write research proposals, because a formal proposal lays the groundwork for a final research report. The next section highlights problems that are unique to the main types of reports: student papers and theses, journal articles, and unpublished reports for specific clients. The chapter's final sections present suggestions for writing and organizing reports, techniques for displaying statistical results, and ethical issues to be considered.

THE RESEARCH PROPOSAL

Be grateful for those who require you to write a formal research proposal and even more for those who give you constructive feedback. Whether your proposal is written for a professor, a thesis committee, an organization seeking practical advice, or a government agency that funds basic research, the proposal will force you to establish a problem statement and a research plan. Too many research projects begin without a clear problem statement or with only the barest of notions about which variables must be measured or what the analysis should look like. Such projects often wander along, lurching from side to side, then collapse entirely or just fizzle out with a report that is given little attention. So even in circumstances when a proposal is not required, you should prepare one and present it to others for feedback. Just writing your ideas down will help you to see how they can be improved, and feedback in almost any form will help you to refine your plans.

A well-designed proposal can go a long way toward shaping the final research report and will make it easier to progress at later research stages. Every research proposal should have at least five sections, defined as follows:

1. *An introductory statement of the research problem.* Clarify the research question(s) that you are interested in studying and how this question was derived from the existing literature.

2. *A literature review.* Explain your problem in greater detail and build on what has already been reported in the literature on this topic.

3. *A methodological plan.* Detail the methods you will employ (including sample, variables, etc.) along with a discussion of how you will respond to the particular mix of opportunities and constraints you face.

4. *A budget.* Present a careful listing of the anticipated costs.

5. *An ethics statement.* Identify human subjects issues in the research and how you will respond to them in an ethical fashion.

If your research proposal will be competitively reviewed, it must present a compelling rationale for funding. It is not possible to overstate the importance of the research problem that you propose to study (see Chapter 2). If you propose to test a hypothesis, be sure it is one with plausible alternatives: "A boring hypothesis is one which, although likely to be correct, has no credible alternatives" (Dawes 1995:93).

Case Study: Criminal Victimization of the Elderly

Particular academic departments, grant committees, and funding agencies will have more specific proposal requirements. As an example, Exhibit 13.1 lists the primary required sections of the "Research Plan" for proposals to the National Institutes of Health (NIH), together with excerpts from a proposal one of your authors submitted in this format to the National Institute of Mental Health (NIMH) with colleagues from the Cornell University Medical College (Mark Lachs, M.D.) and Yale University (Christianna Williams). The Research Plan is limited to 25 pages by NIH guidelines. It must be preceded by an abstract (which has been excerpted), a proposed budget, biographical sketches of project personnel, and a discussion of the available resources for the project. Appendixes may include research instruments, prior publications by the authors, and findings from related work. As you can see from Exhibit 13.1, our proposal had several specific aims, all related to criminal victimization among older adults.

So how does a research proposal like this get evaluated? Typically, grants for all government agencies and many nonprofit agencies also include a peer review process whereby other researchers and practitioners in a particular area of expertise will be asked to review your proposal. Based on their recommendations, your proposal either sinks or swims, so to speak. The NIMH review committee approved this particular project for funding, but required two submissions and reviews before funding was awarded (it often takes several resubmissions before even a worthwhile proposal is funded). The committee members recognized the proposal's strengths, but also identified several problems that they believed had to be overcome before the proposal could be funded. The reviewers pointed out several issues they had with the proposal including the following:

> The outcomes of crime considered are too narrow. . . . Number of physician visits, for example, might be a more appropriate outcome variable than nursing home placement to demonstrate increased health care utilization. Nursing home placement may also be confounded. . . . For example, variables such as weak social networks and support could be associated with both crime victimization and nursing home placement. . . . The authors have not considered "short stay" nursing home placement. Conversely, other types of residential living arrangements (e.g., assisted living) do not appear to be included (i.e., nursing home placement is treated as dichotomous). This research considers only reports of crimes made to police and therefore underestimates its prevalence within the cohort. Similarly, the work assumes no cohort members were victimized prior to inception year 1985; some undoubtedly were and this will confound the analysis. . . . Alcohol is a risk factor for victimization in younger individuals yet it is not considered in the proposed research. . . . There is minimal text on psychosocial aspects of stress theory.

(Text continues on page 454)

A Grant Proposal to the NIMH

Abstract

Little is known about the epidemiology of crime committed against older adults, and virtually nothing is known about its health consequences. The investigators propose linking an established cohort of older adults (The New Haven EPESE cohort, $n = 2,812$) who have been followed annually with standardized measures of medical, functional, and psychosocial health for over a decade, with Police Records from the cohort's catchments area. This would permit the largest community-based study of crime victimization in older adults ever conducted, at a fraction of the expense of assembling a new inception cohort for this purpose. The broad long-term goal of the research is to identify older adults at highest risk for mortality after crime so that preventive strategies may be developed and implemented.

Research Plan

1. Specific Aims

The specific aims of this research are: (1) to identify risk factors for crime victimization in an observational cohort of community-dwelling older adults; (2) to estimate the independent contribution of crime victimization to erosion of quality-of-life in cohort members after victimization is measured by (a) the onset of new or worsening functional disability and (b) a decline in self-perceived health status; (3) to estimate the independent contribution of crime victimization to increased health care utilization in cohort members after victimization as measured by (a) hospitalization and (b) nursing home placement; and (4) to estimate the independent contribution of crime victimization to all-cause mortality in the cohort after the experience of victimization. A secondary aim of this research is to determine if gender is an "effect modifier" in influencing mortality in older adults.

2. Background and Significance

Without question, older adults remain the least studied demographic group with respect to the epidemiology and impact of crime victimization. . . . While older adults have relatively low rates of crime victimization compared to younger adults, when criminal victimization of older adults is examined more closely, some alarming patterns emerge. For example, recent research has revealed that older adults suffer violent death from felony-related homicides perpetrated by strangers at a rate proportionally greater than younger age cohorts. In addition, it has been found that older victims of robbery, particularly women, are more likely to sustain injuries compared to younger victims. Moreover, older victims of violent crime in general are more likely than younger injured victims to suffer a serious medical injury that requires hospital or other medical care. Dr. Lachs's experience in the primary care of older people is consistent with this literature. For example, he has observed that for many older adults who are ostensibly well compensated in all spheres (medical, functional, and psychosocial), a crime victimization can set in motion an inexorable spiral that may ultimately result in a loss of independence.

The notion that a single event might set into motion a progressive spiral of decline in many domains for an older adult has a basis in aging theory. . . . Normal aging is accompanied by a loss of physiologic reserve in various systems that need not lead to phenotypic decline. . . . Rather, when the organism is taxed through stress, illness, or other factors, this loss of physiologic reserve is unmasked. . . . Crime victimization may be such a precipitating event in the life of an older person. It is a stressful experience that may impact on physical health, mental health, and

(Continued)

EXHIBIT 13.1 (Continued)

functional independence. . . . Stress in general, and stressful life events in particular, have been linked to the onset of illness and other maladaptive behavior by several decades of research at both the individual and societal levels. . . . While the geriatric literature is rife with examples of mortality and adverse health outcomes following stressful life events. . . . There is virtually no empirical research which examines the impact of crime on health outcomes, mortality, or general quality of life in older people. . . . The study proposed is the largest and longest longitudinal evaluation of crime, mortality, and quality of life in a well-characterized community-dwelling cohort of older people. It also marries both the medical and criminal justice perspectives bringing together a multidisciplinary team ideally suited to conduct this research and involves a municipal police department at the grassroots level.

Several gaps in our knowledge can be addressed in this research:

- What are the risk factors for crime victimization among older adults? How does medical, functional, and psychosocial health interact with environmental variables (e.g., poverty and social networks) to create a milieu that makes the older person especially vulnerable?
- How does crime victimization influence the need for subsequent nursing home placement? For individuals who escape physical injury from any form of victimization, is there still a quality-of-life toll in the form of new onset functional impairment, social isolation, and subsequent institutionalization months or years after victimization?
- What is the independent contribution of crime victimization to all causes of mortality in older people?
- What role does gender play in influencing the response to crime victimization?

3. Progress Report or Preliminary Studies

To determine whether a records linkage between the New Haven EPESE cohort and the City of New Haven's police records was feasible, the investigators conducted a pilot record match between these disparate epidemiological and administrative data sets. . . . After approval from the institutional review boards at Yale and Cornell, and after development of a carefully devised confidentiality protocol, the investigators received approval from the New Haven Police Department to determine if cohort members had been reported as victims of crime.

A pilot match was undertaken for 200 randomly selected cohort members for a short segment of the follow-up period during which police records were readily available (between 1985 and 1991).... Of the 200 cohort members, there have been 86 police events in 47 cohort members over the 7-year follow-up period. 43% of these were as victims.... These pilot data demonstrate that despite potential technical and bureaucratic obstacles, a records match between the data and police records is indeed feasible....

4. Research Design

Study Sample
The EPESE (Established Population for Epidemiologic Studies of the Elderly) data is an observational cohort of 2,812 older adults living in New Haven who were over the age of 65 in 1982. The study population is based on a stratified cluster sample of 3,337 individuals, 82 percent of whom responded and agreed to participate. . . . Members were interviewed yearly in person or by telephone between 1982 and 1990 with an additional in person interview in 1994.

EXHIBIT 13.1

Additionally, the match rate suggests that ample events have accrued for adequate statistical power.

New Haven Police Records

All cohort members who experienced a crime in New Haven and reported the victimization to police will be included in the records match. Typical information on a crime report includes victim characteristics, type of offense, offender characteristics, injuries sustained, and location of the incident along with other information.

Variables

As the specific aims delineate, crime victimization is both a dependent and an independent variable in this research. For the purpose of this research, a cohort member is deemed to have been a crime victim when he or she has experienced any police contact wherein their role in the crime is that of victim and a specific crime incident type is provided. Several independent variables derived from clinical experience and from the criminological literature will serve as independent variables to identify risk factors for crime victimization (dependent variable for specific aim 1) including sociodemographic (age, income, race, sex, etc.), medical (number and type of chronic conditions), functional (ADL impairment, cognitive impairment), and psychosocial (size and quality of social network, depressive symptoms).

For specific aims 2 and 3, criminal victimization will be used along with the other independent variables to predict the outcomes of custodial nursing home placement and mortality

Strategy of Analysis

The initial analysis involves the construction of time failure or survival curves wherein custodial nursing home placement will be the parameter of interest. . . . Subsequent multivariable analysis will include crime victimization as a time-dependent covariate, with days to nursing home placement after crime victimization as the dependent variable in Cox proportional hazards models. . . . To achieve the third specific aim to predict all cause mortality, Kaplan-Meier curves will be constructed for victimized and non-victimized groups, and their respective survival over the follow-up period will be compared. . . . Subsequent multivariable analysis is intended to determine the independent effects of crime victimization on mortality using hierarchical modeling.

5. Human Subjects

There is no direct contact with subjects in this research as it is a project that involves the linking of two archival data sets. . . . The major risk in this research is violation of confidentiality with respect to being a crime victim for surviving subjects. . . . An extensive protocol has been developed such that in the final merged record, EPESE investigators will have no knowledge of which subjects were victims of crime and conversely, police staff will have no knowledge of which of their many crime victims are EPESE cohort members. Ironically, state statutes with respect to crime are more directly concerned with protecting the identify of the alleged perpetrator of crime. For this reason, in the matching process police staff must first ensure that a crime case has no pending legal disposition before releasing the record to the investigators for anonymous abstraction. . . .

6. Women and Minority Subjects

The EPESE cohort intentionally over-samples minority subjects, over 20% of whom were non-white at cohort inception. Additionally, women outnumber men in the original cohort 1,647 to 1,143, which is typical of the demography of older adults.

If you get the impression that researchers cannot afford to leave any stone unturned in working through procedures in an NIMH proposal, you are right. It is very difficult to convince a government agency that a research project is worth the money requested, particularly when the amount requested is large (over $500,000). And that is as it should be: Your tax dollars should be used only for research that has a high likelihood of yielding findings that are valid and used. But whether you are proposing a smaller project to a more generous funding source or presenting a proposal to your professor, you should scrutinize the proposal carefully before submission and ask others to comment on it. Other people will often think of issues you neglected to consider, and you should allow yourself time to think about these issues and to reread and redraft the proposal. Besides, you will get no credit for having thrown together a proposal as best you could in the face of an impossible submission deadline (or class deadline).

In addition, government funding agencies such as NIMH also ask you to provide many formal statements and certificates: paperwork, paperwork, paperwork! For example, Exhibit 13.2 provides the forms and sections required in a research proposal application to the National Institute of Justice, one of the largest funding agencies for criminological research. Note that the research proposal itself is only one item (Item 12) on this list. As you can see, providing the proper forms in the correct order can almost be as much work as writing the proposal itself!

When you develop a research proposal, it will help to ask yourself a series of questions like those in Exhibit 13.3 (also see Herek 1995). This checklist is important because it is too easy to omit important details and to avoid being self-critical while rushing to put a proposal together. And it is too painful to have a proposal rejected (or to receive a low grade). It is always best to make sure the proposal covers what it should and confronts the tough issues that reviewers (or your professor) will be sure to spot early.

The series of questions in Exhibit 13.3 can serve as a map to preceding chapters in this book as well as a checklist of decisions that must be made throughout any research project. The questions are organized in five sections, each concluding with a checkpoint at which you should consider whether to proceed with the research as planned, modify the plans, or stop the project altogether. The sequential ordering of these questions obscures a bit the way they should be answered: not as single questions, one at a time, but as a unit, first as five separate stages and then as a whole. Feel free to change your answers to earlier questions on the basis of your answers to later questions.

The Methodologist's Toolchest™

You may also find it useful when preparing a research proposal to have a computer program identify and review key decisions that you have to make. The Methodologist's Toolchest (Brent & Thompson 1996) is an expert decision support system containing detailed information on the major components of research proposals. You can review the issues that should be considered for such critical stages as designing a sample, developing measures, and choosing statistics. The software shows you what the alternatives are and asks you which one you prefer. It then can give you feedback to indicate possible consequences of your decision.

This process is represented in Exhibit 13.4. First you see the Toolchest outline of major proposal sections (see Exhibit 13.4a). Then you see the different types of threats to

---

EXHIBIT 13.2   Material and Forms Required for Research Proposal Submitted to the National Institute of Justice

1. Standard Form (SF) 424: application for Federal Assistance

2. Geographic Areas Affected Worksheet

3. Assurances

4. Certifications regarding lobbying, debarment, suspension, and other responsibility matters; and drug-free workplace requirements (one form)

5. Disclosure of lobbying activities

6. Budget detail worksheet

7. Budget narrative

8. Negotiated indirect rate agreement

9. Names and affiliations of all key persons from applicant and subcontractor(s), advisors, consultants, and advisory board members. Include name of principal investigator, title, organizational affiliation, department (if institution of higher education), address, home, and fax.

10. Proposal abstract: meant to serve as a succinct and accurate description of the proposed work

11. Table of contents

12. Program narrative or technical proposal: Applicants must concisely describe the research goals and objectives, research design, and methods for achieving the goals and objectives. The number of pages in the "program narrative" must not exceed 30 double-spaced pages (margins of a minimum of 1," and no less than 11-point font size).

13. Privacy certificate

14. Form 310 (Protection of Human Subjects Assurance Identification/Certification/Declaration)

15. Environmental assessment (if required)

16. References

17. Letters of cooperation from organizations collaborating in the research project

18. Resumes

19. Appendixes (e.g., list of previous NJ awards, their status, and products)

---

validity that you can review, if you are interested (see Exhibit 13.4b). Next you see some background information to help you choose a research design, and then feedback indicating the consequences of your design choice for the validity threat on which you focused (history, in the example) (see Exhibit 13.4c). The last screens provide suggestions for reducing the threat and a quantitative rating of the extent to which your design has addressed the validity threat (see Exhibit 13.4d and Exhibit 13.4e). Of course, this information and feedback is useful only if you treat it as an additional source of suggestions, not as the ultimate authority. You may sometimes receive inappropriate advice from a system like the Toolchest, because it does not understand all the aspects of the research problem. You still must be in charge. Nonetheless, a structured procedure for reviewing decisions and thinking about alternatives can only help improve your proposals.

EXHIBIT 13.3 Decisions in Research

PROBLEM FORMULATION (see Chapters 1–2)

1. Developing a research question
2. Assessing researchability of the problem
3. Consulting prior research
4. Relating to social theory
5. Choosing an approach:
   Deductive? Inductive? Descriptive?
6. Reviewing research guidelines

Checkpoint 1
Alternatives:
- Continue as planned.
- Modify the plan.
- STOP. Abandon the plan.

RESEARCH VALIDITY (see Chapters 3–5)

7. Establishing measurement validity:
   - How are concepts defined?
   - Choose a measurement strategy.
   - Assess available measures or develop new measures.
   - What evidence of reliability and validity is available or can be collected?

8. Establishing generalizability:
   - Was a representative sample used?
   - Are the findings applicable to particular subgroups?
   - Does the population sampled correspond to the population of interest?

9. Establishing causality:
   - What is the possibility of experimental or statistical controls?
   - How to assess the causal mechanism?
   - Consider the causal context.

10. Data required: Longitudinal or cross-sectional?
11. Units of analysis: Individuals or groups?
12. What are major possible sources of causal invalidity?

Checkpoint 2
Alternatives:
- Continue as planned.
- Modify the plan.
- STOP. Abandon the plan.

RESEARCH DESIGN (see Chapters 5–10)

13. Choosing a research design:
    - Experimental? Survey? Participant observation?
    - Historical, comparative? Multiple methods?

14. Secondary analysis? Availability of suitable data sets?
15. Assessing ethical concerns

EXHIBIT 13.3 (Continued)

Checkpoint III
Alternatives:
- Continue as planned.
- Modify the plan.
- STOP. Abandon the plan.

DATA ANALYSIS (see Chapter 12)

18. Choosing a statistical approach:
- Statistics and graphs for describing data
- Identifying relationships between variables
- Deciding about statistical controls
- Testing for interaction effects
- Evaluating inferences from sample data to the population

Checkpoint 4
Alternatives:
- Continue as planned.
- Modify the plan.
- STOP. Abandon the plan.

REVIEWING, PROPOSING, REPORTING RESEARCH (see Chapter 13)

19. Clarifying research goals
20. Identifying the intended audience
21. Searching the literature and the Web
22. Organizing the text
23. Reviewing ethical and practical constraints

Checkpoint 5
Alternatives:
- Continue as planned.
- Modify the plan.
- STOP. Abandon the plan.

RESEARCH REPORT GOALS

The research report will present research findings and interpretations in a way that reflects some combination of the researcher's goals, the research sponsor's goals, the concerns of the research subjects, and perhaps the concerns of a wider anticipated readership. Understanding the goals of these different groups will help the researcher begin to shape the final report even at the start of the research. In designing a proposal and in negotiating access to a setting for the research, commitments often must be made to produce a particular type of report, or at least cover certain issues in the final report. As the research progresses, feedback about the research from its subjects, sponsoring agencies, collaborators, or other interested parties may suggest the importance of focusing on particular issues in the final report. Social researchers traditionally have tried to distance themselves from the

EXHIBIT 13.4a Methodologist's Toolchest™

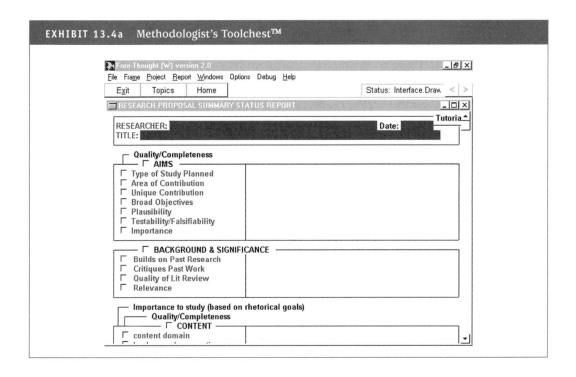

EXHIBIT 13.4b

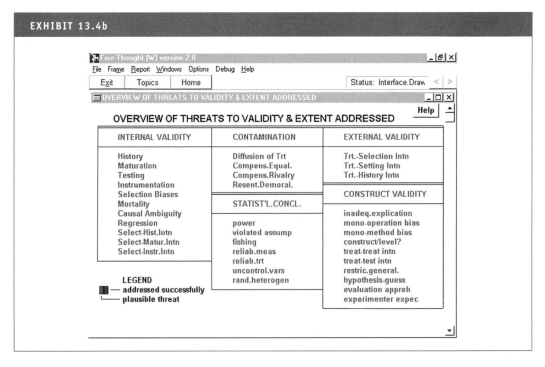

EXHIBIT 13.4c

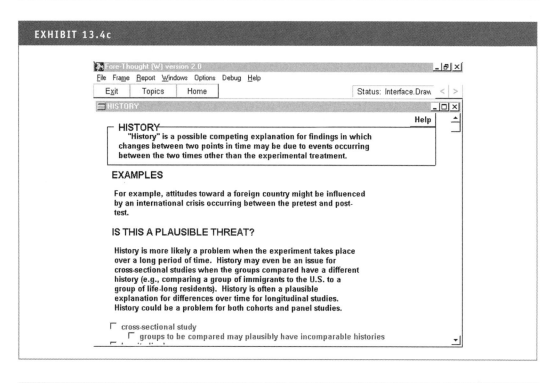

EXHIBIT 13.4d

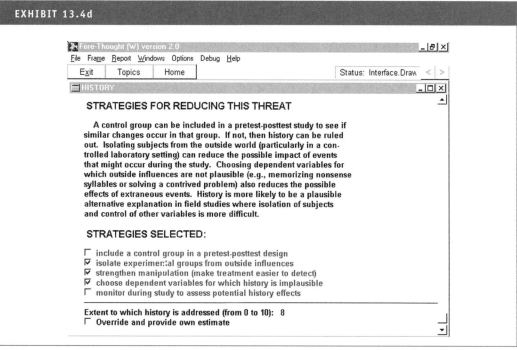

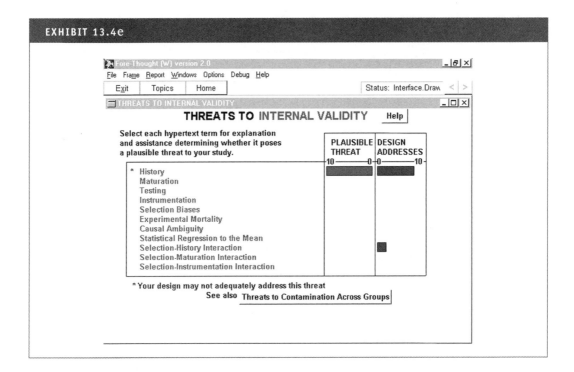

EXHIBIT 13.4e

concerns of such interested parties, paying attention only to what is needed to advance scientific knowledge. But in recent years, some social scientists have recommended bringing these interested parties into the research and reporting process itself.

Advance Scientific Knowledge

To test your understanding of research report goals, go to the Reporting and Reviewing Research Interactive Exercises on the Student Study Site.

Most social science research reports are directed to other social scientists working in the area of study, so they reflect orientations and concerns that are shared within this community of interest. The traditional scientific approach encourages a research goal to advance scientific knowledge by providing reports to other scientists. This approach also treats value considerations as beyond the scope of science: "An empirical science cannot tell anyone what he should do but rather what he can do and under certain circumstances what he wishes to do" (Weber 1949:54).

The idea is that developing valid knowledge about how society is organized or how we live our lives does not tell us how society should be organized or how we should live our lives. There should, as a result, be a strict separation between the determination of empirical facts and the evaluation of these facts as satisfactory of unsatisfactory (Weber 1949:11). Social scientists must not ignore value considerations, which are viewed as a legitimate basis for selecting a research problem to study. After the research is over and a report has been written, many scientists also consider it acceptable to encourage government officials or private organizations to implement the findings. During a research project, however, value considerations are to be held in abeyance.

Shape Social Policy

Many social scientists seek to influence social policy through their writing. You have been exposed to several examples in this text, including all the evaluation research cited in Chapter 11. These particular studies, like much policy-oriented social science research, are similar to those that aim strictly to increase knowledge. In fact, these studies might even be considered contributions to knowledge first, and to social policy debate second. What distinguishes the reports of these studies from strictly academic reports is their attention to policy implications.

Other social scientists who seek to influence social policy explicitly reject the traditional scientific, rigid distinction between facts and values (Sjoberg & Nett 1968). Bellah et al. (1985) have instead proposed a model of "social science as public philosophy," in which social scientists focus explicit attention on achieving a more just society:

> Social science makes assumptions about the nature of persons, the nature of society, and the relation between persons and society. It also, whether it admits it or not, makes assumptions about good persons and a good society and considers how far these conceptions are embodied in our actual society.
>
> Social science as public philosophy, by breaking through the iron curtain between the social sciences and the humanities, becomes a form of social self-understanding or self-interpretation. . . . By probing the past as well as the present, by looking at "values" as much as at "facts," such a social science is able to make connections that are not obvious and to ask difficult questions. (P. 301)

This perspective suggests more explicit concern with public policy implications when reporting research results. But it is important to remember that we all are capable of distorting our research and our interpretations of research results to correspond to our own value preferences. The temptation to see what we want to see is enormous, and research reports cannot be deemed acceptable unless they avoid this temptation.

Organize Social Action

For the same reasons that value questions are traditionally set apart from the research process, many social scientists consider the application of research a nonscientific concern. William Foote Whyte, whose *Street Corner Society* (1943) study you encountered in Chapter 8, has criticized this belief and proposed an alternative research and reporting strategy he calls **participatory action research**. Whyte (1991:285) argues that social scientists must get "out of the academic rut" and engage in applied research to develop better understanding of social phenomena.

In participatory action research, the researcher involves some organizational members as active participants. Both the organizational members and the researcher are assumed to want to develop valid conclusions, to bring unique insights, and to desire change.

Dialogue With Research Subjects

Guba and Lincoln (1989:44–45) have carried the notion of involving research subjects and others in the design and reporting of research one step further. What they call the **constructivist paradigm** is a methodology that emphasizes the importance of exploring how different stakeholders in a social setting construct their beliefs. This approach rejects the assumption that there is a reality around us to be studied and reported on. Instead, social scientists operating in the constructivist paradigm try to develop a consensus among participants in some social process about how to understand the focus of inquiry, a program that is often evaluated. A research report will then highlight different views of the social program and explain how a consensus can be reached.

Constructivist inquiry uses an interactive research process, in which a researcher begins an evaluation in some social setting by identifying the different interest groups in that setting and the stakeholder groups. The researcher goes on to learn what each group thinks and then gradually tries to develop a shared perspective on the problem being evaluated (Guba & Lincoln 1989:42). This process involves four steps that can each be repeated many times in a given study:

To test your understanding of research report goals, go to the Reporting and Reviewing Research Interactive Exercises on the Student Study Site.

1. Identify stakeholders and solicit their claims, concerns, and issues.
2. Introduce the claims, concerns, and issues of each stakeholder group to the other stakeholder groups and ask for their reactions.
3. Focus further information collection on claims, concerns, and issues about which there is disagreement among stakeholder groups.
4. Negotiate with stakeholder groups about the information collected and attempt to reach consensus on the issues about which there is disagreement.

These steps are diagrammed as a circular process in Exhibit 13.5 (originally presented in Chapter 1). In this process,

> The constructions of a variety of individuals deliberately chosen so as to uncover widely variable viewpoints are elicited, challenged, and exposed to new information and new, more sophisticated ways of interpretation, until some level of consensus is reached (although there may be more than one focus for consensus). (Guba & Lincoln 1989:181)

The researcher conducts an open-ended interview with the first respondent (R1) to learn about her thoughts and feelings on the subject of inquiry, her "construction" (C1). The researcher then asks this respondent to nominate a second respondent (R2), who feels very differently. The second respondent is then interviewed in the same way, but also is asked to comment on the themes raised by the previous respondent. The process continues until all major perspectives are represented, and then may be repeated again with the same set of respondents.

The final product is a case report:

> A case report is very unlike the technical reports we are accustomed to seeing in positivist inquiries. It is not a depiction of a "true" or "real" state of affairs. . . . It does not culminate in judgments, conclusions, or recommendations except insofar as these are concurred on by relevant respondents.
>
> The case report helps the reader come to a realization (in the sense of making real) not only of the states of affairs that are believed by constructors [research respondents] to exist but also of the underlying motives, feelings, and rationales leading to those beliefs. The case report is characterized by a thick description that not only clarifies the all-important context, but that makes it possible for the reader vicariously to experience it. (Guba & Lincoln 1989: 180–181)

EXHIBIT 13.5 The Circular Process of Uncovering Attitudes and Beliefs

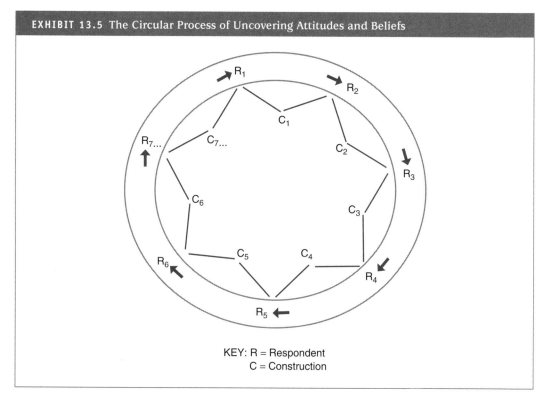

KEY: R = Respondent
C = Construction

Source: Adapted from Guba & Lincoln 1989:152.

Although this is off the beaten social scientific track, the constructivist approach provides a useful way of thinking about how to best make sense of the complexity and subjectivity of the social world. Other researchers write reports intended to influence public policy and often their findings are ignored. Such neglect would be less common if social researchers gave more attention to the different meanings attached by participants to the same events, in the spirit of constructivist case reports. The philosophy of this approach is also similar to the utilization-based evaluation research approach advanced by Patton (see Chapter 11) that involves all stakeholders in the research process.

RESEARCH REPORT TYPES

Research projects designed to produce student papers and theses, applied research reports, and academic articles all have unique features that will influence the final research report. For example, student papers are written for a particular professor or for a thesis committee and often are undertaken with almost no financial resources and in the face of severe time constraints. Applied research reports are written for an organization or agency that usually also has funded the research and has expectations for a particular type of report. Journal articles are written for the larger academic community and will not be published until they are judged acceptable by some representatives of that community (e.g., after the article has gone through extensive peer review).

These unique features do not really match up so neatly with the specific types of research products. For example, a student paper that is based on a research project conducted in collaboration with a work organization may face some constraints for a project designed to produce an applied research report. An academic article may stem from an applied research project conducted for a government agency. An applied research report often can be followed by an academic article on the same topic. In fact, one research study may lead to all three types of research reports, as students write course papers or theses for professors who write both academic articles and applied research reports.

Student Papers and Theses

What is most distinctive about a student research paper or thesis is the audience for the final product: a professor or, for a thesis, a committee of professors. In light of this, it is important for you to seek feedback early and often about the progress of your research and about your professor's expectations for the final paper. Securing approval of a research proposal is usually the first step, but it should not be the last occasion for seeking advice prior to writing the final paper. Do not become too anxious for guidance, however. Professors require research projects in part so that their students can work through, at least somewhat independently, the many issues they confront. A great deal of insight into the research process can be gained this way. So balance your requests for advice with some independent decision making.

Most student research projects can draw on few resources beyond the student's own time and effort, so it is important that the research plan not be overly ambitious. Keep the paper deadline in mind when planning the project, and remember that almost every researcher tends to underestimate the time required to carry out a project.

Group Projects

Pooling your resources with those of several students in a group project can make it possible to collect much more data, but can lead to other problems. Each student's role should be clarified at the outset and written into the research proposal as a formal commitment. Group members should try to help each other out, rather than competing to do the least work possible or to receive the most recognition. Complaints about other group members should be made to the professor when things just cannot be worked out among group members. Each group member should have a clear area of responsibility in the final report, and one may want to serve as the final editor.

The Thesis Committee

Students who are preparing a paper for a committee, usually at the M.A. or Ph.D. level, must be prepared to integrate the multiple perspectives and comments of committee members into a plan for a coherent final report. (The thesis committee chair should be the primary guide in this process; careful selection of faculty to serve on the committee is also important.) As much as possible, committee members should have complementary areas of expertise that are each important for the research project: perhaps one methodologist, one specialist in the primary substantive area of the thesis, and one specialist in a secondary area. Theses using data collected by service agencies or other organizations often benefit if an organizational representative is on the committee.

It is very important that you work with your committee members in an ongoing manner, both individually and collectively. In fact, it is vitally important to have a group meeting with all committee members at the beginning of the project to ensure everyone on the committee supports the research plan. Doing this will avoid obstacles that arise due to miscommunication later in the research process.

Journal Articles

It is the **peer review** process that makes preparation of an academic journal article most unique. Similar to a grant review, the journal's editor sends submitted articles to two or three experts, peers, who are asked whether the paper should be accepted more or less as is, revised and then resubmitted, or rejected. Reviewers also provide comments, sometimes quite lengthy, to explain their decision and to guide any required revisions. The process is an anonymous one at most journals; reviewers are not told the author's name, and the author is not told the reviewers' names. Although the journal editor has the final say, editors' decisions are normally based on the reviewers' comments.

This peer review process must be anticipated in designing the final report. Peer reviewers are not pulled out of a hat. They are expert in the field or fields represented in the paper and usually have published articles themselves in that field. It is critical that the author be familiar with the research literature and be able to present the research findings as a unique contribution to that literature. In most cases, this hurdle is much harder to jump with journal articles than with student papers or applied research reports. In fact, most leading journals have a rejection rate of over 90%, so that hurdle is quite high indeed. Of course, there is also a certain luck of the draw in peer review. One set of two or three reviewers may be inclined to reject an article that another set of reviewers would accept (see the next case study). But in general, the anonymous peer review process results in higher quality research reports because articles are revised prior to publication in response to the suggestions and criticisms of the experts.

Criminological and criminal justice research is published in a myriad of journals within several disciplines, including criminology, law, sociology, psychology, and economics. As a result, there is no one formatting style that all criminological literature abides by. If, for example, you are submitting your paper to a psychology-related journal, you must abide by the formatting style dictated by the *Publication Manual of the American Psychological Association*. The easiest way to determine how to format a paper for a particular journal is to examine recent volumes of the journal and format your paper accordingly. To give you a general idea of what a journal article looks like, an article in its entirety has been reprinted in Appendix C, along with an illustration of how to read a journal article. There are also numerous article available on the Student Study Site for this text.

Despite the slight variations in style across journals, there are typically seven standard sections within a journal article in addition to the title page:

1. *Abstract*. This should be a concise and non-evaluative summary of your research paper (no more than 120 words) that describes the research problem, the sample, the method, and the findings.

2. *Introduction.* The body of a paper should open with an introduction that presents the specific problem under study and describes the research strategy. Before writing this section, you should consider the following questions: What is the point of the study? How do the hypotheses and the research design relate to the problem? What are the theoretical implications of the study, and how does the study relate to previous work in the area? What are the theoretical propositions tested, and how were they derived? A good introduction answers these questions in a few paragraphs by summarizing the relevant argument and the data, giving the reader a sense of what was done and why.

3. *Literature Review.* Discuss the relevant literature in a way that relates each previous study cited to your research, not in an exhaustive historical review. Citation of and specific credit to relevant earlier works is part of the researchers' scientific and scholarly responsibility. It is essential for the growth of cumulative science. This section should demonstrate the logical continuity between previous research and the research at hand. At the end of this section, you are ready to conceptually define your variables and formally state your hypotheses.

4. *Method.* Describe in detail how the study was conducted. Such a description enables the reader to evaluate the appropriateness of your methods and the reliability and the validity of your results. It also permits experienced investigators to replicate the study if they so desire. In this section, you can include subsections that describe the sample, the independent and dependent variables, and the analytical or statistical procedure you will use to analyze the data.

5. *Results.* Summarize the results of the statistical or qualitative analyses performed on the data. It can include tables and figures that summarize findings. If statistical analyses are performed, tests of significance should also be highlighted.

6. *Discussion.* Take the opportunity to evaluate and interpret your results, particularly with respect to your original hypotheses and previous research. Here, you are free to examine and interpret your results as well as draw inferences from them. In general, this section should answer the following questions: What have I contributed to the literature here? How has my study helped to resolve the original problem? What conclusions and theoretical implications can I draw from my study? What are the limitations of my study? What are the implications for future research?

7. *References.* All citations in the manuscript must appear in the reference list, and all references must be cited in the text.

Case Study: Up Close and Personal With Peer Review

In all honesty, your first experience with the peer review process is generally not a pleasant one. In fact, your one-hundredth experience with peer review may not be a very pleasant experience. By this time, however, you will likely have experienced the positive contributions this process can have on your work.

In this section, we will give you an up-close and personal experience with the peer review process. The paper we will use as a case study was written by one of your authors

(Bachman 1996) and was published in the *Journal of Quantitative Criminology*. The title of the paper was "Victims' Perceptions of Initial Police Responses to Robbery and Aggravated Assault: Does Race Matter?"

After many revisions of the paper based on comments from colleagues (both informally and at a professional conference), the paper was submitted to the *Journal of Quantitative Criminology* in the fall of 1994. In early 1995, the editor of the journal, John Laub, gave me an opportunity to revise and resubmit the article based on three anonymous reviewers' comments. A condensed version of these comments by each of the three reviewers is provided here:

> *Reviewer A:* The paper is a well-framed and well-designed investigation of the relevance of race at a seldom-examined stage in official processing. It pays close attention to theory and methodological issues. Particularly well anticipated were the limitations of the data, namely, victim perceptions of police response. . . . The following should tighten the paper's focus and increase its contribution to the literature: Expand the theoretical relevance of the findings. . . . Moderate conclusions based on findings for the victim's perception of arrest. This variable, more than the others, tells us more about the police communication with victims than it does about their actual behavior. It's also more time dependent on the others; an arrest may be in the offing. . . .

> *Reviewer B:* This manuscript is flawed in two related ways: (1) its reliance on respondent perceptions of police actions; (2) conceptual ambiguity of police response. Each flaw stems from the inappropriate use of National Crime Victimization Survey (NCVS) data. The NCVS data are poorly suited to address detailed questions about police actions. . . . the NCVS data is not representative of police or police agencies. The concept of differential police action by race cannot be meaningfully studied with a nationally representative sample of victims. . . . Variation in police responses across departments is more likely than variation across race of victim/offender. . . . What's the point of controlling for sex? Why not control for weapon use, something that might more plausibly affect police response. . . . The first part of the paper is much too long. . . . Presenting six crosstabulation tables is far too cumbersome. . . . The last complete paragraph on page 27 seems a convoluted attempt to fit results into a cherished theory. . . .

> *Reviewer C:* This is a good paper that makes a significant contribution to the literature on the effects of race on criminal justice processing. . . . The author(s) come close to explicitly discussing the importance of differing frequencies of black on white vs. other combinations, they should flesh this out. If most offenders are in intra-racial criminal events, the observed patterns will have weighted effects on the outcomes of the overall patterns of race and police activity. They may wish to discuss this. . . . Finally, I think the author(s) should include a caveat about the data.

One thing that should be immediately apparent from these comments is the diversity of opinion; if you ask three researchers to judge a manuscript, you typically get three different opinions on its quality. This, of course, is an important part of the process. It is the best way to uncover flaws in the research design and the presentation of results. In this case, two reviewers thought the research made an important contribution to the field while the other thought it was little more than toilet paper. Despite the diversity of opinion, all three reviews helped to improve the final version of the manuscript. But would there be a final version?!?

After more statistical analyses and rewriting, another version of the manuscript was resubmitted to the journal in August of 1995. It was then sent out to the same reviewers along with one additional reviewer. In late October, it received a conditional acceptance; but the paper again had to be revised based on the reviewers' comments. Only if the reviewers' comments and concerns were adequately handled would the paper be published. More rewriting! In December, another version of the manuscript was submitted to the journal and in January of 1996, this version of the paper was finally accepted by the journal. It was published in the final 1996 volume of the journal. And you thought you had it bad when it takes a professor a few weeks to grade and provide comments on a paper! It took this research paper almost two years to finally be published, and this does not include the preparation time for the first draft. In the end, however, the final product was significantly better than the first draft originally submitted in 1994. And as a footnote, this was one of the better experiences I had with publishing! A paper often gets rejected by one or two journals before it is finally accepted for publication.

Applied Reports

Unlike journal articles, applied reports are usually commissioned by a particular government agency, corporation, or nonprofit organization. As such, the most important problem that applied researchers confront is the need to produce a final report that meets the funding organization's expectations. This is called the hired gun problem. Of course, the extent to which being a hired gun is a problem varies greatly with the research orientation of the funding organization and with the nature of the research problem posed. The ideal situation is to have few constraints on the nature of the final report, but sometimes research reports are suppressed or distorted because the researcher comes to conclusions that the funding organization does not like.

Applied reports that are written in a less highly charged environment can face another problem—even when they are favorably received by the funding organization their conclusions are often ignored. This problem can be more a matter of the organization not really knowing how to use research findings than not wanting to use them. And this is not just a problem of the funding organization; many researchers are prepared only to present their findings, without giving any thought to how research findings can be translated into organizational policies or programs.

An Advisory Committee

An advisory committee can help the applied researcher avoid the problems of incompatible expectations for the final report and insufficient understanding of how to use the research results, without adopting the more engaged strategy of Whyte's participatory action research or Guba and Lincoln's (1989) constructivist inquiry. An advisory committee should be formed before the start of the project to represent the various organizational segments with stakes in the outcomes of the research. The researcher can use the committee as a source of intelligence about how particular findings may be received and as a sounding board for ideas about how the organization or agency can use research findings. Perhaps most important, an advisory committee can help the researcher work out many problems in research design, implementation, and data collection. Because an advisory committee is meant to comprise all stakeholders, it is inevitable that conflicts will arise between advisory group members. In our experience, however, these conflicts almost invariably can be used to strategize more effectively about the research design and the final product.

Advisory committees are particularly necessary for research investigating controversial issues. For example, after a study conducted in 1999 found that several death row inmates had been wrongly convicted of their crimes, the governor of Illinois placed a moratorium on all death sentences in the state. Other results of the study suggested that the death penalty was handed down unfairly; it found proportionally more minority and poor offenders were sentenced to death than whites and those who could afford hired legal counsel. This caused a great deal of media attention and calls for other states that practice the death penalty to institute similar moratoriums. As a consequence of this attention, other states have begun to examine their implementation of the death penalty as well. Maryland is one such state. In 2001, the state legislature in Maryland commissioned Raymond Paternoster at the University of Maryland to conduct a study of its practice of the death penalty. The primary goal of the study was to determine whether the administration of the death penalty in the state was affected by the race of the defendant or victim.

At this writing, the results of the study were not yet released, but as you can imagine, they will inevitably be very controversial, regardless of what is found. For example, if racial bias is found, prosecutors may question the study's methodology. Similarly, if racial bias is not found, defense attorneys and other interest groups may also question the study's methodology. To make sure all interests are represented, Paternoster set up an advisory committee before undertaking the study. The advisory committee consists of a group of prosecutors and defense counsel who have had experience in capital cases. They will advise Professor Paternoster on several critical issues, including the years that the study should cover, the sources where information can be found, and the particularly important variables related to sentencing outcomes. Not only will the advisory committee provide substantive input into the research, but by having a broad spectrum of the legal community "on board," it also will lend credibility to the study's findings.

A BIT MORE ON WRITING

"Perfectionism is the voice of the oppressor, the enemy of the people. It will keep you cramped and insane your whole life and it is the main obstacle between you and a shitty first draft." (Lamott 1994:28)

We often hear lamentations from students such as "It is impossible to know where to begin" or "I have a hard time getting started." To this, we say, begin wherever you are most comfortable but begin early! You do not have to start with the introduction; start in the method section if you prefer. The main point is to begin somewhere and then keep typing, keep typing, and keep typing! It is always easier to rewrite a paper than it is to write the first draft. The fine art of writing is really in the rewriting!

Those of you who began with a research proposal have a head start; you will find that the final report is much easier to write, and more adequate you write more material for it as you work out issues during the project. It is very disappointing to discover that something important was left out when it is too late to do anything about it. And we do not need to point out that students (and professional researchers) often leave final papers (and reports) until the last possible minute (often for understandable reasons, including other coursework and job or family responsibilities). But be forewarned: The last-minute approach does not work for research reports.

A successful report must be well-organized and clearly written. Getting to such a product is a difficult but not impossible goal. Consider the following principles formulated by experienced writers (Booth, Colomb, & Williams 1995:150–151):

- Start with an outline.
- Respect the complexity of the task, and do not expect to write a polished draft in a linear fashion. Your thinking will develop as you write, causing you to reorganize and rewrite.
- Leave enough time for dead ends, restarts, revisions, and so on, and accept the fact that you will discard much of what you write.
- Write as fast as you comfortably can. Do not worry about spelling, grammar, and so on until you are polishing things up.
- Ask all the people whom you trust for their reactions to what you have written.
- Write as you go along, so you have notes and report segments drafted even before you focus on writing the report.

It is important to remember that no version of a manuscript is ever final. As you write, you will get new ideas about how to organize the report. Try them out. As you review the first draft, you will see many ways to improve your writing. Focus particularly on how to shorten and clarify your statements. Make sure each paragraph concerns only one topic. Remember the golden rule of good writing: Writing is revising!

Another useful tip is called reverse outlining. After you have written a first complete draft, outline it on a paragraph-by-paragraph basis, ignoring the actual section headings you used. See if the paper you wrote actually fits the outline you planned. How could the organization be improved?

And perhaps most important, leave yourself enough time so that you can revise, several times if possible, before turning in the final draft.

A well-written research report requires (to be just a bit melodramatic) blood, sweat, and tears and more time than you will at first anticipate. But the process of writing one will help you to write the next. And the issues you consider, if you approach your writing critically, will be sure to improve your subsequent research projects and sharpen your evaluations of others.

Those of you interested in a more focused discussion of writing in general (e.g., grammar, elements of style, emotional aspects of writing) should see Becker (1986), Booth, Colomb, and Williams (1995), Mullins (1977), Strunk and White (1979), and Turabian (1967).

DATA DISPLAYS

You learned in Chapter 11 about some of the statistics that are useful in analyzing and reporting data, but there are some additional methods of presenting statistical results that can improve research reports. Combined and compressed displays are used most often in applied research reports and government documents, but they can also help to communicate findings more effectively in student papers and journal articles.

In a **combined frequency display**, the distributions for a set of conceptually similar variables with the same response categories are presented together, with common headings for the responses. For example, you could identify the variables in the leftmost column and the value labels along the top. Exhibit 13.6 is a combined display reporting the frequency distributions in percentage form for five variables that indicate the responses of high school seniors to questions about their delinquent behavior. From this table you can infer several pieces of information besides the basic distribution of self-reported delinquency. You can determine the variation in the cohort's involvement in specific types of delinquent behavior and you can also determine whether this behavior increased or decreased over time (1982–1994). For example, although the majority of seniors reported having an argument or fight with their parents, only a small percentage of students (around 20% on average) reported getting into a fight where a group of their friends were against another group, or getting into trouble with police (around 25% on average). You can also see that the rate of fighting behavior did not change much over the 13-year time period examined; however, the percentage of youth getting into trouble with the police actually decreased.

Compressed frequency displays can also be used to present crosstabular data and summary statistics more efficiently, by eliminating unnecessary percentages (such as those corresponding to the second value of a dichotomous variable) and by reducing the need for repetitive labels. Exhibit 13.7 presents a compressed display used to highlight the results of a survey on attitudes toward problems facing the country by several demographic characteristics. For several problem areas, respondents were asked to tell interviewers if they believed the country was "about the same today," "making progress in this area," or "losing ground" on this area. The percentages displayed are for those respondents who said that they believed the country was "losing ground" on a particular problem area.

Combined and compressed statistical displays present a large amount of data in a relatively small space. To the experienced reader of statistical reports, such displays can convey much important information. They should be used with caution, however, because people who are not used to them may be baffled by the rows of numbers. Graphs can also provide an efficient tool for summarizing relationships among variables. Exhibit 13.8 is from the evaluation report of the Children at Risk Program performed by Harrell, Cavanagh, and Sridharan (1999) (see Chapter 9). It presents the percentage of youth who used drugs at different times during the course of the evaluation for both treatment (youth who participated in the program) and control (youth who did not) groups. As you can readily see, compared to youth who did not receive the CAR program, fewer youth who participated in the CAR program used drugs at all times that were measured.

EXHIBIT 13.6 Combined Frequency Display of High School Seniors Reporting Involvement in Selected Delinquent Activities in the Past 12 Months, United States, 1986–1994

Delinquent Activity	Class of 1987 (N = 3,179)	Class of 1988 (N = 3,361)	Class of 1989 (N = 3,350)	Class of 1990 (N = 2,879)	Class of 1991 (N = 2,627)	Class of 1992 (N = 2,569)	Class of 1993 (N = 2,690)
Argued or had a fight with either of your parents?							
Not at all	8.8	9.7	9.6	9.3	10.0	9.3	12.1
Once	8.5	8.2	8.7	8.8	8.9	8.7	9.4
Twice	12.1	11.0	10.2	12.8	12.7	11.7	12.4
3 or 4 times	23.1	23.7	23.6	23.2	24.7	24.7	20.2
5 or more times	47.5	47.5	47.9	45.9	43.6	45.5	45.9
Taken part in a fight where a group of your friends were against another group?							
Not at all	80.4	80.5	79.7	81.1	79.6	78.7	77.8
Once	11.3	11.1	12.1	11.4	11.2	11.5	11.2
Twice	4.4	4.4	3.9	4.4	5.0	4.4	5.8
3 or 4 times	2.6	2.4	2.4	1.9	2.5	3.2	2.9
5 or more times	1.4	1.6	1.8	1.2	1.7	2.2	2.3
Gotten into trouble with police because of something you did?							
Not at all	75.9	77.5	76.6	75.8	77.4	77.8	90.4
Once	15.3	12.8	13.7	13.2	12.4	11.9	5.9
Twice	4.5	6.2	5.5	6.0	6.0	5.2	1.8
3 or 4 times	2.8	2.4	2.6	3.4	2.7	3.0	1.2
5 or more times	1.5	1.1	1.6	1.6	1.5	2.2	0.6

Source: Monitoring the Future Data provided by Monitoring the Future Project, Survey Research Center, Lloyd D. Johnston, Jerald G. Bachman, and Patrick M. O'Malley, Principal Investigators. Table adapted from Sourcebook of Criminal Justice Statistics: 1994 (NCJ-154591). Bureau of Justice Statistics, U.S. Department of Justice, Washington, DC.

EXHIBIT 13.7   Compressed Display of Attitudes Toward Problems Affecting the Country Today by Demographic Characteristics, 1995

Question: "Next, as I read you some problem areas, please tell me how you think each is affecting the country today. First, do you think the problem of . . . is about the same today, is the country making progress in this area, or is the country losing ground?"

	Percentage Responding "Losing Ground"		
	Crime (in %)	Families Split Up (in %)	Drugs (in %)
Sex			
Male	71	73	60
Female	82	77	71
Race			
White	77	75	64
Nonwhite	77	75	79
Age			
Under 30 years	73	75	64
30 to 49 years	79	73	63
50 to 64 years	78	76	69
65 years and older	77	76	70
Education			
College graduate	71	71	64
Some college	84	79	67
High school graduate	80	77	66
Less than high school graduate	69	72	66

Source: Interview data from a nationwide sample of 1,800 adults by Princeton Survey Research Associates for the Pew Research Center for The People & The Press. Adapted from the *Sourcebook of Criminal Justice Statistics: 1995* (NCJ-158900). Bureau of Justice Statistics, U.S. Department of Justice, Washington, DC.

Another good example of the use of graphs to show relationships is provided by a Bureau of Justice Statistics report on age patterns of victims of violent crime (Perkins 1997). Exhibit 13.9, taken from that report, shows how the rates of violent crimes have varied by particular age groups over time. You can see that whereas violent crime victimization rates have remained relatively stable over time for older age cohorts, younger age cohorts, particularly those under the age of 19, experienced increases in their rate of victimization during the late 1980s and early 1990s.

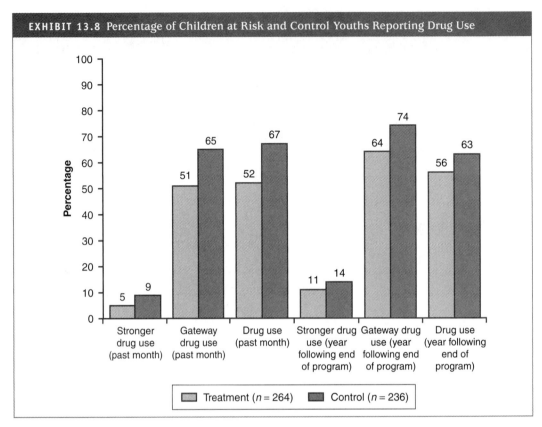

EXHIBIT 13.8 Percentage of Children at Risk and Control Youths Reporting Drug Use

Source: Harrell, A., Cavanagh, S., & Sridharan, S. 1999. *Evaluation of the Children at Risk Program: Results 1 Year After the End of the Program* (NCJ-178914). National Institute of Justice: Research in Brief. Washington, D.C.

ETHICS AND REPORTING

It is at the time of reporting research results that the researcher's ethical duty to be honest becomes paramount. Here are some guidelines:

- *Provide an honest accounting of how the research was carried out and where the initial research design had to be changed.* Readers do not have to know about every change you made in your plans and each new idea you had, but they should be informed about major changes in hypotheses or research design.
- *Maintain a full record of the research project so that questions can be answered if they arise.* Many details will have to be omitted from all but the most comprehensive reports, but these omissions should not make it impossible to

EXHIBIT 13.9 Violent Crime Rates by Age

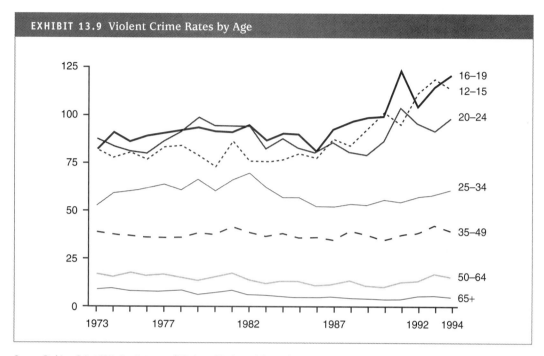

Source: Perkins, C.A. 1997. *Age Patterns of Victims of Serious Violent Crime* (NCJ-162031). Bureau of Justice Statistics. Washington, D.C.

track down answers to specific questions about research procedures that may arise in the course of data analysis or presentation.

- *Avoid "lying with statistics" or using graphs to mislead.* (See Chapter 12.)
- *Acknowledge the sponsors of the research.* In part, this is so that others can consider whether this sponsorship may have tempted you to bias your results in some way; and thank staff who made major contributions.
- *Be sure that the order of authorship for coauthored reports is discussed in advance and reflects agreed-upon principles. Be sensitive to coauthors' needs and concerns.*

Ethical research reporting should not mean ineffective reporting. You need to tell a coherent story in the report and to avoid losing track of the story in a thicket of minuscule details. You do not need to report every twist and turn in the conceptualization of the research problem or the conduct of the research, but be suspicious of reports that do not seem to admit to the possibility of any room for improvement. Social science is an ongoing enterprise in which one research report makes its most valuable contribution by laying the groundwork for another, more sophisticated research project. Highlight important findings in the research report, but use the research report also to point out what are likely to be the most productive directions for future researchers.

CONCLUSION

Good critical skills are essential when evaluating research reports, whether your own or those produced by others. There are always weak points in any research, even published research. It is an indication of strength, not weakness, to recognize areas where one's own research needs to be, or could have been, improved. And it is really not just a question of sharpening our knives and going for the jugular. You need to be able to weigh the strengths and weaknesses of particular research results and to evaluate a study in terms of its contribution to understanding its particular research question, not whether it gives a definitive answer for all time.

But this is not to say that anything goes. Much research lacks one or more of the three legs of validity—measurement validity, causal validity, or generalizability—and sometimes contributes more confusion than understanding about particular issues. Top journals generally maintain very high standards, partly because they have good critics in the review process and distinguished editors who make the final acceptance decisions. But some daily newspapers do a poor job of screening; and research reporting standards in many popular magazines, TV shows, and books are often abysmally poor. Keep your standards high and your view critical when reading research reports, but not so high or so critical that you turn away from studies that make tangible contributions to the literature, even if they do not provide definitive answers. And don't be so intimidated by the need to maintain high standards that you shrink from taking advantage of opportunities to conduct research yourself.

The growth of social science methods from infancy to adolescence, perhaps to young adulthood, ranks as a key intellectual accomplishment of the 20th century. Opinions about the causes and consequences of crime no longer need depend on the scattered impressions of individuals, and criminal justice policies can be shaped by systematic evidence of their effectiveness.

Of course, social research methods are no more useful than the commitment of the researchers to their proper application. Research methods, like all knowledge, can be used poorly or well, for good purposes or bad, when appropriate or not. A claim that a belief is based on social science research in itself provides no extra credibility. As you have learned throughout this book, we must first learn which methods were used, how they were applied, and whether interpretations square with the evidence. To investigate the social world, we must keep in mind the lessons of research methods.

KEY TERMS

Combined frequency display
Compressed display
Constructivist paradigm

Participatory action research
Peer review

HIGHLIGHTS

- Proposal writing should be a time for clarifying the research problem, reviewing the literature, and thinking ahead about the report that will be required.

- Relations with research subjects and consumers should be developed in a manner that achieves key research goals and preparation of an effective research report. The traditional scientific approach of minimizing the involvement of research subjects and consumers in research decisions has been challenged by proponents of participatory action research and adherents of the constructivist paradigm.

- Different types of reports typically pose different problems. Authors of student papers must be guided in part by the expectations of their professor. Thesis writers have to meet the requirements of different committee members, but can benefit greatly from the areas of expertise represented on a typical thesis committee. Applied researchers are constrained by the expectations of the research sponsor; an advisory committee from the applied setting can help to avoid problems. Journal articles must pass a peer review by other social scientists and often are much improved in the process.

- Research reports should include an introductory statement of the research problem, a literature review, a methodology section, a findings section with pertinent data displays, and a conclusions section that identifies any weaknesses in the research design and points out implications for future research and theorizing. This basic report format should be modified according to the needs of a particular audience.

- All reports should be revised several times and critiqued by others before they are presented in final form.

- Some of the data in many reports can be displayed more efficiently by using combined and compressed statistical displays.

- The central ethical concern in research reporting is to be honest. This honesty should include providing a truthful accounting of how the research was carried out, maintaining a full record about the project, using appropriate statistics and graphs, acknowledging the research sponsors, and being sensitive to the perspectives of coauthors.

EXERCISES

1. Select a recent article published in a peer-reviewed criminological journal and answer the following questions: How effective is the article in conveying the design and findings of the research? Could the article's organization be improved at all? Are there bases for disagreement about the interpretation of the findings?

2. Formulate a research problem, review the decision checklist (see Exhibit 13.3), and write a research proposal. You will find that this is a major project, so you should attempt it only if you will have a month or more to work on it. Alternatively, just answer selected questions from the decision checklist about a research question you would like to pursue.

3. Call a local criminal justice official and arrange for an interview. Ask the official about his or her experience with applied research reports and his conclusions about the value of social research and the best techniques for reporting to practitioners.

4. Rate four criminological journal articles for overall quality of the research and for effectiveness of the writing and data displays. Discuss how each could have been improved.

5. How firm a foundation do social research methods provide for understanding the social world? Stage an in-class debate, with the pro and con arguments focusing on the variability of social research findings across different social contexts and the difficulty of understanding human subjectivity.

DEVELOPING A RESEARCH PROPOSAL

Now, it is time to bring all the elements of your proposal together.

1. Organize the proposal material you wrote for previous chapters in a logical order. Based on your research question, select the most appropriate research method as your primary method (see Chapters 5–10).

2. To add a multiple component to your research design, select another research method that could add knowledge about your research question.

3. Rewrite the entire proposal, adding an introduction. Also add sections that outline a budget and state the limitations of your study.

Student Study Site

The companion Web site for *The Practice of Research in Criminology and Criminal Justice*, Third Edition
   http://www.sagepub.com/prccj3
   Visit the Web-based Student Study Site to enhance your understanding of the chapter content and to discover additional resources that will take your learning one step further. You can enhance your understanding of the chapters by using the comprehensive study material, which includes e-flashcards, Web exercises, practice self-tests, and more. You will also find special features, such as Learning from Journal Articles, which incorporates SAGE's online journal collection.

WEB EXERCISES

1. Go to the National Science Foundation's Law and Social Science Program Web site at www.nsf.gov/sbe/ses/law/start.htm. What are the components that the National Science Foundation's Law and Social Science Program look for in a proposed piece of research? Write a detailed outline for a research proposal to study a subject of your choice to be submitted to the National Science Foundation for funding.

2. Go to the Ethics, Public Policy, and Public Administration site at http://plsc.uark.edu/book/books/ethics/index.htm. Choose "Cases" and then "Integrity and Separations" (No. 14). Read the case "Integrity and Separations." Briefly describe the issues surrounding social science research, methods, reporting research results, and ethics in the case. Now return to the home page of the Ethics site. Choose "Set of Questions" and answer the "Questions for Consideration" in terms of the case "Integrity and Separations."

3. Using the Web, find five different examples of criminological research projects that have been completed. Briefly describe each. How does each differ in its approach to reporting the research results? To whom do you think the author(s) of each is "reporting" (i.e., who is the "audience")? How do you think the predicted audience has helped to shape the author's approach to reporting the results? Be sure to note the Web sites at which you located each of your five examples.

ETHICS EXERCISES

1. You learned in Chapter 2 about the controversy resulting from Ehrlich's (1975) failure to make his data on the impact of capital punishment available for reanalysis by other researchers. Would you recommend legal regulations about the release of research data? What would those regulations be? Would they differ depending on the researcher's source of funding? Would you allow researchers exclusive access to their own data for some period of time after they have collected it?

2. Full disclosure of sources of research as well as of other medically-related funding has become a major concern for medical journals. Should researchers publishing in criminology and criminal justice journals also be required to fully disclose all sources of funding? Should full disclosure of all previous funds received by criminal justice agencies be required in each published article? What about disclosure of any previous jobs or paid consultations with criminal justice agencies? Write a short justification of the regulations you propose.

SPSS EXERCISES

1. How do friends' opinions and support for delinquent activities influence levels of delinquency among youth? A combined frequency display of the distributions of a series of YOUTH.POR variables will help you to answer this question.
   a. Obtain a frequency distribution and descriptive statistics for friends' attitudes toward delinquent acts index (FROPINION), index of friends' engagement in delinquent acts (FRBEHAVE), and the delinquency index at time 11 (DELINQ1).
   b. Using the mean as the measure of center, recode the three indexes to measure low and high levels of each variable (e.g., all values below the mean represent low levels of a given variable and all values above the mean represent high levels of the given variable).
   c. Use the percentages in these distributions to prepare a combined frequency display.
   d. Discuss what you have learned about the influence of friends' delinquent tendencies on delinquency levels among youth.

2. Repeat Exercise 1 using the parental index (PARNT2), certainty of punishment index (CERTAIN), and the morality index (MORAL). Note: The scale for the variables of interest in this exercise are measured opposite the delinquency scale. For example, high scores on the parental index indicate high parental supervision, which in theory should correspond with low levels of delinquency. Be sure to place the percentages for DELINQ1 accordingly (make note of the placement of the percentiles for this variable or create more intuitive column labels).

Conducting Literature Reviews and Finding Information

Social Research Foundations—
How to Conduct a Literature Review

Case Study: Arrest and Domestic Violence

How do we find prior research on questions of interest? You may already know some of the relevant material from prior coursework or your independent reading, but that would not be enough. When you are about to launch an investigation of a new research question, you must apply a very different standard than when you are studying for a test or just seeking to learn about domestic violence. You need to find reports of previous investigations that sought to answer the same research question that you wish to answer, not just those that were about a similar topic. If there are no prior studies of exactly the same research question on which you wish to focus, you should seek to find reports from investigations of very similar research questions. Once you have located reports from prior research similar to the research you wish to conduct, you may expand your search to include investigations about related topics or studies that used similar methods.

Sometimes, you will find that someone else has already reviewed the literature on your research question in a special review article or book chapter. For example, Chalk and Garner (2001) published an excellent review of the research on arrest and domestic violence in the journal *New Directions for Evaluation*. Most of the research articles you find will include a literature review. These reviews can help a lot, but they are no substitute for reviewing the literature yourself. No one but you can decide what is relevant for your research question and the research circumstances you will be facing: the setting you will study, the timing of your study, the new issues that you want to include in your study, and your specific methods. And you cannot depend on any published research review for information on the most recent work. New research results about many questions appear continually in scholarly journals and books, in research reports from government agencies and other organizations, and on Web sites all over the world; you will need to check for new research such as this yourself.

Finding Information

Conducting a thorough search of the research literature and then reviewing critically what you have found is an essential foundation for any research project. Fortunately, much of this information can be identified online, without leaving your desktop, and an increasing number of published journal articles can be downloaded directly to your own computer (depending on your particular access privileges). But just because there is a lot available online does not mean that you need to find it all. Keep in mind that your goal is to find reports of prior research investigations, which means that you should focus on scholarly journals that choose articles for publication after they have been reviewed by other social scientists: "refereed" journals. Newspaper and magazine articles just will not do, although you may find some that raise important issues or even that summarize social science research investigations.

Every year, the Web offers more and more useful material, including indexes of the published research literature. You may find copies of particular rating scales, reports from research in progress, papers that have been presented at professional conferences, and online discussions of related topics. This section reviews the basic procedures for finding relevant research information in both the published literature and on the Web, but keep in mind that the primary goal is to identify research articles published in refereed journals.

Searching the Literature

The social science literature should be consulted at the beginning and end of an investigation. Even while an investigation is in progress, consultations with the literature may help to resolve methodological problems or facilitate supplementary explorations. As with any part of the research process, the method you use will affect the quality of your results. You should try to ensure that your search method includes each of the steps below.

Specify your research question. Your research question should be neither so broad that hundreds of articles are judged relevant nor so narrow that you miss important literature. "Is informal social control effective?" is probably too broad. "Does informal social control reduce rates of burglary in large cities?" is probably too narrow. "Is informal social control more effective in reducing crime rates than policing?" provides about the right level of specificity.

Identify appropriate bibliographic databases to search. Criminal Justice Abstracts and *Sociological Abstracts* may meet many of your needs, but if you are studying a question about medical consequences or other health issues, you should also search in *Medline*, the database for searching the medical literature. If your focus is on mental health, you will also want to include a search in the online *Psychological Abstracts* database, *PsycINFO*, or the version that also contains the full text of articles since 1985, *PsycARTICLES*. To find articles that refer to a previous publication, such as Sherman and Berk's (1984) study of the police response to domestic violence, the *Social Science Citation Index* would be helpful. In addition, the search engine Google now offers anyone with Web access Google Scholar (which indexes and searches the full text of selected journals) and Google Print (which digitizes and searches the full text of the books that are owned by selected research libraries). (At the time this book went to press, the Google Print project was on hold due to copyright concerns raised by some publishers, while the search engine and directory Yahoo! was starting a similar venture that focused only on older books that are no longer covered by copyright law; Hafner, 2005:C1.)

Choose a search technology. For most purposes, an online bibliographic database that references the published journal literature will be all you need. However, searches for unpopular topics or very recent literature may require that you also search Web sites or bibliographies of relevant books.

Create a tentative list of search terms. List the parts and subparts of your research question and any related issues that you think are important: "informal social control," "policing," "influences on crime rates," and perhaps "community cohesion and crime." List the authors of relevant studies. Specify the most important journals that deal with your topic.

Narrow your search. The sheer number of references you find can be a problem. For example, searching for "social capital" resulted in 2,293 citations in *Sociological Abstracts*. Depending on the database you are working with and the purposes of your search, you may want to limit your search to English-language publications, to journal articles rather than conference papers or dissertations (both of which are more difficult to acquire), and to materials published in recent years.

Refine your search. Learn as you go. If your search yields too many citations, try specifying the search terms more precisely. If you have not found much literature, try using more general terms. Whatever terms you search first, do not consider your search complete until you have tried several different approaches and have seen how many articles you find. A search for "domestic violence" in *Sociological Abstracts* on September 11, 2005, yielded 1,569 hits; adding "effects" or "influences" as required search terms dropped the number of hits to 370.

Use Boolean search logic. It is often a good idea to narrow down your search by requiring that abstracts contain combinations of words or phrases that include more of the specific details of your research question. Using the Boolean connector "and" allows you to do this, whereas using the connector "or" allows you to find abstracts containing different words that mean the same thing. Exhibit A.1 provides an example.

Use appropriate subject descriptors. Once you have found an article that you consider to be appropriate, take a look at the "descriptors" field in the citation (see Exhibit A.2). You can then redo your search after requiring that the articles be classified with some or all of these descriptor terms.

Check the results. Read the titles and abstracts you have found, and identify the articles that appear to be most relevant. If possible, click on these article titles and generate a list of their references. See if you find more articles that are relevant to your research question but that you have missed so far. You will be surprised (I always am) at how many important articles your initial online search missed.

Read the articles. Now, it is time to find the full text of the articles of interest. If you are lucky, some of the journals you need will be available to patrons of your library in online versions, and you will be able to link to the full text just by clicking on a "full text" link. But many

EXHIBIT A.1   Use of Boolean Connectors in a Literature Search

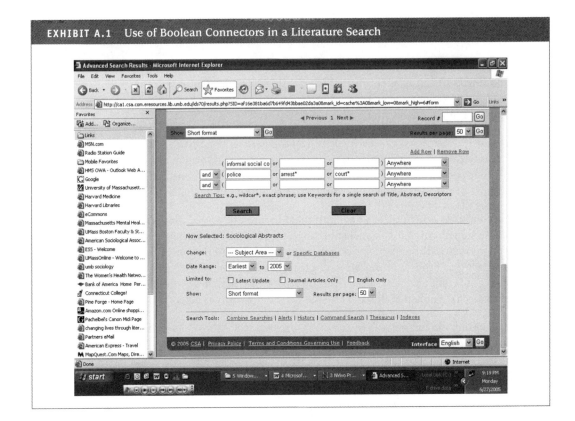

journals, specific issues of some journals, or both will be available only in print; in this case, you will have to find them in your library or order a copy through interlibrary loan.

Refer to a good book for even more specific guidance. Fink's (2004) *Conducting Research Literature Reviews: From the Internet to Paper* is an excellent guide.

You may be tempted to write up a "review" of the literature based on reading the abstracts or using only those articles available online, but you will be selling yourself short. Many crucial details about methods, findings, and theoretical implications will be found only in the body of the article, and many important articles will not be available online. To understand, critique, and really learn from previous research studies, you must read the important articles, no matter how you have to retrieve them.

If you have done your job well, you will now have more than enough literature as background for your own research, unless it is on a very obscure topic (see Exhibit A.3). (Of course, ultimately your search will be limited by the library holdings you have access to and by the time you have to order or find copies of journal articles, conference papers, and perhaps dissertations that you cannot obtain online.) At this point, your main concern is to construct a coherent framework in which to develop your research question, drawing as many lessons as you can from previous research. You may use the literature to

EXHIBIT A.2   Checking Standard Subject Matter Descriptors

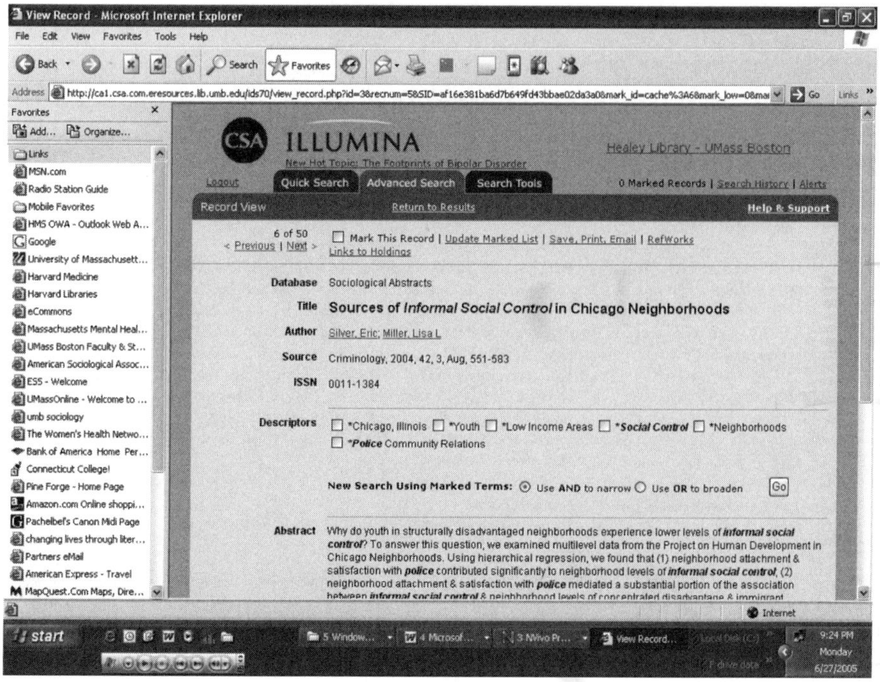

identify a useful theory and hypotheses to be reexamined, to find inadequately studied specific research questions, to explicate the disputes about your research question, to summarize the major findings of prior research, and to suggest appropriate methods of investigation.

Be sure to take notes on each article you read, organizing your notes into standard sections: Theory, Methods, Findings, Conclusions. In any case, write the literature review so that it contributes to your study in some concrete way; do not feel compelled to discuss an article just because you have read it. Be judicious. You are conducting only one study of one issue; it will only obscure the value of your study if you try to relate it to every tangential point in related research.

Don't think of searching the literature as a one-time-only venture, something that you leave behind as you move on to your "real" research. You may encounter new questions or unanticipated problems as you conduct your research or as you burrow deeper into the literature. Searching the literature again to determine what others have found in response to these questions or what steps they have taken to resolve these problems can yield substantial improvements in your own research. There is so much literature on so many topics that often it is not possible to figure out in advance every subject you should search the literature for or what type of search would be most beneficial.

EXHIBIT A.3   A Search in *Sociological Abstracts* on Informal Control

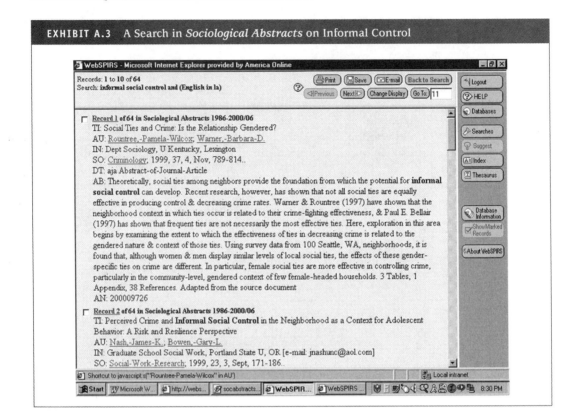

Another reason to make searching the literature an ongoing project is that the literature is always growing. During the course of one research study, whether it takes only one semester or several years, new findings will be published and relevant questions will be debated. Staying attuned to the literature and checking it at least when you are writing up your findings may save your study from being outdated.

Searching the Web

The World Wide Web provides access to vast amounts of information of many different sorts (Ó Dochartaigh 2002). You can search the holdings of other libraries and download the complete text of government reports, some conference papers, and newspaper articles. You can find policies of local governments, descriptions of individual social scientists and particular research projects, and postings of advocacy groups. It is also hard to avoid finding a lot of information in which you have no interest, such as commercial advertisements, third-grade homework assignments, or college course syllabi. In 1999, there were already about 800 million publicly available pages of information on the Web (Davis 1999). Today, there may be as many as 15 billion pages on the Web (Novak 2003).

After you are connected to the Web with a browser such as Microsoft Internet Explorer or Netscape Navigator, you can use three basic strategies for finding information: direct addressing (i.e., typing in the address, or uniform resource locator [URL], of a specific site); browsing (i.e., reviewing online lists of Web sites); and searching (i.e., Google is currently the most popular search engine for searching the Web). For some purposes, you will need to use only one strategy; for other purposes, you will want to use all three.

Exhibit A.4 illustrates the first problem that you may encounter when searching the Web: the sheer quantity of resources that are available. It is a much bigger problem than when searching bibliographic databases. On the Web, less is usually more. Limit your inspection of Web sites to the first few pages that turn up in your list (they are ranked by relevance). See what those first pages contain, and then try to narrow your search by including some additional terms. Putting quotation marks around a phrase that you want to search will also help to limit your search; for example, searching for "informal social control" on Google (on September 11, 2005) produced 31,100 sites, compared to the roughly 15,500,000 sites retrieved when we omitted the quotes wherein Google searched "informal" and "social" and "control."

EXHIBIT A.4   Google Search Results for "Informal Social Control"

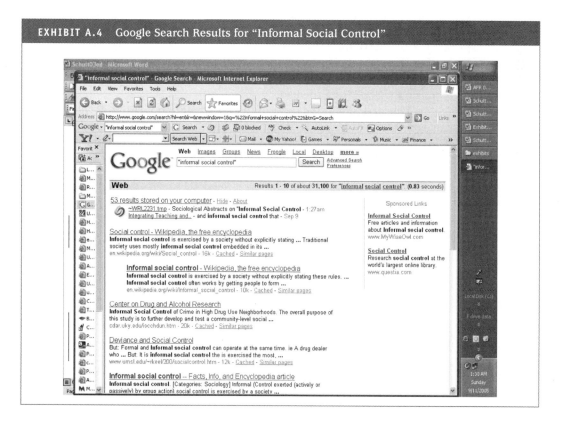

Remember the following warnings when you conduct searches on the Web:

- *Clarify your goals.* Before you begin the search, jot down the terms that you think you need to search for as well as a statement of what you want to accomplish with your search. This will help to ensure that you have a sense of what to look for and what to ignore.
- *Quality is not guaranteed.* Anyone can post almost anything, so the accuracy and adequacy of the information you find are always suspect. There is no journal editor or librarian to evaluate quality and relevance.
- *Anticipate change.* Web sites that are not maintained by stable organizations can come and go very quickly. Any search will result in attempts to link to some URLs that no longer exist.
- *One size does not fit all.* Different search engines use different procedures for indexing Web sites. Some attempt to be all-inclusive, whereas others aim to be selective. As a result, you can get different results from different search engines (e.g., Google or Yahoo!) even though you are searching for the same terms.
- *Be concerned about generalizability.* You might be tempted to characterize police department policies by summarizing the documents you find at police department Web sites. But how many police departments are there? How many have posted their policies on the Web? Are these policies representative of all police departments? To answer all these questions, you would have to conduct a research project just on the Web sites themselves.
- *Evaluate the sites.* There is a lot of stuff out there, so how do you know what is good? Some Web sites contain excellent advice and pointers on how to differentiate the good from the bad.
- *Avoid Web addiction.* Another danger of the enormous amount of information available on the Web is that one search will lead to another and to another and so on. There are always more possibilities to explore and one more interesting source to check. Establish boundaries of time and effort to avoid the risk of losing all sense of proportion.
- *Cite your sources.* Using text or images from Web sources without attribution is plagiarism. It is the same as copying someone else's work from a book or article and pretending that it is your own. Record the Web address (URL), the name of the information provider, and the date on which you obtain material from the site. Include this information in a footnote to the material that you use in a paper.

Reviewing Research

Effective review of the prior research you find is an essential step in building the foundation for new research. You must assess carefully the quality of each research study, consider the implications of each article for your own plans, and expand your thinking about your research question to take account of new perspectives and alternative arguments. It is through reviewing the literature and using it to extend and sharpen your own ideas and methods that you become a part of the social science community. Instead of being just one individual studying an issue that interests you, you are building on an ever-growing body of knowledge that is being constructed by the entire community of scholars.

The research information you find on various Web sites comes in a wide range of formats and represents a variety of sources. *Caveat emptor* (buyer beware) is the watchword when you search the Web; following review guidelines such as those we have listed will minimize, but not eliminate, the risk of being led astray. By contrast, the published scholarly journal literature that you find in databases such as *Sociological Abstracts* and *Psychological Abstracts* follows a much more standard format and has been subject to a careful review process. There is some variability in the contents of these databases: some journals publish book reviews, comments on prior articles, dissertation abstracts, and conference papers. However, most literature you will find on a research topic in these databases represents peer-reviewed articles reporting analyses of data collected in a research project. These are the sources on which you should focus. This section concentrates on the procedures you should use for reviewing these articles. These procedures also can be applied to reviews of research monographs: books that provide much more information from a research project than that contained in a journal article.

Reviewing the literature is really a two-stage process. In the first stage, you must assess each article separately. This assessment should follow a standard format such as that represented by the "Questions to Ask About a Research Article" in Appendix B. However, you should keep in a mind that you cannot adequately understand a research study if you just treat it as a series of discrete steps, involving a marriage of convenience among separate techniques. Any research project is an integrated whole, so you must be concerned with how each component of the research design influenced the others, for example, how the measurement approach might have affected the causal validity of the researcher's conclusions and how the sampling strategy might have altered the quality of measures.

The second stage of the review process is to assess the implications of the entire set of articles (and other materials) for the relevant aspects of your research question and procedures and then to write an integrated review that highlights these implications. Although you can find literature reviews that consist simply of assessments of one published article after another—that never get beyond stage one in the review process—your understanding of the literature and the quality of your own work will be much improved if you make the effort to write an integrated review.

In the next two sections, we will show how you might answer many of the questions in Appendix B as we review a research article about domestic violence. We will then show how the review of a single article can be used within an integrated review of the body of prior research on this research question. Because at this early point in the text you will not be familiar with all the terminology used in the article review, you might want to read through the more elaborate article review in Appendix C later in the course.

A Single-Article Review: Formal and Informal Deterrents to Domestic Violence

Anthony Pate and Edwin Hamilton at the National Police Foundation designed one of the studies funded by the U.S. Department of Justice to replicate the Minneapolis Domestic Violence Experiment. This section examines the article that resulted from that replication, which was published in the *American Sociological Review* (Pate & Hamilton 1992). The numbers in brackets refer to the article review questions in Appendix B.

The Research Question. Like Sherman and Berk's (1984) original Minneapolis study, Pate and Hamilton's (1992) Metro-Dade spouse assault experiment sought to test the deterrent effect of arrest in domestic violence cases, but with an additional focus on the role of informal social control [1]. The purpose of the study was explanatory because the goal was to explain variation in the propensity to commit spouse abuse [2]. Deterrence theory provided the theoretical framework for the study, but this framework was broadened to include the proposition by Williams and Hawkins (1986) that informal sanctions such as stigma and the loss of valued relationships augment the effect of formal sanctions such as arrest [3]. Pate and Hamilton's literature review referred, appropriately, to the original Sherman and Berk research, to the other studies that attempted to replicate the original findings, and to research on informal social control [4].

There is no explicit discussion of ethical guidelines in the article, although reference is made to a more complete unpublished report [6]. Clearly, important ethical issues had to be considered, given the experimental intervention in the police response to serious assaults, but the adherence to standard criminal justice procedures suggests attention to the welfare of victims as well as the rights of suspects. We will consider these issues in more detail later in this chapter.

The Research Design. Developed as a follow-up to the original Minneapolis experiment, the Metro-Dade experiment exemplified the guidelines for scientific research that were presented in Chapter 2 [5]. It was designed systematically, with careful attention to specification of terms and clarification of assumptions, and focused on the possibility of different outcomes rather than certainty about one preferred outcome. The major concepts in the study, formal and informal deterrence, were defined clearly [7] and then measured with straightforward indicators: arrest or nonarrest for formal deterrence and marital status and employment status for informal deterrence. However, the specific measurement procedures for marital and employment status were not discussed, and no attempt was made to determine whether they captured adequately the concept of informal social control.

Three hypotheses were stated and also related to the larger theoretical framework and prior research [8]. The study design focused on the behavior of individuals [11] and collected data over time, including records indicating subsequent assault up to 6 months after the initial arrest [12]. The project's experimental design was used appropriately to test for the causal effect of arrest on recidivism [13]. The research project involved all eligible cases, rather than a sample of cases, but there were a number of eligibility criteria that narrowed the ability to generalize these results to the entire population of domestic assault cases in the Metro-Dade area or elsewhere [14]. There is a brief discussion of the 92 eligible cases that were not given the treatment to which they were assigned, but it does not clarify the reasons for the misassignment [15].

The Research Findings and Conclusion. Pate and Hamilton's (1992) analysis of the Metro-Dade experiment was motivated by concern with effect of social context because the replications in other cities of the original Minneapolis domestic violence experiment had not had consistent results [19]. Their analysis gave strong support to the expectation that informal social control processes are important: As they had hypothesized, arrest had a deterrent effect on suspects who were employed but not on those who were unemployed (see Exhibit A.5).

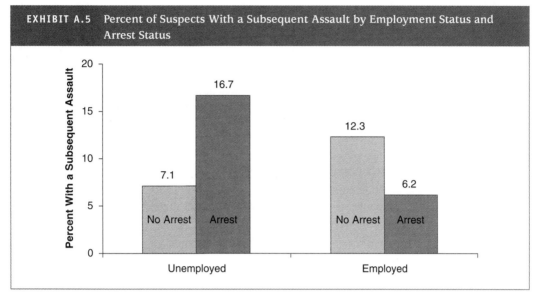

EXHIBIT A.5 Percent of Suspects With a Subsequent Assault by Employment Status and Arrest Status

Source: Pate & Hamilton 1992:695. Reprinted with permission.

However, marital status had no such effect [20]. The subsequent discussion of these findings gives no attention to the implications of the lack of support for the effect of marital status [21], but the study represents an important improvement over earlier research that had not examined informal sanctions [22]. The need for additional research is highlighted, and the importance of the findings for social policy are discussed: Pate and Hamilton suggest that their finding that arrest deters only those who have something to lose (e.g., a job) must be taken into account when policing policies are established [23].

Overall, the Pate and Hamilton (1992) study represents an important contribution to understanding how informal social control processes influence the effectiveness of formal sanctions such as arrest. Although the use of a population of actual spouse assault cases precluded the use of very sophisticated measures of informal social control, the experimental design of the study and the researchers' ability to interpret the results in the context of several comparable experiments distinguishes this research as exceptionally worthwhile. It is not hard to understand why these studies continue to stimulate further research and ongoing policy discussions.

An Integrated Literature Review: When Does Arrest Matter?

The goal of the second stage of the literature review process is to integrate the results of your separate article reviews and develop an overall assessment of the implications of prior research. The integrated literature review should accomplish three goals: (1) summarize prior research, (2) critique prior research, and (3) present pertinent conclusions (Hart 1998:186–187).

Summarize prior research. Your summary of prior research must focus on the particular research questions that you will address, but you also may need to provide some more general background. Hoyle and Sanders (2000:14) begin their *British Journal of Criminology* research article about mandatory arrest policies in domestic violence cases with what they term a "provocative" question: What is the point of making it a crime for men to assault their female partners and ex-partners? They then review the different theories and supporting research that has justified different police policies: the "victim choice" position, the "pro-arrest" position, and the "victim empowerment" position. Finally, they review the research on the "controlling behaviors" of men that frames the specific research question on which they focus: how victims view the value of criminal justice interventions in their own cases (p. 15).

Ask yourself three questions about your summary of the literature:

1. *Have you been selective?* If there have been more than a few prior investigations of your research question, you will need to narrow your focus to the most relevant and highest quality studies. Do not cite a large number of prior articles "just because they are there."

2. *Is the research up-to-date?* Be sure to include the most recent research, not just the "classic" studies.

3. *Have you used direct quotes sparingly?* To focus your literature review, you need to express the key points from prior research in your own words. Use direct quotes only when they are essential for making an important point (Pyrczak 2005:51–59).

Critique prior research. Evaluate the strengths and weakness of the prior research. In addition to all the points you develop as you answer the Article Review Questions in Appendix B, you should also select articles for review that reflect work published in peer-reviewed journals and written by credible authors who have been funded by reputable sources. Consider the following questions as you decide how much weight to give each article:

1. *How was the report reviewed prior to its publication or release?* Articles published in academic journals go through a rigorous review process, usually involving careful criticism and revision. Top refereed journals may accept only 10% of submitted articles, so they can be very selective. Dissertations go through a lengthy process of criticism and revision by a few members of the dissertation writer's home institution. A report released directly by a research organization is likely to have had only a limited review, although some research organizations maintain a rigorous internal review process. Papers presented at professional meetings may have had little prior review. Needless to say, more confidence can be placed in research results that have been subject to a more rigorous review.

2. *What is the author's reputation?* Reports by an author or team of authors who have published other work on the research question should be given somewhat greater credibility at the outset.

3. *Who funded and sponsored the research?* Major federal funding agencies and private foundations fund only research proposals that have been evaluated carefully and ranked highly by a panel of experts. They also often monitor closely the progress of the research. This does not guarantee that every such project

report is good, but it goes a long way toward ensuring some worthwhile products. On the other hand, research that is funded by organizations that have a preference for a particular outcome should be given particularly close scrutiny (Locke, Silverman, & Spirduso 1998:37–44).

Present pertinent conclusions. Do not leave the reader guessing about the implications of the prior research for your own investigation. Present the conclusions you draw from the research you have reviewed. As you do so, follow several simple guidelines:

- Distinguish clearly your own opinion of prior research from conclusions of the authors of the articles you have reviewed.
- Make it clear when your own approach is based on the theoretical framework you are using rather than on the results of prior research.
- Acknowledge the potential limitations of any empirical research project. Do not emphasize problems in prior research that you cannot avoid either (Pyrczak 2005:53–56).
- Explain how the unanswered questions raised by prior research or the limitations of methods used in prior research make it important for you to conduct your own investigation (Fink 2004:190–192).

A good example of how to conclude an integrated literature review is provided by an article based on the replication in Milwaukee of the Minneapolis Domestic Violence Experiment. For this article, Paternoster et al. (1997) sought to determine whether police officers' use of fair procedures when arresting assault suspects would lessen the rate of subsequent domestic violence. Paternoster et al. conclude that there has been a major gap in the prior literature: "Even at the end of some seven experiments and millions of dollars, then, there is a great deal of ambiguity surrounding the question of how arrest impacts future spouse assault" (p. 164) Specifically, they note that each of the seven experiments focused on the effect of arrest itself but ignored the possibility that "particular kinds of police procedure might inhibit the recurrence of spouse assault" (p. 165).

So Paternoster et al. (1997) ground their new analysis in additional literature on procedural justice and conclude that their new analysis will be "the first study to examine the effect of fairness judgments regarding a punitive criminal sanction (arrest) on serious criminal behavior (assaulting one's partner)" (p. 172).

SEARCHING THE WEB

To find useful information on the Web, you have to be even more vigilant than when you search the literature directly. With billions of Web pages on the Internet, there is no limit to the amount of time you can squander and the volume of useless junk you can find as you conduct your research on the Web. However, we can share with you some good ways to avoid the biggest pitfalls.

Direct Addressing

Knowing the exact address (i.e., URL) of a useful Web site is the most efficient way to find a resource on the Web.

Professional Organizations

- American Society of Criminology (http://www.asc41.com)
- American Sociological Association (http://www.asanet.org)
- American Psychological Association (http://www.apa.org)

Government Sites

- U.S. Office of Justice Programs (http://www.ojp.usdoj.gov)
- U.S. Bureau of the Census (http://www.census.gov)

Journals and Newspapers

- *Annual Review of Sociology* (http://www/annualreviews.org)
- *The New York Times* (http://www.nytimes.com)

Bibliographic Formats for Citing Electronic Information

- Electronic reference formats suggested by the American Psychological Association (http://www.apastyle.org/elecref.html)
- Karla Tonella's Guide to Citation Style Guides (http://bailiwick.lib.uiowa.edu/journalism/cite.html) contains more than a dozen links to online style guides
- Style Sheets for Citing Resources (print and electronic) (http://www.lib.berkeley.edu/TeachingLib/Guides/Internet/Style.html)

When you find Web sites that you expect you will return to often, you can save their addresses as "bookmarks" or "favorites" in your Web browser. However, since these can very quickly multiply, you should try to be selective.

Browsing Subject Directories

Subject directories (also called guides, indexes, or clearinghouses) contain links to other Web resources that are organized by subject. They vary in quality and authoritativeness, but a good one can be invaluable to your research and save you much time. The main advantage to using subject directories is that they contain links to resources that have been selected, evaluated, and organized by human beings and thus present a much more manageable number of resources. If the person managing the guide is an expert in the field of concern, or just a careful and methodological evaluator of Web resources, the guide can help you to identify good sites that contain useful and trustworthy information, and you can avoid wading through thousands of "hits" and evaluating all the sites yourself.

There are general and specialized directories. The following are three examples of general directories:

- Yahoo! (http://www.yahoo.com) is often mistaken for a search engine, but it is actually a subject directory, and a monster one at that. It also functions as a portal or a gateway for a collection of resources that can be customized by the user. Unlike search engines, when you search Yahoo!, you are not searching across the Web but rather just within the Web pages that Yahoo! has cataloged. Yahoo! has a subject directory for the social sciences with more specific listings, including one for social work (http://dir.yahoo.com/social_science/social_work/). Yahoo! also links to versions of its site in about 20 countries, which would be good to go to when conducting extensive research on one of those countries (http:world .yahoo.com/).
- Open Directory (http://dmoz.org) is the largest Web directory with four million sites (Hock 2004), and unlike Yahoo!, it is not a portal. In fact, other directories and search engines such as Yahoo! and Google use it. It has 16 top-level categories, including Social Sciences.
- Librarians' Index to the Internet (http://lii.org) is a small and highly selective Web directory produced by the Library of California.

The following are some examples of specialized subject directories:

- Argus Clearinghouse (http://www.clearinghouse.net/searchbrowse.html) is a guide to subject directories on the Internet, and it classifies them under subject headings.
- BUBL INK (http://bubl.ac/uk/link) contains over 12,000 links covering all academic areas.
- Social Sciences Virtual Library (http://www.clas.ufl.edu/users/gthursby/socsci/) includes listings for anthropology, demographics, psychology, social policy and evaluation, sociology, women's studies, and other areas.
- INFOMINE: Scholarly Internet Resource Collections (http://infomine.ucr.edu/) is produced by librarians across several campuses of the University of California system, and it includes a subject directory for the social sciences.
- SOSIG, Social Science Information Gateway (http://www.sosig.ac/uk), is a British site that aims to be comprehensive. It is classified according to the Dewey Decimal System, the classification system used by most public libraries.

Many other Internet subject directories are maintained by academic departments, professional organizations, and individuals. It is often hard to determine whether a particular subject directory such as this is up-to-date and reasonably comprehensive, but you can have some confidence in subject directories published by universities or government agencies. *The Internet Research Handbook* is an excellent source for more information on subject directories (Ó Dochartaigh 2002).

EXHIBIT A.6   The Results of a Google Title Search

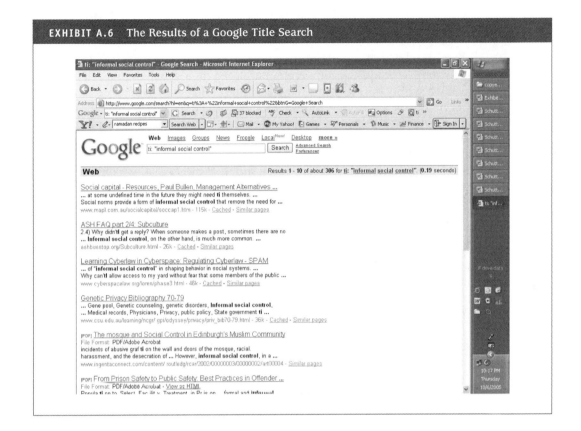

Search Engines

Search engines are powerful Internet tools. It is already impossible to imagine life without them. The biggest problem is the huge number of results that come back to you. If the number of results is still unmanageable, you can try a title search. Exhibit A.6 shows the results of typing the following into the Google search box: ti: "informal social control." This search will retrieve those pages that have that phrase in their title as opposed to anywhere in the page. This practice usually results in a dramatically smaller yield of results. If you are looking for graphical information such as a graph or a chart, you can limit your search to those pages that contain an image. On Google, this just requires clicking on the "Images" link located above the search box.

There are many search engines, and none will give you identical results when you use them to search the Web. Different search engines use different strategies to find Web sites and offer somewhat different search options for users. Due to the enormous size of the Web and its constantly changing content, it simply is not possible to identify one search engine that will give you completely up-to-date and comprehensive results. You can find the latest information about search engines at http://searchenginewatch.com. Hock's (2004)

The Extreme Searcher's Internet Handbook contains a wealth of information on specific search engines. Although there are many search engines, you may find the following to be particularly useful for general searching:

- Google (http://www.google.com) has become the leading search engine for many users in recent years. Its coverage is relatively comprehensive, and it does a good job of ranking search results by their relevancy (based on the terms in your search request). Google also allows you to focus your search just on images, discussions, or directories.
- AlltheWeb (http://www.alltheweb.com) is a more recent comprehensive search engine that also does a good job of relevancy ranking and allows searches restricted to images and so on.
- Microsoft's search engine (http://search.msn.com) adds a unique feature: Editors review and pick the most popular sites. As a result, your search request may result in a Popular Topics list that can help you to focus your search.
- Teoma (http://teoma.com) is one of the newest search engines and has a unique Resources section that links users to specialized directories.

In conclusion, use the appropriate tool for your searches. Do not use a search engine in place of searching literature that is indexed in tools such as *Sociological Abstracts*. Bookmark the key sites that you find in your area of interest. Become familiar with subject directories that cover your areas of interest, and look there before going to a search engine. And when you do use a search engine, take a moment to learn about how it works and what steps you should take to get the best results in the least amount of time.

Questions to Ask About a Research Article

1. What is the basic research question or problem? Try to state it in just one sentence. (Chapter 2)

2. Is the purpose of the study explanatory, evaluative, exploratory, or descriptive? Did the study have more than one purpose? (Chapter 1)

3. Was the theoretical framework presented? What was it? Did it seem appropriate for the research question addressed? Can you think of a different theoretical perspective that might have been used? (Chapter 2)

4. What prior literature was reviewed? Was it relevant to the research problem? To the theoretical framework? Does the literature review appear to be adequate? Are you aware of (or can you locate) any important omitted studies? (Chapter 2)

5. How well did the study live up to the guidelines for science? Do you need additional information in any areas to evaluate the study? To replicate it? (Chapter 2)

6. Did the study seem consistent with current ethical standards? Were any trade-offs made between different ethical guidelines? Was an appropriate balance struck between adherence to ethical standards and use of the most rigorous scientific practices? (Chapter 2 and in each methods chapter)

7. What were the major concepts in the research? How, and how clearly, were they defined? Were some concepts treated as unidimensional that you think might best be thought of as multidimensional? (Chapter 3)

8. Were any hypotheses stated? Were these hypotheses justified adequately in terms of the theoretical framework? In terms of prior research? (Chapter 2)

9. What were the independent and dependent variables in the hypothesis or hypotheses? Did these variables reflect the theoretical concepts as intended? What direction of association was hypothesized? Were any other variables identified as potentially important? (Chapter 2)

How to Read
a Research Article

The discussions of research articles throughout the text may provide all the guidance you need to read and critique research on your own. But reading about an article in bits and pieces in order to learn about particular methodologies is not quite the same as reading an article in its entirety in order to learn what the research discovered. The goal of this appendix is to walk you through an entire research article, answering the review questions introduced in Appendix B. Of course, this is only one article, and our "walk" will take different turns from one taken by a review of other articles, but after this review you should feel more confident when reading other research articles on your own.

For this example, we will use an article by Alarid, Burton, and Cullen (2000) on gender and crime among felony offenders (reprinted in this volume on pp. C-8–C-31). It focuses on a topic important to explaining the similarities and differences between male and female offending patterns as well as to questions of the generalizability of central criminological theoretical perspectives. Moreover, the article is published in a reputable criminological journal, the *Journal of Research in Crime and Delinquency,* indicating the article makes an important contribution to what is known about the etiology of criminal offending across gender groups.

We have reproduced below each of the article review questions from Appendix B, followed by our answers to them. After each question, we indicate the article page or pages that we are referring to. You can also follow our review by reading through the article itself and noting our comments.

> 1. *What is the basic research question or problem? Try to state it in just one sentence.*
> (Chapter 2)

The clearest statement of the research question actually consists of two questions: "[Can] social control and differential association theories . . . help account for participation in crime among young adult male and female felons?" (p. 174 in the original article). Prior to this point, the authors focus on the research problem, describing the limitations of past research and why exploring social control and differential association theories across gender groups among felony offenders is important.

2. *Is the purpose of the study explanatory, evaluative, exploratory, or descriptive? Did the study have more than one purpose?* (Chapter 1)

This study is exploratory in nature. The literature review (pp. 171–174) makes it clear that the primary purpose of this research is exploratory since new ground is being covered based on the limitations of past research. In addition, the problem statement indicates the exploratory character of the study: It "attempts to test the generality of social control and differential association theories to a population that has been infrequently studied" (p. 173).

3. *Was the theoretical framework presented? What was it? Did it seem appropriate for the research question addressed? Can you think of a different theoretical perspective that might have been used?* (Chapter 2)

This study uses two theoretical perspectives, "social control" and "differential association." This framework is very appropriate for the research questions addressed because this study explores the applicability of these theories to adult populations and their generalizability across gender groups. Because the research question focuses specifically on the applicability and generalizability of these specific theoretical perspectives, no other theoretical framework would be applicable.

4. *What prior literature was reviewed? Was it relevant to the research problem? To the theoretical framework? Does the literature review appear to be adequate? Are you aware of (or can you locate) any important omitted studies?* (Chapter 2)

Literature is reviewed from the article's first page until the "method" section (pp. 171–174). All the literature seems relevant to the particular problem as well as the general theoretical framework. The first few paragraphs generalize the findings of several studies that measures of both theories are related to criminal involvement (pp. 171–172). The authors then discuss the limited focus of previous studies and their rationale for selecting social control and differential association as their focus for this study (pp. 172–174). We leave it to you to find out whether any important studies were omitted.

5. *How well did the study live up to the guidelines for science? Do you need additional information in any areas to evaluate the study? To replicate it?* (Chapter 2)

It would be best to return to this question after reading the whole article. The study clearly involves a test of ideas against empirical reality to the extent that reality could be measured; it was carried out systematically and disclosed, as far as we can tell. Because the authors used a sample of newly incarcerated felons, it may be difficult for others to replicate the authors' findings. Drawing from previous research, the authors make some general assumptions about what factors best measure social control and differential association (p. 177). However, in previous studies, those factors measuring social control and differential association have been applied primarily to juvenile populations (p. 172). Intuitively, the same measures or indicators may not be applicable to adult populations (e.g., parental attachment and peer attachment), but we have no empirical evidence of this. The authors also assume respondents' reports of past criminal behavior are valid (pp. 177–178). It appears these are reasonable assumptions, but they are unproved assumptions that could be challenged. This is not in itself a criticism of the research, since some assumptions must be made in any study. The authors specified the meaning of key terms, as required in scientific research. The authors did search for regularities or patterns in their data, thus living up to another guideline. A skeptical stance toward current knowledge is apparent in the literature review and in the authors' claim that "social control and differential association theory appear to have general effects" on young adult felons and across gender groups (pp. 189–190). They encourage the development of new theories but emphasize the importance of "older" theoretical perspectives and their application to underresearched topics (p. 192). The study thus seems to exemplify adherence to basic scientific guidelines and to be very replicable.

6. *Did the study seem consistent with current ethical standards? Were any trade-offs made between different ethical guidelines? Was an appropriate balance struck between adherence to ethical standards and use of the most rigorous scientific practices?* (Chapter 2)

To the best of our knowledge, the study was consistent with current ethical standards. "All data were collected directly from the participants themselves" (p. 175). In accordance with the ASA code of ethics, participation in the study was voluntary, and the participants gave their informed consent to participate. The reporting seems honest and open. The original survey used by the authors is not likely to have violated any ethical guidelines concerning the treatment of human subjects, although it would be necessary to inspect the survey instrument itself to evaluate this.

7. *What were the major concepts in the research? How, and how clearly, were they defined? Were some concepts treated as unidimensional that you think might best be thought of as multidimensional?* (Chapter 3)

There are two key concepts in this study, social control and differential association. These concepts are the focus of the study and are discussed at length. These concepts, however, were not defined so that a layperson would understand what these concepts mean.

8. *Were any hypotheses stated? Were these hypotheses justified adequately in terms of the theoretical framework? In terms of prior research?* (Chapter 2)

Three hypotheses are stated, although they are labeled as theoretical predictions, which are derived from previous research (p. 173). The first hypothesis is that social control will account for variation in offending across gender groups. The second hypothesis is that differential association will account for variation in offending across gender groups. Finally, the third hypothesis is that the effect of the competing theory will be spurious.

9. *What were the independent and dependent variables in the hypothesis(es)? Did these variables reflect the theoretical concepts as intended? What direction of association was hypothesized? Were any other variables identified as potentially important?* (Chapter 2)

The dependent variable for this study is past criminal activity. For purposes of analysis, past criminal behavior is divided into three subscales resulting in four dependent variables: general crime (the total of all criminal activities), violent crime, property crime, and drug crime (p. 177). There are eight independent variables included in this study. The independent variables for the first hypothesis include five social control indicators. These include marital/partner attachment (0 = not married/unattached and 1 = married/attached), attachment to parents (a three-item composite index, higher values represent stronger attachment), attachment to friends (a two-item index, higher values represent stronger attachment), involvement in conventional activities (a two-item index measuring how much free time individual has, higher values represent more free time), and moral belief in the law (measured by a two-item index with higher values representing stronger belief) (pp. 178–180).

The independent variables for the second hypothesis include three indicators of differential association. These include individual definitions toward crime (a five-item index, higher values represent greater tolerance of criminal behavior), others' definitions toward crime (a three-item index, higher values represent greater association with individuals who hold favorable definition toward violating the law), and criminal friends (continuous variable indicating the number of close friends who have broken the law) (pp. 180–182).

Can you identify the variables in the final hypothesis?

10. *Did the instruments used—the measures of the variables—seem valid and reliable? How did the author attempt to establish this? Could any more have been done in the study to establish measurement validity?* (Chapter 3)

The measurement of the dependent variable was straightforward. The authors report the scale used has been identified as a valid and reliable self-report assessment of antisocial behavior (p. 176). In addition, reliability was assessed using Cronbach's alpha method, which indicated the general scale and the subscales are a reliable measure of criminal behavior. The measurement of the independent variables was also straightforward. The authors report the scales used have been identified as valid measures of social control and differential association indicators. In addition, Cronbach's alpha reliability testing shows the measures are reliable.

11. *What were the units of analysis? Were they appropriate for the research question? If some groups were the units of analysis, were any statements made at any point that are open to the ecological fallacy? If individuals were the units of analysis, were any statements made at any point that suggest reductionist reasoning?* (Chapter 4)

The authors sampled young adult, first-time felony offenders sentenced to a residential court-ordered boot camp program. The units of analysis are appropriate for the research question. No statements were made at any point that suggest reductionist reasoning.

12. *Was the study design cross-sectional or longitudinal, or did it use both types of data? If the design was longitudinal, what type of longitudinal design was it? Could the longitudinal design have been improved in any way, as by collecting panel data rather than trend data, or by decreasing the dropout rate in a panel design? If cross-sectional data were used, could the research question have been addressed more effectively with the longitudinal data?* (Chapter 5)

The survey was cross-sectional. The research question could have been addressed more effectively with longitudinal data that followed people from childhood into adulthood, since many of the authors' interpretations reflect their interest in how individuals' past experiences shape their propensity for criminal behavior.

13. *Were any causal assertions made or implied in the hypotheses or in subsequent discussion? What approach was used to demonstrate the existence of causal effects? Were all three criteria for establishing causal relationships addressed? What, if any, variables were controlled in the analysis to reduce the risk of spurious relationships? Should any other variables have been measured and controlled? How satisfied are you with the internal validity of the conclusions?* (Chapters 5, 6)

The explanatory hypotheses indicate that the authors were concerned with causality. However, the lack of previous research on their chosen population suggests an exploratory approach to causality. In order to reduce the risk of spuriousness in the presumed causal relationships, three control variables were included in the analysis: age, race/ethnicity, and gender (p. 188); these variables have been shown to be strongly related to criminal activity. There are other variables that might have created a spurious relationship, but at least several of the most likely contenders have been controlled. For example, some of the differences in social control and differential association measures may be due to social class—that is, individuals who were raised in more affluent families are more likely to have had greater social control in their lives and less opportunity to be involved with individuals who favor, or at least justify, criminal behavior. Income has also been shown to be negatively related to criminal activity—that is, the higher an individual's income, the less likely he or she is to engage in criminal activity.

14. *Was a sample or the entire population of elements used in the study? What type of sample was selected? Was a probability sampling method used? Did the authors think the sample was generally representative of the population from which it was drawn? Do you? How would you evaluate the likely generalizability of the findings to other populations?* (Chapter 4)

The sample was a nonrandom (nonprobability) sample of young, first-time felony offenders. A purposive sampling technique was used when sampling women, the authors "focused [their] efforts on obtaining responses from the entire [female] population . . . in the program over a 15-month period" (p. 175). The authors did not focus on sampling the entire male population; therefore, the authors obtained an availability sample of males sentenced to the program over the same time period. The authors admit the sample is not representative of "felony crimes committed by men and women in prison, [but] closely resembles first-time felony offenders in other shock incarceration programs as an alternative to jail or prison incarceration" (p. 176). Do you believe the authors can generalize their findings to similar populations? Why or why not?

15. *Was the response rate or participation rate reported? Does it appear likely that those who did not respond or participate were markedly different from those who did participate? Why or why not? Did the author(s) adequately discuss this issue?* (Chapters 4, 6)

The response rate was reported. The authors obtained responses from 122 of 124 (or 98.4%) female offenders and 1,031 (or 85.9%) of male offenders (p. 175). With the exception of female offenders, the authors did not discuss differences among those who did and did not participate with the study. Of the two female offenders who did not participate in the study, one did not speak English and the other woman escaped from the facility (p. 175). In general, it is reasonable to believe those who did not participate in the study were not markedly different from those who did participate. The large sample of males and females obtained for this study should have provided good variability.

16. *Was an experimental, survey, participant observation, or some other research design used? How well was this design suited to the research question posed and the specific hypotheses tested, if any? Why do you suppose the author(s) chose this particular design? How was the design modified in response to research constraints? How was it modified in order to take advantage of research opportunities?* (Chapters 6–10)

Survey research was the method of choice and probably was used for this article because it provided the most cost-effective means of gathering data. Survey research seems appropriate for the research questions posed.

17. *Was an evaluation research design used? Which type was it? What was the primary purpose of the evaluation?* (Chapter 10)

This study did not use any type of evaluation research.

18. *Were multiple methods used? Were findings obtained with different methods complementary?* (Chapter 10)

This study used only survey methods.

19. *Was any attention given to social context? To biological processes? If so, what did this add? If not, would it have improved the study? Explain.* (Chapters 5, 10)

Social context is given attention in this study on an individual level. The majority of social control and differential association indicators are measures of social conditions and experiences of the respondents. However, no attention is given to the potential importance of larger social contexts, such as neighborhood or region of the country. Biological factors that may increase the propensity for criminality are not considered in this study.

20. *Summarize the findings. How clearly were statistical and/or qualitative data presented and discussed? Were the results substantively important?* (Chapters 9, 12)

Statistical data are presented clearly using descriptive statistics, examination of bivariate relationships, and multiple regression analysis (a multivariate statistical technique) that highlight the most central findings. No qualitative data are presented. The findings seem substantively important, since they identify social control and differential association theories that can be applied to adult populations across gender groups (pp. 184–188).

21. *Did the author(s) adequately represent the findings in the discussion and/or conclusion sections? Were conclusions well grounded in the findings? Are any other interpretations possible?* (Chapter 12)

The findings are well represented in the discussion and conclusion section (pp. 188–192). The authors point out in their literature review that the focus for testing social control and differential association theories data is often limited to juvenile delinquent populations and frequently fails to consider possible differences between males and females. The findings suggest these theoretical perspectives can be applied to adult criminal populations and are important in explaining variation in criminal behavior across gender groups. You might want to consider what other interpretations of the findings might be possible. Remember that other interpretations are always possible for particular findings—it is a question of the weight of the evidence, the persuasiveness of the theory used, and the consistency of the findings with other research.

22. *Compare the study to others addressing the same research question. Did the study yield additional insights? In what ways was the study design more or less adequate than the design of previous research?* (Chapters 2, 10, 12)

This study investigated the relationship between social control and differential association on a population that had not previously received much attention (young adult felony offenders). This helped the authors to gain additional insights into the influence of social control and differential association on young adults across gender groups.

23. *What additional research questions and hypotheses are suggested by the study's results? What light did the study shed on the theoretical framework used? On social policy questions?* (Chapters 2, 10, 12)

The authors suggest no additional research questions or hypotheses. However, the study does shed light on the importance of "traditional" theoretical perspectives in explaining variation in criminal behavior among adults across gender groups. The authors note the importance of new theoretical paradigms but emphasize "that the value of 'older' theoretical perspectives should be remembered and . . . applied to underresearched topics" (p. 192).

Gender and Crime Among Felony Offenders

Assessing the Generality of
Social Control and Differential Association Theories

Leanne Fiftal Alarid

Velmer S. Burton, Jr.

Francis T. Cullen

Although often tested empirically on high school samples, differential association and social control theories have only infrequently been used to explain offending by felons. Based on a sample of 1,153 newly incarcerated felons, the authors examine the ability of differential association and social control theories to explain self-reported offending across types of crime and gender groups. Overall, the analyses lend support to both perspectives and suggest that they are "general" theories of crime. It also appears, however, that differential association theory has more consistent effects, especially for men. Parental attachment is a significantly stronger predictor of female than male participation in violent crime. These results indicate that future studies of criminal behavior risk being misspecified if they do not include measures of these "traditional" theories of crime.

Since the 1969 publication of Hirschi's *Causes of Delinquency*, social control or control theory and differential association or "cultural deviance" theory have competed vigorously against one another for the status of the preeminent microlevel sociological theory of crime (see also Kornhauser 1978). Even today, spirited exchanges occur in which the logical and empirical

---

NOTE: Reprinted from the *Journal of Research in Crime and Delinquency*, Vol. 37, No. 2, May 2000, 171-199. © Sage Publications, Inc. A previous version of this article was presented at the 1995 meeting of the Academy of Criminal Justice Sciences.

adequacy of the theories is debated (see Akers 1996, 1998; Costello 1997; Hirschi 1996; Matsueda 1988, 1997). These paradigms continue to guide research (see, e.g., Akers 1998; Sampson and Laub 1993) not only because of their theoretical parsimony and elegance but also because they accrue a measure of empirical support (Akers 1996; Krohn 1995).

Although noteworthy exceptions exist, most of the tests of social control and differential association theory have been conducted using self-report surveys on juveniles. Particularly instructive are empirical studies in which measures of both theories have been incorporated in the analysis. In general, this research suggests that measures of both theories are related to delinquent or adult criminal involvement, although in some studies, the strength of the differential association variables is greater (Benda 1994; Conger 1976; Kandel and Davies 1991; Macdonald 1989; Matsueda 1982; Matsueda and Heimer 1987; McGee 1992; see also Akers 1998; Burton 1991; Krohn 1995).

A key concern, however, is the generality of social control and differential association theories. These perspectives typically have been seen as having the ability to explain various forms of crime for all people (see, e.g., Akers 1998; Sutherland and Cressey 1955; Hirschi 1969; Sampson and Laub 1993). Clearly, the theoretical power of each of these perspectives hinges on their ability to achieve the status of a general theory. Accordingly, a central empirical and thus theoretical issue is whether social control and differential association theories can explain criminal behavior that extends beyond the delinquency found among community samples of youth.

There have been studies conducted on adult samples that have found support for differential association/social learning theory (Akers 1998; Akers and LaGreca 1988, 1991; Akers et al. 1989; Boeringer, Shehan, and Akers 1991; Dull 1983; Orcutt 1987; Tittle, Burke, and Jackson 1986). Research on social control theory with adult samples also suggests that adult social bonds may reduce criminal behavior (Horney, Osgood, and Marshall 1995; Lasley 1988; Sampson and Laub 1993). Even so, the empirical literature assessing whether social control and differential association theories can explain adult criminal behavior remains limited, especially when one searches for studies that include in their analyses measures of both theoretical perspectives.[1] Among the few adult studies testing social control versus differential association theory, the results show mixed support for both perspectives. Thus, two studies using community samples of adults found stronger support for differential association theory (Burton 1991; Macdonald 1989), while another study revealed that social control theory was better able to account for variation in arrests for men and women drug offenders (Covington 1985).

Even with this research, however, there are typically two limitations. First, especially with adult samples, studies often do not explore whether the effects of variables measuring traditional criminological theories vary by gender. Traditional theories have been criticized for their perceived inability to explain female criminality (Adler 1975; Daly and Chesney-Lind 1988; Leonard 1982) and for ignoring how gender-related factors—such as patriarchal power relations—differentially shape the involvement of gender groups in crime (Heidensohn 1985; Messerschmidt 1993) and victimization (Chesney-Lind and Shelden 1992). In turn, it is argued that scholars should formulate separate or different theories of crime for women (Leonard 1982; Messerschmidt 1993; Naffine and Gale 1989; Smart 1976).

Second, tests of social control and differential association theory have only infrequently been conducted on samples that include a high base rate of serious offending. As a result, we have only limited information on whether these theories explain more than relatively minor

TABLE C.1: Sample Characteristics

Characteristic	Women (n = 122)	Men (n = 1,031)
Age (17-28 years)	20.6	19.5
Race/ethnicity (percentage)		
African American	44.3	44.4
Caucasian	45.1	31.4
Hispanic	10.7	23.8
Formal education (years)	10.5	10.3
Marital/partner attachment		
(percentage attached)	32.0	35.1
Number of children (percentage)		
None	46.7	69.8
One	30.3	21.3
Two	9.8	6.1
Three or more	13.1	2.9
Median household income (in dollars)	16,800	27,500
Current conviction (percentage)		
Drug	41.8	28.9
Person	19.7	19.5
Property	38.5	51.6

Sample Characteristics

The sample characteristics of the 1,031 men and 122 women indicated that the vast majority lived in the inner city or surrounding urban area at the time of arrest for their most recent conviction. Men and women were both between the ages of 17 and 28 years, with the average age calculated at 19.5 for men and 20.6 years for women. The African American composition of both samples was identical (44 percent). However, there were more Caucasian women (45.1 percent) than men (31.4 percent), and there were more Hispanic men than women (23.8 percent and 10.7 percent, respectively). The amount of formal education completed ranged between 6th grade and three years of higher education, with a mean of a 10th-grade education. Single/unattached women (68.0 percent) and men (64.9 percent) comprised the majority of the sample relative to married and cohabiting women and men.

The demographic differences between the men and women in the sample were number of children, median household income, and current conviction. First, about 46.7 percent of women and 69.8 percent of men had no dependent children. Of all the fathers and mothers, the women more often reported being the primary caretaker and sole financial respondent for the children. As primary caretakers, the women lived in households that generated much lower incomes than men. The median household income for women (most with children) was $16,800, while men's reported annual household income was $27,500.

As for the most recent conviction, crimes of violence were comparably represented by gender; in the sample, 19.7 percent of women and 19.5 percent of men were violent offenders. More women (41.8 percent) than men (28.9 percent) were convicted of a drug offense. Bear in mind that the offense type of each convicted offender (violent, property, or drug) does not represent felony crimes committed by men and women in prison. Even though women are increasingly being sentenced for felony drug offenses as a result of the war on drugs

(U.S. Bureau of Justice Statistics 1994; Immarigeon and Chesney-Lind 1993), our sample more closely resembles first-time felony offenders in other shock incarceration programs as an alternative to jail or prison incarceration (MacKenzie and Brame 1994).

Dependent Variables

Dependent variable measures for the analysis were originally derived from Elliott and Ageton's (1980) scale of delinquent behaviors offenses found in the National Youth Survey (NYS). The NYS is currently "the most highly respected self-report assessment of antisocial behavior" (Caspi et al. 1994). We adjusted certain measures to apply to our young adult sample. Respondents were asked whether they had ever committed 35 different deviant and criminal acts (Hindelang, Hirschi, and Weis 1981). Our prevalence dimension of crime measures offense participation or the number of crime types committed. Other researchers have preferred the use of the prevalence measure of variety:

> Variety scores are useful for individual-differences research for several reasons. First, they show the extent of involvement in different types of crimes. . . . Second, they are less skewed than frequency scores. Third, they give equal weight to all delinquent acts, unlike frequency scores, which give more weight to minor crimes that are committed more frequently (e.g., underage drinking) and less weight to serious, less frequent crimes (e.g., rape). (Caspi et al. 1994:170-71)[4]

In the current study, one point was assigned if a respondent admitted involvement in a specific act; if respondents did not admit involvement, no points were assigned. The general crime scale was summed to obtain a composite score, which ranged from 0 to 35. Items in the general crime scale have been employed in previous tests of criminological theories using samples of adults (Burton et al. 1993, 1994, 1998). In addition, we distinguished type of criminal behavior by dividing the offenses into three subscales: violent, property, and drug. Table C.2 shows the means, standard deviations, and reliability for all dependent and independent variables used in the analysis.

Using Cronbach's (1951) alpha method, the 35-item general crime scale was reliable at .82 for women and .90 for men; the 15-item property crime subscale had a reliability level for women and men of .72 and .85, respectively; the 11-item drug subscale was .76 for women and .81 for men; and the 9-item violent crime subscale was reliable for females at .61 and .66 for males (see the appendix for individual offenses included in each subscale and accompanying frequencies separated by gender). Table C.2 shows that men admitted past involvement in an average of nine crimes: three drug offenses, four property crimes, and two violent crimes. Women engaged in an average of nine total crimes: four drug offenses, three property crimes, and two violent crimes. Although the property crime rate differed slightly for men and women, the types of drug offenses and violent crimes committed by both men and women were similar (see English 1993).[5]

Independent Variables

The items used in this study to measure social control and differential association variables were all drawn from previous theoretical tests (see Burton 1991; Burton et al. 1993,

TABLE C.2: Individual Dependent and Independent Variables by Gender

Variable	Number of Items	Women Reliability	Women Mean (SD)	Men Reliability	Men Mean (SD)	Sample F-Ratio
Dependent						
General crime	35	.82	8.57 (5.12) (range: 0-26)	.90	9.49 (6.85) (range: 0-31)	2.06
Violent crime	9	.61	1.64 (1.25) (range: 0-6)	.66	1.72 (1.70) (range: 0-9)	0.27
Property crime	15	.72	2.98 (2.60) (range: 0-11)	.85	4.19 (3.76) (range: 0-17)	11.99**
Drug crime	11	.76	3.96 (2.55) (range: 0-11)	.81	3.62 (2.72) (range: 0-11)	1.68
Independent						
Social control						
Marital/partner attachment	1	—	0.32 (0.47)	—	0.35 (0.48)	0.48
Parental attachment	3	.75	15.25 (2.98)	.64	14.87 (3.02)	1.77
Peer attachment	2	.81	6.84 (2.91)	.73	7.68 (2.79)	9.58**
Involvement	2	.69	6.44 (2.89)	.56	6.09 (2.69)	1.82
Belief	2	.63	8.88 (2.41)	.51	8.11 (2.40)	11.23**
Differential association						
Individual definitions	5	.77	10.97 (4.95)	.62	11.98 (4.52)	5.30*
Others' definitions	3	.71	9.50 (3.79)	.66	10.11 (3.61)	3.11
Criminal friends	1	—	2.87 (1.66)	—	2.77 (1.85)	0.30

*$p < .05$. **$p < .01$.

1994, 1998; Fiftal 1993; Matsueda 1982).[6] Participants in our sample responded to each item by using a 6-point Likert scale that ranged from *strongly agree* (6) to *strongly disagree* (1). Some responses were recoded so that a high score on an item indicated the presence of differential association or social control. The reliability of each composite independent theory measure was checked by Cronbach's (1951) alpha method. Finally, factor analysis was performed on each composite measure to ensure that each of the individual measures loaded on one factor. The mean, standard deviation, and reliability of each independent variable, separated by gender, can be found in Table C.2.

Social Control Theory

Hirschi's (1969) social control theory was originally formulated to rival differential association or "cultural deviance" theory as an explanation of why some juveniles conform and others engage in delinquency. More recent research has applied this perspective to adult criminality. Social control theory asserts that individuals with strong ties to family and friends (Covington 1985; Sampson and Laub 1990, 1993) will be protected from criminal involvement. In this study, we measure the adult social bond by marital attachment, attachment to parents, attachment to friends, involvement, and belief.[7]

Marital/partner attachment. Attachment to a significant other of the opposite sex by cohabiting or marriage is perceived as providing informal social control by protecting individuals from participation in criminal and other antisocial activities (Rand 1987; Sampson and Laub 1993). In our analysis, we use marital/partner attachment status as one of the adult social bond measures (0 = not married/unattached and 1 = married/ attached).

Attachment to parents. We include three items to measure attachment to parents to determine the effect of longstanding connections to parents by gender: "Throughout my life, I have had a lot of respect for my mother and father" (LaGrange and White 1985; Rosenbaum 1987). The second item asks whether respondents have "gotten along" well with their parents throughout their lives. This measure is modified from previous tests of social control (Eve 1978; Simons et al. 1980). The third component states, "My family is the most important thing in my life." The level of reliability for the attachment to parents scale is .75 for women and .64 for men.

Attachment to friends. The bond to friends is an important part of conventional attachment. The items used in the questionnaire yielded a reliability level of .81 for women and .73 for men. Respondents were asked to respond to the following two items: "It is important for me to spend time with my friends," and "My friends are a very important part of my life" (Agnew 1985; Burton 1991; Canter 1982; Friedman and Rosenbaum 1988; Hindelang 1973; Johnson, Marcos, and Bahr 1987; Matsueda 1982; Paternoster and Triplett 1988; Rosenbaum 1987).

Involvement. Involvement in conventional activities is measured by how much free time individuals have (Agnew 1985; Burton 1991; Canter 1982; Johnson 1979; Rosenbaum 1987). The following two items (reliable at .69 for women and .56 for men) were borrowed from Burton's (1991) study: "Between work, family, and community activities, I don't have much free time," and "Before coming here, I had a lot of free time on my hands."

Belief. Finally, the conventional bond is measured by moral belief in the law, specifically in the police, as measured by the following statement: "I have a lot of respect for the police" (Hindelang 1973; Hirschi 1969). The second item, "It is all right to get around the law if you can get away with it," was also used by Hirschi (1969) in previous tests. The two items have a reliability level of .63 for women and .51 for men.

Differential Association Theory

According to Sutherland's (1947) theory of differential association, individuals develop internalized definitions that are favorable or nonfavorable toward violating the law. As individual exposure to procriminal values, patterns, and associates increases, the likelihood of criminal involvement also increases. As in previous studies, we use three differential association variables to measure the likelihood of criminal exposure: individual definitions toward the law, others' definitions toward the law, and number of criminal friends.

Individual definitions toward crime. The five-item scale measuring individual definitions toward crime examined an individual's degree of tolerance for criminal behavior, the moral validity of violating the law, and the level of agreement with committing criminal acts.

Cronbach's (1951) alpha reliability for the scale for women is .77 and .62 for men. To measure disapproval toward violating the law, we asked for a response to the following item: "No matter how small the crime, breaking the law is a serious matter" (Akers et al. 1989; Jackson, Tittle, and Burke 1986; Short 1960). A second item, "It is morally wrong to break the law," also assessed anticriminal definitions (see Jackson et al. 1986; Matsueda 1989; Silberman 1976; Tittle et al. 1986). In contrast, to measure adults' willingness to violate the law, three items were included: "Sometimes you just don't have any choice but to break the law" (Krohn, Lanza-Kaduce, and Akers 1984; Short 1960), "If someone insulted me, I would be likely to hit or slap them," and "If breaking the law really doesn't hurt anyone, and you can make a quick buck doing it, then it's really not all that wrong" (Burton 1991).

Others' definitions toward crime. According to Sutherland (1947), criminal behavior is influenced primarily through exposure to other individuals holding definitions favorable toward violating the law. Relying on previous tests of differential association (Akers et al. 1979; Cressey 1953; Dull 1983; Griffen and Griffen 1978; Jaquith 1981; Johnson et al. 1987; Short 1960; Tittle et al. 1986), we included the following item: "Many of the people I associate with think it's okay to break the law if you can get away with it." The second item is designed to determine the type of people (criminal or noncriminal) with whom an individual associates. Thus, respondents were presented with the following item: "Most of the people I associate with would never break the law." In previous research, items similar to this have been related to delinquent involvement (Akers et al. 1989; Krohn et al. 1984; Jackson et al. 1986; Orcutt 1987). Finally, the third item assesses the extent to which individuals are "often in situations where people encourage [them] to do something illegal." This three-item scale has a reliability coefficient of .71 for women and .66 for men.

Criminal friends. Many empirical tests of differential association theory have relied on the number of criminal associates as evidence of interaction with criminal members within an individual's primary group (see, e.g., Akers et al. 1979; Dull 1983; Johnson et al. 1987; Warr and Stafford 1993; Winfree, Griffith, and Sellers 1989). In our study, we apply a measure of the actual number of criminal friends in our model by asking, "In the last 12 months, how many of your five closest friends have done something they could have gotten arrested for?"

FINDINGS

We assessed the bivariate relationships between the eight independent variables and crime by zero-order correlations. Tables C.3 and C.4 show that for the sample as a whole, all three of the differential association variables (individual definitions, others' definitions, and number of criminal friends) were significantly and directly correlated with property, violent, and drug offenses.[8] Similarly, these tables reveal that the social control variables of attachment to parents, involvement in conventional activities, and belief in the law were significantly and inversely correlated with criminal behavior. On the other hand, marriage reduced involvement only in property crime, and attachment to peers was positively correlated with all types of crime.[9]

TABLE C.3: Zero-Order Correlations of Theoretical Variables and Crime Scales (men)

	General Crime	Violent Crime	Property Crime	Drug Crime	Peer Attachment	Parental Attachment	Marital/ Partner Attachment	Involvement	Individual Belief	Other Definitions	Criminal Definitions	Friends
General crime	1.00											
Violent crime	.75**	1.00										
Property crime	.91**	.60**	1.00									
Drug crime	.80**	.44**	.53**	1.00								
Peer attachment	.15**	.09**	.14**	.13**	1.00							
Parent attachment	-.23**	-.16**	-.22**	-.18**	-.05	1.00						
Marital/partner attachment	-.05	-.01	-.10**	-.00	-.07*	.04	1.00					
Involvement	-.19**	-.12**	-.20*	-.14**	-.10**	-.01	.10**	1.00				
Belief	-.25**	-.24**	-.20*	-.18**	-.02	.21**	-.02	.18**	1.00			
Individual definitions	.27**	.29**	.24**	.18**	.04	-.22**	-.02	-.17**	-.54**	1.00		
Others' definitions	.37**	.32**	.34**	.24**	.02	-.18**	-.07*	-.28**	-.29**	.28**	1.00	
Criminal friends	.39**	.30**	.37**	.29**	.02	-.12**	-.09**	-.22**	-.27**	.27**	.42**	1.00

*p < .05. **p < .01.

TABLE C.4: Zero-Order Correlations of Theoretical Variables and Crime Scales (women)

	General Crime	Violent Crime	Property Crime	Drug Crime	Peer Attachment	Parental Attachment	Marital/ Partner Attachment	Involvement	Individual Belief	Other Definitions	Criminal Definitions	Friends
General crime	1.00											
Violent crime	.61**	1.00										
Property crime	.86**	.41**	1.00									
Drug crime	.83**	.32**	.50**	1.00								
Peer attachment	.32**	.13	.31**	.27**	1.00							
Parent attachment	-.35**	-.24**	-.33**	-.25**	-.12	1.00						
Marital/partner attachment	-.30**	-.09	-.29**	-.26**	-.20*	.08	1.00					
Involvement	-.26**	-.38**	-.13	-.19*	-.10	-.12	-.07	1.00				
Belief	-.22*	-.17*	-.19*	-.16	-.13	.27**	.10	.09	1.00			
Individual definitions	.28**	.23*	.21*	.24**	.19*	-.20*	-.01	-.16	-.50**	1.00		
Others' definitions	.47**	.35**	.35**	.42**	.17	-.24**	-.00	-.32**	-.18*	.33**	1.00	
Criminal friends	.31**	.22**	.28**	.23**	.03	-.10	-.00	-.13	-.12	.14	.28**	1.00

*p < .05. **p < .01.

General Criminal Behavior and Gender

Following the bivariate analysis, we used ordinary least squares (OLS) regression to examine the effects of independent variables on dependent variables. The first dependent variable (general crime) was regressed on three control variables (age, race, and gender), five social control variables, and three differential association variables.[10] Race was dummy coded (White vs. non-White), while age was a continuous variable. Regression coefficients were determined with all variables entered simultaneously into the equation. Table C.5 shows the standardized and unstandardized regression coefficients for men and women when both theories were regressed against each other. Gender groups were analyzed both independently and together.

Social control. Table C.5 shows that three out of five social control variables were significantly related to the overall measure of criminal behavior for the sample. Attachment to peers was significantly and positively related to criminal behavior. Attachment to parents and involvement in conventional activities were significantly and inversely related to involvement in criminal behavior. Those who got along well with their parents and considered them an important part of their lives were less likely to engage in criminal behavior. Furthermore, respondents who were immersed in time-consuming activities (i.e., work, family, and community activities) were also less likely to commit crimes.

We conducted a difference of slopes test on the significant unstandardized regression coefficients and their corresponding standard errors to determine if any significant differences existed between men and women. When conducting a hypothesis test of two samples of this nature, we relied on the formula suggested by Paternoster et al. (1998).

TABLE C.5: Effects of Social Control and Differential Association Variables on General Crime: Standardized Betas (unstandardized *B* coefficients) Reported

Variable Name	Women (n = 122)		Men (n = 1,031)		Women and Men (N = 1,153)	
Social control						
Marital/partner attachment (0 = not attached)	.21	(2.27)**	.01	(.08)	.02	(.33)
Parental attachment	−.31	(−.53)**	−.09	(−.21)**	−.12	(−.32)**
Peer attachment	.07	(.12)	.06	(.16)*	.07	(.16)*
Involvement	−.24	(−.43)**	−.06	(−.16)*	−.07	(−.18)*
Belief	.03	(.05)	−.06	(−.17)	−.05	(−.15)
Differential association						
Individual definitions	.11	(.11)	.12	(.19)**	.12	(.17)**
Others' definitions	.18	(.31)*	.16	(.31)**	.16	(.30)**
Criminal friends	.16	(.50)*	.26	(.98)**	.26	(.94)**
Control variables						
Age	−.08	(−.15)	.03	(.08)	.01	(.04)
Gender (0 = male)	—	—	—	—	−.05	(−.93)
Race (0 = non-White)	.22	(2.23)**	.19	(2.75)**	.19	(2.64)**
R^2		.51		.29		.30

*$p < .05$. **$p < .01$.

We found that attachment to parents ($z = -2.214$) was a significantly stronger predictor of female ($b = -.533$; $SE = .129$) as opposed to male ($b = -.208$; $SE = .070$) crime participation. On the other hand, there were no significant differences between women and men and the level of involvement as a predictor of criminal participation. The effect, then, of involvement on participation in crime is similar for men and women.

One of the strongest predictors for a woman's involvement in crime was whether she lived with a mate or was married. The significance of this relationship was not in the predicted direction of social control theory, which would hypothesize that family relationships, including marriage, insulate against criminal involvement. On the other hand, marital attachment was not significant for predicting men's criminal behavior.

One factor that was significantly related to men's participation in crime was peer attachment. Young men who were attached to their friends were more likely to engage in criminal behavior—a relationship in the direction opposite to that predicted by social control theory. Other studies have interpreted peer attachment and increased criminal involvement as indirect support for differential association theory, depending on whether the friends were delinquent (Conger 1976).[11]

Differential association. In Table C.5, others' definitions and criminal friends significantly predicted both male and female involvement in crime. A difference of slopes test on criminal friends indicated there were no significant differences between women and men; the effect of criminal friends on participation in crime is similar for men and women.

Individual definitions, although in the expected direction for women, were significant only for men. Furthermore, for men, differential association variables had stronger effects than social control theory variables in predicting their participation in crime. For women, there was some tendency for the social control variables to have stronger effects. The overall explanatory power for both theories in the model for women was $R^2 = .51$ and for men was .29.

Control variables. Race/ethnicity significantly predicted criminal involvement for women and men. Specifically, Anglo men and women were more likely to have been involved in crime than non-White individuals. Age was not a significant predictor of overall rates of criminal behavior. To more precisely measure the nature of gender differences and similarities for overall crime, a separate t-test for independent samples was conducted using unstandardized partial regression coefficients while using gender as the grouping variable. Results indicated that the mean for the men (9.88) was significantly higher than for women (8.56) for the general crime scale.

Type of Criminal Behavior and Gender

Table C.6 shows the contribution of social control and differential association theories to individually and simultaneously explain young men's and women's involvement in drug, property, and violent crimes.

Social control. A consistent and significant predictor of all three crime types for women was parental attachment. Parental attachment was also significant in the direction predicted by

TABLE C.6: Effects of Social Control and Differential Association Variables on Drug, Property, and Violent Crime: Standardized Betas (and unstandardized coefficients) Reported

Variable Name	Drug Crime			Property Crime			Violent Crime		
	Women	Men	Sample	Women	Men	Sample	Women	Men	Sample
Social control									
Marital/partner attachment	.16 (.89)*	.04 (.24)	.06 (.33)*	.21 (1.17)**	-.04 (-.32)	-.02 (-.18)	.08 (.21)	.02 (.07)	.02 (.08)
Parental attachment	-.18 (-.15)*	-.06 (-.05)	-.06 (-.07)*	-.29 (-.26)**	-.10 (-.12)**	-.12 (-.18)**	-.29 (-.12)**	-.06 (-.04)*	-.10 (-.07)**
Peer attachment	-.02 (-.02)	.06 (.06)	.06 (.06)*	.15 (.14)	.06 (.08)*	.07 (.09)*	.01 (.00)	.06 (.04)	.05 (.03)
Involvement	-.22 (-.20)**	-.07 (-.07)*	-.09 (-.08)**	-.10 (-.09)	-.07 (-.10)*	-.07 (-.09)*	-.34 (-.15)**	.01 (.01)	-.02 (-.01)
Belief	.01 (.01)	-.08 (-.09)*	-.07 (-.08)*	.02 (.02)	-.02 (-.03)	-.02 (-.03)	.01 (.01)	-.05 (-.04)	-.04 (-.03)
Differential association									
Individual definitions	.11 (.06)	.08 (.05)*	.09 (.05)*	.06 (.03)	.10 (.08)**	.09 (.08)**	.09 (.02)	.15 (.06)**	.14 (.05)**
Others' definitions	.15 (.18)*	.08 (.06)*	.10 (.08)**	.14 (.21)*	.16 (.16)**	.15 (.15)**	.10 (.03)	.17 (.08)**	.16 (.07)**
Criminal friends	.15 (.22)*	.23 (.34)**	.22 (.33)**	.17 (.23)*	.24 (.50)**	.24 (.48)**	.05 (.04)	.18 (.17)**	.17 (.16)**
Control variables									
Age	.14 (.14)*	.16 (.20)**	.16 (.19)**	-.18 (-.18)*	-.06 (-.11)*	-.08 (-.13)**	-.23 (-.11)**	-.03 (-.02)	-.04 (-.03)
Gender (0 = male)	—	—	.01 (.06)	—	—	-.09 (-1.04)**	—	—	-.01 (.07)
Race (0 = non-White)	.44 (2.25)**	.26 (1.51)**	.28 (1.62)**	.03 (.14)	.13 (1.07)**	.12 (.96)**	-.07 (-.17)	.02 (.08)	-.00 (-.00)
R^2	.49	.23	.25	.40	.25	.26	.36	.18	.19

$*p < .05.$ $**p < .01.$

social control theory for men's property and violent crime. Attachment to parents served as an effective insulator for men and women against criminal involvement, a finding consistent in delinquency studies (Burton et al. 1995). A difference of slopes test indicated that the effect of parental attachment ($z = -2.136$) was a significantly stronger predictor for women ($b = -.122$; $SE = .036$) than for men ($b = -.036$; $SE = .018$) for violent crime participation. The effect of parental attachment on property crime ($z = 1.62$) is similar for males and females since the difference was found to be not statistically significant.

Lack of involvement in conventional activities significantly predicted men's and women's participation in drug crime, as well as women's violent crime and men's property crime. Difference of slopes tests verified that lack of involvement in activities has similar effects for men's and women's involvement in drug crime and overall crime participation.

Marital attachment significantly predicted women's involvement in drug and property crimes but not in the direction predicted by social control theory. In other words, women who were married or living with a mate or boyfriend were more likely to be involved in drugs and/or property offenses. Marital attachment for men, however, did not insulate against or foster crime.

Differential association. For men, all three measures of differential association theory were significantly and positively related to all three types of crime. Thus, when criminal behavior was separated by type, differential association theory more consistently explained men's participation in drug, property, and violent criminal behaviors than did social control theory. For women, though, both differential association and social control theory were related to explaining female drug and property offenses, while social control variables were stronger predictors of women's participation in violent crime. Difference of slopes tests were conducted to detect significant effects on others' definitions and criminal friends between men and women. None of the differences was statistically significant for participation in property, drug, or overall crime. In other words, the effects of criminal friends and others' definitions on criminal participation were similar for men and women.

Control variables. Of the three control variables, age and race/ethnicity were the most consistent and strongest predictors of differences in rates of offending for men and women. As the age of male and female adults increased, involvement in drug crimes also significantly increased, while male and female property and female violent crimes significantly decreased. Anglo men and women were more likely to have been involved in drug crime than non-White men and women. Also, Anglo men were more likely than were African American and Hispanic men to have been involved in property crime.

The only type of crime that varied by gender was property offenses. Men were more likely than women to have committed crimes against property. For independent samples, *t*-tests were conducted for each of the three subscales, using unstandardized partial regression coefficients. Gender was used in each *t*-test as the grouping variable to more precisely measure the nature of gender differences and similarities for drug, property, and violent offenses. Results confirmed that the mean for the men (4.37) was significantly higher than for women (2.98) for the property crime subscale. However, for drug offenses, there was no difference between men (3.75) and women (3.94). The same held true for the violent crime subscale, in which men (1.77) were found to have a mean similar to that of women (1.64).

DISCUSSION

If psychology is sometimes referred to as the "science of college sophomores," then criminology might earn the label of the "science of high school sophomores"; while psychologists run experiments on their students, criminologists rely, over and over again, on samples of adolescents in school. To be sure, such self-report research is important and has considerably advanced our understanding of crime causation. Nonetheless, the skewness in focusing on adolescents and on minor delinquencies means that the applicability of criminological theory to serious and adult offending for men and women has been examined too infrequently.

A strategy employed episodically to rectify this omission has been to test theories on samples of convicted offenders, who have higher base rates of serious criminality (see, e.g., Covington 1985; Horney et al. 1995; Longshore et al. 1996; Sampson and Laub 1993). Building on this research, we have attempted to assess the generality of social control and differential association theories and to explain self-report crime across gender among first-time convicted felons serving a three-month sentence in a shock probation program, as an alternative to jail or prison incarceration. Several salient results were found.

First, both social control theory and differential association theory appear to have general effects. For the sample as a whole, the support for differential association is especially strong and consistent: The three differential association variables are significantly related to the general crime scale and to the drug, property, and violent crime subscales. Note that many studies that assess differential association use only the single measure of "number of delinquent peers." In a cross-sectional study, the peer-delinquency relationship is subject to the claim of spuriousness—that it is merely a case of "birds of a feather flocking together" (see Hirschi 1969; Gottfredson and Hirschi 1990). In our analysis, however, it is noteworthy that the attitudinal measures of others' definitions and one's own individual definitions also are related to all the crime scales for the sample. Some additional support for differential association theory can be drawn from the positive relationships of peer attachment to all the crime measures for the sample. This result is contrary to Hirschi's (1969) prediction. Furthermore, to the extent that the members of the sample are likely to have delinquent friends, then such attachment might be viewed as an indicator of differential association.

The findings for the social control measures are less impressive but nonetheless offer support for the perspective. Parental attachment had a consistent negative relationship to all of the crime measures for the sample. It appears, therefore, that bonds to parents have continuing effects on serious felons into early adulthood (Sampson and Laub 1993). Furthermore, with the exception of violent crime, involvement in conventional activities was negatively and significantly associated with the crime measures. Previous research on social control theory exploring the relationship between involvement and criminality has revealed mixed findings. While some studies have reported that involvement has few effects on offending (see, e.g., Hirschi 1969), others have found that involvement in conventional activities has a significant and inverse relationship both to minor forms of deviance (Agnew 1985; Friedman and Rosenbaum 1988; Rosenbaum 1987) and to adult offending behavior (Fiftal 1993; Lasley 1988).

Second, the analyses suggest that the "traditional" criminological perspectives can account for offending across the genders. This is not to say, however, that gender-specific theories could not expand our understanding not only of female crime but also of male criminal participation (see, e.g., Chesney-Lind 1989; Hagan 1989; Messerschmidt 1993). Still, the data reveal that differential association and social control variables—albeit with some variation—have similar effects across male and female felons. In fact, for every crime measure, the amount of explained variation is higher for the female sample than it is for the male sample. Furthermore, our results are consistent with a recent meta-analysis of predictors of criminal behavior, which reports that antisocial attitudes/associates and parental/family factors have similar effects for males and females (Andrews and Bonta 1994:68).

Some differences do emerge, however, when analyzing the effects of theoretical variables by gender. We should caution that making interpretations is difficult because the large sample size for the male offenders (1,031 men vs. 122 women) means that the analyses of the males have more power to detect statistically significant relationships. With this caveat stated, it appears that the differential association variables have consistent effects not only across crime types but also across male and female offenders. Only with individual definitions are any gender differences apparent; even here, the relationships, though not significant for women, are in the same direction.

In contrast, the effects of social control variables were stronger for the women than for the men in our sample. We found that the effects of parental attachment significantly and inversely affected males' offending, but the impact of parental attachment on females was significantly stronger for participation in violent offenses. This difference between men and women was not found for property crime, and parental attachment was not significant for male drug crime participation. The effect of a lack of parental attachment as a significant predictor of female violent crime is especially noteworthy for furthering our understanding of female offending.

Covington's (1985) analysis revealed that measures of social control theory were stronger predictors of adult women's criminal involvement. She suggests that as adolescents, females generally are exposed to more parental supervision than male adolescents. Thus, females may be more negatively influenced when familial relations are problematic or when parental supervision is reduced or nonexistent. Covington raises the additional possibility that these findings indicate differential perception of female and male adolescent offending by the juvenile justice system. While male adolescent offending may stem from a variety of causes, "many of the early female [juvenile] arrests may have been based on their detached and unsupervised status as juveniles rather than serious criminality" (Covington 1985:350).

We should note that with the exception of violent crime, being married or attached to a male partner was positively and significantly related to criminal involvement for women. For males, female partner attachment had no effects on any crime scale. The result for males is consistent with Sampson and Laub (1993), who found that marriage only had effects on their sample of former delinquents when marital quality is taken into account (see also Laub, Nagin, and Sampson 1998). Unfortunately, we do not have detailed information on the nature of the marital relationship to disentangle why marriage or attachment fosters criminality.

From a differential association perspective, it may be that the female felons in the sample are involved with men who reinforce their criminality. Research on battered women offenders found that being involved in intimate violent relationships was a main pathway to increasing the risk of offending (Richie 1996). Another possibility is that women from abusive, dysfunctional homes—those who are at a high risk for offending (see, e.g., Chesney-Lind 1989)—attempt to escape family victimization through early marriages and attachments (Miller 1986). These attachments, in turn, influenced the roles these women played in criminal behavior in relation to men (see Alarid et al. 1996). Future research should take into account the quality of the marital relationship, as well as the role that battering and intimate abuse have on subsequent involvement with the criminal justice system.

Another noteworthy finding is that for both males and females, being Caucasian was associated with higher rates of self-reported crime for every scale with the exception of violent offending. These differences could reflect racial differences in criminal justice processing: Compared to minorities, Whites had to have more involvement in self-reported crime before being arrested and incarcerated.

An added finding that may have resulted from decisions made during prosecutorial and judicial case processing was the offending patterns of men and women. While men reported having committed more overall crime than women, particularly more property offenses, women and men did not differ significantly in the type and number of drug and violent crimes. These similarities may mean that judges who sentenced young, first-time felony offenders to shock incarceration as a prison alternative were impartial to gender.

In closing, the analyses reported here lend continuing support to the traditional theories of social control and, especially, differential association/social learning. Importantly, the data support the claim that these are "general" theories, explaining variations in self-report criminality among felony offenders and across men and women. As Cole (1975) demonstrates, however, the fate of crime and deviance theories does not always or usually hinge on their empirical vitality but on whether newer "paradigms" offer fresh research opportunities. New theories, of course, should be welcomed and their explanatory power assessed. But in doing so, we would caution that the value of "older" perspectives should be remembered and, as in our research, applied to underresearched topics.

APPENDIX   Prevalence of Self-Reported Criminal Behavior: "Have You Ever . . . ?"

	Women n (%)		Men n (%)	
Property crime subscale (15 items)				
1. Avoided paying at restaurants, the movie theater, etc.?	29	(23.8)	346	(33.7)*
2. Knowingly bought, held, sold stolen property?	50	(41)	531	(51.8)*
3. Taken someone else's vehicle without their permission?	33	(27)	379	(36.9)*
4. Taken anything ($5 or less) from your job?	11	(9)	152	(14.8)
5. Taken anything (between $5 and $50) from your job?	11	(9)	118	(11.5)
6. Taken anything (over $50) from your job?	3	(2.5)	102	(10.0)**
7. Taken anything ($5 or less) from someone (other than work)?	46	(37.7)	317	(31.0)
8. Taken anything (between $5 and $50) from someone (other than work)?	25	(20.5)	306	(30.0)*
9. Taken anything (over $50) from someone else (other than work)?	31	(25.4)	346	(33.7)

10. Damaged/destroyed property belonging to relative/family of origin?	15	(12.3)	166	(16.2)
11. Purposely damaged/destroyed property belonging to an employer?	2	(1.6)	59	(5.7)*
12. Purposely damaged/destroyed property belonging to spouse/partner/friend (someone other than family of origin or an employer)?	22	(18.0)	327	(31.9)**
13. Broken into a building/vehicle?	25	(20.5)	423	(41.2)**
14. Stolen or attempted to steal a motor vehicle?	19	(15.6)	314	(30.6)**
15. Thrown objects at cars/property?	41	(33.6)	419	(40.9)
Drug crime subscale (11 items)				
1. Drank alcoholic beverages before age 21?	92	(75.4)	788	(76.8)
2. Drove a car while drunk?	51	(41.8)	471	(46.0)
3. Bought/provided beer/liquor for someone under 21?	51	(41.8)*	340	(33.2)
4. Had marijuana/hashish?	86	(70.5)**	572	(55.8)
5. Used hallucinogens/LSD?	26	(21.3)	226	(22.0)
6. Had amphetamines?	17	(13.9)	129	(12.6)
7. Had barbiturates?	14	(11.5)	91	(8.9)
8. Had heroin?	6	(4.9)	32	(3.1)
9. Had cocaine?	60	(49.2)**	278	(27.1)
10. Sold marijuana?	34	(27.9)	359	(35.2)
11. Sold hard drugs (cocaine, crack, heroin)?	46	(37.7)	435	(42.5)
Violent crime subscale (9 items)				
1. Been involved in a gang fight?	25	(20.5)	232	(22.6)
2. Used physical force to get money from a relative/family of origin?	4	(3.3)	48	(4.7)
3. Used physical force to get money from someone at work?	1	(0.8)	35	(3.4)
4. Used physical force to get money from someone, a family member, or someone at work?	12	(9.8)	151	(14.7)
5. Hit or threatened to hit a relative/member of family of origin?	42	(34.4)	337	(32.9)
6. Hit or threatened to hit someone at work?	5	(4.1)	144	(14.0)**
7. Hit or threatened to hit friend/partner/spouse?	80	(65.6)**	528	(51.7)
8. Had or tried to have sex with someone against their will?	1	(0.8)	31	(3.0)
9. Attacked someone with the idea of seriously hurting/killing them?	30	(24.6)	264	(25.7)

NOTE: Women, $N = 122$; men, $N = 1,031$.

*$p < .05$. **$p < .01$.

NOTES

1. There have been several adult tests of social control and differential association measures assessed against deterrence, low self-control, social learning, strain, or subcultural theory measures (see, e.g., Burton et al. 1994; Ginsberg and Greenley 1978; Makkai and Braithwaite 1991; Tittle 1980).

2. Covington's (1985) tests of social control and differential association theories differed from the variables presented in this study. For example, involvement was not included as a social control variable, and individual definitions toward crime were not included as a differential association variable.

3. Strain theory, a fundamental perspective that also offers explanations on criminal behavior, was not included in the final analysis. Although Agnew's (1985) general strain theory has received some empirical support, Merton's social strain theory has received less support in previous studies. We originally included social strain variables that tested relative deprivation and blocked opportunities for both men and women in the data set. The variables measuring blocked opportunities were not reliable (.44) and were not used in the analysis. The variables measuring relative deprivation (reliable at .74) were regressed separately with all dependent crime variables. Relative deprivation was not significant for women for any type of crime and was significant only for male property crime but not for male drug or violent offenses. When relative deprivation was regressed against social control theory and/or differential association theory, relative deprivation had no effect for any type of crime.

4. Incidence data (how many times in the past year) were collected on all criminal behaviors but were not used due to a low reliability of the incidence subscales. The 35-item general crime incidence scale was, however, reliable at .61.

5. We are not suggesting that women are becoming more violent in their criminal involvement, especially as they relate to the women's self-reported involvement of violent crime, particularly the three "hit or threatened to hit" measures in the violent crime subscale (see appendix). The similarities in self-reported crime participation by women and men in our study can be accounted for in part by English (1993:373), who found that incarcerated adult women and men prisoners had similar participation rates for assault, robbery, motor vehicle theft, fraud, and dealing drugs. Our data show, however, that women and men participate in a wide variety of criminal behaviors, from less serious to more serious, which may or may not be related to their conviction offense.

6. A potential weakness of our measures is the varying time frame of our independent and dependent variables (e.g., the time frame for "peer attachment" and "marital attachment" is current, the time frame for "criminal friends" is in the past year, while the time frame for the dependent measure of prevalence is lifetime). We are largely assuming that our measures of our independent variables are relatively stable over time. Furthermore, we believe that our analysis has merit despite these issues. Future research would benefit from longitudinal analysis to determine the stability of theoretical variables over time.

7. Commitment, an element of Hirschi's (1969) social bond, was not used in the analysis due to its low reliability.

8. Men and women were also analyzed separately using correlations between differential association variables with the dependent variables. Our analyses revealed that a significant correlation existed between all variables. For the sample as a whole, the three differential association measures are more highly correlated with each other than are the five measures of social control. This may increase the likelihood of finding greater consistency with differential association measures.

9. When bivariate correlations were analyzed separately by gender, a mixed portrayal emerged for social control variables. Attachment to parents for both men and women was significantly and inversely related to all types of crime. For men, attachment to friends, involvement, and belief were significantly related to all crime types, with the exception of marital attachment, which was significant only for property crime. For women, attachment to friends and marital attachment were related to drug and property crimes but not violent crime. On the other hand, involvement was significantly related to women's drug and violent crimes but not women's property offenses. Finally, belief was significantly related to women's property crime and violent crime but was not significantly related to women's drug crime.

10. We separately analyzed the control variables with each of the theories before regressing them together as shown in Table C.5. The results of the separate regressions show a significant relationship between age, race, and self-reported offending behavior. Age was inversely related to crime—that is, as age increased, less crime was committed. Men and women of color reported they committed significantly less crime than did Caucasian men and women. When social control theory was introduced into the equation with the three control variables, all social control variables, except for marital attachment, were significantly related to criminal offending. All social control variables were significant in the predicted direction, except for attachment to friends. When social control variables were removed and differential association variables were regressed with the three control variables, all three differential association variables were significantly and directly related to crime in the predicted direction.

11. To detect the presence of an interaction between peer attachment and criminal friends, we introduced a multiplicative term into each regression equation. A hierarchical F-test indicated that an interaction effect was not present for women, but an interaction existed for men for drug, property, and general crime scales (Jaccard, Turrisi, and Wan 1990). The difference between the R^2 values in the main effect and interactive models indicated that the interaction contributed less than 1 percent to each model (.63 percent for drug, .89 percent for general crime, and .90 percent for property offenses).

REFERENCES

Adler, Freda. 1975. *Sisters in Crime: The Rise of the New Female Offender.* New York: McGraw-Hill.

Agnew, Robert. 1985. "Social Control Theory and Delinquency: A Longitudinal Test." *Criminology* 23:47-61.

Akers, Ronald L. 1996. "Is Differential Association/Social Learning Cultural Deviance Theory?" *Criminology* 34:229-47.

_____. 1998. *Social Learning and Social Structure: A General Theory of Crime and Deviance.* Boston: Northeastern University Press.

Akers, Ronald L., Marvin Krohn, Lonn Lanza-Kaduce, and Marcia Radosevich. 1979. "Social Learning and Deviant Behavior: A Specific Test of a General Theory." *American Sociological Review* 44:636-55.

Akers, Ronald L., and Anthony LaGreca. 1988. "Alcohol, Contact with the Legal System, and Illegal Behavior among the Elderly." In *Older Offenders,* edited by Belinda Mccarthy and Robert Langworthy. New York: Praeger.

_____. 1991. "Alcohol Use among the Elderly: Social Learning, Community Context, and Life Events." In *Society, Culture, and Drinking Patterns Re-Examined,* edited by David J. Pittman and Helene Raskin White. New Brunswick, NJ: Rutgers Center of Alcohol Studies.

Akers, Ronald L., Anthony LaGreca, John Cochran, and Christine Sellers. 1989. "Social Learning Theory and Alcohol Behavior among the Elderly." *Sociological Quarterly* 30:625-38.

Alarid, Leanne Fiftal, Velmer S. Burton, James W. Marquart, Francis T. Cullen, and Steven J. Cuvelier. 1996. "Women's Roles in Serious Offenses: A Study of Adult Felons." *Justice Quarterly* 13:431-54.

Andrews, D. A., and James Bonta. 1994. *The Psychology of Criminal Conduct.* Cincinnati, OH: Anderson.

Benda, Brent B. 1994. "Testing Competing Theoretical Concepts: Adolescent Alcohol Consumption." *Deviant Behavior* 15:375-96.

Boeringer, Scot, Constance L. Shehan, and Ronald L. Akers. 1991. "Social Contexts and Social Learning in Sexual Coercion and Aggression: Assessing the Contribution of Fraternity Membership." *Family Relations* 40:558-64.

Burton, Velmer S. 1991. "Explaining Adult Criminality: Testing Strain, Differential Association, and Control Theories." Ph.D. dissertation, University of Cincinnati.

Burton, Velmer S., Francis T. Cullen, T. David Evans, Leanne Fiftal Alarid, and R. Gregory Dunaway. 1998. "Gender, Self-Control, and Crime." *Journal of Research in Crime and Delinquency* 35:123-47.

Burton, Velmer S., Francis T. Cullen, T. David Evans, and R. Gregory Dunaway. 1994. "Reconsidering Strain Theory: Operationalization, Rival Theories, and Adult Criminality." *Journal of Quantitative Criminology* 10:213-40.

Burton, Velmer S., Francis T. Cullen, T. David Evans, R. Gregory Dunaway, Sesha R. Kethineni, and Gary L. Payne. 1995. "The Impact of Parental Controls on Delinquency." *Journal of Criminal Justice* 23:111-26.

Burton, Velmer S., James W. Marquart, Steven J. Cuvelier, Robert J. Hunter, and Leanne Fiftal. 1993. "The Harris County CRIPP Program: An Outline for Evaluation, Part 1." *Texas Probation* 8:1-8.

Canter, Rachelle J. 1982. "Family Correlates of Male and Female Delinquency." *Criminology* 20:149-67.

Caspi, Avshalom, Terrie E. Moffitt, Phil A. Silva, Magda Stouthamer-Loeber, Robert F. Krueger, and Pamela S. Schmutte. 1994. "Are Some People Crime Prone? Replications of the Personality-Crime Relationship across Countries, Genders, Races, and Methods." *Criminology* 32:163-95.

Chesney-Lind, Meda. 1989. "Girls' Crime and Woman's Place: Toward a Feminist Model of Female Delinquency." *Crime and Delinquency* 35:5-29.

Chesney-Lind, Meda, and Randall G. Shelden. 1992. Girls, *Delinquency, and Juvenile Justice.* Pacific Grove, CA: Brooks/Cole.

Cole, Stephen. 1975. "The Growth of Scientific Knowledge: Theories of Deviance as a Case Study." In *The Idea of Social Structure: Papers in Honor of Robert K. Merton,* edited by Lewis A. Coser. New York: Harcourt Brace.

Conger, Rand. 1976. "Social Control and Social Learning Models of Delinquency: A Synthesis." *Criminology* 14:17-40.

Costello, Barbara. 1997. "On the Logical Adequacy of Cultural Deviance Theories." *Theoretical Criminology* 1:403-28.

Covington, Jeanette. 1985. "Gender Differences in Criminality among Heroin Users." *Journal of Research in Crime and Delinquency* 22:329-53.

Cressey, Donald. 1953. *Other People's Money.* Glencoe, IL: Free Press.

Cronbach, Lee J. 1951. "Coefficient Alpha and the Internal Structure of Tests." *Psychometrika* 16:297-334.

Daly, Kathleen, and Meda Chesney-Lind. 1988. "Feminism and Criminology." *Justice Quarterly* 5:497-538.

Dull, R. Thomas. 1983. "Friend's Use and Adult Drug and Drinking Behavior: A Further Test of Differential Association Theory." *Journal of Criminal Law and Criminology* 74:1608-19.

Elliott, Delbert S., and Suzanne S. Ageton. 1980. "Reconciling Race and Class Differences in Self-Reported and Official Estimates of Delinquency." *American Sociological Review* 45:95-110.

English, Kim. 1993. "Self Reported Crime Rates of Women Prisoners." *Journal of Quantitative Criminology* 9:357-82.

Eve, Raymond. 1978. "A Study of the Efficacy of Interactions of Several Theories for Explaining Rebelliousness among High School Students." *Journal of Criminal Law and Criminology* 69:115-25.

Fiftal, Leanne E. 1993. "Assessing Adult Female Criminality: Testing Differential Association and Social Control Theories." Master's thesis, Sam Houston State University.

Friedman, Jennifer, and Dennis P. Rosenbaum. 1988. "Social Control Theory: The Salience of Components of Age, Gender, and Type of Crime." *Journal of Quantitative Criminology* 4:363-81.

Ginsberg, Irving J., and James R. Greenley. 1978. "Competing Theories of Marijuana Use: A Longitudinal Study." *Journal of Health and Social Behavior* 19:22-34.

Gottfredson, Michael R., and Travis Hirschi. 1990. *A General Theory of Crime.* Stanford, CA: Stanford University Press.

Griffin, Brenda S., and Charles T. Griffin. 1978. "Marijuana Use among Students and Peers." *Drug Forum* 7:155-65.

Hagan, John. 1989. *Structural Criminology.* New Brunswick, NJ: Rutgers University Press.

Heidensohn, Frances M. 1985. *Women and Crime.* New York: Macmillan.

Hindelang, Michael J. 1973. "Causes of Delinquency: A Partial Replication and Extension." *Social Problems* 20:471-87.

Hindelang, Michael J., Travis Hirschi, and Joseph G. Weis. 1981. *Measuring Delinquency.* Beverly Hills, CA: Sage.

Hirschi, Travis. 1969. *Causes of Delinquency.* Berkeley: University of California Press.

_____. 1996. "Theory without Ideas: Reply to Akers." *Criminology* 34:249-56.

Horney, Julie, and Ineke Haen Marshall. 1992. "Risk Perceptions among Serious Offenders: The Role of Crime and Punishment." *Criminology* 30:575-94.

Horney, Julie, D. Wayne Osgood, and Ineke Haen Marshall. 1995. "Criminal Careers in the Short-Term: Intra-Individual Variability in Crime and Its Relation to Local Life Circumstances." *American Sociological Review* 60:655-73.

Immarigeon, Russ, and Meda Chesney-Lind. 1993. "Women's Prisons: Overcrowded and Overused." In *It's a Crime: Women and Justice,* edited by Roslyn Muraskin and Ted Alleman. Englewood Cliffs, NJ: Regents/Prentice Hall.

Jaccard, James, Robert Turrisi, and Choi K. Wan. 1990. *Interaction Effects in Multiple Regression.* Newbury Park, CA: Sage.

Jackson, Elton F., Charles R. Tittle, and Mary Jean Burke. 1986. "Offense-Specific Models of the Differential Association Process." *Social Problems* 33:335-56.

Jaquith, Susan M. 1981. "Adolescent Marijuana and Alcohol Use: An Empirical Test of Differential Association Theory." *Criminology* 19:271-80.

Johnson, Richard. 1979. *Juvenile Delinquency and Its Origins: An Integrated Theoretical Approach.* Cambridge, MA: Cambridge University Press.

Johnson, Richard E., Anastasios C. Marcos, and Stephen J. Bahr. 1987. "The Role of Peers in the Complex Etiology of Adolescent Drug Use." *Criminology* 25:323-40.

Kandel, Denise, and Mark Davies. 1991. "Friendship Networks, Intimacy, and Illicit Drug Use in Young Adulthood: A Comparison of Two Competing Theories." *Criminology* 29:441-69.

Kornhauser, Ruth. 1978. *Social Sources of Delinquency.* Chicago: University of Chicago Press.

Krohn, Marvin. 1995. "Control and Deterrence Theories of Criminality." In *Criminology: A Contemporary Handbook,* 2nd ed., edited by Joseph F. Sheley. Belmont, CA: Wadsworth.

Krohn, Marvin, Lonn Lanza-Kaduce, and Ronald Akers. 1984. "Community Context and Theories of Deviant Behavior: An Examination of Social Learning and Social Bonding Theories." *The Sociological Quarterly* 25:353-71.

LaGrange, Randy L., and Helene Raskin White. 1985. "Age Differences in Delinquency: A Test of Theory." *Criminology* 23:19-45.

Lasley, James R. 1988. "Toward a Control Theory of White-Collar Offending." *Journal of Quantitative Criminology* 4:347-62.

Laub, John H., Daniel S. Nagin, and Robert J. Sampson. 1998. "Trajectories of Change in Criminal Offending: Good Marriages and the Desistance Process." *American Sociological Review* 63:225-38.

Leonard, Eileen B. 1982. *Women, Crime, and Society: A Critique of Theoretical Criminology.* New York: Longman.

Longshore, Douglas, Susan Turner, and Judith A. Stein. 1996. "Self-Control in a Criminal Sample: An Examination of Construct Validity." *Criminology* 34:209-28.

Macdonald, Patrick T. 1989. "Competing Theoretical Explanations of Cocaine Use: Differential Association Versus Control Theory." *Journal of Contemporary Criminal Justice* 5:73-88.

MacKenzie, Doris Layton, and Robert Brame. 1994. "Shock Incarceration and Positive Adjustment during Community Supervision: A Multi-Site Evaluation." Presented at the annual meeting of the Academy of Criminal Justice Sciences, June, Chicago.

Makkai, Toni, and John Braithwaite. 1991. "Criminological Theories and Regulatory Compliance." *Criminology* 29:191-220.

Matsueda, Ross L. 1982. "Testing Control Theory and Differential Association: A Causal Modeling Approach." *American Sociological Review* 47:489-504.

_____. 1988. "The Current State of Differential Association Theory." *Crime and Delinquency* 34:277-306.

_____. 1989. "The Dynamics of Moral Beliefs and Minor Deviance." *Social Forces* 68:428-57.

_____. 1997. "'Cultural Deviance Theory': The Remarkable Persistence of a Flawed Term." *Theoretical Criminology* 1:429-52.

Matsueda, Ross L., and Karen Heimer. 1987. "Race, Family Structure, and Delinquency: A Test of Differential Association and Social Control Theories." *American Sociological Review* 52:826-40.

McGee, Zina T. 1992. "Social Class Differences in Parental and Peer Influence on Adolescent Drug Use." *Deviant Behavior* 13:349-72.

Merton, Robert. 1938. "Social Theory and Anomie." *American Sociological Review* 3:672-82.

Messerschmidt, James W. 1993. *Masculinities and Crime.* Lanham, MD: Rowman & Littlefield.

Messner, Steven F., and Richard Rosenfeld. 1994. *Crime and the American Dream.* Belmont, CA: Wadsworth.

Miller, Eleanor M. 1986. *Street Woman.* Philadelphia, PA: Temple University Press.

Naffine, Ngaire, and Fay Gale. 1989. "Testing the Nexus: Crime, Gender, and Unemployment." *British Journal of Criminology* 29:144-56.

Orcutt, James. 1987. "Differential Association and Marijuana Use: A Closer Look at Sutherland (with a Little Help from Becker)." *Criminology* 25:341-58.

Paternoster, Raymond, Robert Brame, Paul Mazerolle, and Alex Piquero. 1998. "Using the Correct Statistical Test for the Equality of Regression Coefficients." *Criminology* 36:859-66.

Paternoster, Raymond, and Ruth Triplett. 1988. "Disaggregating Self-Reported Delinquency and Its Implications for Theory." *Criminology* 26:591-625.

Rand, Alicia. 1987. "Transitional Life Events and Desistance from Delinquency and Crime." In *From Boy to Man: From Delinquency to Crime,* edited by Marvin Wolfgang, Terence P. Thornberry, and Robert M. Figlio. Chicago: University of Chicago Press.

Richie, Beth E. 1996. *Compelled to Crime: The Gender Entrapment of Battered Black Women.* New York: Routledge Kegan Paul.

Rosenbaum, James L. 1987. "Social Control, Gender and Delinquency: An Analysis of Drug, Property and Violent Offenders." *Justice Quarterly* 4:117-32.

Sampson, Robert J., and John H. Laub. 1990. "Crime and Deviance over the Life Course: The Salience of Adult Social Bonds." *American Sociological Review* 55:609-27.

_____. 1993. *Crime in the Making.* Cambridge, MA: Harvard University Press.

Short, James F. 1960. "Differential Association as a Hypothesis: Problems of Empirical Testing." *Social Problems* 8:14-24.

Silberman, Matthew. 1976. "Toward a Theory of Deterrence." *American Sociological Review* 41:442-61.

Simons, Ronald L., Martin G. Miller, and Stephen M. Aigner. 1980. "Contemporary Theories of Deviance and Female Delinquency: An Empirical Test." *Journal of Research in Crime and Delinquency* 17:42-57.

Smart, Carol. 1976. *Women, Crime and Criminology: A Feminist Critique.* Boston: Routledge Kegan Paul.

Sutherland, Edwin H. 1947. *Principles of Criminology.* 4th ed. Philadelphia, PA: Lippincott.

Sutherland, Edwin H., and Donald Cressey. 1955. *Principles of Criminology.* Philadelphia, PA: Lippincott.

Tittle, Charles R. 1980. *Sanctions and Social Deviance.* New York: Praeger.

Tittle, Charles, Mary Jean Burke, and Elton Jackson. 1986. "Modeling Sutherland's Theory of Differential Association: Toward an Empirical Clarification." *Social Forces* 65:405-32.

U.S. Bureau of Justice Statistics. 1994. *Special Report, March, 1994.* Washington DC: Government Printing Office.

Warr, Mark, and Mark Stafford. 1993. "Age, Peers, and Delinquency." *Criminology* 31:17-40.

Winfree, L. Thomas Curt, T. Griffiths, and Christine Sellers. 1989. "Social Learning Theory, Drug Use and American Indian Youths: A Cross-Cultural Test." *Justice Quarterly* 6:395-418.

Table of Random Numbers

Appendix D

Line/Col.	(1)	(2)	(3)	(4)	(5)	(6)	(7)	(8)	(9)	(10)	(11)	(12)	(13)	(14)
1	10480	15011	01536	02011	81647	91646	69179	14194	62590	36207	20969	99570	91291	90700
2	22368	46573	25595	85393	30995	89198	27982	53402	93965	34095	52666	19174	39615	99505
3	24130	48360	22527	97265	76393	64809	15179	24830	49340	32081	30680	19655	63348	58629
4	42167	93093	06243	61680	07856	16376	39440	53537	71341	57004	00849	74917	97758	16379
5	37570	39975	81837	16656	06121	91782	60468	81305	49684	60672	14110	06927	01263	54613
6	77921	06907	11008	42751	27756	53498	18602	70659	90655	15053	21916	81825	44394	42880
7	99562	72905	56420	69994	98872	31016	71194	18738	44013	48840	63213	21069	10634	12952
8	96301	91977	05463	07972	18876	20922	94595	56869	69014	60045	18425	84903	42508	32307
9	89579	14342	63661	10281	17453	18103	57740	84378	25331	12566	58678	44947	05585	56941
10	85475	36857	43342	53988	53060	59533	38867	62300	08158	17983	16439	11458	18593	64952
11	28918	69578	88231	33276	70997	79936	56865	05859	90106	31595	01547	85590	91610	78188
12	63553	40961	48235	03427	49626	69445	18663	72695	52180	20847	12234	90511	33703	90322
13	09429	93969	52636	92737	88974	33488	36320	17617	30015	08272	84115	27156	30613	74952
14	10365	61129	87529	85689	48237	52267	67689	93394	01511	26358	85104	20285	29975	89868
15	07119	97336	71048	08178	77233	13916	47564	81056	97735	85977	29372	74461	28551	90707
16	51085	12765	51821	51259	77452	16308	60756	92144	49442	53900	70960	63990	75601	40719
17	02368	21382	52404	60268	89368	19885	55322	44819	01188	65255	64835	44919	05944	55157
18	01011	54092	33362	94904	31273	04146	18594	29852	71585	85030	51132	01915	92747	64951
19	52162	53916	46369	58586	23216	14513	83149	98736	23495	64350	94738	17752	35156	35749
20	07056	97628	33787	09998	42698	06691	76988	13602	51851	46104	88916	19509	25625	58104
21	48663	91245	85828	14346	09172	30168	90229	04734	59193	22178	30421	61666	99904	32812
22	54164	58492	22421	74103	47070	25306	76468	26384	58151	06646	21524	15227	96909	44592
23	32639	32363	05597	24200	13363	38005	94342	28728	35806	06912	17012	64161	18296	22851
24	29334	27001	87637	87308	58731	00256	45834	15398	46557	41135	10367	07684	36188	18510
25	02488	33062	28834	07351	19731	92420	60952	61280	50001	67658	32586	86679	50720	94953
26	81525	72295	04839	96423	24878	82651	66566	14778	76797	14780	13300	87074	79666	95725
27	29676	20591	68086	26432	46901	20849	89768	81536	86645	12659	92259	57102	80428	25280
28	00742	57392	39064	66432	84673	40027	32832	61362	98947	96067	64760	64584	96096	98253
29	05366	04213	25669	26422	44407	44048	37937	63904	45766	66134	75470	66520	34693	90449
30	91921	26418	64117	94305	26766	25940	39972	22209	71500	64568	91402	42416	07844	69618
31	00582	04711	87917	77341	42206	35126	74087	99547	81817	42607	43808	76655	62028	76630
32	00725	69884	62797	56170	86324	88072	76222	36086	84637	93161	76038	65855	77919	88006
33	69011	65797	95876	55293	18988	27354	26575	08625	40801	59920	29841	80150	12777	48501

Line/Col.	(1)	(2)	(3)	(4)	(5)	(6)	(7)	(8)	(9)	(10)	(11)	(12)	(13)	(14)
34	25976	57948	29888	88604	67917	48708	18912	82271	65424	69774	33611	54262	85963	03547
35	09763	83473	73577	12908	30883	18317	28290	35797	05998	41688	34952	37888	38917	88050
36	91567	42595	27958	30134	04024	86385	29880	99730	55536	84855	29080	09250	79656	73211
37	17955	56349	90999	49127	20044	59931	06115	20542	18059	02008	73708	83317	36103	42791
38	46503	18584	18845	49618	02304	51038	20655	58727	28168	15475	56942	53389	20562	87338
39	92157	89634	94824	78171	84610	82834	09922	25417	44137	48413	25555	21246	35509	20468
40	14577	62765	35605	81263	39667	47358	56873	56307	61607	49518	89656	20103	77490	18062
41	98427	07523	33362	64270	01638	92477	66969	98420	04880	45585	46565	04102	46880	45709
42	34914	63976	88720	82765	34476	17032	87589	40836	32427	70002	70663	88863	77775	69348
43	70060	28277	39475	46473	23219	53416	94970	25832	69975	94884	19661	72828	00102	66794
44	53976	54914	06990	67245	68350	82948	11398	42878	80287	88267	47363	46634	06541	97809
45	76072	29515	40980	07391	58745	25774	22987	80059	39911	96189	41151	14222	60697	59583
46	90725	52210	83974	29992	65831	38857	50490	83765	55657	14361	31720	57375	56228	41546
47	64364	67412	33339	31926	14883	24413	59744	92351	97473	89286	35931	04110	23726	51900
48	08962	00358	31662	25388	61642	34072	81249	35648	56891	69352	48373	45578	78547	81788
49	95012	68379	93526	70765	10593	04542	76463	54328	02349	17247	28865	14777	62730	92277
50	15664	10493	20492	38391	91132	21999	59516	81652	27195	48223	46751	22923	32261	85653
51	16408	81899	04153	53381	79401	21438	83035	92350	36693	31238	59649	91754	72772	02338
52	18629	81953	05520	91962	04739	13092	97662	24822	94730	06496	35090	04822	86772	98289
53	73115	35101	47498	87637	99016	71060	88824	71013	18735	20286	23153	72924	35165	43040
54	57491	16703	23167	49323	45021	33132	12544	41035	80780	45393	44812	12515	98931	91202
55	30405	83946	23792	14422	15059	45799	22716	19792	09983	74353	68668	30429	70735	25499
56	16631	35006	85900	98275	32388	52390	16815	69298	82732	38480	73817	32523	41961	44437
57	96773	20206	42559	78985	05300	22164	24369	54224	35083	19687	11052	91491	60383	19746
58	38935	64202	14349	82674	66523	44133	00697	35552	35970	19124	63318	29686	03387	59846
59	31624	76384	17403	53363	44167	64486	64758	75366	76554	31601	12614	33072	60332	92325
60	78919	19474	23632	27889	47914	02584	37680	20801	72152	39339	34806	08930	85001	87820
61	03931	33309	57047	74211	63445	17361	62825	39908	05607	91284	68833	25570	38818	46920
62	74426	33278	43972	10119	89917	15665	52872	73823	73144	88662	88970	74492	51805	99378
63	09066	00903	20795	95452	92648	45454	09552	88815	16553	51125	79375	97596	16296	66092
64	42238	12426	87025	14267	20979	04508	64535	31355	86064	29472	47689	05974	52468	66834
65	16153	08002	26504	41744	81959	65642	74240	56302	00033	67107	77510	70625	28725	34191
66	21457	40742	29820	96783	29400	21840	15035	34537	33310	06116	95240	15957	16572	06004

(Continued)

Appendix D (Continued)

Line/Col.	(1)	(2)	(3)	(4)	(5)	(6)	(7)	(8)	(9)	(10)	(11)	(12)	(13)	(14)
67	21581	57802	02050	89728	17937	37621	47075	42080	97403	48626	68995	43805	33386	21597
68	55612	78095	83197	33732	05810	24813	86902	60397	16489	03264	88525	42786	05269	92532
69	44657	66999	99324	51281	84463	60563	79312	93454	68876	25471	93911	25650	12682	73572
70	91340	84979	46949	81973	37949	61023	43997	15263	80644	43942	89203	71795	99533	50501
71	91227	21199	31935	27022	84067	05462	35216	14486	29891	68607	41867	14951	91696	85065
72	50001	38140	66321	19924	72163	09538	12151	06878	91903	18749	34405	56087	82790	70925
73	65390	05224	72958	28609	81406	39147	25549	48542	42627	45233	57202	94617	23772	07896
74	27504	96131	83944	41575	10573	08619	64482	73923	36152	05184	94142	25299	84387	34925
75	37169	94851	39117	89632	00959	16487	65536	49071	39782	17095	02330	74301	00275	48280
76	11508	70225	51111	38351	19444	66499	71945	05422	13442	78675	84081	66938	93654	59894
77	37449	30362	06694	54690	04052	53115	62757	95348	78662	11163	81651	50245	34971	52924
78	46515	70331	85922	38329	57015	15765	97161	17869	45349	61796	66345	81073	49106	79860
79	30986	81223	42416	58353	21532	30502	32305	86482	05174	07901	54339	58861	74818	46942
80	63798	64995	46583	09765	44160	78128	83991	42865	92520	83531	80377	35909	81250	54238
81	82486	84846	99254	67632	43218	50076	21361	64816	51202	88124	41870	52689	51275	83556
82	21885	32906	92431	09060	64297	51674	64126	62570	26123	05155	59194	52799	28225	85762
83	60336	98782	07408	53458	13564	59089	26445	29789	85205	41001	12535	12133	14645	23541
84	43937	46891	24010	25560	86355	33941	25786	54990	71899	15475	95434	98227	21824	19585
85	97656	63175	89303	16275	07100	92063	21942	18611	47348	20203	18534	03862	78095	50136
86	03299	01221	05418	38982	55758	92237	26759	86367	21216	98442	08303	56613	91511	75928
87	79626	06486	03574	17668	07785	76020	79924	25651	83325	88428	85076	72811	22717	50585
88	85636	68335	47539	03129	65651	11977	02510	26113	99447	68645	34327	15152	55230	93448
89	18039	14367	61337	06177	12143	46609	32989	74014	64708	00533	35398	58408	13261	47908
90	08362	15656	60627	36478	65648	16764	53412	09013	07832	41574	17639	82163	60859	75567
91	79556	29068	04142	16268	15387	12856	66227	38358	22478	73373	88732	09443	82558	05250
92	92608	82674	27072	32534	17075	27698	98204	63863	11951	34648	88022	56148	34925	57031
93	23982	25835	40055	67006	12293	02753	14827	22235	35071	99704	37543	11601	35503	85171
94	09915	96306	05908	97901	28395	14186	00821	80703	70426	75647	76310	88717	37890	40129
95	50937	33300	26695	62247	69927	76123	50842	43834	86654	70959	79725	93872	28117	19233
96	42488	78077	69882	61657	34136	79180	97526	43092	04098	73571	80799	76536	71255	64239
97	46764	86273	63003	93017	31204	36692	40202	35275	57306	55543	53203	18098	47625	88684
98	03237	45430	55417	63282	90816	17349	88298	90183	36600	78406	06216	95787	42579	90730
99	86591	81482	52667	61583	14972	90053	89534	76036	49199	43716	97548	04379	46370	28672
100	38534	01715	94964	87288	65680	43772	39560	12918	86537	62738	19636	51132	25739	56947

Source: Beyer 1968.

References

Abbott, Andrew. 1992. "From Causes to Events: Notes on Narrative Positivism." *Sociological Methods* 20 (May): 428–55.

Abbott, Andrew. 1994. "History and Sociology: The Lost Synthesis." Pp. 77–112 in *Engaging the Past: The Uses of History Across the Social Sciences*. Durham, NC: Duke University Press.

Abel, David. 2000. "Census May Fall Short at Colleges." *The Boston Sunday Globe*, March 26, pp. B1, B4.

Abma, Tineke A. 2005. "Responsive Evaluation: Its Meaning and Special Contribution to Health Promotion." *Evaluation and Program Planning* 28:279–289.

Abrams, Philip. 1982. *Historical Sociology*. Ithaca, NY: Cornell University Press.

Adair, G., T. W. Dushenko, and R. C. L. Lindsay. 1985. "Ethical Regulations and Their Impact on Research Practice." *American Psychologist* 40:59–72.

Agnew, R. 1986. "Work and Delinquency Among Juveniles Attending School." *Journal of Criminal Justice* 9:19–41.

Alarid, Leanne Fiftal, Velmer S. Burton, Jr., and Francis T. Cullen. 2000. "Gender and Crime Among Felony Offenders: Assessing the Generality of Social Control and Differential Association Theories." *Journal of Research in Crime and Delinquency* 37:171–99.

Alfred, Randall. 1976. "The Church of Satan." Pp. 180–202 in *The New Religious Consciousness*, edited by Charles Glock and Robert Bellah. Berkeley: University of California Press.

Altheide, David L. and John M. Johnson. 1994. "Criteria for Assessing Interpretive Validity in Qualitative Research." Pp. 485–99 in *Handbook of Qualitative Research*, edited by Norman K. Denzin and Yvonna S. Lincoln. Thousand Oaks, CA: Sage.

Alvarez, Alex. 2000. *Governments, Citizens, and Genocide*. Bloomington: Indiana University Press.

Amenta, E. and J. D. Poulsen. 1994. "Where to Begin: A Survey of Five Approaches to Selecting Independent Variables for Qualitative Comparative Analysis." *Sociological Methods and Research* 23 (1): 22–53.

American Psychiatric Association. 1994. *Diagnostic and Statistical Manual of Mental Disorders (DSM IV)*. 4th ed. Washington, DC: American Psychiatric Association.

American Sociological Association. 1997. *Code of Ethics*. Washington, DC: American Sociological Association.

Anderson, Elijah. 1990. *Streetwise: Race, Class, and Change in an Urban Community*. Chicago: University of Chicago Press.

Anderson, Elijah. 2003. "Jelly's Place: An Ethnographic Memoir." *Symbolic Interaction* 26:217–37.

Anderson, Tammy. Forthcoming. Post-Rave Dance Scenes and Cultural Change. New York: Oxford University Press.

Anderson, Elijah. 1999. *Code of the Street: Decency, Violence, and the Moral Life of the Inner City*. New York: Norton.

Anderton, Douglas L., Richard E. Barrett, and Donald J. Bogue. 1997. *The Population of the United States*. New York: Free Press.

Anglin, M. D., D. Longshore, and S. Turner. 1999. "Treatment Alternatives to Street Crime: An Evaluation of Five Programs." *Criminal Justice and Behavior* 26 (2): 168–95.

Archer, Dane and Rosemary Gartner. 1984. *Violence and Crime in Cross-National Perspective*. New Haven, CT: Yale University Press.

Asmussen, Kelly J. and John W. Creswell. 1995. "Campus Response to a Student Gunman." *Journal of Higher Education* 66 (5): 575–91.

Babbie, Earl and Fred Halley. 1995. *Adventures in Social Research: Data Analysis Using SPSS for Windows*. Thousand Oaks, CA: Pine Forge.

Babbie, Earl, Fred Halley, and Jeanne Zaino. 2000. *Adventures in Social Research: Data Analysis Using SPSS v. 9.0 and 10.0 for Windows 95/98*. Thousand Oaks, CA: Pine Forge.

Babor, Thomas F., Robert S. Stephens, and G. Alan Marlatt. 1987. "Verbal Report Methods in Clinical Research on Alcoholism: Response Bias and Its Minimization." *Journal of Studies on Alcohol* 48 (5): 410–24.

Bachman, Ronet. 1992. *Death and Violence on the Reservation: Homicide, Family Violence, and Suicide in American Indian Populations*. Westport, CT: Auburn House.

Bachman, Ronet. 1996. "Victim's Perceptions of Initial Police Responses to Robbery and Aggravated Assault: Does Race Matter?" *Journal of Quantitative Criminology* 12:363–90.

Bachman, Ronet. 2000. "A Comparison of Annual Incidence Rates and Contextual Characteristics of Intimate Perpetrated Violence Against Women From the National Crime Victimization Survey (NCVS) and the National Violence Against Women Survey (NVAWS)." *Violence Against Women* 6 (8): 839–67.

Bachman, Ronet and Raymond Paternoster. 2003. *Statistics for Criminology and Criminal Justice*. 2nd ed. New York: McGraw-Hill.

Bachman, Ronet, Raymond Paternoster, and Sally Ward. 1992. "Rationality of Sexual Offending: Testing a Deterrence/Rational Choice Conception of Sexual Assault." *Law and Society Review* 26:401–32.

Bachman, Ronet and Linda Saltzman. 1995. "Violence Against Women: Estimates From the Redesigned National Crime Victimization Survey." *Bureau of Justice Statistics Special Report* (NCJ 154348). Washington, DC: U.S. Department of Justice.

Bachman, Ronet and Bruce Taylor. 1994. "The Measurement of Rape and Family Violence by the Redesigned National Crime Victimization Survey." *Justice Quarterly* 11:702–14.

Bailey, William C. 1990. "Murder, Capital Punishment, and Television: Execution Publicity and Homicide Rates." *American Sociological Review* 55 (October): 628–33.

Bainbridge, William Sims. 1989. *Survey Research: A Computer-Assisted Introduction*. Belmont, CA: Wadsworth.

Bandura, Albert, Dorothea Ross, and Sheila A. Ross. 1963. "Imitation of Film-Mediated Aggressive Models." *Journal of Abnormal and Social Psychology* 66:3–11.

Barringer, Felicity. 1993. "Majority in Poll Back Ban on Handguns." *The New York Times*, Junes 4, p. A14.

Baskin, Deborah R. and Ira B. Sommers. 1998. *Casualties of Community Disorder: Women's Careers in Violent Crime*. Boulder, CO: Westview.

Bayley, David H. 1994. *Police for the Future*. Oxford, UK: Oxford University Press.

Becker, Howard S. 1958. "Problems of Inference and Proof in Participant Observation." *American Sociological Review* 23:652–60.

Becker, Howard S. 1963. *The Outsiders: Studies in the Sociology of Deviance*. New York: Free Press.

Becker, Howard S. 1986. *Writing for Social Scientists*. Chicago: University of Chicago Press. (This may be ordered directly from the American Sociological Association: 1722 N Street NW, Washington, DC, 20036; 202-833-3410.)

Bellah, Robert N., Richard Madsen, William M. Sullivan, Ann Swidler, and Steven M. Tipton. 1985. *Habits of the Heart: Individualism and Commitment in American Life*. New York: Harper & Row.

Bench, Lawrence L. and Terry D. Allen. 2003. "Investigating the Stigma of Prison Classification: An Experimental Design." *The Prison Journal* 83 (4): 367–82.

Bennet, Lauren, Lisa Goodman, and Mary Ann Dutton. 1999. "Systematic Obstacles to the Criminal Prosecution of a Battering Partner: A Victim Perspective." *Journal of Interpersonal Violence* 14:761–72.

Berk, Richard A., Alice Campbell, Ruth Klap, and Bruce Western. 1992. "The Deterrent Effect of Arrest in Incidents of Domestic Violence: A Bayesian Analysis of Four Field Experiments." *American Sociological Review* 57:698–708.

Best, Joel and David F. Luckenbill. 1990. "Male Dominance and Female Criminality: A Test of Harris's Theory of Deviant Type-Scripts." *Sociological Inquiry* 60:71–86.

Beyer, William H., ed. 1968. *CRC Handbook for Tables for Probability and Statistics*. 2nd ed. Boca Raton, FL: CRC Press.

Binder, Arnold and James W. Meeker. 1993. "Implications of the Failure to Replicate the Minneapolis Experimental Findings." *American Sociological Review* 58 (December): 886–88.

Black, Donald J., ed. 1984. *Toward a General Theory of Social Control*. Orlando, FL: Academic Press.

Blackwell, Brenda Sims and Alex R. Piquero. 2005. "On the Relationships Between Gender, Power Control, Self-Control, and Crime." *Journal of Criminal Justice* 33:1–17.

Blau, Peter M. and Judith R. Blau. 1982. "The Cost of Inequality: Metropolitan Structure and Violence Crime." *American Sociological Review* 47:114–29.

Bollen, Kenneth A., Barbara Entwisle, and Arthur S. Alderson. 1993. "Macrocomparative Research Methods." *Annual Review of Sociology* 19:321–51.

Booth, Wayne C., Gregory G. Colomb, and Joseph M. Williams. 1995. *The Craft of Research*. Chicago: University of Chicago Press.

Boruch, Robert F. 1997. *Randomized Experiments for Planning and Evaluation: A Practical Guide*. Thousand Oaks, CA: Sage.

Botein, B. 1965. "The Manhattan Bail Project: Its Impact in Criminology and the Criminal Law Process." *Texas Law Review* 43:319–31.

Bourgois, Philippe, Mark Lettiere, and James Quesada. 1997. "Social Misery and the Sanctions of Substance Abuse: Confronting HIV Risk Among Homeless Heroin Addicts in San Francisco." *Social Problems* 44:155–73.

Boyer, Barbara and Peter Mucha. 2006. "Violent Crime Up in the City." *Philadelphia Figuira*, Sept. 20, p. B01.

Braga, Anthony A., David L. Weisburd, Elin J. Waring, Lorraine Green Mazerolle, William Spelman, and Francis Gajewski. 1999. "Problem-Oriented Policing in Violent Crime Places: A Randomized Controlled Experiment." *Criminology* 37 (4): 541–80.

Brame, Robert and Doris L. MacKenzie. 1996. "Shock Incarceration and Positive Adjustment During Community Supervision: A Multisite Evaluation." In *Correctional Boot Camps: A Tough Intermediate Sanction*, edited by D. L. MacKenzie and E. E. Herbert (NCJ 57639). Washington, DC: National Institute of Justice, U.S. Department of Justice.

Brener, Nancy D., Thomas R. Simon, Etienne G. Krug, and Richard Lowry. 1999. "Recent Trends in Violence-Related Behaviors Among High School Students in the United States." *Journal of the American Medical Association* 282 (5): 133–47.

Brennan, T. and J. Austin. 1997. *Women in Jail: Classification Issues*. Washington, DC: Department of Justice, National Institute of Corrections.

Brent, Edward and Alan Thompson. 1996. *Methodologist's Toolchest™ for Windows: User's Guide and Reference Manual*. Columbia, MO: Idea Works.

Bridges, George S. and Joseph G. Weis. 1989. "Measuring Violent Behavior: Effects of Study Design on Reported Correlates of Violence." Pp. 14–34 in *Violent Crime, Violent Criminals*, edited by Neil Alan Weiner and Marvin E. Wolfgang. Newbury Park, CA: Sage.

Broder, David S. 2000. "Don't Toy With the Census." *Washington Post*, April 4. p. A29.

Bureau of Justice Statistics. 1994. *Sourcebook of Criminal Justice Statistics: 1994* (NCJ 154591). Washington, DC: U.S. Department of Justice.

Bureau of Justice Statistics. 1995. *Sourcebook of Criminal Justice Statistics: 1995* (NCJ 158900). Washington, DC: U.S. Department of Justice.

Bureau of Justice Statistics. 1998. *Alcohol and Crime* (NCJ 168632). Washington, D.C.: U.S. Department of Justice.

Bureau of Justice Statistics. 2002. "Nation's Violent Crime Victimization Rate Falls 10 Percent." Retrieved September 9, 2002 (www.ojp.usdoj.gov/bjs/pub/press/cv01pr.htm).

Bureau of Justice Statistics. 2006. *Prison Statistics*. Retrieved June 26, 2006 (http://www.ojp .usdoj.gov/bjs/prisons.htm).

Burt, Martha R. 1996. "Homelessness: Definitions and Counts." Pp. 15–23 in *Homelessness in America*, edited by Jim Baumohl. Phoenix, AZ: Oryx.

Bushman, Brad J. 1995. "Moderating Role of Trait Aggressiveness in the Effects of Violent Media on Aggression." *Journal of Personality and Social Psychology* 69 (5): 950–60.

Bushway, S. and P. Reuter. 1997. "Labor Markets and Crime Risk Factors." In *Preventing Crime: What Works, What Doesn't, What's Promising*, edited by L. W. Sherman, D. Gottfredson, D. MacKenzie, J. Eck, P. Reuter, and S. Bushway. Unpublished Report by the Department of Criminology and Criminal Justice, University of Maryland.

Butterfield, Fox. 1996a. "After 10 Years, Juvenile Crime Begins to Drop." *The New York Times*, August 9, pp. A1, A25.

Butterfield, Fox. 1996b. "Gun Violence May Be Subsiding, Studies Find." *The New York Times*, October 14, p. A10.

Butterfield, Fox. 1997. "Serious Crime Decreased for Fifth Year in a Row." *The New York Times*, January 5, p. 10.

Buzawa, Eva S. and Carl G. Buzawa. 1996. *Domestic Violence: The Criminal Justice Response*. 2nd ed. Thousand Oaks, CA: Sage.

Campbell, Donald T. and M. Jean Russo. 1999. *Social Experimentation*. Thousand Oaks, CA: Sage.

Campbell, Donald T. and Julian C. Stanley. 1996. *Experimental and Quasi-Experimental Designs for Research*. Chicago: Rand McNally.

Campbell, Richard T. 1992. "Longitudinal Research." Pp. 1146–58 in *Encyclopedia of Sociology*, edited by Edgar F. Borgatta and Marie L. Borgatta. New York: Macmillan.

Campbell, T. T. and D. W. Fiske. 1959. "Convergent and Discriminant Validity by the Multi-Trait, Multi-Method Matrix." *Psychological Bulletin* 56:126–39.

Campbell, Wilson. 2002. "A Statement from The Governmental Accounting Standards Board and Performance Measurement Staff." American Society for Public Administration. Retrieved July 20, 2002 (http://www.aspanet.org/cap/forum_statement.html#top).

Cantor, David. 1984. *Comparing Bounded and Unbounded Three and Six Month Reference Periods in Rate Estimation*. Washington, DC: Bureau of Social Science Research.

Cantor, David. 1985. "Operational and Substantive Differences in Changing the NCS Reference Period." Pp. 125–37 in *Proceedings of the American Statistical Association, Social Statistics Section*. Washington, DC: American Statistical Association.

Cao, L., A. Adams, and V. J. Jensen. 1997. "A Test of the Black Subculture of Violence Thesis: A Research Note." *Criminology* 35 (2): 367–79.

Catalano. Shannan. 2006. *Criminal Victimization, 2005*. (NCJ 214644.) Washington, D.C.: U.S. Department of Justice, Bureau of Justice Statistics.

Cavender, Gray and Lisa Bond-Maupin. 2000. "Fear and Loathing on Reality Television: An Analysis of *America's Most Wanted* and *Unsolved Mysteries*." Pp. 51–57. In *Criminology: Perspectives*, edited by Steven Cooper. Bellevue, WA: Coursewise.

Chalk, Rosemary and Joel H. Garner. 2001. "Evaluating Arrest for Intimate Partner Violence: Two Decades of Research and Reform." *New Directions for Evaluation* 90:9–23.

Chamlin, Mitchell B. and John K. Cochran. 1997. "Social Altruism and Crime." *Criminology* 35 (2): 203–28.

Cloward, Richard and Lloyd E. Ohlin. 1960. *Delinquency and Opportunity: A Theory of Delinquent Gangs*. New York: Free Press.

Coffey, Amanda and Paul Atkinson. 1996. *Making Sense of Qualitative Data: Complementary Research Strategies*. Thousand Oaks, CA: Sage.

Cohen, Lawrence E. and Marcus Felson. 1979. "Social Change and Crime Rate Trends: A Routine Activity Approach." *American Sociological Review* 44:588–608.

Coleman, James S. and Thomas Hoffer. 1987. *Public and Private High Schools: The Impact of Communities*. New York: Basic Books.

Coleman, James S., Thomas Hoffer, and Sally Kilgore. 1982. *High School Achievement: Public, Catholic, and Private Schools Compared*. New York: Basic Books.

Converse, Jean M. 1984. "Attitude Measurement in Psychology and Sociology: The Early Years." Pp. 3–40 in *Surveying Subjective Phenomena*, Vol. 2, edited by Charles F. Turner and Elizabeth Martin. New York: Russell Sage.

Cook, Philip J. and John H. Laub. 1998. "The Epidemic in Youth Violence." *In Youth Violence: Crime and Justice*, Vol. 24, edited by Michael Tonry and Mark H. Moore. Chicago: University of Chicago Press.

Cook, Thomas D. and Donald T. Campbell. 1979. *Quasi-Experimentation: Design and Analysis Issues for Field Settings*. Chicago: Rand McNally.

Cooper, Harris and Larry V. Hedges. 1994. "Research Synthesis as a Scientific Enterprise." Pp. 3–14 in *The Handbook of Research Synthesis*, edited by Harris Cooper and Larry V. Hedges. New York: Russell Sage.

Core Institute. 1994. "Core Alcohol and Drug Survey: Long Form." Carbondale: FIPSE Core Analysis Grantee Group, Core Institute, Student Health Programs, Southern Illinois University.

Corse, Sara J., Nancy B. Hirschinger, and David Zanis. 1995. "The Use of the Addiction Severity Index With People With Severe Mental Illness." *Psychiatric Rehabilitation Journal* 19 (1): 9–18.

Costner, Herbert L. 1989. "The Validity of Conclusions in Evaluation Research: A Further Development of Chen and Rossi's Theory-Driven Approach." *Evaluation and Program Planning* 12:345–53.

Crossen, Cynthia. 1994. "How 'Tactical Research' Muddied Diaper Debate." *The Wall Street Journal*, May 17, pp. B1, B9.

Czaja, Ronald and Bob Blair. 1995. *Survey Research*. Newbury Park, CA: Pine Forge.

D'Amico, Elizabeth J. and Kim Fromme. 2002. "Brief Prevention for Adolescent Risk-Taking Behavior." *Addiction* 97:563–574.

Dannefer, W. Dale and Russell K. Schutt. 1982. "Race and Juvenile Justice Processing in Court and Police Agencies." *American Journal of Sociology* 87 (March): 1113–32.

Davies, Paul. 1993. "The Holy Grail of Physics." *The New York Times Book Review*, March 7, pp. 11–12.

Davis, James A. and Tom W. Smith. 1992. *The NORC General Social Survey: A User's Guide*. Newbury Park, CA: Sage.

Davis, Ryan. 1999. "Study: Search Engines Can't Keep Up With Expanding Net." *The Boston Globe*, July 8, pp. C1 C3.

Dawes, Robyn. 1995. "How Do You Formulate a Testable Exciting Hypothesis?" Pp. 93–96 in *How to Write a Successful Research Grant Application: A Guidance for Social and Behavioral Scientists*, edited by Willo Pequegnat and Ellen Stover. New York: Plenum.

Decker, Scott H. and Barrik Van Winkle. 1996. *Life in the Gang: Family, Friends, and Violence*. Cambridge, UK: Cambridge University Press.

Dentler, Robert A. 2002. *Practicing Sociology: Selected Fields*. Westport, CT: Praeger.

Denzin, Norman K. 2002. "The Interpretive Process." Pp. 349–368 in *The Qualitative Researcher's Companion*, edited by A. Michael Huberman and Matthew B. Miles. Thousand Oaks, CA: Sage.

Denzin, Norman K. and Yvonna S. Lincoln. 1994. "Introduction: Entering the Field of Qualitative Research." Pp. 1–17 in *Handbook of Qualitative Research*, edited by Norman K. Denzin and Yvonna S. Lincoln. Thousand Oaks, CA: Sage.

Denzin, Norman K. and Yvonna S. Lincoln, ed. 2000. *The Handbook of Qualitative Research*. 2nd ed. Thousand Oaks, CA: Sage.

Dewan, Shaila K. 2004a. "As Murders Fall, New Tactics Are Tried Against Remainder." *The New York Times*, December 31, pp. A24–A25.

Dewan, Shaila K. 2004b. "New York's Gospel of Policing by Data Spreads Across U.S." *The New York Times*, April 26, pp. A1, C16.

Diamond, Timothy. 1992. *Making Gray Gold: Narratives of Nursing Home Care.* Chicago: University of Chicago Press.

Dillman, Don A. 1978. *Mail and Telephone Surveys: The Total Design Method.* New York: Wiley.

Dillman, Don A. 1982. "Mail and Other Self-Administered Questionnaires." Chapter 12 in *Handbook of Survey Research*, edited by Peter Rossi, Kames Wright, and Andy Anderson. New York: Academic Press. Reprinted on pp. 637–38 in Delbert C. Miller. 1991. *Handbook of Research Design and Social Measurement.* 5th ed. Newbury Park, CA: Sage.

Dillman, Don A. 2000. Mail and Internet Surveys: *The Tailored Design Method.* 2nd ed. New York: John Wiley.

Dillman, Don A., James A. Christenson, Edwin H. Carpenter, and Ralph M. Brooks. 1974. "Increasing Mail Questionnaire Response: A Four-State Comparison." *American Sociological Review* 39 (October): 744–56.

Dixon, J. and Al J. Lizotte. 1987. "Gun Ownership and the Southern Subculture of Violence." *American Journal of Sociology* 93:383–405.

Dolnick, Edward. 1984. "Why Have the Pollsters Been Missing the Mark?" *The Boston Globe*, July 16, pp. 27–28

Duggan, Paul, Michael D. Shear, and Marc Fisher. 1999. "Killers Fused Violent Fantasy, Reality." *The Washington Post*, April 22, p. A1.

Dunford, Franklyn W., David Huizinga, and Delbert Elliott. 1990. "The Role of Arrest in Domestic Assault: The Omaha Police Experiment." *Criminology* 28:183–206.

Durkheim, Emile. [1951] 1987. *Suicide.* New York: Free Press.

Ehrlich, Isaac 1975. "The Deterrent Effect of Capital Punishment: A Question of Life and Death." *American Economic Review* 65:397–417.

Elliott, Delbert and Suzanne Ageton. 1980. "Reconciling Race and Class Differences in Self-Reported and Official Estimates of Delinquency." *American Sociological Review* 45:95–110.

Elliott, Delbert A., David Huizinga, and Suzanne S. Ageton. 1985. *Explaining Delinquency and Drug Use.* Newbury Park, CA: Sage.

Ellison, C. G. 1991. "An Eye for an Eye? A Note on the Southern Subculture of Violence Thesis." *Social Forces* 69:1223–39.

Emerson, Robert M., ed. 1983. *Contemporary Field Research.* Prospect Heights, IL: Waveland.

Emerson, Robert M., Rachel I. Fretz, and Linda L. Shaw. 1995. *Writing Ethnographic Fieldnotes.* Chicago: University of Chicago Press.

Erikson, Kai T. 1966. *Wayward Puritans: A Study in the Sociology of Deviance.* New York: Wiley.

Erikson, Kai T. 1967. "A Comment on Disguised Observation in Sociology." *Social Problems* 12:366–73.

Farrington, David P. 1977. "The Effects of Public Labeling." *British Journal of Criminology* 17 (2): 112–25.

Felson, Richard B., Allen E. Liska, Scott J. South, and Thomas L. McNulty. 1994. "The Subculture of Violence and Delinquency: Individual Versus School Context Effects." *Social Forces* 73 (1): 155–74.

Fenno, Richard F., Jr. 1978. *Home Style: House Members in Their Districts.* Boston: Little, Brown.

Fink, Arlene. 1998. *Conducting Research Literature Reviews: From Paper to the Internet.* Thousand Oaks, CA: Sage.

Fink, Arlene. 2004. *Conducting Research Literature Reviews: From the Internet to Paper.* Thousand Oaks, CA: Sage.

Fischer, Constance T. and Frederick J. Wertz. 2002. "Empirical Phenomenological Analyses of Being Criminally Victimized." Pp. 275–304 in *The Qualitative Researcher's Companion*, edited by A. Michael Huberman and Matthew B. Miles. Thousand Oaks, CA: Sage.

Fleury-Steiner, Benjamin. 2003. *Jurors' Stories of Death: How America's Death Penalty Invests in Inequality*. Ann Arbor: University of Michigan Press.

Forero, Juan. 2000. "Census Takers Top '90 Efforts in New York City, With More to Go." *The New York Times*, June 12, p. A29.

Fowler, Floyd J. 1988. *Survey Research Methods*. Rev. ed. Newbury Park, CA: Sage.

Fowler, Floyd J. 1995. *Improving Survey Questions: Design and Evaluation*. Thousand Oaks, CA: Sage.

Fraker, Thomas and Rebecca Maynard. 1987. "Evaluating Comparison Group Designs With Employment-Related Programs." *Journal of Human Resources* 22 (2): 194–227.

Freedman, David A. 1991. "Statistical Models and Shoe Leather." Pp. 291–313 in *Sociological Methodology*, Vol. 21, edited by Peter V. Marsden. Oxford, UK: Basil Blackwell.

Gallup, George 1986. *Public Opinion, 1985*. Wilmington, DE: Scholarly Resources Press.

Garner, Joel, Jeffrey Fagan, and Christopher Maxwell. 1995. "Published Findings From the Spousal Assault Replication Program: A Critical Review." *Journal of Quantitative Criminology* 11:3–28.

Geertz, Clifford. 1973. "Thick Description: Toward an Interpretive Theory of Culture." Pp. 3–30 in *The Interpretation of Cultures*, edited by Clifford Geertz. New York: Basic Books.

"'Get Tough' Youth Programs Are Ineffective, Panel Says." 2004. *The New York Times*, October 17, p. 25.

Gilligan, Carol. 1988. "Adolescent Development Reconsidered." Pp. vii–xxxix in *Mapping the Moral Domain*, edited by Carol Gilligan, Janie Victoria Ward, and Jill McLean Taylor. Cambridge, MA: Harvard University Press.

Glaser, Barney G. and Anselm L. Strauss. 1967. *The Discovery of Grounded Theory: Strategies for Qualitative Research*. London: Weidenfeld and Nicholson.

Glashow, Sheldon. 1989. "We Believe That the World Is Knowable." *The New York Times*, October 22, p. E24.

Gleick, James. 1990. "The Census: Why We Can't Count." *The New York Times*, July 15, pp. 22–26, 54.

Glover, Judith. 1996. "Epistemological and Methodological Considerations in Secondary Analysis." Pp. 28–38 in *Cross-National Research Methods in the Social Sciences*, edited by Linda Hantrais and Steen Mangen. New York: Pinter.

Glueck, Sheldon and Elenor Glueck. 1950. *Unraveling Juvenile Delinquency*. New York: Commonwealth Fund.

Goffman, Erving. 1961. *Asylums: Essays on the Social Situation of Mental Patients and Other Inmates*. Garden City, NY: Doubleday.

Goldenberg, Sheldon. 1992. *Thinking Methodologically*. New York: HarperCollins.

Goldfinger, Stephen M., Russell K. Schutt, Larry J. Seidman, Winston M. Turner, Walter E. Penk, and George S. Tolomiczenko. 1996. "Self-Report and Observer Measures of Substance Abuse Among Homeless Mentally Ill Persons in the Cross-Section and Over Time." *The Journal of Nervous and Mental Disease* 184 (11): 667–72.

Goleman, Daniel. 1993a. "Placebo Effect Is Shown to Be Twice as Powerful as Expected." *The New York Times*, August 17, p. C3.

Goleman, Daniel. 1993b. "Pollsters Enlist Psychologists in Quest for Unbiased Results." *The New York Times*, September 7, pp. C1, C11.

Goleman, Daniel. 1995. *Emotional Intelligence*. New York: Bantam.

Gordon, Raymond. 1992. *Basic Interviewing Skills*. Itasca, IL: Peacock.

Gregg v. Georgia, 428 U.S. 153 (1976).

Griffin, Larry J. 1993. "Narrative, Event-Structure Analysis, and Causal Interpretation in Historical Sociology." *American Journal of Sociology* 98 (March): 1094–133.

Grinnell, Frederick. 1992. *The Scientific Attitude*. 2nd ed. New York: Guilford.

Grossman, David C., Jolly J. Neckerman, Thomas D. Koepsell, Ping-Yu Liu, Kenneth N. Asher, Kathy Beland, Darin Frey, and Frederick P. Rivara. 1997. "Effectiveness of a Violence Prevention Curriculum Among Children in Elementary School: A Randomized Controlled Trial." *The Journal of the American Medical Association* 277 (20): 1605–12.

Groves, Robert M. 1989. Survey Errors and Survey Costs. New York: Wiley.

Groves, Robert M. and Mick P. Couper. 1998. *Nonresponse in Household Interview Surveys*. New York: Wiley.

Groves, Robert M. and Robert L. Kahn. 1979. *Surveys by Telephone: A National Comparison With Personal Interviews*. New York: Academic Press. Adapted in Delbert C. Miller. 1991. *Handbook of Research Design and Social Measurement*. 5th ed. Newbury Park, CA: Sage.

Gruenewald, Paul J., Andrew J. Treno, Gail Taff, and Michael Klitzner. 1997. *Measuring Community Indicators: A Systems Approach to Drug and Alcohol Problems*. Thousand Oaks, CA: Sage.

Guba, Egon G. and Yvonna S. Lincoln. 1989. *Fourth Generation Evaluation*. Newbury Park, CA: Sage.

Guba, Egon G. and Yvonna S. Lincoln. 1994. "Competing Paradigms in Qualitative Research." Pp. 105–17 in *Handbook of Qualitative Research*, edited by Norman K. Denzin and Yvonna S. Lincoln. Thousand Oaks, CA: Sage.

Gubrium, Jaber F. and James A. Holstein. 1997. *The New Language of Qualitative Method*. New York: Oxford University Press.

Gubrium, Jaber F. and James A. Holstein. 2000. "Analyzing Interpretive Practice." Pp. 487–508 in *The Handbook of Qualitative Research*, 2nd ed., edited by Norman Denzin and Yvonna S. Lincoln. Thousand Oaks, CA: Sage.

Hadaway, C. Kirk, Penny Long Marler, and Mark Chaves. 1993. "What the Polls Don't Show: A Closer Look at U.S. Church Attendance." *American Sociological Review* 58 (December): 741–52.

Hafner, Katie. 2005. "In Challenge to Google, Yahoo Will Scan Books." *The New York Times*, October 3, pp. C1, C4.

Hagan, J., A. R. Gillis, and J. Simpson. 1985. "The Class Structure of Gender and Delinquency: Toward a Power-Control Theory." *American Journal of Sociology* 90:1151–76.

Hagan, John. 1994. *Crime and Disrepute*. Thousand Oaks, CA: Pine Forge.

Hage, Jerald and Barbara Foley Meeker. 1988. *Social Causality*. Boston: Unwin Hyman.

Hagedorn, John. 1988. *People and Folks*. Chicago: Lake View.

Hamilton, V. Lee and Joseph Sanders. 1983. "Universals in Judging Wrongdoing: Japanese and Americans Compared." *American Sociological Review* 48 (April) 199–211.

Haney, C., C. Banks, and Philip G. Zimbardo. 1973. "International Dynamics in a Simulated Prison." *International Journal of Criminology and Penology* 1:69–97.

Hantrais, Linda and Steen Mangen. 1996. "Method of Management of Cross-National Social Research." Pp 1–12 in *Cross-National Research Methods in the Social Sciences*, edited by Linda Hantrais and Steen Mangen. New York: Pinter.

Harding, Sandra. 1989. "Value-Free Research Is a Delusion." *The New York Times*, October 22, p. E24.

Harrell, Adele, S. Cavanagh, and S. Sridharan. 1999. *Evaluation of the Children at Risk Program: Results 1 Year After the End of the Program* (NCJ 178914). Washington, DC: National Institute of Justice, U.S. Department of Justice.

Harris, Anthony R., Stephen H. Thomas, Gene A. Fisher, and David J. Hirsch. 2002. "Murder and Medicine: The Lethality of Criminal Assault 1960–1999." *Homicide Studies*, 6:128–166.

Hart, Chris. 1998. *Doing a Literature Review: Releasing the Social Science Research Imagination*. London: Sage.

Heckathorn, Douglas D. 1997. "Respondent-Driven Sampling: A New Approach to the Study of Hidden Populations." *Social Problems* 44:174–99.

Heimer, Karen and Stacy De Coster. 1999. "The Gendering of Violent Delinquency." *Criminology* 37 (2): 277–318.

Herek, Gregory. 1995. "Developing a Theoretical Framework and Rationale for a Research Proposal." Pp. 85–91 in *How to Write a Successful Research Grant Application: A Guide for Social and Behavioral Scientists*, edited by Willo Pequegnat and Ellen Stover. New York: Plenum.

Hesse-Biber, Sharon. 1989. "Eating Problems and Disorders in a College Population: Are College Women's Eating Problems a New Phenomenon?" *Sex Roles* 20:71–89.

Hirsch, Kathleen. 1989. *Songs From the Alley*. New York: Doubleday.

Hirschel, J. David, Ira W. Hutchison, III, and Charles W. Dean. 1992. "The Failure of Arrest to Deter Spouse Abuse." *Journal of Research in Crime and Delinquency* 7–33.

Ho, D. Y. R. 1996. "Filial Piety and Its Psychological Consequences." Pp. 155–65 in *Handbook of Chinese Psychology*, edited by M. H. Bond. Hong Kong: Oxford University Press.

Hock, Randolph. 2004. *The Extreme Searcher's Internet Handbook: A Guide for the Serious Searcher*. Medford, NJ: CyberAge.

Holmes, Steven A. 1994. "Census Officials Plan Big Changes in Gathering Data." *The New York Times*, May 16, pp. A1, A13.

Hoover, Kenneth R. 1980. *The Elements of Social Scientific Thinking*. 2nd ed. New York: St. Martin's.

Horney, Julie, D. Wayne Osgood, and Ineke Haen Marshall. 1995. "Criminal Careers in the Short-Term: Intra-Individual Variability in Crime and Its Relation to Local Life Circumstances." *American Sociological Review* 60:655–73.

Horney, Julie and Cassia Spohn. 1991. "Rape Law Reform and Instrumental Change in Six Urban Jurisdictions." *Law and Society Review* 25:117–53.

Hoyle, Carolyn and Andrew Sanders. 2000. "Police Response to Domestic Violence: From Victim Choice to Victim Empowerment." *British Journal of Criminology* 40:14–26.

Huberman, A. Michael and Matthew B. Miles. 1994. "Data Management and Analysis Methods." Pp. 428–44 in *Handbook of Qualitative Research*, edited by Norman K. Denzin and Yvonna S. Lincoln. Thousand Oaks, CA: Sage.

Humphrey, Nicholas. 1992. *A History of the Mind: Evolution and the Birth of Consciousness*. New York: Simon & Schuster.

Humphreys, Laud. 1970. *Tearoom Trade: Impersonal Sex in Public Places*. Chicago: Aldine.

Hunt, Morton. 1985. *Profiles of Social Research: The Scientific Study of Human Interactions* New York: Russell Sage.

Irwin, John. 1970. *The Felon*. Englewood Cliffs, NJ: Prentice Hall.

James, Thomas S. and Jeanne M. Granville. 1984. "Practical Issues in Vocational Education for Serious Juvenile Offenders." *In Violent Juvenile Offenders: An Anthology*, edited by R. Mathias, P. DeMuro, and R. Allison. San Francisco, CA: National Council on Crime and Delinquency.

Janesick, Valerie J. 1994. "The Dance of Qualitative Research Design: Metaphor, Methodolatry, and Meaning." Pp. 209–19 in *Handbook of Qualitative Research*, edited by Norman K. Denzin and Yvonna S. Lincoln. Thousand Oaks, CA: Sage.

Jervis, Robert. 1996. "Counterfactuals, Causation, and Complexity." Pp. 309–16 in *Counterfactual Thought Experiments in World Politics: Logical, Methodological, and Psychological Perspectives*, edited by Philip E. Tetlock and Aaron Belkin. Princeton, NJ: Princeton University Press.

Johnson, Dirk. 1997. "Party Animals in Fraternities Face the Threat of Extinction." *The New York Times*, May 15, pp. A1, A29.

Johnson, Holly. 1996. *Dangerous Domains: Violence Against Women in Canada*. Toronto: Nelson Canada.

Johnston, Lloyd D., Jerald G. Bachman, and Patrick M. O'Malley. 1994. *Sourcebook of Criminal Justice Statistics: 1994* (NCJ 154591). Washington, DC: Bureau of Justice Statistics, U.S. Department of Justice.

Kamin, Leon. 1974. *The Science and Politics of IQ*. Potomac, MD: Erlbaum.

Kandakai, Tina L., James H. Price, Susan K. Telljohann, and Carter A. Wilson. 1999. "Mothers' Perceptions of Factors Influencing Violence in Schools." *Journal of School Health* 69 (5): 189–205.

Kaplan, Fred. 2002. "NY Continues to See Plunge in Number of Felonies." *The Boston Globe*, April 15, p. A3.

Kaufman, Sharon R. 1986. *The Ageless Self: Sources of Meaning in Late Life*. Madison: University of Wisconsin Press.

Kennedy, David M., Anne M. Piehl, and Anthony A. Braga. 1996. "Youth Violence in Boston: Gun Markets, Serious Youth Offenders, and a Use-Reduction Strategy." *Law and Contemporary Problems* 59:147–196.

Kershaw, Sarah. 2000. "In a Black Community, Mistrust of the Government Hinders Census." *The New York Times*, May 16, p. A20.

Kershaw, Sarah. 2002. "Report Shows Serious Crime Rose in 2001." *The New York Times*, June 24, p. A10.

Kifner, John. 1994. "Pollster Finds Error on Holocaust Doubts." *The New York Times*, May 20, p. A12.

Kifner, John. 1999. "Killers Fused Violent Fantasy, Reality." *The Washington Post*, April 22, p. A1.

Kincaid, Harold. 1996. *Philosophical Foundations of the Social Sciences: Analyzing Controversies in Social Research*. Cambridge, UK: Cambridge University Press.

King, Gary, Robert O. Keohane, and Sidney Verba. 1994. *Scientific Inference in Qualitative Research*. Princeton, NJ: Princeton University Press.

Kobelarcik, E. L., C. A. Alexander, R. P. Singh, and G. M. Shapiro. 1983. "Alternative Reference Periods for the National Crime Survey." In *Proceedings of the American Statistical Association: Section on Survey Methods*. Washington, DC: American Statistical Association.

Koegel, Paul. 1987. *Ethnographic Perspectives on Homeless and Homeless Mentally Ill Women*. Washington, DC: Alcohol, Drug Abuse, and Mental Health Administration; Public Health Service; U.S. Department of Health and Human Services.

Kohn, Melvin L. 1987. "Cross-National Research as an Analytic Strategy." *American Sociological Review* 52:713–31.

Kohut, Andrew. 1988. "Polling: Does More Information Lead to Better Understanding?" *The Boston Globe*, November 7, p. 25.

Kraemer, Helena Chmura and Sue Thiemann. 1987. *How Many Subjects? Statistical Power Analysis in Research*. Newbury Park, CA: Sage.

Krauss, Clifford. 1996. "New York Crime Rate Plummets to Levels Not Seen in 30 Years." *The New York Times*, December 20, pp. A1, B4.

Krueger, Richard A. 1988. *Focus Groups: A Practical Guide for Applied Research*. Newbury Park, CA: Sage.

Kubey, Robert. 1990. "Television and the Quality of Family Life." *Communication Quarterly* 38 (Fall): 312–24.

Kuhn, Thomas S. 1970. *The Structure of Scientific Revolutions*. 2nd ed. Chicago: University of Chicago Press.

Kvale, Steinar. 1996. *Interviews: An Introduction to Qualitative Research Interviewing*. Thousand Oaks, CA: Sage.

Labaw, Patricia J. 1980. *Advanced Questionnaire Design*. Cambridge, MA: ABT Books.

LaLonde, Robert J. 1986. "Evaluating the Economic Evaluations of Training Programs With Experimental Data." *The American Economic Review* 76:604–20.

Lamott, Anne. 1994. *Bird by Bird: Some Instructions on Writing and Life*. New York: Anchor.

Langford, Terri. 2000. "Census Workers in Dallas Find the Well-off Hard to Count." *The Boston Globe*, June 1, p. A24.

Larson, Calvin J. 1993. *Pure and Applied Sociological Theory: Problems and Issues*. New York: Harcourt Brace Jovanovich.

Larson, Calvin J. and Gerald R. Garrett. 1996. *Crime, Justice and Society*. 2nd ed. Dix Hills, NY: General Hall.

Latour, Francie. 2002. "Marching Orders: After 10 Years, State Closes Prison Boot Camp." *Boston Sunday Globe*, June 16, pp. B1, B7.

Lavin, Danielle and Douglas W. Maynard. 2001. "Standardization vs. Rapport: Respondent Laughter and Interviewer Reaction During Telephone Surveys." *American Sociological Review* 66:453–479.

Lavrakas, Paul J. 1987. *Telephone Survey Methods: Sampling, Selection, and Supervision*. Newbury Park, CA: Sage.

Lazarsfeld, Paul F. and Anthony R. Oberschall. 1965. "Max Weber and Empirical Research." *American Sociological Review* (April): 185–99.

Lempert, Richard. 1989. "Humility Is a Virtue: On the Publicization of Policy-Relevant Research." *Law and Society Review* 23:146–61.

Lempert, Richard and Joseph Sanders. 1986. *An Invitation to Law and Social Science: Desert, Disputes, and Distribution*. New York: Longman.

Lewin, Tamar. 2001. "Income Education Is Found to Lower Risk of New Arrest." *The New York Times*, November 16, p. A18.

Lichtblau, Eric. 2000. "Crime Dip Levels Off; Assault, Rape Up." *The New York Times*, December 19, p. A2.

Lieberson, Stanley. 1985. *Making It Count: The Improvement of Social Research and Theory*. Berkeley: University of California Press.

Lieberson, Stanley. 1991. "Small *N*'s and Big Conclusions: An Examination of the Reasoning in Comparative Studies Based on a Small Number of Cases." *Social Forces* 70:307–20.

Lieberson, Stanley. 1992. "Einstein, Renoir, and Greeley: Some Thoughts About Evidence in Sociology." *American Sociological Review* 57 (February): 1–15.

Linsky, Arnold, Ronet Bachman, and Murray Straus. 1995. *Stress Culture and Aggression in the United States*. New Haven, CT: Yale University Press.

Lipsky, Michael. 1980. *Street-Level Bureaucracy*. New York: Russell Sage.

Liptak, Adam. 2004. "Long Term in Drug Case Fuels Debate on Sentencing." *The New York Times*, September 12, p. 16.

Litwin, Mark S. 1995. *How to Measure Survey Reliability and Validity*. Thousand Oaks, CA: Sage.

Locke, Lawrence F., Stephen J. Silverman, and Waneen Wyrick Spirduso. 1998. *Reading and Understanding Research*. Thousand Oaks, CA: Sage.

Lofland, John and Lyn H. Lofland. 1984. *Analyzing Social Settings: A Guide to Qualitative Observation and Analysis*. 2nd ed. Belmont, CA: Wadsworth.

Lofland, John and Lyn H. Lofland. 1995. *Analyzing Social Settings: A Guide to Qualitative Observation and Analysis*. 3rd ed. Belmont, CA: Wadsworth.

Loth, Renee. 1992. "Bush May Be Too Far Back, History of Polls Suggests." *The Boston Globe*, October 25, p. 19.

Lynch, James P. 1996. "Clarifying Divergent Estimates of Rape From Two National Surveys." *Public Opinion Quarterly* 60:410–30.

Lynch, Michael and David Bogen. 1997. "Sociology's Asociological 'Core': An Examination of Textbook Sociology in Light of the Sociology of Scientific Knowledge." *American Sociological Review* 62:481–93.

MacKenzie, Doris L. 1994. "Results of a Multisite Study of Boot Camp Prisons." *Federal Probation* 58 (2): 60–66.

MacKenzie, Doris L., Robert Brame, David McDowall, and C. Souryal. 1995. "Boot Camp, Prisons and Recidivism in Eight States." *Criminology* 33 (3): 401–30.

MacKenzie, Doris L. and Alex Piquero. 1994. "The Impact of Shock Incarceration Programs on Prison Crowding." *Crime and Delinquency* 40 (2): 222–49.

MacKenzie, Doris L. and C. Souryal. 1995. "Inmate Attitude Change during Incarceration: A Comparison of Boot Camp With Traditional Prison." *Justice Quarterly* 12 (2): 125–50.

MacKenzie, Doris L. and C. Souryal. 1996. "Multisite Study of Correctional Boot Camps." *In Correctional Boot Camps: A Tough Intermediate Sanction*, edited by D. L. MacKenzie and E. E. Herbert (NCJ 157639). Washington, DC: National Institute of Justice, U.S. Department of Justice.

Mangione, Thomas W. 1995. *Mail Surveys: Improving the Quality*. Thousand Oaks, CA: Sage.

Margolis, Eric. 2004. "Looking at Discipline, Looking at Labour: Photographic Representations of Indian Boarding Schools." *Visual Studies* 19:72–96.

Marini, Margaret Mooney and Burton Singer. 1988. "Causality in the Social Sciences." Pp. 347–409 in *Sociological Methodology*, Vol. 18, edited by Clifford C. Clogg. Washington, DC: American Sociological Association.

Martin, Lawrence L. and Peter M. Kettner. 1996. *Measuring the Performance of Human Service Programs*. Thousand Oaks, CA: Sage.

Martin, Linda G. and Kevin Kinsella. 1995. "Research on the Demography of Aging in Developing Countries." Pp. 356–403 in *Demography of Aging*, edited by Linda G. Martin and Samuel H. Preston. Washington, DC: National Academy Press.

Matt, Georg E. and Thomas D. Cook. 1994. "Threats to the Validity of Research Syntheses." Pp. 503–20 in *The Handbook of Research Synthesis*, edited by Harris M. Cooper and Larry V. Hedges. New York: Russell Sage.

Maxwell, Joseph A. 1996. *Qualitative Research Design: An Interactive Approach*. Thousand Oaks, CA: Sage.

Mayr, Ernst. 1982. *The Growth of Biological Thought: Diversity, Evolution, and Inheritance*. Cambridge, MA: Harvard University Press.

McDowall, David, Colin Loftin, and Brian Wiersema. 1992. "A Comparative Study of the Preventive Effects of Mandatory Sentencing Laws for Gun Crimes." *The Journal of Criminal Law and Criminology* 83 (2): 378–91.

McLellan, A. Thomas, Lester Luborsky, John Cacciola, Jeffrey Griffith, Frederick Evans, Harriet L. Barr, and Charles P. O'Brien. 1985. "New Data From the Addiction Severity Index: Reliability and Validity in Three Centers." *The Journal of Nervous and Mental Disease* 173 (7): 412–23.

Menand, Louis. 2001. *The Metaphysical Club: A Story of Ideas in America*. New York: Farrar, Straus, and Giroux.

Merton, Robert K. 1938. "Social Structure and Anomie." *American Sociological Review* 3:672–82.

Merton, Robert K., Marjorie Fiske, and Patricia L. Kendall. 1956. *The Focused Interview*. Glencoe, IL: Free Press.

Messner, Steven F. and Richard Rosenfeld. 1994. *Crime and the American Dream*. Belmont, CA: Wadsworth.

Miczek, Klaus A., Joseph F. DeBold, Margaret Haney, Jennifer Tidey, Jeffrey Vivian, and Elise M. Weerts. 1994. "Alcohol, Drugs of Abuse, Aggression, and Violence." Pp. 377–570 in *Understanding and Preventing Violence: Vol. 3, Social Influences*, edited by Albert J. Reiss, Jr., and Jeffrey A. Roth. Washington, DC: National Academy Press.

Miethe, Terance D., Wendy C. Regoeczi, and Kriss A. Drass. 2004. *Rethinking Homicide: Exploring the Structure and Process Underlying Deadly Situations*. Cambridge, UK: Cambridge University Press.

Milgram, Stanley. 1965. "Some Conditions of Obedience and Disobedience to Authority." *Human Relations* 18:57–75.

Milgram, Stanley. 1974. *Obedience to Authority: An Experimental View*. New York: Harper & Row.

Mill, John Stuart. 1872. *A System of Logic: Ratiocinative and Inductive*. 8th ed., Vol. 2. London: Longmans, Green, Reader, & Dyer.

Miller, Walter B. 1992. *Crime by Youth Gangs and Groups in the United States*. Washington, DC: Office of Juvenile Justice and Delinquency Prevention.

Miller, Delbert C. 1991. *Handbook of Research Design and Social Measurement*. 5th ed. Newbury Park, CA: Sage.

Miller, Jody. 2000. *One of the Guys: Girls, Gangs, and Gender*. New York: Oxford University Press.

Miller, Susan. 1999. *Gender and Community Policing: Walking the Talk*. Boston: Northeastern University Press.

Miller, Susan L. and C. Burack. 1993. "A Critique of Gottfredson and Hirschi's General Theory of Crime: Selective (in)Attention to Gender and Power Positions." *Women and Criminal Justice* 4:115–34.

Miller, William L. and Benjamin F. Crabtree. 1999a. "Clinical Research: A Multimethod Typology and Qualitative Roadmap." Pp. 3–30 in *Doing Qualitative Research*, 2nd ed., edited by Benjamin F. Crabtree and William L. Miller. Thousand Oaks, CA: Sage.

Miller, William L. and Benjamin F. Crabtree. 1999b. "The Dance of Interpretation." Pp. 127–143 in *Doing Qualitative Research*, 2nd ed., edited by Benjamin F. Crabtree and William L. Miller. Thousand Oaks, CA: Sage.

Miller, H. G., Gribble, J. N., Mazade, L. C., and Turner, C. F. 1998. "Abortion and Breast Cancer: Facto or Artifact?" In Stone, A. (Ed.) *Science of Self Report*. Mahwah, NJ.: Lawrence Erlbaum.

Mills, C. Wright. 1959. *The Sociological Imagination*. New York: Oxford University Press.

Mohr, Lawrence B. 1992. *Impact Analysis for Program Evaluation*. Newbury Park, CA: Sage.

Monkkonen, Eric H. 1994. "Introduction." Pp. 1–8 in *Engaging the Past: The Uses of History Across the Social Sciences*. Durham, NC: Duke University Press.

Mooney, Christopher Z. and Mei Hsien Lee. 1995. "Legislating Morality in the American States: The Case of Abortion Regulation Reform." *American Journal of Political Science* 39:599–627.

Moore, Joan W. 1978. *Homeboys: Gangs, Drugs, and Prison in the Barrios of Los Angeles*. Philadelphia: Temple University Press.

Moore, Joan W. 1991. *Going Down to the Barrio: Homeboys and Homegirls in Change*. Philadelphia: Temple University Press.

Morin, Richard. 1999. "Unconventional Wisdom: New Facts and Hot Stats From the Social Sciences." *Washington Post*, September 5, p. B5 (Outlook section).

Morrill, Calvin, Christine Yalda, Madeleine Adelman, Michael Musheno, and Cindy Bejarano. 2000. "Telling Tales in School: Youth Culture and Conflict Narratives." *Law & Society Review* 34:521–65.

Mosher, Clayton. J., Terance D. Miethe, and Dretha M. Phillips. 2002. *The Mismeasure of Crime*. Thousand Oaks, CA: Sage.

Mueser, Kim T., Paul R. Yarnold, Douglas F. Levinson, Hardeep Singhy, Alan S. Bellack, Kimmy Kee, Randall L. Morrison, and Kashinath G. Yadalam. 1990. "Prevalence of Substance Abuse in Schizophrenia: Demographic and Clinical Correlates." *Schizophrenia Bulletin* 16 (1): 31–56.

Mullins, Carolyn J. 1977. *A Guide to Writing and Publishing in the Social and Behavioral Sciences*. New York: Wiley.

National Institute of Alcohol Abuse and Alcoholism (NIAAA). 1994. "Alcohol-Related Impairment." *Alcohol Alert* 25 (July): 1–5.

National Institute of Alcohol Abuse and Alcoholism (NIAAA). 1995. "College Students and Drinking." *Alcohol Alert* 29 (July): 1–6.

National Institute of Alcohol Abuse and Alcoholism (NIAAA). 1997. "Alcohol Metabolism." *Alcohol Alert* 35 (January): 1–4.

National Opinion Research Center. 1992. *The NORC General Social Survey: Questions and Answers*. Mimeographed. Chicago: National Data Program for the Social Sciences.

National Victim Center and the Crime Victims Research and Treatment Center. 1992. *Rape in America: A Report to the Nation*. Arlington, VA: National Victim Center and the Crime Victims Research and Treatment Center.

Navarro, Mireya. 1990. "Census Questionnaire: Link to Democracy and Source of Data." *The New York Times*, March 25, p. 36.

Needleman, Carolyn. 1981. "Discrepant Assumptions in Empirical Research: The Case of Juvenile Court Screening." *Social Problems* 28 (February): 24–262.

Neuendorf, Kimberly A. 2002. *The Content Analysis Guidebook*. Thousand Oaks, CA: Sage.

Newport, Frank. 2000. "Popular Vote in Presidential Race Too Close to Call." Princeton, NJ: The Gallup Organization. Retrieved December 12, 2000 (www.gallup.com/poll/releases/pr001107.asp).

Nie, Norman H. and Lutz Erbring. 2000. *Internet and Society: A Preliminary Report.* Palo Alto, CA: Stanford Institute for the Quantitative Study of Society.

Nisbett, Richard E. and Dov Cohen. 1996. Culture of Honor: *The Psychology of Violence in the South.* Boulder, CO: Westview.

Novak, David. 2003. "The Evolution of Internet Research: Shifting Allegiances." *Online* 27:21.

Ó Dochartaigh, Niall. 2002. *The Internet Research Handbook: A Practical Guide for Students and Researchers in the Social Sciences.* Thousand Oaks, CA: Sage.

Oberschall, Anthony. 1972. "The Institutionalization of American Sociology." Pp. 187–251 in *The Establishment of Empirical Sociology: Studies in Continuity, Discontinuity, and Institutionalization,* edited by Anthony Oberschall. New York: Harper & Row.

Olzak, Susan, Suzanne Shanahan, and Elizabeth H. McEneaney. 1996. "Poverty, Segregation, and Race Riots: 1960 to 1993." *American Sociological Review* 61:590–613.

Orcutt, James D. and J. Blake Turner. 1993. "Shocking Numbers and Graphic Accounts: Quantified Images of Drug Problems in the Print Media." *Social Problems* 49 (May): 190–206.

Orshansky, Mollie. 1977. "Memorandum for Daniel P. Moynihan. Subject: History of the Poverty Line." Pp. 232–37 in *The Measure of Poverty. Technical Paper I: Documentation of Background Information and Rationale for Current Poverty Matrix,* edited by Mollie Orshansky. Washington, DC: U.S. Department of Health, Education, and Welfare.

Ousey, Graham C. and Matthew R. Lee. 2004. "Investigating the Connections Between Race, Illicit Drug Markets, and Legal Violence, 1984–1997." *Journal of Research in Crime and Delinquency* 41:352–383.

Padilla, Felix M. 1992. *The Gang as an American Enterprise.* New Brunswick, NJ: Rutgers University Press.

Papineau, David. 1978. *For Science in the Social Sciences.* London: Macmillan.

Parlett, Malcolm and David Hamilton. 1976. "Evaluation as Illumination: A New Approach to the Study of Innovative Programmes." Pp. 140–157 in *Evaluation Studies Review Annual,* Vol. 1, edited by G. Glass. Beverly Hills, CA: Sage.

Pate, Anthony M. and Edwin E. Hamilton. 1992. "Formal and Informal Deterrents to Domestic Violence: The Dade County Spouse Assault Experiment." *American Sociological Review* 57:691–97.

Paternoster, Raymond. 1991. *Capital Punishment in America.* New York: Lexington.

Paternoster, Raymond, Robert Brame, Ronet Bachman, and Lawrence W. Sherman. 1997. "Do Fair Procedures Matter? The Effect of Procedural Justice on Spouse Assault." *Law & Society Review* 31 (1): 163–204.

Patton, Michael Quinn. 1997. *Utilization Focused Evaluation: The New Century Text.* 3rd ed. Thousand Oaks, CA: Sage.

Patton, Michael Quinn. 2002. *Qualitative Research & Evaluation Methods.* 3rd ed. Thousand Oaks, CA: Sage.

Pepinsky, Harold E. 1980. "A Sociologist on Police Patrol." Pp. 223–34 in *Fieldwork Experience: Qualitative Approaches to Social Research,* edited by William B. Shaffir, Robert A. Stebbins, and Allan Turowetz. New York: St. Martin's.

Perkins, Craig. 1997. *Age Patterns of Victims of Serious Violent Crime.* Washington, DC: U.S. Department of Justice, Bureau of Justice Statistics.

Phillips, Derek L. 1971. *Knowledge From What? Theories and Methods in Social Research.* Chicago: Rand McNally.

Phoenix, Ann. 2003. "Neoliberalism and Masculinity: Racialization and the Contradictions of Schooling for 11- to 14-Year-Olds." *Youth & Society* 36:227–46.

Piquero, Nicole L., Angela R. Gover, J. Mohn MacDonald, and Alex R. Piquero. 2005. "The Influence of Delinquent Peers on Delinquency: Does Gender Matter?" *Youth & Society* 36:251–75.

Plessy v. Ferguson, 163 U.S. 537 (1896).

Posavac, E. J. and R. G. Carey. 1997. *Program Evaluation: Methods and Case Studies*. Upper Saddle River, NJ: Prentice Hall.

Powell, Kenneth E., Lois Muir-McClain, and Lakshmi Halasyamani. 1995. "A Review of Selected School-Based Conflict Resolution and Peer Mediation Projects." *Journal of School Health* 65 (10): 426–32.

Presley, Cheryl A., Philip W. Meilman, and Rob Lyerla. 1994. "Development of the Core Alcohol and Drug Survey: Initial Findings and Future Directions." *Journal of American College Health* 42:248–55.

Punch, Maurice. 1994. "Politics and Ethics in Qualitative Research." Pp. 83–97 in *Handbook of Qualitative Research*, edited by Norman K. Denzin and Yvonna S. Lincoln. Thousand Oaks, CA: Sage.

Purdy, Matthew. 1994. "Bronx Mystery: 3d-Rate Service for 1st-Class Mail." *The New York Times*, March 12, pp. 1, 3.

Putnam, Israel. 1977. "Poverty Thresholds: Their History and Future Development." Pp. 272–83 in *The Measure of Poverty. Technical Paper I: Documentation of Background Information and Rationale for Current Poverty Matrix*, edited by Mollie Orshansky. Washington, DC: U.S. Department of Health, Education, and Welfare.

Pyrczak, Fred. 2005. *Evaluating Research in Academic Journals: A Practical Guide to Realistic Evaluation*. 3rd ed. Glendale, CA: Pyrczak.

Radin, Charles A. 1997. "Partnerships, Awareness Behind Boston's Success." *The Boston Globe*, February 19, pp. A2, B7.

Ragin, Charles C. 1987. *The Comparative Method: Moving beyond Qualitative and Quantitative Strategies*. Berkeley: University of California Press.

Ragin, Charles C. 1994. *Constructing Social Research*. Thousand Oaks, CA: Pine Forge.

Rashbaum, William K. 2002. "Reasons for Crime Drop in New York Elude Many." *The New York Times*, November 29, p. A28.

Raudenbush, Stephen W. and Robert J. Sampson. 1999. "Ecometrics: Toward a Science of Assessing Ecological Settings, With Application to the Systematic Social Observation of Neighborhoods." *Sociological Methodology* 29:1–41.

Regoli, R. M. and J. D. Hewitt. 1994. "Delinquency in Society: A Child-Centered Approach." New York: McGraw-Hill.

Reiss, Albert J., Jr. 1971. *The Police and the Public*. New Haven, CT: Yale University Press.

Reiss, Albert J. and Jeffrey A. Roth. 1993. *Understanding and Preventing Violence*. Washington, DC: National Academy Press.

Reynolds, Paul Davidson. 1979. *Ethical Dilemmas and Social Science Research*. San Francisco: Jossey-Bass.

Richards, Thomas J. and Lyn Richards. 1994. "Using Computers in Qualitative Research." Pp. 445–62 in *Handbook of Qualitative Research*, edited by Norman K. Denzin and Yvonna S. Lincoln. Thousand Oaks, CA: Sage.

Richardson, Laurel. 1995. "Narrative and Sociology." Pp. 198–221 in *Representation in Ethnography*, edited by John Van Maanen. Thousand Oaks, CA: Sage.

Riedel, Marc. 2000. *Research Strategies for Secondary Data: A Perspective for Criminology and Criminal Justice*. Thousand Oaks, CA: Sage.

Riessman, Catherine Kohler. 2002. "Narrative Analysis." Pp. 217–270 in *The Qualitative Researcher's Companion*, edited by A. Michael Huberman and Matthew B. Miles. Thousand Oaks, CA: Sage.

Ringwalt, Christopher L., Jody M. Greene, Susan T. Ennett, Ronaldo Iachan, Richard R. Clayton, and Carl G. Leukefeld. 1994. *Past and Future Directions of the D.A.R.E. Program: An Evaluation Review*. Research Triangle, NC: Research Triangle Institute.

Rise, Eric W. 1995. *The Martinsville Seven: Race, Rape, and Capital Punishment*. Charlottesville: University Press of Virginia.

Rives, Norfleet W., Jr., and William J. Serow. 1988. *Introduction to Applied Demography: Data Sources and Estimation Techniques*. Sage University Paper Series on Quantitative Applications in the Social Sciences, Series No. 07–039. Newbury Park, CA: Sage.

Rosen, Lawrence. 1995. "The Creation of the Uniform Crime Report: The Role of Social Science." *Social Science History* 19:215–38.

Rosenbaum, David E. 2000. "Seeking Answers, Census Is Raising Privacy Questions." *The New York Times*, April 1, pp. A1, A9.

Rosenberg, Morris. 1968. *The Logic of Survey Analysis*. New York: Basic Books.

Rosenfeld, Richard, Timothy M. Bray, and Arlen Egley. 1999. "Facilitating Violence: A Comparison of Gang-Motivated, Gang-Affiliated, and Nongang Youth Homicides." *Journal of Quantitative Criminology* 15 (4): 496–516.

Rossi, Peter H. 1989. *Down and Out in America: The Origins of Homelessness*. Chicago: University of Chicago Press.

Rossi, Peter H. and Howard E. Freeman. 1989. *Evaluation: A Systematic Approach*. 4th ed. Newbury Park, CA: Sage.

Rossman, Gretchen B. and Sharon F. Rallis. 1998. *Learning in the Field: An Introduction to Qualitative Research*. Thousand Oaks, CA: Sage.

Ruback, R. Barry and Paula J. Vardaman. 1997. "Decision Making in Delinquency Cases: The Role of Race and Juveniles' Admission/Denial of the Crime." *Law and Human Behavior* 21 (1): 47–69.

Rubin, Herbert J. and Irene S. Rubin. 1995. *Qualitative Interviewing: The Art of Hearing Data*. Thousand Oaks, CA: Sage.

Rueschemeyer, Dietrich, Evelyne Huber Stephens, and John D. Stephens. 1992. *Capitalist Development and Democracy*. Chicago: University of Chicago Press.

Ruggles, Patricia. 1990. *Drawing the Line: Alternative Poverty Measures and Their Implications for Public Policy*. Washington, DC: The Urban Institute Press.

Sampson, Robert J. 1987. "Urban Black Violence: The Effect of Male Joblessness and Family Disruption." *American Journal of Sociology* 93 (September): 348–82.

Sampson, Robert J. and John H. Laub. 1990. "Crime and Deviance Over the Life Course: The Salience of Adult Social Bonds." *American Sociological Review* 55 (October): 609–27.

Sampson, Robert J. and John H. Laub. 1993. "Structural Variations in Juvenile Court Processing: Inequality, the Underclass, and Social Control." *Law and Society Review* 27 (2): 285–311.

Sampson, Robert J. and Janet L. Lauritsen. 1994. "Violent Victimization and Offending: Individual-, Situational-, and Community-Level Risk Factors." Pp. 1–114 in *Understanding and Preventing Violence: Vol. 3, Social Influences*, edited by Albert J. Reiss, Jr., and Jeffrey A. Roth. Washington, DC: National Academy Press.

Sampson, Robert J. and Stephen W. Raudenbush. 1999. "Systematic Social Observation of Public Spaces: A New Look at Disorder in Urban Neighborhoods." *American Journal of Sociology* 105:603–51.

Sampson, Robert J. and Stephen W. Raudenbush. 2001. "Disorder in Urban Neighborhoods—Does It Lead to Crime?" *Research in Brief*. Washington, DC: National Institute of Justice, U.S. Department of Justice.

Sampson, Robert J., Stephen W. Raudenbush, and Felton Earls. 1997. "Neighborhoods and Violent Crime: A Multilevel Study of Collective Efficacy." *Science* 277:918–24.

Sanchez-Jankowski, Martin. 1991. *Islands in the Street*. Berkeley: University of California Press.

Saretsky, H. 1972. "The OEO P. C. Experiment and the John Henry Effect." *Phi Delta Kappan* 53:579–81.

Savelsberg, Joachim L., Ryan King, and Lara Cleveland. 2002. "Politicized Scholarship? Science on Crime and the State." *Social Problems* 49:327–348.

Schnelle, J. F., R. E. Kirchner, J. W. Macrae, M. P. McNess, R. H. Eck, S. Snodgrass, J. D. Casey, and R. H. Uselton, Jr. 1978. "Police Evaluation Research: An Experimental and Cost-Benefit Analysis of a Helicopter Patrol in a High Crime Area." *Journal of Applied Behavior Analysis* 11:11–21.

Schorr, Lisbeth B. and Daniel Yankelovich. 2000. "In Search of a Gold Standard for Social Programs." *The Boston Globe*, February 18, p. A19.

Schuman, Howard and Stanley Presser. 1981. *Questions and Answers in Attitude Surveys: Experiments on Question Form, Wording, and Context*. New York: Academic Press.

Schutt, Russell K. and M. L. Fennell. 1992. "Shelter Staff Satisfaction With Services, the Service Network and Their Jobs." *Current Research on Occupations and Professions* 7:177–200.

Schutt, Russell K., Stephen M. Goldfinger, and Walter E. Penk. 1997. "Satisfaction With Residence and With Life: When Homeless Mentally Ill Persons Are Housed." *Evaluation and Program Planning* 20(2):185–194.

Schutt, Russell K., Suzanne Gunston, and John O'Brien. 1992. "The Impact of AIDS Prevention Efforts on AIDS Knowledge and Behavior Among Sheltered Homeless Adults." *Sociological Practice Review* 3 (1): 1–7.

Schwandt, Thomas A. 1994. "Constructivist, Interpretivist Approaches to Human Inquiry." Pp. 118–37 in *Handbook of Qualitative Research*, edited by Norman K. Denzin and Yvonna S. Lincoln. Thousand Oaks, CA: Sage.

Scull, Andrew T. 1988. "Deviance and Social Control." Pp. 667–93 in *Handbook of Sociology*, edited by Neil J. Smelser. Newbury Park, CA: Sage.

Sechrest, Lee and Souraya Sidani. 1995. "Quantitative and Qualitative Methods: Is There an Alternative?" *Evaluation and Program Planning* 18:77–87.

Seidman, Larry J. 1997. "Neuropsychological Testing." Pp. 498–508 in *Psychiatry*, Vol. 1, edited by Allan Tasman, Jerald Kay, and Jeffrey Lieberman. Philadelphia: W. B. Saunders.

Seltzer, Richard A. 1996. *Mistakes That Social Scientists Make: Error and Redemption in the Research Process*. New York: St. Martin's.

Shadish, William R. 1995. "Philosophy of Science and the Quantitative-Qualitative Debates: Thirteen Common Errors." *Evaluation and Program Planning* 18:63–75.

Shadish, William R., Thomas D. Cook, and Laura C. Leviton, eds. 1991. *Foundations of Program Evaluation: Theories of Practice*. Thousand Oaks, CA: Sage.

Shaw, Clifford R. and Henry D. McKay. 1942. *Juvenile Delinquency and Urban Areas*. Chicago: University of Chicago Press.

Shepherd, Jane, David Hill, Joel Bristor, and Pat Montalvan. 1996. "Converting an Ongoing Health Study to CAPI: Findings From the National Health and Nutrition Study." Pp. 159–64 in *Health Survey Research Methods Conference Proceedings*, edited by Richard B. Warnecke. Hyattsville, MD: U.S. Department of Health and Human Services.

Sherman, Lawrence W. 1992. *Policing Domestic Violence: Experiments and Dilemmas*. New York: Free Press.

Sherman, Lawrence W. 1997. "Thinking About Crime Prevention." *In Preventing Crime: What Works, What Doesn't, What's Promising*, edited by L. W. Sherman, D. Gottfredson, D. MacKenzie, J. Eck, P. Reuter, and S. Bushway. Unpublished Report by the Department of Criminology and Criminal Justice, University of Maryland.

Sherman, Lawrence W. and Richard A. Berk. 1984. "The Specific Deterrent Effects of Arrest for Domestic Assault." *American Sociological Review* 49:261–72.

Sherman, Lawrence W. and Ellen G. Cohn. 1989. "The Impact of Research on Legal Policy: The Minneapolis Domestic Violence Experiment." *Law and Society Review* 23:117–44.

Sherman, Lawrence W., P. Gartin, and M. Buerger. 1989. "Hot Spots of Predatory Crime: Routine Activities and the Criminology of Place." *Criminology* 27:27–56.

Sherman, Lawrence W., D. Gottfredson, Doris MacKenzie, J. Eck, P. Reuter, and S. Bushway. 1997. *Preventing Crime: What Works, What Doesn't, What's Promising.* Unpublished Report by the Department of Criminology and Criminal Justice, University of Maryland.

Sherman, Lawrence W., Douglas A. Smith, Janell D. Schmidt, and Dennis P. Rogan. 1992. "Crime, Punishment, and Stake in Conformity: Legal and Informal Control of Domestic Violence." *American Sociological Review* 57:680–90.

Short, James F., Jr., and Ivan Nye. 1957–58. "Reported Behavior as a Criterion of Deviant Behavior." *Social Problems* 5:207–13.

Simon, Rita J. and Sandra Baxter. 1989. "Gender and Violent Crime." Pp. 171–97 in *Violent Crime, Violent Criminals*, edited by Neil Alan Weiner and Marvin E. Wolfgang. Newbury Park, CA: Sage.

Sjoberg, Gideon, ed. 1967. *Ethics, Politics, and Social Research*. Cambridge, MA: Schenkman.

Sjoberg, Gideon and Roger Nett. 1968. *A Methodology for Social Research*. New York: Harper & Row.

Skinner, Harvey A. and Wen-Jenn Sheu. 1982. "Reliability of Alcohol Use Indices: The Lifetime Drinking History and the MAST." *Journal of Studies on Alcohol* 43 (11): 1157–70.

Skocpol, Theda. 1984. "Emerging Agendas and Recurrent Strategies in Historical Sociology." Pp. 356–91 in *Vision and Method in Historical Sociology*, edited by Theda Skocpol. New York: Cambridge University Press.

Snow, David L., Jacob Kraemer Tebes, and Michael W. Arthur. 1992. "Panel Attrition and External Validity in Adolescent Substance Use Research." *Journal of Consulting and Clinical Psychology* 60:804–807.

Sobell, Linda C., Mark B. Sobell, Diane M. Riley, Reinhard Schuller, D. Sigfrido Pavan, Anthony Cancilla, Felix Klajner, and Gloria I. Leo. 1988. "The Reliability of Alcohol Abusers' Self-Reports of Drinking and Life Events That Occurred in the Distant Past." *Journal of Studies on Alcohol* 49 (2): 225–32.

Sosin, Michael R., Paul Colson, and Susan Grossman. 1988. *Homelessness in Chicago: Poverty and Pathology, Social Institutions and Social Change*. Chicago: Chicago Community Trust.

Specter, Michael. 1994. "Census-Takers Come Calling and Get a Scolding." *The New York Times*, March 3, p. A4.

Stake, Robert E. 1995. *The Art of Case Study Research*. Thousand Oaks, CA: Sage.

Stewart, David W. 1984. *Secondary Research: Information Sources and Methods*. Beverly Hills, CA: Sage.

Stone, Christopher 2006. "A Tale of Two Cities: Boston, New York and Crime." *The Boston Globe*, February 12, p. E12.

Stout, David. 1997. "Officials Are Starting Early in Their Defense of the 2000 Census." *The New York Times*, March 23, p. 37.

Straus, Murray. 1979. "Measuring Intrafamily Conflict and Violence: The Conflict Tactics (TC) Sale." *Journal of Marriage and the Family* 41:75–88.

Straus, Murray A. and Richard J. Gelles. 1990. *Physical Violence in American Families: Risk Factors and Adaptations to Violence in 8,145 Families*. New Brunswick, NJ: Transaction Publishers.

Strunk, William, Jr., and E. B. White. 1979. *The Elements of Style*. 3rd ed. New York: Macmillan.

Sudman, Seymour. 1976. *Applied Sampling*. New York: Academic Press.

"Survey on Adultery: 'I Do' Means 'I Don't.'" 1993. *The New York Times*, October 19, p. A20.

Tannenbaum, F. 1938. *Crime and the Community*. New York: Columbia University Press.

Tarnas, Richard. 1991. *The Passion of the Western Mind: Understanding the Ideas That Have Shaped Our World View*. New York: Ballantine.

Thorne, Barrie. 1993. *Gender Play: Girls and Boys in School*. New Brunswick, NJ: Rutgers University Press.

Thrasher, Frederic. 1927. *The Gang: A Study of 1,313 Gangs in Chicago*. Chicago: University of Chicago Press.

Tjaden, Patricia and Nancy Thoennes. 2000. "Extent, Nature, and Consequences of Intimate Partner Violence." In *Findings From the National Violence Against Women Survey* (NCJ 181867). Washington, DC: National Institute of Justice, U.S. Department of Justice.

Toby, Jackson. 1957. "Social Disorganization and Stake in Conformity: Complementary Factors in the Predatory Behavior of Hoodlums." *Journal of Criminal Law, Criminology and Police Science* 48:12–17.

Tonry, Michael and Mark H. Moore. 1998. *Youth Violence: Crime and Justice*. Vol. 24. Chicago: University of Chicago Press.

Tourangeau, R. and Smith, T. W. 1996. "Asking Sensitive Questions: The Impact of Data Collection Mode, Question Format, and Question Context." *Public Opinion Quarterly* 60: 275–301.

Tufte, Edward R. 1983. *The Visual Display of Quantitative Information*. Cheshire, CT: Graphics.

Turabian, Kate L. 1967. *A Manual for Writers of Term Papers, Theses, and Dissertations*. Rev. 3rd ed. Chicago: University of Chicago Press.

Turner, Stephen P. 1980. *Sociological Explanation as Translation*. Cambridge, UK: Cambridge University Press.

Turner, C. F., L. Ku, S. M. Rogers, L. D. Lindberg, J. H. Pleck, and F. L. Sonenstein. 1998. "Adolescent Sexual Behavior, Drug Use, and Violence: Increased Reporting With Computer Survey Technology." *Science* 280:867–73.

Turner, Charles F. and Elizabeth Martin, eds. 1984. *Surveying Subjective Phenomena*. Vols. I and II. New York: Russell Sage.

Tyler, T. 1990. *Why People Obey the Law*. New Haven, CT: Yale University Press.

U.S. Bureau of the Census. 1999. *United States Census 2000, Updated Summary: Census 2000 Operational Plan*. Washington, DC: U.S. Department of Commerce, Bureau of the Census.

U.S. Bureau of the Census. 2000a. "Census Bureau Directory Says 92 Percent of U.S. Households Accounted For: Thanks President and Vice President for Message to Census Workers." In *United States Department of Commerce News*. Retrieved May 31, 2000 (www.census.gov/Press-Release/www/2000/cb00cn41.html).

U.S. Bureau of the Census. 2000b. "U.S. Commerce Secretary William M. Daley Delegates Decision to Census Bureau on Adjusting Census 2000." In *United States Department of Commerce News*. Retrieved June 14, 2000 (www.census.gov/Press-Release/www/2000).

Vaillant, George E. 1995. *The Natural History of Alcoholism Revisited*. Cambridge, MA: Harvard University Press.

van de Vijver, Fons and Kwok Leung. 1997. *Methods and Data Analysis for Cross-Cultural Research*. Thousand Oaks, CA: Sage.

Van Maanen, John. 1982. "Fieldwork on the Beat." Pp. 103–51 in *Varieties of Qualitative Research*, edited by John Van Maanen, James M. Dabbs, Jr., and Robert R. Faulkner. Beverly Hills: Sage.

Venkatesh, Sudhir Alladi. 1997. "The Social Organization of Street Gang Activity in an Urban Ghetto." *American Journal of Sociology* 103:82–102.

Venkatesh, Sudhir Alladi. 2000. *American Project: The Rise and Fall of a Modern Ghetto*. Cambridge, MA: Harvard University Press.

Vidich, Arthur J. and Stanford M. Lyman. 1994. "Qualitative Methods: Their History in Sociology and Anthropology." Pp. 23–59 in *Handbook of Qualitative Research*, edited by Norman K. Denzin and Yvonna S. Lincoln. Thousand Oaks, CA: Sage.

Vigil, James Diego. 1988. *Barrio Gangs*. Austin: University of Texas Press.

Wallace, Walter L. 1971. *The Logic of Science in Sociology*. Chicago: Aldine.

Wallace, Walter L. 1983. *Principles of Scientific Sociology*. New York: Aldine.

Wallerstein, J. S. and Wyle, C. J. 1947. "Our Law-Abiding Law Breakers." *Probation* 25:107–12.

Walters, Pamela Barnhouse, David R. James, and Holly J. McCammon. 1997. "Citizenship and Public Schools: Accounting for Racial Inequality in Education for the Pre- and Post-Disfranchisement South." *American Sociological Review* 62:34–52.

Watson, Charles G., Curt Tilleskjor, E. A. Hoodecheck-Schow, John Pucel, and Lyle Jacobs. 1984. "Do Alcoholics Give Valid Self-Reports?" *Journal of Studies on Alcohol* 45 (4): 344–48.

Watson, Roy E. L. 1986. "The Effectiveness of Increased Police Enforcement as a General Deterrent." *Law and Society Review* 20 (2): 293–99.

Weatherby, Norman L., Richard Needle, Helen Cesari, Robert Booth, Clyde B. McCoy, John K. Waters, Mark Williams, and Dale D. Chitwood. 1994. "Validity of Self-Reported Drug Use Among Injection Drug Users and Crack Cocaine Users Recruited Through Street Outreach." *Evaluation and Program Planning* 17 (4): 347–55.

Webb, Eugene, Donald T. Campbell, Richard D. Schwartz, and Lee Sechrest. [1966] 2000. *Unobtrusive Measures: Nonreactive Research in the Social Sciences*. Chicago: Rand McNally.

Weber, Max. 1949. *The Methodology of the Social Sciences*. Translated and edited by Edward A. Shils and Henry Finch. New York: Free Press.

Weber, Robert Philip. 1985. *Basic Content Analysis*. Beverly Hills, CA: Sage.

Weber, Robert Philip. 1990. *Basic Content Analysis*, 2nd ed. Newbury Park, CA: Sage.

Wechsler, Henry, Andrea, Davenport, George Dowdall, Barbara Moeykens, and Sonia Castillo. 1994. "Health and Behavioral Consequences of Binge Drinking in College: A National Survey of Students at 140 Campuses." *JAMA: The Journal of the American Medical Association* 272 (21): 1672–77.

Weisburd, David, L. Maher, and Lawrence Sherman. 1992. "Contrasting Crime General and Crime Specific Theory: The Case of Hot Spots of Crime." *Advances in Criminological Theory* 4:45–69.

Weisburd, David, Stanton Wheeler, Elin Waring, and Nancy Bode. 1991. *Crimes of the Middle Class: White-Collar Offenders in the Federal Courts*. New Haven, CT: Yale University Press.

Weiss, Carole H. 1977. "Introduction." In *Using Social Research in Public Policy Making*, edited by C. H. Weiss. Lexington, MA: D. C. Heath.

Weiss, Carol H. 1993. "Where Politics and Evaluation Research Meet." *Evaluation Practice* 14: 93–106.

Wells, L. Edward and Joseph H. Rankin. 1991. "Families and Delinquency: A Meta-Analysis of the Impact of Broken Homes." *Social Problems* 38 (February): 71–93.

Whyte, William Foote. 1943. *Street Corner Society: The Social Structure of an Italian Slum*. Chicago: University of Chicago Press.

Whyte, William Foote. 1955. *Street Corner Society*. Chicago: University of Chicago Press.

Whyte, William Foote. 1991. *Social Theory for Social Action: How Individuals and Organizations Learn to Change*. Newbury Park, CA: Sage.

Widom, Cathy Spatz. 2000. "Childhood Victimization: Early Adversity, Later Psychopathology." In *National Institute of Justice Journal* (NCJ JR000242). Washington, DC: National Institute of Justice, U.S. Department of Justice.

Williams, Kirk R. and Richard Hawkins. 1986. "Perceptual Research on General Deterrence: A Critical Review." *Law and Society Review* 20:545–572.

Williams, Christine and E. Joel Heikes. 1993. "The Importance of Researcher's Gender in the In-depth Interview: Evidence From Two Case Studies of Male Nurses." *Gender and Society* 7:280–91.

Wilson, William Julius. 1987. *The Truly Disadvantaged: The Inner City, the Underclass, and Public Policy*. Chicago: University of Chicago Press.

Wines, Michael. 2006. "To Fill Notebooks, and Then a Few Bellies." *The New York Times*, August 27, Week in Review.

Wolcott, Harry F. 1995. *The Art of Fieldwork*. Walnut Creek, CA: AltaMira.

Wolf, N., C. L. Blitz, J. Shi, R. Bachman, and J. Siegel. 2006. "Sexual Violence Inside Prisons: Rates of Victimization." *Journal of Urban Health* 83 (5): 835–851.

Wolfgang, Marvin E. and F. Ferracuti. 1967. *The Subculture of Violence*. London: Tavistock.

Wright, Richard and Scott Decker. 1994. *Burglars on the Job: Streetlife and Residential Break-ins*. Boston: Northeastern University Press.

Zaret, David. 1996. "Petitions and the 'Invention' of Public Opinion in the English Revolution." *American Journal of Sociology* 101:1497–555.

Zelditch, M., Jr. 1962. "Some Methodological Problems of Field Studies." *American Journal of Sociology* 67:566.

Zielbauer, Paul. 2000. "2 Cities Lag Far Behind the U.S. in Heeding the Call of the Census." *The New York Times*, April 21, p. A21.

Glossary/Index

Good Morning America, 122
Gore, A., 113
Governmental Accounting Standards
 Board (GASB), 376
Graphs
 bar chart, 408–409e, 410e
 frequency polygon, 410, 412e
 guidelines for using, 411–412
 histogram, 409–410, 411e
 misuse of, 410–411, 413e
 use of, 408
Gregg v. Georgia, 60, 232
Grounded research, 38–39
Grounded theory Systematic theory developed
 inductively, based on observations that are
 summarized into conceptual categories,
 reevaluated in the research setting, and
 gradually refined and linked to other
 conceptual categories, 315e
Group-administered survey A survey that is
 completed by individual respondents who
 are assembled in a group, 238
 compared to other survey designs,
  236e, 247–249, 248e
Group units of analysis, 133–134
Grouped frequency distribution A frequency
 distribution in which the data are
 organized into categories, either because
 there are more values than can be easily
 displayed or because the distribution of the
 variable will be clearer or more meaningful,
 414, 416–418e
GSS (General Social Survey), 128, 208–209, 327
GTC (General Theory of Crime), 160
Guba, E. G., 24–25, 380, 461–462, 468
Gun crime, 17, 389–390

Hawthorne effect A type of contamination in
 experimental and quasi-experimental
 designs that occurs when members of the
 treatment group change in terms of the
 dependent variable because their
 participation in the study makes them
 feel special, 191
Head Start, 180
Health Research Extension Act of 1985, 394
Hermeneutic circle, 23–24e
Hermeneutic perspective, 296
Hirschi, T., 160, C–8, C–14
Histogram A graphic for quantitative variables in
 which the variable's distribution is displayed
 with adjacent bars, 409–410, 411e

Historical events research Research in which
 social events of only one time period in the
 past are studied, 18, 328, 330–331
 case study, 330–331
History effect A source of causal invalidity
 that occurs when something other than the
 treatment influences outcome scores; also
 called an effect of external events, 188–189
Homeless people
 AIDS prevention education, 224
 mental illnesses, 377
 population definition, 103–105
 substance abuse by, 121
 validity of data on, 80
Homicide rates
 among American Indians, 352–355, 353e
 cross-national comparisons, 332–333, 334e
 declines in, 141
 gang homicides in St. Louis, 342–345, 343e, 344e
Honesty, 60–61
Horney, J., 183, 184e, 367, 389
Human subjects, research on, 62–65
 children, 287–288
 distribution of benefits, 187, 198–199
 informed consent, 64, 197, 198
 institutional review boards, 63
 potential harm, 63–64, 355
 withholding beneficial treatment, 64
 See also Ethical issues
HyperRESEARCH, 317–318e, 319
Hypothesis A tentative statement about
 empirical reality involving the relationship
 between two or more variables, 45–50
 direction of association and, 46e–50e
 example of, 45, 46e

ICPSR (Inter-University Consortium for Political
 and Social Research), 18, 209, 326, 356
Identity disclosure, 267, 288–289
Idiosyncratic variation Variation in responses to
 questions that is caused by individuals'
 reactions to particular words or ideas in the
 question instead of by variation in the concept
 that the question is intended to measure, 221–222
Illogical reasoning Prematurely jumping to conclusions and arguing on the basis of invalid
 assumptions, 6, 9, 16
Impact evaluation (or analysis) Analysis of the
 extent to which a treatment or other service
 has an effect, 373–375e
Improvement-oriented evaluations Evaluations that
 seek to improve a program rather than simply